Seismological Algorithms

Seismological Algorithms

Computational Methods and Computer Programs

Edited by

DURK J. DOORNBOS

Institute of Geophysics
University of Oslo, Norway

1988

ACADEMIC PRESS
Harcourt Brace Jovanovich, Publishers

London San Diego New York
Boston Sydney Tokyo Toronto

ACADEMIC PRESS LIMITED
24/28 Oval Road,
London NW1

United States Edition published by
ACADEMIC PRESS INC.
San Diego, CA 92101

British Library Cataloguing in Publication Data

Doornbos, D. J. (Durk J)
Seismological algorithms
1. Seismology. Applications of computer systems. Algorithms
I. Title
551.2′2′0285

ISBN 0-12-220770-X

Printed in Northern Ireland at The Universities Press (Belfast) Ltd.

Contributors

Uri Ascher Department of Computer Science, The University of British Columbia, 6356 Agricultural Road, Vancouver BC V6T 1W5, Canada

E. Buforn Catedra de Geofisica, Facultad de Ciencias Fisicas, Universidad Complutense de Madrid, 28040 Madrid, Spain

Ray Buland US Geological Survey, Denver Federal Center, Box 25046, MS 967, Colorado 80225, USA

V. Červený Institute of Geophysics, Charles University, Ke Karlovu 3, 121 16 Praha 2, Czechoslovakia

Chris Chapman Department of Earth Sciences, University of Cambridge, Bullard Laboratories, Madingley Rise, Madingley Road, Cambridge CB3 0EZ, UK

Vernon Cormier University of Connecticut, Department of Geology and Geophysics, Storrs, Connecticut 06268, USA

D. J. Doornbos Institute of Geophysics, University of Oslo, P.O. Box 1022, Blindern, 0315 Oslo 3, Norway

L. Neil Frazer Hawaii Institute of Geophysics, University of Hawaii, 2525 Correa Road, Honolulu, Hawaii 96822, USA

Andrzej Hanyga Institute of Geophysics, Polish Academy of Sciences, P1–00 973 Warszawa, Pasteura 3, Poland

Chu Jen-yi Department of Geophysics, Peking University, Beijing, China

B. L. N. Kennett Research School of Earth Sciences, Australian National University, GPO Box 4, Canberra ACT 2601, Australia

L. Klimeš Institute of Geology and Geotechnics, Czechoslovak Academy of Sciences, V. Hollesovickach 41, 18209 Praha 4, Czechoslovakia

D. G. Lyness Department of Earth Sciences, University of Cambridge, Bullard Laboratories, Madingley Rise, Madingley Road, Cambridge CB3 0EZ, UK

Gary Pavlis Department of Geology, Indiana University, 1005 East Tenth Street, Bloomington, Indiana 47405, USA

I. Pšenčík Geophysical Institute, Czechoslovak Academy of Sciences, Bocni II, 141 31 Praha 4, Czechoslovakia

Paul G. Richards Lamont-Doherty Geological Observatory, Columbia University, Palisades, New York 10964, USA

Masanori Saito Department of Applied Physics, Tokyo Institute of Technology, 2–12–1 Ookayama, Meguro-Ku, Tokyo 152, Japan

A. Udias Cátedra de Geofísica, Facultad de Ciencias Fisicas, Universidad Complutense de Madrid, 28040 Madrid, Spain

J. H. Woodhouse Department of Earth and Planetary Sciences, Harvard University, 20 Oxford Street, Cambridge, Massachusetts 02138, USA

Preface

The plans for publication of this book and of the associated program package developed in response to signals from within the seismological community, indicating the need to bring together in an organized and surveyable way a selection of basic and advanced methods and algorithms. During the IASPEI General Assembly in Tokyo, in August 1985, the commissions on Seismological Theory and on Practice supported these plans, and the Joint Subcommission on Algorithms was charged with the task of sponsoring publication of a book. In a way, this was a natural follow-up and a synthesis of two rather different earlier projects. (1) The subcommission, on the initiative of its former chairman, Dr Engdahl, was already sponsoring an ongoing documentation program of earthquake algorithms. Report SE-35 of World Data Center A summarizes the status of this program in 1984. (2) In their series *Methods in Computational Physics,* Academic Press published in 1972 two volumes on seismology (Volumes 11 and 12 were edited by Professor Bolt). An update was considered timely, although of course the philosophy behind the present work is different from that of previous publications. It is an important aspect of this work that algorithms are also made available in the form of computer routines. The complete collection of these routines is available on tape or disk and can be obtained from World Data Center A for Solid Earth Geophysics. An index to the programs is included with this book.

Close cooperation with many colleagues around the world was essential for the completion of this project. I learned to appreciate the often substantial amount of extra work it takes an author to present text and programs in combination. I thank the contributors, and those contributing with advice, comments and reviews. This IASPEI project was supported by a grant from UNESCO.

DURK J. DOORNBOS

Chairman, Joint Subcommission on Algorithms

Contents

Contributors v

Preface vii

Introduction xv

I RAYS AND BODY WAVES IN ONE-DIMENSIONAL MEDIA

I.1 Spectral Synthesis of Body Waves in Earth Models Specified by Vertically Varying Layers **3**

VERNON F. CORMIER and PAUL G. RICHARDS

1 Introduction 3
2 Example body-wave problem 5
3 Synthesis of seismograms 15
4 Extensions and special applications 31
5 Conclusions 41
Acknowledgements 43
References 43

I.2 The WKBJ Seismogram Algorithm **47**

C. H. CHAPMAN, CHU JEN-YI and D. G. LYNESS

1 Introduction 47
2 WKBJ seismogram theory 49
3 Description of subroutines 57
4 Conclusions 69
Acknowledgements 69
Appendix: WKBJ seismograms with attenuation 69
References 73

I.3 Aspehricity and Ellipticity Corrections **75**

D. J. DOORNBOS

1 Introduction 75
2 Velocity models 76
3 Ray-tracing algorithm 77
4 Spherical harmonics in rotated coordinates 79
5 Ellipticity corrections 81
6 A test case: the core–mantle boundary 82
References 85

II RAYS AND BODY WAVES IN THREE-DIMENSIONAL MEDIA

II.1 Complete Seismic Ray Tracing in Three-Dimensional Structures **89**

V. ČERVENÝ, L. KLIMEŠ and I. PŠENČÍK

1 Introduction 89
2 Coordinate system 92
3 Model of the medium 94
4 Codes for elementary waves 105
5 Complete ray tracing 106
6 Initial points of rays 140
7 Applications and processing of the results of complete ray tracing 143
References 167

II.2 Numerical Methods for Tracing Rays and Wavefronts **169**

ANDRZEJ HANYGA

1 Basic assumptions and methods 169
2 Initial-value ray tracing 176
3 Two-point ray tracing 180
4 Dynamic ray tracing 195
5 Computation of body-wave pulses 213
Acknowledgements 217
Appendices 218
References 231

III REFLECTIVITY METHODS FOR STRATIFIED MEDIA

III.1 Systematic Approximations to the Seismic Wavefield **237**

B. L. N. KENNETT

1 Introduction 237
2 Reflection and transmission of seismic waves 239

3 Response of a medium to a source 243
4 Split stratification 244
5 The construction of approximations to the response 247
6 Examples of approximations 252
Appendix: Recursive approach for a stack of uniform layers 256
References 258

III.2 Hybrid Methods in Reflectivity Calculations **261**

URI ASCHER

1 Introduction 261
2 Methods for solving two-point problems 264
3 A hybrid method 271
4 An example 273
References 276

III.3 Quadrature of Wavenumber Integrals **279**

L. NEIL FRAZER

1 Introduction 279
2 Trapezoidal Filon 280
3 Alternative methods 286
Acknowledgements 288
References 288

IV MODE METHODS FOR STRATIFIED MEDIA

IV.1 Disper80: A Subroutine Package for the Calculation of Seismic Normal-Mode Solutions **293**

MASANORI SAITO

1 Introduction 293
2 Transversely isotropic Earth model 293
3 Methods of integration and specification of the model 294
4 Torsional oscillations 296
5 Spheroidal oscillations of a non-gravitating Earth 300
6 Spheroidal oscillations of a self-gravitating Earth 307
7 Love waves and Rayleigh waves 317
Acknowledgements 319
References 319

IV.2 The Calculation of Eigenfrequencies and Eigenfunctions of the Free Oscillations of the Earth and the Sun **321**

J. H. WOODHOUSE

1 Introduction 321
2 The algebra of minors 326

3 Linear transformations of vector subspaces 337
4 Linear systems of ordinary differential equations 343
5 The differential equations governing the Earth's free oscillations 354
6 Description of program 367
Acknowledgements 370
References 370

IV.3 Variational Methods **371**

RAY BULAND

1 Rayleigh's Principle 371
2 The Rayleigh–Ritz algorithm 372
3 Natural basis functions 377
4 Finite-element basis functions 381
5 Frequency-dependent problems 393
6 Instabilities 394
7 Natural basis functions revisited 395
8 Comparison with other methods 395
9 Closing remarks 398
References 398

V UTILITY ALGORITHMS AND SOURCES

V.1 Vector and Matrix Manipulation Algorithms **403**

GARY L. PAVLIS

1 Introduction 403
2 Basic vector operations 404
3 Manipulation of non-sparse matrices 407
4 Sparse-matrix inversion algorithms 420
Acknowledgements 424
References 424

V.2 Radiation from a Moment-Tensor Source **427**

B. L. N. KENNETT

1 Introduction 427
2 Point sources 428
3 Body-wave radiation patterns 431
4 Surface-wave radiation patterns 437
Appendix: Principal values of the moment tensor 439
References 440

V.3 Single and Joint Fault-Plane Solutions from First-Motion Data **443**

A. UDÍAS and E. BUFORN

1 Introduction 443
2 Geometry of the problem 443
3 Probability model and likelihood function 445
4 Joint solutions 449
5 Computer programs 452
Acknowledgements 453
References 453

Program Index 455

Subject Index 463

Introduction

PURPOSE

The algorithms presented in this book and in the associated program package deal with a wide range of seismological problems. Nevertheless, we realize that every new research project involves computational problems that require new programs, or a modification of existing ones. Thus we do not expect the algorithms and programs in this book to be tailored to the specific needs of every user. Rather, we hope to provide the reader with "building blocks" and with the background information to use these blocks intelligently in the construction of his own programs. Briefly, this book has the dual purpose of presenting adequate background information about the seismological theory and methods, and of providing the resulting algorithms for practical use. Some of these methods are new; others are based on theoretical developments documented in recent textbooks and research papers referenced in the individual contributions to this book. The reader is encouraged to consult these works.

SCOPE

Some algorithms are special-purpose routines, but many have been developed to deal with such ever-recurring questions as how to calculate body-wave travel times, surface-wave dispersion curves and free-oscillation eigenfrequencies, how to calculate amplitudes, waveforms, and complete seismograms or approximations thereof, and—once the forward problem is solved—how to apply matrix methods in the inverse problem. The algorithms dealing with wave-propagation problems in this book are based on ray and generalized ray methods, reflectivity methods and normal mode methods. The part and chapter division reflects the differences in methods and purposes. However, it frequently happens that a computational method developed within one area or chapter finds applications in other areas or chapters as well. The book contains many examples of this. To locate "hidden" algorithms, the reader is advised to consult not only the program index and the list of contents, but also the subject index.

APPLICABILITY

When considering the applicability of the different methods in Parts I–IV, it may be useful to keep in mind the following points:

(1) Some methods are formulated for plane stratified media, others for spherically stratified media. Algorithms for both types of media are included. In practice these algorithms are often applicable to either type of media owing to the possibility of applying a so-called Earth-flattening transformation to the velocity-density model. An accessible source of information on this point is Aki and Richards' (1980) textbook.

(2) Some routines calculate the seismic response for a specified point source. Others calculate just the Green functions or normal-mode terms, and the evaluation of the source-excitation factors is left to the user. These factors are readily obtained for a moment-tensor point-source representation, and this is pointed out in several of the contributions. For a more detailed exposition, the reader is referred to recent textbooks (e.g. Aki and Richards, 1980; Kennett, 1983).

(3) Although the algorithms of Parts I–IV solve forward problems of seismology, the same algorithms often form the main ingredient of inverse methods as well. Several examples are given in Part V.

OTHER SOURCES

Obviously not all computational methods of interest to seismology could be included in this volume. Within the selected subject areas, an attempt has been made to present alternative algorithms where this was possible, but we concede that several important algorithms are not included. Some of these are contained in World Data Center A report SE-35 (Engdahl,

1984). Subject areas not or only marginally covered include, among others, signal-processing methods, and finite-difference and finite-element methods for three-dimensional wave problems. Published algorithms within these areas are spread throughout the engineering literature, and useful examples are also often contained in books on numerical analysis (e.g. Press *et al.*, 1986). Within the geophysics literature, a number of textbooks contain examples of algorithms in some of these areas (e.g. Robinson, 1983; Menke, 1984; Claerbout, 1985).

PROGRAM LIBRARY

All contributors have submitted Fortran routines, and these together form a comprehensive software package. The programs are arranged in files (which may contain more than one program). The file names are listed in the program index at the end of this book. The software package can be obtained on tape or disk from World Data Center A for Solid Earth Geophysics in Boulder, Colorado. It should be realized that the programming style, level of Fortran and machine portability differ considerably from one contribution to the next. Relevant information concerning these points is included with the routines. It should also be noted that the contributors make no claims that the routines are error-free, and responsibility for correctness of numerical results rests with the user. The contributors would of course appreciate suggestions for improvement of algorithms.

REFERENCES

Aki, K. and Richards, P. G. (1980). *Quantitative Seismology,* Vols I and II. W. H. Freeman, San Fransisco.

Claerbout, J. F. (1985). *Imaging the Earth's Interior.* Blackwell, Oxford.

Engdahl, E. R. (1984). Documentation of earthquake algorithms. World Data Center A Report SE-35.

Kennett, B. L. N. (1983). *Seismic Wave Propagation in Stratified Media.* Cambridge University Press.

Menke, W. (1984). *Geophysical data analysis: Discrete Inverse Theory.* Academic Press, Orlando.

Press, W. H., Flannery, B. P., Teukolsky, S. A. and Vetterling, W. T. (1986). *Numerical Recipes.* Cambridge University Press.

Robinson, E. A. (1983). *Multichannel Time Series Analysis,* 2nd edn. Goose Pond Press, Houston.

I

RAYS AND BODY WAVES IN ONE-DIMENSIONAL MEDIA

I.1

Spectral Synthesis of Body Waves in Earth Models Specified by Vertically Varying Layers

VERNON F. CORMIER
Department of Geology and Geophysics, University of Connecticut, USA

PAUL G. RICHARDS
Lamont-Doherty Geological Observatory and Department of Geological Sciences, Columbia University, New York, USA

1 INTRODUCTION

Computational algorithms for synthesizing seismograms are intimately affected by the way in which an Earth model is parametrized. This chapter considers algorithms appropriate for a spherically symmetric laterally homogeneous Earth, parametrized by velocity functions that are analytic in depth and smoothly varying over thick regions. Examples of two Earth models that have been parametrized in this fashion are PEM (Dziewonski *et al.*, 1975) and PREM (Dziewonski and Anderson, 1981). In these models seismic velocities are given by a small number of polynomials in radius over large ranges of depth, separated by a small number of first- and second-order discontinuities. PREM (Fig. 1), for example, specifies velocity polynomials for the outer core, inner core, lowermost mantle, lower mantle, upper-mantle transition zones, low velocity zone, upper-mantle lid and crust. An advantage of working with such models is that the physical significance of geophysical quantities related to the first and second derivatives of velocities with respect to depth is not obscured by the parametrization. These quantities include the stratification parameter, the pressure derivative of the bulk modulus and the Gruneisen parameter. By using modelling algorithms that work

SEISMOLOGICAL ALGORITHMS
ISBN 0-12-220770-X

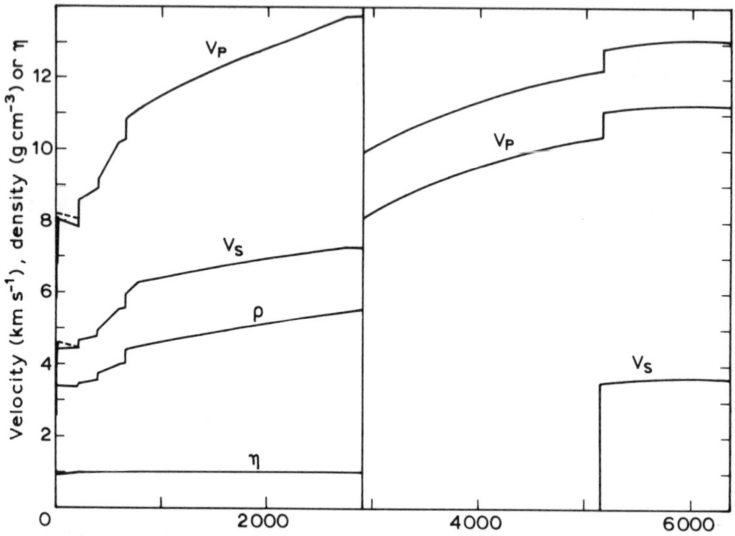

Fig. 1 The Preliminary Reference Earth Model (PREM) of Dziewonski and Anderson (1981) is an example of a model specified by a small number of layers separated by first- or second-order discontinuities in velocity and density. In each layer, velocity is specified by a polynomial in radius.

directly with velocity models that are analytic over thick regions, it is possible to directly investigate the sensitivity of a particular set of waveform data to these quantities.

Synthesis techniques that can handle this type of model parametrization include the collocation method, described by Ascher (Chapter III.2) and the WKBJ or slowness method, described by Chapman *et al.* (Chapter I.2). The algorithms for computing the delay time τ and the complex turning points described here for isotropic and transversely isotropic models complement those described by Chapman and can be combined with subroutines used for the computation of body waves by the WKBJ/slowness technique.

The WKBJ approximation, however, does not accurately describe diffraction and tunnelling of body waves at and near grazing incidence to a boundary (Fig. 2). In these instances a uniform asymptotic approximation is needed that smoothly patches between the travelling wave and exponential forms of the solution to the radially separated part of the wave equation. This uniform approximation has been described by Langer (1932, 1949) and extended by Olver (1954). At non-grazing angles it reproduces the non-uniform WKBJ approximation and at grazing angles it reproduces the non-uniform Airy approximation. The

Langer approximation can be incorporated in a body-wave synthesis problem by simple substitution of the uniform approximation for the WKBJ approximation to the radial wavefunctions appearing in the fundamental matrices for vertically propagating waves. When reflection/transmission coefficients are then calculated using the modified form of the fundamental matrices, it is seen that the Langer approximation reduces to the substitution of generalized cosines and ratios of vertical wavefunctions defined in terms of Airy functions (Richards, 1976). This chapter will describe algorithms for evaluating the generalized cosines and vertical wavefunctions of the Langer approximation. An outline is also given of how the Langer approximation can be incorporated into algorithms that calculate the reflectivity response of a stack of n layers (see Chapter III.1 by Kennett). In this case, the Langer approximation allows each layer to have a vertically varying velocity profile.

2 EXAMPLE BODY-WAVE PROBLEM

2.1 Representation of the complex spectrum

In order to rapidly introduce notation and motivate the discussion, we shall begin by considering a particular problem. Consider the asymptotic representation of the complex spectrum of a P wave of the form

$$u_r(r, \Delta_0, \omega) = \frac{-\mathrm{i}\omega M_0(\omega)\mathrm{i}\omega}{4\pi\rho_s\alpha_s^4}\int_\Gamma \omega^2 pF^{\mathrm{P}}\pi^{(1)}(r_s)\grave{P}\acute{P}\alpha(r)\xi(r)\pi^{(1)}(r)Q_l^{(2)}(\cos\Delta_0)\,\mathrm{d}p. \quad (1)$$

The P wave has been radiated by an earthquake or explosion having a moment $M_0(\omega)$ and radiation pattern F^{P}. The source is located at a point at radius r_s. Its radiated P wave is reflected as a P wave by a single discontinuity or zone of several discontinuities, and the radial component of displacement u_r has been recorded by a station at radius r and distance Δ_0. Γ denotes a contour in the complex ray-parameter (p) plane. $l = \omega p - \frac{1}{2}$ is a complex angular order number. The derivation of the representation given by (1) is described by Scholte (1956), Richards (1973) and in Chapter 9 of Aki and Richards (1980).

The function $Q_l^{(2)}(\cos\Delta)$ is one of the travelling-wave components of the Legendre function. The functions $\pi^{(1)}$ for upgoing P waves and $\pi^{(2)}$ for downgoing P waves are proportional to solutions of the radially separated component R of the equation of motion; e.g., to first order in frequency, R satisfies the equation

$$\frac{\mathrm{d}^2}{\mathrm{d}r^2}(rR) + \omega^2\left(\frac{1}{\alpha^2} - \frac{p^2}{r^2}\right)rR = 0. \tag{2}$$

$\acute{\xi}(r)$ is defined by $\acute{\xi} = (\mathrm{d}\pi^{(1)}/\mathrm{d}r)/(\mathrm{i}\omega\pi^{(1)})$. Unless the P wave is close to grazing incidence, $\acute{\xi}$ can be simply taken as the vertical slowness of an incident plane wave of ray parameter p at r, i.e. $\acute{\xi}(r) \sim 1/\alpha^2 - p^2/r^2$.

The source may be parametrized either by the elements of a moment tensor M_{ij} or by the strike, dip and rake of a fault plane describing the orientation of a point shear dislocation. In either case, these parameters determine the weights w_i of three different functions of the ray parameter $f_i(p)$ that comprise F^{P}:

$$F^{\mathrm{P}}(\phi, p) = w_1(\phi)f_1(p) + w_2(\phi)f_2(p) + w_3(\phi)f_3(p). \tag{3}$$

The integration over p is performed for each separate f_i. The weighting functions w_i, which are functions of the azimuthal take-off angle ϕ at the source, are fully determined from input consisting of either the elements of a moment tensor or the orientation of a double couple. For each function f_i the complex spectrum is inverted into the time domain by fast Fourier transform and three fundamental seismograms weighted by w_i are summed.

2.2 An outline of the algorithm

The algorithm and related programs can be summarized in a brief outline as follows:

Given: analytic velocity and density in several layers, a description of the layers, interfaces traversed by the ray, and the integration contour Γ for the ray parameter integral,

Compute: displacement $u(\boldsymbol{x}, t)$:

Compute $u(\boldsymbol{x}, \omega)$ at many real frequencies ω and then transform to $u(\boldsymbol{x}, t)$ by FFT:

Compute $u(p, \omega)$ at many complex ray parameters p and then transform to $u(x, \omega)$ by integration along Γ:

Assemble $u(p, \omega)$ from the ξ, π and reflection/transmission coefficients C at various layer interfaces:

Compute vertical slownesses ξ and η

Compute vertical wavefunctions π and/or σ by

WKBJ approximation if $\omega\tau > 5$
Langer approximation if $\omega\tau \leq 5$

Compute reflection/transmission coefficients $C = f(\xi, \eta, \pi, \sigma)$

Assemble $u(p, \omega) = f(\xi, \eta, \pi, \sigma, C)$

Do contour integral using the trapezoidal rule, $u(p, \omega) \to u(\boldsymbol{x}, \omega)$

Do inverse Fourier transform $u(\boldsymbol{x}, \omega) \to u(x, t)$ by FFT

The computation of the vertical wavefunctions π and σ, the generalized vertical slownesses π and η, and generalized reflection/transmission coefficients are described in Subsections 2.3–2.8. Example problems and the choice of integration contours are discussed in Section 3. Section 4 describes extensions and applications to problems that include intrinsic attenuation, transverse isotropy, n-layered reflectivity and antipodal receivers.

2.3 The WKBJ approximation

The WKBJ approximation to the solution of (2) gives for upgoing, $R = g^{(1)}$, and downgoing, $R = g^{(2)}$, travelling waves the result

$$g^{(1),(2)} \sim e^{\mp i\pi/4} \frac{\alpha^{1/2}}{\omega r} \frac{e^{\pm i\omega\tau}}{\lambda^{1/2}}, \tag{4a}$$

where

$$\lambda = \left[\frac{1}{\alpha^2(r)} - \frac{p^2}{r^2}\right]^{1/2}, \qquad \tau = \int_{r_p}^{r_b} \lambda \, dr.$$

In (4a) (and similar expressions below) the upper/lower signs correspond to the upgoing (1)/downgoing (2) waves. The lower bound r_p of the integral for τ is the turning-point radius or the radius at which λ vanishes, i.e. $\lambda(r_p) = 0$. The upper bound r_b is the radius of a discontinuity at which either a source or receiver is located or at which a reflection/transmission coefficient is evaluated.

The functions $\pi^{(1),(2)}$ are proportional to $g^{(1),(2)}$ through $\pi^{(1),(2)} = (\rho_s/\rho)^{1/2}\alpha_s(g^{(1),(2)}/\alpha)$. The vertical wavefunctions for up- and downgoing S waves, $\sigma^{(1),(2)}$, are the same but with S velocities β and β_s substituted for P velocities α and α_s. The normalization in (4a) has been chosen so that in a homogeneous layer, $g^{(1),(2)}$ are exactly spherical Hankel functions.

2.4 The Langer approximation

The uniform Langer approximation to the solution of (2) gives the result

$$g^{(1),(2)} \sim \frac{2}{\omega r}\left(\frac{\pi\alpha_s}{\xi}\right)^{1/2} \zeta^{1/4}\,\mathrm{Ai}\,(-\mathrm{e}^{\pm 2\mathrm{i}\pi/3}\zeta), \tag{4b}$$

where $\zeta = (\frac{3}{2}\omega\tau)^{2/3}$. (Equation (4b) can also be written in terms of Hankel functions of order $\frac{1}{3}$ as in Richards (1976).) When ζ is large, the Langer approximation (4b) reproduces the WKBJ approximation (4a). In practice, when $\omega\tau$ is equal to 5, the Langer approximation (4b) equals (4a) to within about 1%. Evaluation of $\omega\tau$ for a particular domain of the ray parameter p can be useful for determining whether the WKBJ approximation is adequate for a particular range of incidence angles and frequencies.

2.5 The horizontal wavefunction

For large ωp and Δ not near the source or the antipode, $Q_l^{(1),(2)}$ is accurately given by the WKBJ approximation of the solution to Legendre's equation. This is

$$Q_l^{(1),(2)} \sim \frac{\exp\,[\mp\mathrm{i}(\omega p\Delta_0 - \frac{1}{4}\pi)]}{2\pi\omega p \sin\Delta_0}. \tag{5}$$

If ωp is small or Δ_0 is close to zero or a multiple of π then it is best not to divide the Legendre function into travelling-wave components $P_l = Q_l^{(1)} + Q_l^{(2)}$. The calculation of P_l for antipodal problems is described in a later section of this chapter.

2.6 Generalized reflection–transmission coefficients

The function $\grave{P}\acute{P}$ in the integrand of (1) is a generalized reflection coefficient or reflectivity function for a P wave incident on a single boundary or a zone of several layers separated by first- or second-order discontinuities in velocity (Fig. 3). For a single boundary at radius $r = r_b$,

$\grave{P}\acute{P}$ can be found by solving a linear equation expressing the continuity of a stress–displacement vector in the transformed domain of (ω, p):

$$(U, V, R, W)_{r_b^+} = (U, V, R, W)_{r_b^-}. \tag{6a}$$

If a P wave is incident from above on the r_b^+ side then the boundary condition (6a) can be written in terms of fundamental matrices $\boldsymbol{F}_2(r_b^+)$ and $\boldsymbol{F}_1(r_b^-)$, weighted by generalized coefficients of reflection, transmission and conversion:

$$\boldsymbol{F}_2(r_b^+)\begin{bmatrix} \grave{P}\acute{P} \\ 1 \\ \grave{P}\acute{S} \\ 0 \end{bmatrix} = \boldsymbol{F}_1(r_b^-)\begin{bmatrix} 0 \\ \grave{P}\grave{P} \\ 0 \\ \grave{P}\grave{S} \end{bmatrix}, \tag{6b}$$

where

$$\boldsymbol{F}(r) = r^{-1}\left(\frac{\mathrm{i}}{\rho}\right)^{1/2}\begin{bmatrix} \acute{\xi}\pi^{(1)} & -\grave{\xi}\pi^{(2)} & r^{-1}p\sigma^{(1)} & r^{-1}p\sigma^{(2)} \\ -\mathrm{i}r^{-1}p\pi^{(1)} & -\mathrm{i}r^{-1}p\pi^{(2)} & \mathrm{i}\acute{\eta}\sigma^{(1)} & -\mathrm{i}\grave{\eta}\sigma^{(2)} \\ -\mathrm{i}A\pi^{(1)} & -\mathrm{i}A\pi^{(2)} & \mathrm{i}B\acute{\eta}\sigma^{(1)} & -\mathrm{i}B\grave{\eta}\sigma^{(2)} \\ B\acute{\xi}\pi^{(1)} & -B\grave{\xi}\pi^{(2)} & A\sigma^{(1)} & A\sigma^{(2)} \end{bmatrix} \tag{7}$$

(Woodhouse, 1978; Cormier, 1980). A and B are used in defining the fundamental matrix in (7) and are functions of shear modulus μ, density ρ and flattened horizontal slowness p/r defined by

$$A = 2\frac{p^2}{r^2}\mu - \rho, \tag{8a}$$

$$B = 2\frac{p}{r}\mu. \tag{8b}$$

$\acute{\xi}$, $\grave{\xi}$, $\acute{\eta}$, $\grave{\eta}$ are generalized vertical slownesses for upgoing and downgoing P and S waves respectively, given in terms of the vertical wavefunctions and their radial derivatives by

$$\acute{\xi} = \frac{\mathrm{d}\pi^{(1)}}{\mathrm{d}r}\bigg/(\mathrm{i}\omega\pi^{(1)}), \tag{9a}$$

$$\grave{\xi} = \frac{\mathrm{d}\pi^{(2)}}{\mathrm{d}r}\bigg/(-\mathrm{i}\omega\pi^{(2)}), \tag{9b}$$

$$\acute{\eta} = \frac{\mathrm{d}\sigma^{(1)}}{\mathrm{d}r}\bigg/(\mathrm{i}\omega\sigma^{(1)}), \tag{9c}$$

$$\grave{\eta} = \frac{\mathrm{d}\sigma^{(2)}}{\mathrm{d}r}\bigg/(-\mathrm{i}\omega\sigma^{(2)}), \tag{9d}$$

$\boldsymbol{F}_2$ and $\boldsymbol{F}_1$ are evaluated in the analytic velocity profiles on the r_b^+ and r_b^- sides of the boundary respectively.

In a vertically inhomogeneous layer, $\boldsymbol{F}$ satisfies to first order in frequency the linear equation

$$\frac{\partial \boldsymbol{F}}{\partial r} = \omega \boldsymbol{A}\boldsymbol{F}, \tag{10}$$

where the 4×4 matrix $\boldsymbol{A}$ for P–SV problems is given in Chapman (1973, 1974).

Solving (6b) for a single boundary between two solids, $\grave{P}\acute{P}$ is found to be

$$\grave{P}\acute{P} = -\left\{\left[(b\grave{\xi}_1 - c\grave{\xi}_2)F + (\alpha + d\grave{\xi}_2\grave{\eta}_1)\frac{Gp^2}{r_1^2}\right]\Big/ D\right\}\frac{\pi_2^{(1)}}{\pi_2^{(1)}}, \tag{11}$$

with

$$D = EF + \frac{GHp^2}{r_1^2},$$

$$E = b\grave{\xi}_1 + c\acute{\xi}_2, \qquad F = b\grave{\eta}_1 + c\acute{\eta}_2,$$

$$G = \alpha - d\grave{\xi}_1\acute{\eta}_2, \qquad H = \alpha - d\acute{\xi}_2\grave{\eta}_1,$$

and a, b, c and d defined by

$$\alpha = A_2 + A_1, \qquad b = -A_2 + \frac{B_1 p}{r},$$

$$c = -A_1 + \frac{B_2 p}{r}, \qquad d = \frac{(B_2 - B_1)p}{r}.$$

The subscripts 2 and 1 denote the solid on the upper and lower sides of the boundary respectively.

The Langer approximation to the vertical slownesses and wavefunctions corrects for frequency dependence of the generalized reflection–transmission coefficients at grazing incidence. An important phenomena related to this frequency dependence is that of tunnelling. In tunnelling, rays that turn just above or below a boundary can interact with the boundary for finite frequency. For rays bottoming just above a boundary, transmitted real rays can be excited below the boundary. This transmitted or tunnelled energy also acts to reduce the magnitude of the reflection coefficient at ray parameters close to grazing incidence. Examples of these effects are shown in Fig. 2.

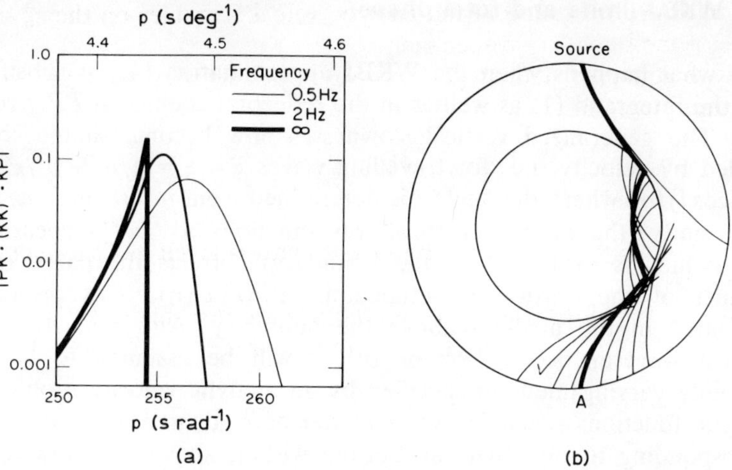

Fig. 2 (a) The product of reflection and transmission coefficients relevant to P4KP is plotted against ray parameter for two frequencies (after Richards, 1973). (b) The ray drawn heaviest is core-grazing. There is an exponential decay of energy below a turning point. Thus for finite frequency, energy can leak into the outer core from turning rays in the mantle. This contributes to non-zero energy shown in (a) for ray parameters greater than grazing at finite frequency.

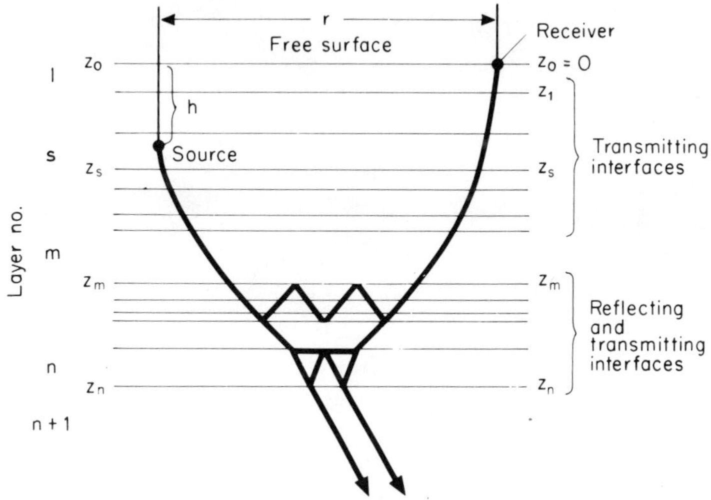

Fig. 3 $\grave{P}\acute{P}$ is used to describe the reflection coefficient of a P wave incident downwards on z_m. Thus $\grave{P}\acute{P}$ includes both the upward reflection (from z_m) as a P wave, and the P-wave ray that turns in the layer between z_m and the transmitting interfaces. In a reflectivity calculation $\grave{P}\acute{P}$ denotes this wave as well as all possible multiples in the layers between z_m and z_n that are returned as P waves in the layer above z_m.

2.7 WKBJ limits and total phase

Note what happens when the WKBJ approximation (4a) is substituted into the integrand (1) as well as in the reflection coefficient $\grave{P}\acute{P}$ given by (11). The generalized vertical slownesses now become simple cosines divided by velocity, i.e. for travelling waves $\acute{\xi} = \grave{\xi} = (1/\alpha^2 - p^2/r^2)^{1/2} = \lambda = (\cos i)/\alpha$, where the angle i is determined from $p = (r \sin i)/\alpha$. Most importantly, the ratio of vertical wavefunctions $\pi_2^{(2)}/\pi_2^{(1)}$ appearing in (11) (which is evaluated on the boundary) corrects the phase of the product of the vertical wavefunctions $\pi^{(1)}(r)\pi^{1}(r_s)$. This product is calculated in the medium above the boundary. Any evaluation of a vertical wavefunction $\pi^{(1),(2)}$ or $\sigma^{(1),(2)}$ will be assumed to be in a smoothly varying medium specified by an analytic velocity profile. The analytic functions specifying velocity can be evaluated at complex radii corresponding to rays that can bottom within, above or below discontinuities bounding the vertically inhomogeneous layer (Fig. 4). Hence, for low-enough ray parameters the product $\pi^{(1)}(r)\pi^{(1)}(r_s)$ may describe in the WKBJ limit the phasor $\exp[i\omega(\int_{r_p}^{r} \lambda_2 \, dr + \int_{r_p}^{r_s} \lambda_2 \, dr)]$. This is a ray that

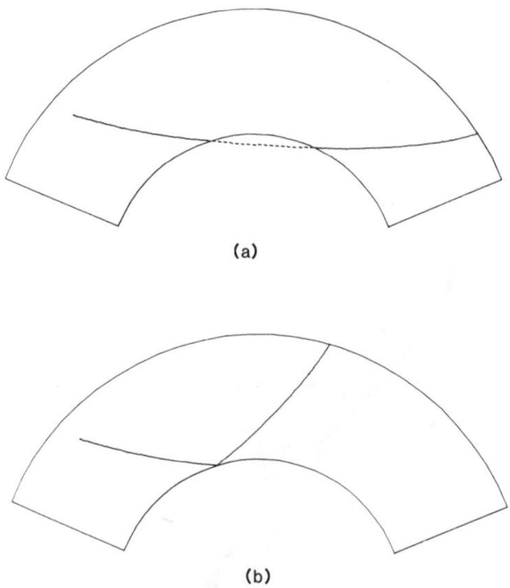

Fig. 4 The ratio $\pi_2^{(2)}/\pi_2^{(1)}$, which appears in the generalized reflection coefficient $\grave{P}\acute{P}$, acts in the product $\grave{P}\acute{P}\pi_2^{(1)}(r_s)\pi_2^{(1)}(r)$ to subtract the phase of rays that turn in the analytic continuation of the velocity profile in layer 2 below the boundary separating layers 2 and 1. Such a turning ray is shown dashed in (a). The resulting phase given by the product $\grave{P}\acute{P}\pi_2^{(1)}(r_s)\pi_2^{(1)}(r)$ describes the reflected ray shown in (b).

bottoms at a radius r_p below the discontinuity r_b. The definition of this r_p requires analytic continuation of the velocity function $\alpha(r)$ in medium 2 to values of radius below r_b, i.e. outside the range in which the layer is physically defined. If the WKBJ limit of the ratio $\pi_2^{(2)}/\pi_2^{(1)}$ appearing in (11) is evaluated, it is seen to produce the phasor $\exp[-2i\omega(\int_{r_p}^{r_b}\lambda_2\,dr)]$. The product of these two phasors, however, is seen to be simply the phase of a plane P wave reflected upward in medium 2 by the boundary at r_b, namely $\exp[i\omega(\int_{r_b}^{r_s}\lambda_2\,dr + \int_{r_b}^{r}\lambda_2\,dr)]$. Although the ray described by the product $\pi_2^{(1)}(r_s)\pi_2^{(1)}(r_s)$ may bottom below r_b, the phase of the generalized reflection coefficient acts to subtract the phase of the turning ray in the analytic extension of the velocity profile in medium 2 below r_b.

Substitution of WKBJ approximations for all vertical and horizontal wavefunctions in (1) gives

$$u_r(r, \Delta_0, \omega) = \frac{i\omega M_0(\omega)e^{-i\pi/4}\omega^{1/2}}{(2\pi\rho\rho_s \sin\Delta_0)^{1/2}4\pi r r_s \alpha_s^2}\int_\Gamma B(p)e^{i\omega J(p)}\,dp, \tag{12}$$

where

$$B(p) = p^{1/2}\left(\frac{1}{\alpha^2} - \frac{p^2}{r^2}\right)^{1/4} \Big/ \left(\frac{1}{\alpha_s^2} - \frac{p^2}{r_s^2}\right)^{1/4} \grave{P}\acute{P}.$$

$\grave{P}\acute{P}$ in (11) is now the WKBJ limit of the generalized coefficient in (11) without the ratio $\pi^{(2)}/\pi^{(1)}$. The phase given by the ratio $\pi^{(2)}/\pi^{(1)}$ has been absorbed into the total phase factor $J(p)$. For a ray reflected by the boundary, $J(p)$ is calculated by

$$J(p) = \int_{r_b}^{r_s}\frac{\cos i}{\alpha}\,dr + \int_{r_b}^{r}\frac{\cos i}{\alpha}\,dr + p\Delta_0. \tag{13}$$

If the ray bottoms above the boundary then the plane-wave reflection coefficient becomes equal to 1 and the lower limits of the integrals in the $J(p)$ calculation are taken to be the turning-point radius r_p in the upper medium.

2.8 Calculation of total phase in provided subroutines

2.8.1 Total phase when the Langer approximation is invoked

It is important to remember that the subroutines for evaluating generalized reflection–transmission–conversion coefficients, which will be described in the following section, return ratios of vertical wavefunctions of the type $\pi^{(2)}/\pi^{(1)}$ described for the $\grave{P}\acute{P}$ coefficient. These act to subtract or add phase necessary to convert the phase of rays that turn in the medium above a boundary into the phase of rays that are reflected by, transmitted through, or converted to another wave type at the boundary. The main program SCHAIRY and the subroutines VALUES and TRAPEZ provided in the companion package of Fortran programs for this chapter assemble integrands of the type given in (1). When reflection–transmission coefficients are specified using the Langer approximation option given by flags equal to 2 or 3 in the ILNGVS array, ratios of vertical wavefunctions of the type $\pi^{(2)}/\pi^{(1)}$ act to subtract or add phase.

2.8.2 Total phase when the Langer approximation is not invoked

If the Langer approximation is not invoked by setting flags ILNGVS equal to 1 or 0 in the ILNGVS array, then plane-wave reflection–transmission coefficients are invoked, in which $\acute{\xi}$, $\grave{\xi}$, $\acute{\eta}$, $\grave{\eta}$ are cosines of angles of incidence divided by velocity. Also with this option, the ratios of vertical wavefunctions of the type $\pi^{(2)}/\pi^{(1)}$ are returned as complex constants $e^{\pm i\pi}$, $e^{\pm i\pi/4}$ or 1.

2.8.3 The assembly of the total phase

In either case of ILNGVS options, the total phase is assembled by calls to subroutine ROMTAU from subroutine VALUES. The total phase depends on the input requested for turning rays as well as the ratios of vertical wavefunctions appearing in generalized reflection–transmission coefficients whenever the Langer approximation is invoked. The example input for PKP waves provided with this package illustrates the interaction of the phase of turning rays with phase given by ratios of vertical wavefunctions. This input should be carefully understood before proceeding to assemble integrands for other wavetypes. A rule of thumb is to have rays always turn above the first discontinuity at which the Langer approximation is invoked. At subsequent lower boundaries all vertical wavefunctions of a desired wavetype should be either (i) calculated using the Langer approximation or (ii) the generalized reflection/transmission coefficients need not be calculated because they have simple limits of 1 or 0.

2.8.4 When should the Langer approximation be used?

Whether it is necessary to use the Langer approximation for a particular domain of (ω, p) for a wavetype depends on how close the wave is to grazing incidence to the boundary. The lower the frequency, the greater the angular range in which it is necessary to use the Langer approximation. A test for whether it is necessary to invoke the Langer approximation can be made by calculating the quantities $\omega\tau = \int_{r_p}^{r_b} (\cos i)/v \, dr$ for each P and S velocity function on either side of the boundary r_b. If $|\omega\tau|$ is smaller than 5 then the errors in the use of the WKBJ approximation become significant.

3 SYNTHESIS OF SEISMOGRAMS

3.1 Main programs and subroutines

Equation (1) can be (*a*) evaluated at the discrete frequencies needed to invert the complex spectrum into the time domain with a fast Fourier transform (spectral method) or (*b*) written as a series of convolutions in the time domain after performing the p integration along the real axis (slowness method). The program WIGGLER provided in the subroutine package for this chapter performs the inversion to the time domain by fast Fourier transform. Representations for the complex spectrum of the type given by (1) are assembled by the program SCHAIRY and are integrated by trapezoidal integration along a contour Γ specified in the complex p plane.

3.2 Integration contours

Example integration contours for S, ScS, SKS and PKP phases are provided with the programs. Discussion of the contours appropriate for different phases are given in Richards (1973), Choy (1977), Cormier and Richards (1977), and Chapter 9 of Aki and Richards (1980). A more rigorous derivation of contours is given in Nussenzweig (1965) and Richards (1973).

3.2.1 Saddle points

A first step in defining an integration contour is to estimate, if possible, the values of ray parameter corresponding to the ray theoretical stationary phases or saddle points of the waves to be synthesized. A particular contour can often be used for a range of distances without modification, provided that segments along the real p axis include or are sufficiently close to all of the saddle points. A precise estimate of these saddle points is not usually necessary, and values can be estimated from

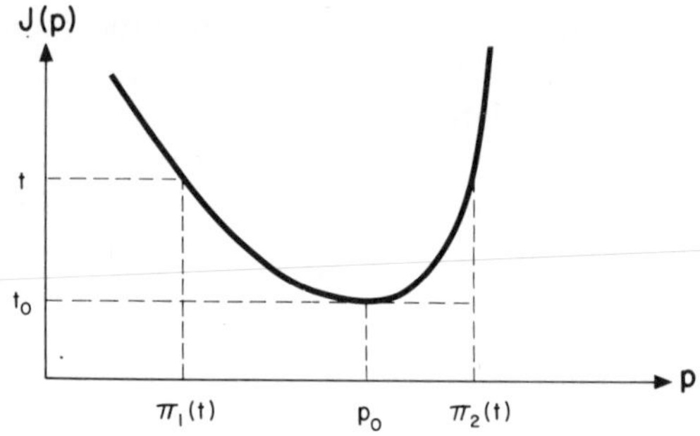

Fig. 5 The total phase function $J(p)$ is a minimum at the ray theoretical stationary phase p_0 for a simple turning ray.

the $\mathrm{d}T/\mathrm{d}\Delta$ values listed in the Jeffreys and Bullen (1958) table of travel times. The units of ray parameter assumed in the program are $\mathrm{s\,rad^{-1}}$. Thus to convert a value of $\mathrm{d}T/\mathrm{d}\Delta$ from $\mathrm{s\,deg^{-1}}$ to $\mathrm{s\,rad^{-1}}$, multiply by $180/\pi$. The saddle points can also be estimated by running SCHAIRY with an option that reproduces the results of geometrical ray theory. The option attempts to find the point of stationary phase by searching for extrema in the phase function $T(p) - \Delta(p) + p\Delta_0$ (Fig. 5). In this case a range of p must be specified along the real axis, which includes the desired saddle points.

3.2.2 Diffraction poles

A second preliminary step is to estimate the approximate position of all important diffraction poles in the vicinity of the integration contour. These poles will be close to the values of ray parameter corresponding to grazing incidence of P or S waves on the upper and lower sides of first-order velocity discontinuities. They will emanate from the values $p = r_b/\alpha(r_b^+)$, $r_b/\alpha(r_b^-)$, $r_b/\beta(r_b^+)$ and $r_b/\beta(r_b^-)$ (Figs 6 and 7), which can be calculated from the values of radius r_b and P and S velocities on both sides of first-order velocity discontinuities. When the wave or waves to be synthesized have saddle points close to a diffraction pole, care should be taken to place the integration contour such that its residue contribution will be included when the ends of the contour are extended to enclose the upper half-plane (positive imaginary p).

3.2.3 Orientation of saddles

The third step is to estimate the orientation of saddle points. These are usually $\pm 45°$ with respect to the real p axis. Generally, phases having

$d\Delta/dp$ positive, such as PcP, ScS and PKP–AB, have saddles oriented −45°, and phases having $d\Delta/dp$ negative, such as simple turning rays and directly transmitted waves, have saddles oriented +45°. If there is some doubt about the orientation of saddles, they can be determined from the complete phase of the integrand given by substituting the WKBJ representations (4a) for the vertical wavefunctions into the integral representation appropriate for the wavetype. The saddle orientation can then be found by the usual technique of maximizing the negative imaginary part of the phase ϕ, where ϕ is defined by the complex phasor $\rho \exp[i\omega\phi(p)]$ = integrand. Another method is simply to plot the integrand for a trial complex p path and examine whether the integrand is exponentially decaying as $\mathrm{Im}(p)$ increases.

3.2.4 Length of contours off the real p axis

The subroutines supplied with this chapter are not automated to estimate how far off the real axis a contour needs to be extended in order to reach a region of an exponentially small integrand. Generally, the lower the frequency, the more the contour needs to be extended off the real axis. Currently, testing of the contours and plotting of the integrand must be performed to ensure that exponential decay is reached. Analysis of the total phase given by WKBJ approximations, however, makes it possible to modify the programs to either estimate the length of contours off the real axis needed to achieve exponential decay or to estimate the contribution from the point of contour truncation to a point at complex infinity along the linear extension of the contour (see e.g. Thomson and Chapman, 1986).

3.2.5 Example: P + PcP and P diffracted

As examples, the integration contours for two different problems are shown in Figs 6 and 7. Figure 6 illustrates the contours for the P + PcP phase and P diffracted phases at distances corresponding to grazing incidence on the core mantle boundary. In the lit zone at distances at which separate P and PcP phases can be observed, the contour traverses saddle points oriented at −45° and +45° respectively and ends in regions of exponential decay on either side of a line of diffraction poles emanating from the ray parameter corresponding to core grazing incidence. The saddles need not be crossed precisely by the contour as long as the contour is sufficiently close to the saddle points. In practice, this means that the effects of the saddle points and the exponential decay are easily observed in plots of the integrand. The contour shown in Fig. 6

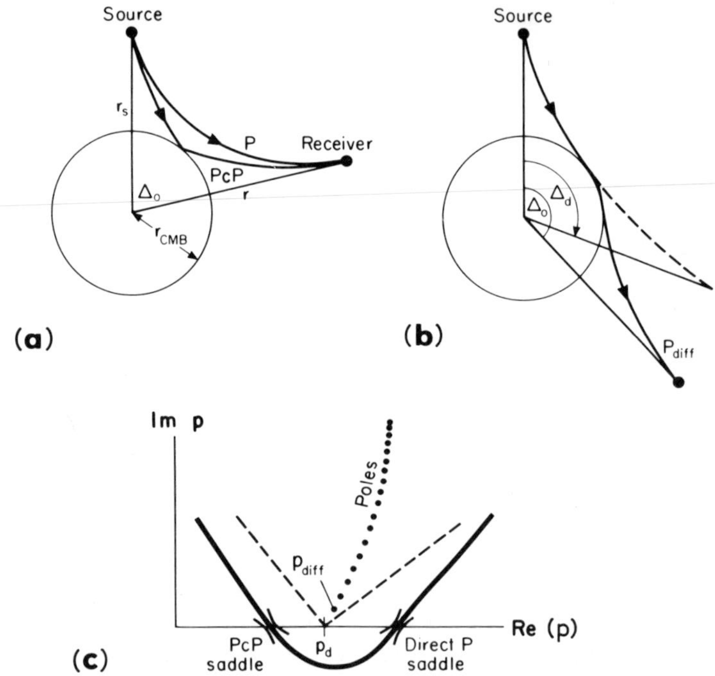

Fig. 6 (a) Rays for direct P and PcP. (b) Beyond Δ_d lies a shadow zone into which P waves are diffracted. (c) The orientation of poles, saddle points and integration contours for P + PcP and P diffracted. A single contour is adequate for a range of distances on either side of the shadow boundary.

is also adequate in the shadow zone at the distance ranges at which only a diffracted phase is observed. The integrand decays quickly as $\text{Im}(p)$ increases on portions of the contour off the real axis. This decay is more rapid as the frequency increases. As frequency increases, the amplitude of higher frequency components will eventually approach a nearly constant small value, which will be a function of machine precision. At these values, it is appropriate not to perform the calculation for P diffracted and simply to set the complex spectral amplitude to zero.

3.2.6 Example: PKiKP + PKIKP + PKIIKP + . . .

Figure 7 illustrates integration paths for the PKP–DF + PKP–CD branches in the distance range 110°–135°. Integration along these contours describes the phase assemblage PKiKP + PKIKP + PKIIKP + PKIIIKP + . . . (Cormier and Richards, 1977). After inverting the complex spectrum into the time domain, the pulses are quite simple in this

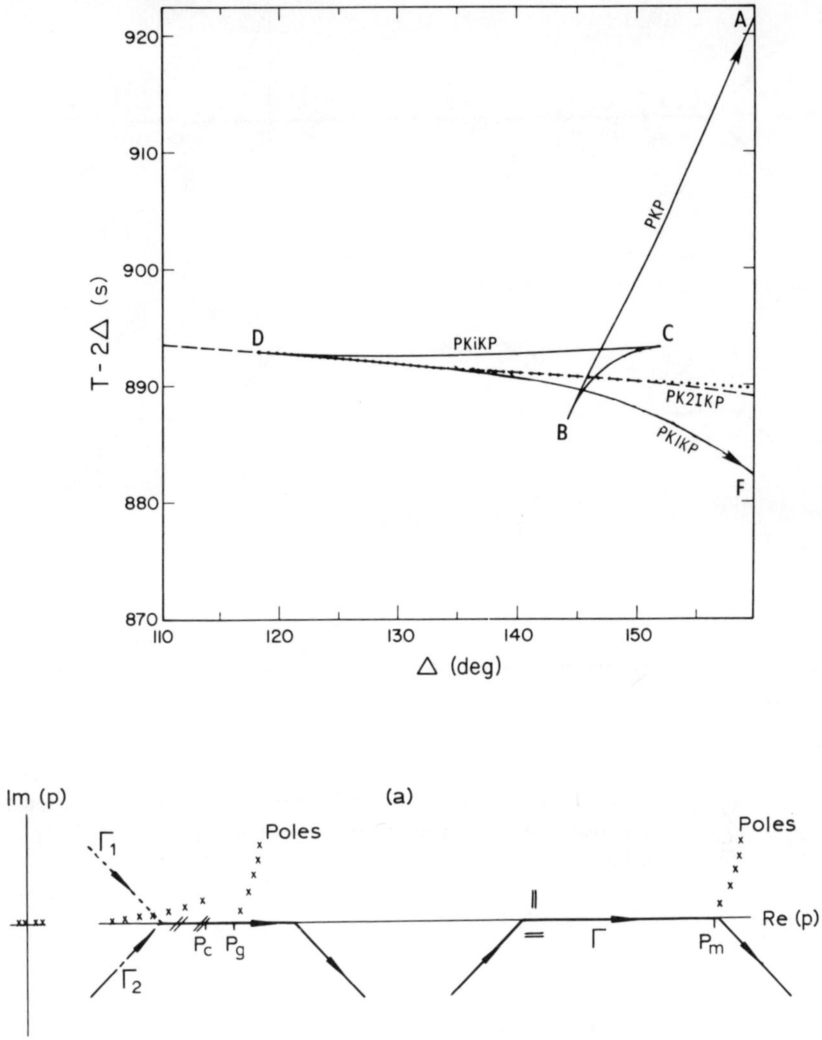

Fig. 7 The travel-time curve for PKP waves and associated integration paths in the complex p plane for computing synthetic seismograms. With $\grave{K}\acute{K}$ in (1), the integration is conducted over Γ_1. With $(\grave{K}\grave{I} - \acute{I}\acute{K})/(1 - \acute{I}\hat{I})$ in (1), the integration is conducted over Γ_2. For the remainder of the path Γ the two integrands are combined into one term proportional to $K(\check{I})K$. In the $K(\check{I})K$ term, poles associated with $\grave{\xi}_1$ (the poles asymptotically parallel to the imaginary p axis) are replaced with poles associated with $\check{\xi}_1$ (the poles asymptotically parallel to the real p axis). (a)–(c) show modifications of paths and motions of saddle points in the distance ranges 110°–135° (a), 135°–152° (b), and >152° (c).

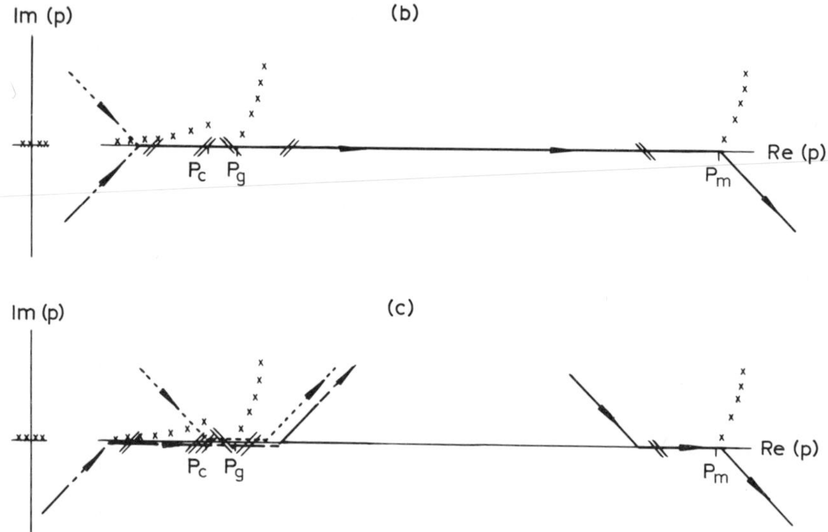

Fig. 7 (*continued*)

distance range. The separation of PKIKP (PKP–DF) from PKiKP (PKP–CD) only begins to be visible in high-frequency and broadband recordings at around 130°.

In this example, the integrand contains a product of generalized reflection/transmission coefficients of the form $\grave{P}\grave{K} \cdot \grave{K}(\check{I})\acute{K} \cdot \acute{K}\acute{P}$. $\grave{P}\acute{K}$ and $\acute{K}\acute{P}$ are the coefficients for downward and upward trsnsmission respectively through the outer core–mantle boundary. The coefficient $\grave{K}(\check{I})\acute{K}$ can be expanded into terms

$$\grave{K}(\check{I})\acute{K} = \grave{K}\acute{K} + \frac{\grave{K}\grave{I} \cdot \acute{I}\acute{K}}{1 - \acute{I}\hat{I}} \tag{14}$$

where the individual terms are generalized reflection–transmission coefficients calculated as in (6b). These terms contain both generalized vertical slownesses $\acute{\xi}$, $\grave{\xi}$ etc. and ratios of vertical wavefunctions $\pi^{(2)}/\pi^{(1)}$ that provide phase information.

From (14) it can be seen that the coefficient $\grave{K}(\check{I})\acute{K}$ describes the PKiKP wave reflected from the upper side of the inner core boundary (the coefficient $\grave{K}\acute{K}$) as well as PKIKP (the product $\grave{K}\grave{I} \cdot \acute{I}\acute{K}$) and all of the underside reflections PKIIKP + PKIIIKP + . . . comprising an interference head wave (the expansion of the factor $1/(1 - \acute{I}\hat{I})$). Similar coefficients and ray expansions can be found for other problems

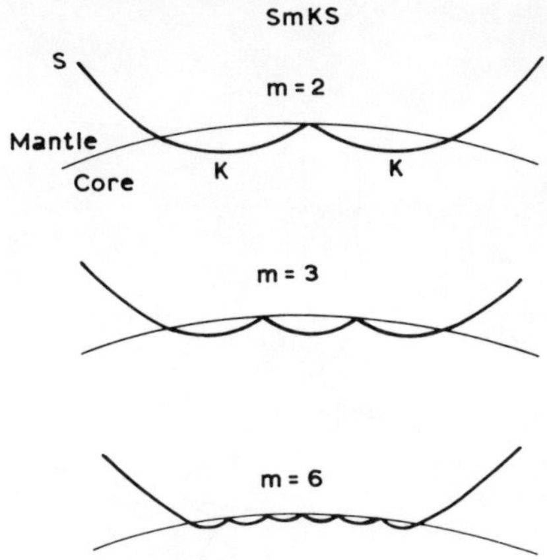

Fig. 8 An interference head wave is set up along the underside of the medium having the faster velocity. The uppermost part of the faster medium acts as a waveguide similar to a "whispering gallery". Examples of interference head waves include the Pn phase associated with turning rays in the mantle beneath the Moho, the PKIKP + PKIIKP + PKIIIKP + . . . waves on the underside of the inner-core boundary, and the SKS + SKKS + SKKKS + . . . waves on the underside of the outer-core boundary.

associated with waves incident on a boundary having a discontinuous velocity increase. Examples include the S to K interactions on the underside of the outer core (Fig. 8, from Choy (1977)), the multiple P reflections on the underside of the Moho (Menke and Richards, 1980) comprising the Pn phase, and the S to converted multiple P waves on the underside of the Moho comprising the PL phase (Frazer, 1977). These expansions and contours were first described by Ludwig (1970) in a paper on the interaction of electromagnetic waves with a cylinder having a discontinuous increase in phase velocity.

The integration contours for the inner-core problem (Fig. 7) illustrate the use of taking separate contours for different terms in the expansion (14). In effect a contour can be treated like a string that is split apart, with the separate parts, representing different parts of the integrand, taken into different parts of the complex p plane. It is important to note that the separate terms in the ray expansion of the integrand may have different singularities in the complex p plane. It is valid to allow a split portion of a contour to pass through a line of poles as long as that line of poles does not represent the poles of the particular integrand taken for that split portion of the contour. This happens for the segment taken for the $\grave{K}\acute{K}$ term in Fig. 7, which passes through a sequence of Rayleigh

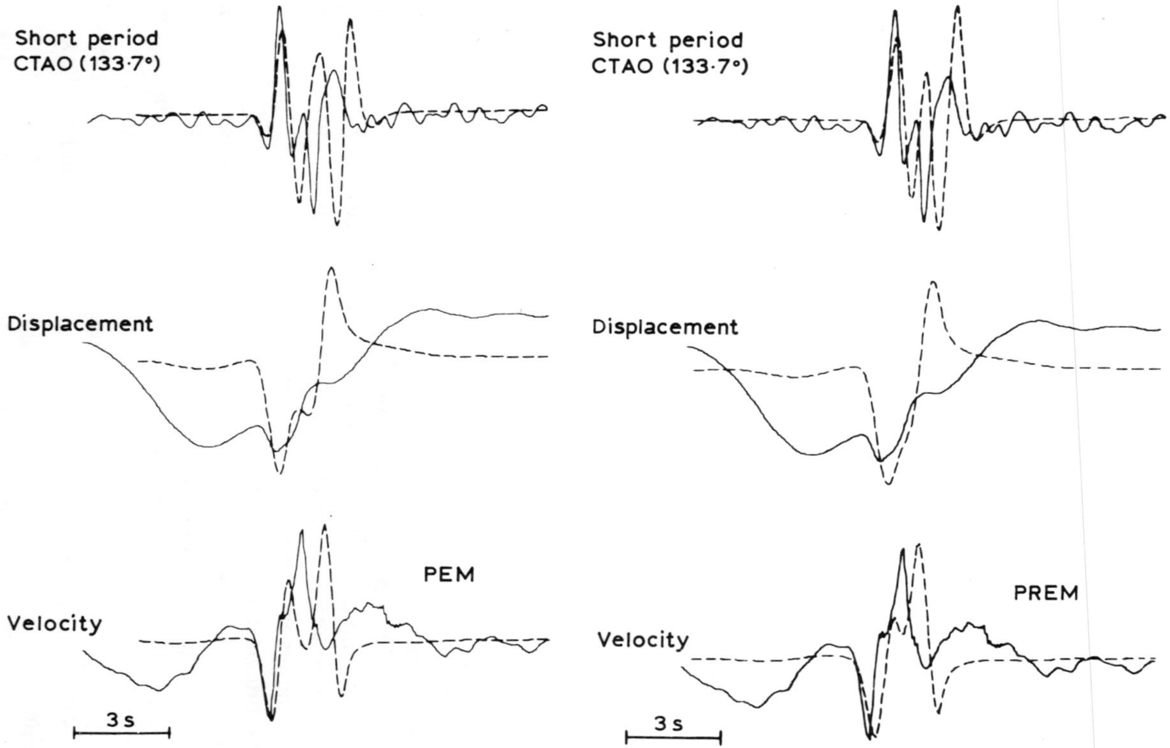
Short period
CTAO (133·7°)
Displacement
Velocity
PEM
3 s
Short period
CTAO (133·7°)
Displacement
Velocity
PREM
3 s

(a)

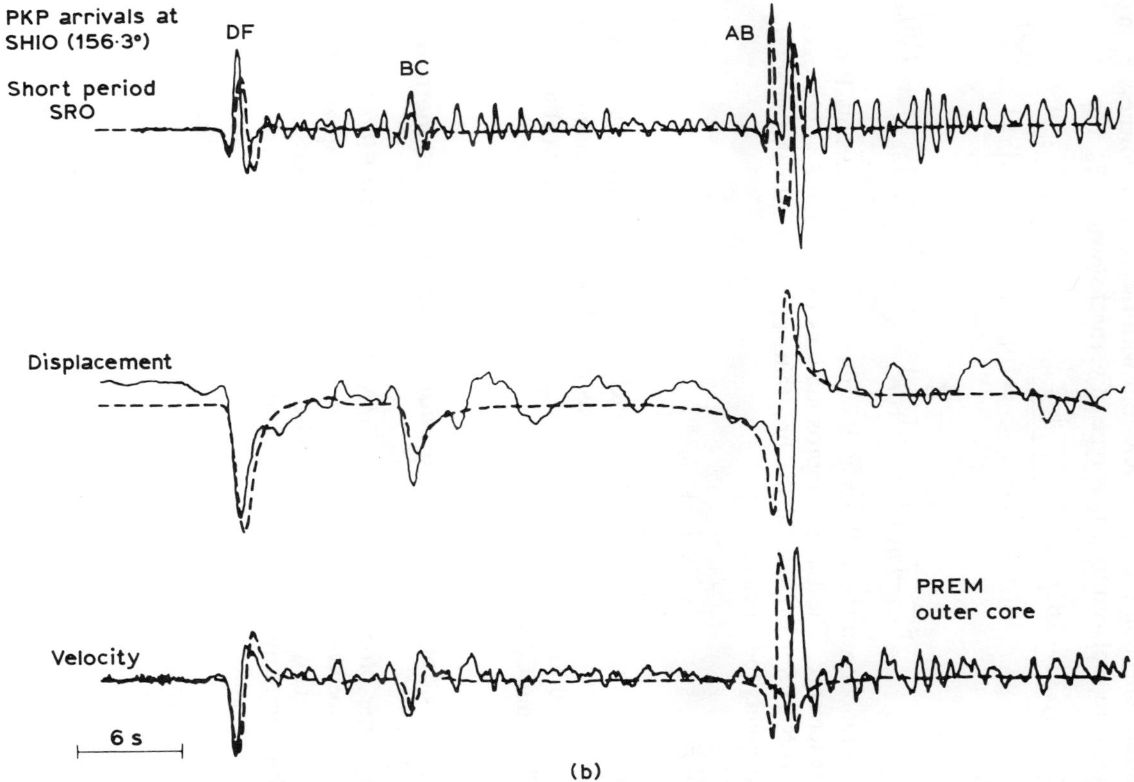

Fig. 9 (a) Synthetic seismograms for PKiKP + PKIKP + PKIIKP + . . . for a deep-focus earthquake recorded by the GDSN station CTAO at 133.7°. Results are shown for two Earth models PEM and PREM (dashed) and comparison with observed waveforms at CTAO (solid). (After Choy and Cormier (1983).) (b) Observed (solid) and synthetic (dashed) waveforms for PKP from a deep-focus earthquake observed at GDSN station SHIO at 156.3°. (After Choy and Cormier (1983).)

poles. These Rayleigh poles are poles for the $\grave{K}\grave{I} \cdot \acute{I}\acute{K}/(1-\acute{I}\grave{I})$ term but not for the $\grave{K}\acute{K}$ term.

The unexpanded $\grave{K}(\check{I})\acute{K}$ coefficient has the same algebraic form as the generalized reflection coefficient $\grave{K}\acute{K}$, but with the vertical slowness $\grave{\xi}_1$ in the inner core replaced by a new type of vertical slowness $\check{\xi}_1$, i.e.

$$K(I)K = \frac{[\rho_1(1-2\beta_1^2p^2/r_1^2)^2\grave{\xi}_2 - \rho_2\check{\xi}_1 + 4\rho_1\beta_1^4(p^2/r_1^2)\grave{\xi}_2\grave{\eta}_1\check{\xi}_1]\pi_2^{(2)}}{[\rho_1(1-2\beta_1^2p^2/r_1^2)^2\acute{\xi}_2 + \rho_2\check{\xi}_1 + 4\rho_1\beta_1^4(p^2/r_1^2)\acute{\xi}_2\grave{\eta}_1\check{\xi}_1]\pi_2^{(1)}}, \tag{15}$$

where

$$\check{\xi} = \frac{1}{-\mathrm{i}\omega(\pi^{(1)}+\pi^{(2)})}\frac{\mathrm{d}}{\mathrm{d}r}(\pi^{(1)}+\pi^{(2)}). \tag{16}$$

The wavefunction $\pi^{(3)} = \pi^{(1)} + \pi^{(2)}$ describes the phase of turning P rays in the inner core. The Langer approximations of $\pi^{(3)}$ and $\check{\xi}$ are invoked by the ILNGVS = 3 option in the programs provided with this chapter, and represents the calculation

$$\pi^{(3)} = \left(\frac{\pi}{2}\right)^{1/2} \omega^{1/6} \frac{\tau^{1/4}}{Q^{1/2}} \mathrm{Ai}\,(-\zeta). \tag{17}$$

Input for integration contours for PKP phases in the distance ranges 110°–156° is provided with the programs included with this chapter. This input was used to synthesize the seismograms shown in Figs 9(a) and (b) from Choy and Cormier (1983).

3.3 Modification for Filon integration and the slowness method

Although the program packages SCHAIRY and WIGGLER are built around a spectral method of synthesis with simple trapezoidal p integration, many subroutines can also be used in a slowness approach and with alternative forms of p integration. For example, Filon integration (Frazer and Gettrust, 1984, and Chapter III.3) can instead be used, provided that one isolates all of the factors contributing to a phase that depends approximately linearly on frequency. If the Langer approximation is used to calculate the coefficient $\grave{P}\acute{P}$ then one could write

$$\grave{P}\acute{P} = R(\omega, p)\exp\left(-2\mathrm{i}\omega\int_{r_\mathrm{p}}^{r_\mathrm{b}} \frac{\cos i}{\alpha}\,\mathrm{d}r\right), \tag{18a}$$

where

$$R(\omega, p) = \grave{P}\acute{P}\exp\left(2\mathrm{i}\omega\int_{r_\mathrm{p}}^{r_\mathrm{b}} \frac{\cos i}{\alpha}\,\mathrm{d}r\right). \tag{18b}$$

The phase $-2i\omega \int_{r_p}^{r_b} (\cos i)/\alpha \, dr$ should be then included in the total phase $J(p)$ used in Filon integration. For large $\omega\tau$, the WKBJ approximations (4a) can be substituted for the vertical wavefunctions appearing in $\grave{P}\acute{P}$. These will act to cancel the phasor appearing in (18b), returning a function $R(\omega, p)$ that is independent of frequency and that is simply equal to the phaseless plane-wave reflection coefficient. For small $\omega\tau$, the Langer approximations (4b) must be used for the vertical wavefunctions appearing in $\grave{P}\acute{P}$, and $R(\omega, p)$ becomes frequency-dependent. In a slowness approach to synthesis, the frequency-dependent function $R(\omega, p)$ can be transformed either analytically (if possible) or by fast Fourier transform into the time domain for a sequence of slownesses.

3.4 Specification of the Earth model

The Earth model is stored in a BLOCK DATA routine in the form of densities at layer boundaries and the polynomial coefficients of third-order polynomials in radius for the P and S velocities. In the package provided with this chapter, BLOCK DATA contains the coefficients of the isotropic form of PREM, referenced at 1 Hz. Given a complex radius, the subroutines CALPHA and CBETA evaluate velocities in thick layers from these polynomial coefficients. The velocities are generally complex and may also be frequency-dependent if the option for dispersive attenuation is invoked (IAD = 2). Subroutines CBETP and CALP evaluate the radial derivatives of velocities α and β respectively for a complex radius. The program package can be easily modified for velocities specified by any function that is analytic in radius by modifying these routines. An example is provided for a velocity profile consisting of a product of a polynomial in radius and an exponential.

Both the Langer approximation and its associated WKBJ limit represent solutions for body waves having single turning points. Layers that induce two turning points in a p domain of interest should be avoided. Low-velocity zones should be parametrized by several inhomogeneous layers having either uniformly positive or a uniformly negative gradient rather than as a continuous layer. Interactions with the boundaries of the low-velocity zone can be handled by the reflectivity response of the layered zone or by partial-ray expansions of that response incorporating generalized reflection–transmission coefficients. A rule of thumb to use when parametrizing an inhomogeneous layer by a polynomial or other analytic function is to not introduce pronounced kinks or inflections in the velocity profile. These inflections can generate multiple turning points and make the turning-point search converge on the wrong turning point for a particular complex p. The situations in which this happens are usually quite obvious. The plot of the integrand of (1) will either exhibit unexpected growth or decay in inappropriate regions of the complex p plane or have unexpected discontinuities. The arrival time of the calculated phase will also be in error from that predicted using ray theory. This has never been the case with many examples calculated using the PREM velocity polynomials.

A model T7 (Burdick and Helmberger, 1978) was calculated by parametrizing velocity with sixth-order polynomials in radius (Cormier and Choy, 1981). The velocity profile in the thick layer below the lid exhibited some problems with multiple turning points in a narrow range of complex p. These could be eliminated by dividing the layer into two layers at the position of the transition from a strong to a weaker gradient near 200 km (Fig. 10). The situation in which this happens becomes obvious through discontinuities and inappropriate exponential growth when the integrand of (1) is plotted.

The calculations of the upper mantle triplications of model T7 were performed using integration paths and integrands analogous to those described for PKP phases, in which a product of two functions analogous to $\grave{K}(\check{I})\acute{K}$ were constructed using the Langer option ILNGVS = 3 for

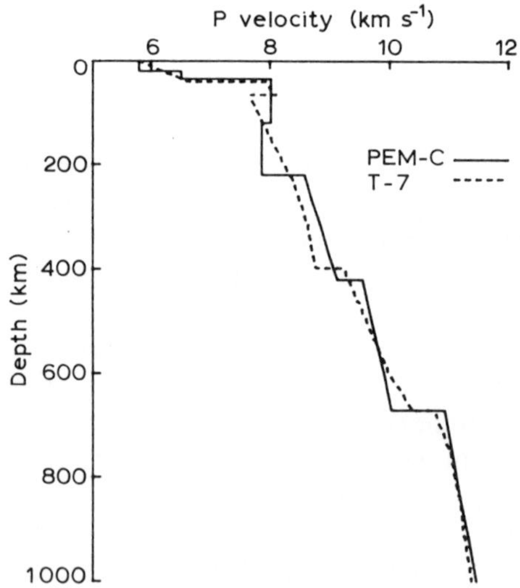

Fig. 10 A comparison of the P velocity profiles for the upper mantle of the isotropic 1 Hz PEM-C (Dziewonski *et al.*, 1975) with model T7 (Burdick and Helmberger, 1978). T7 was parametrized by a single polynomial between the base of the lid of the low-velocity zone to 400 km depth.

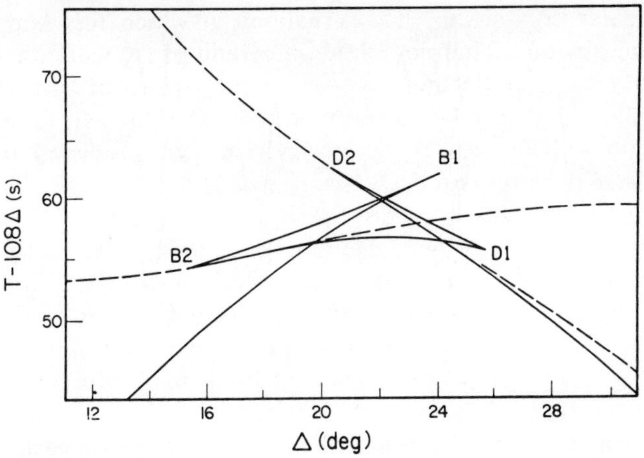

Fig. 11 Reduced travel-time curve for model T7 with weaker amplitude arrivals dashed. Interactions with the low-velocity-zone lid are not shown.

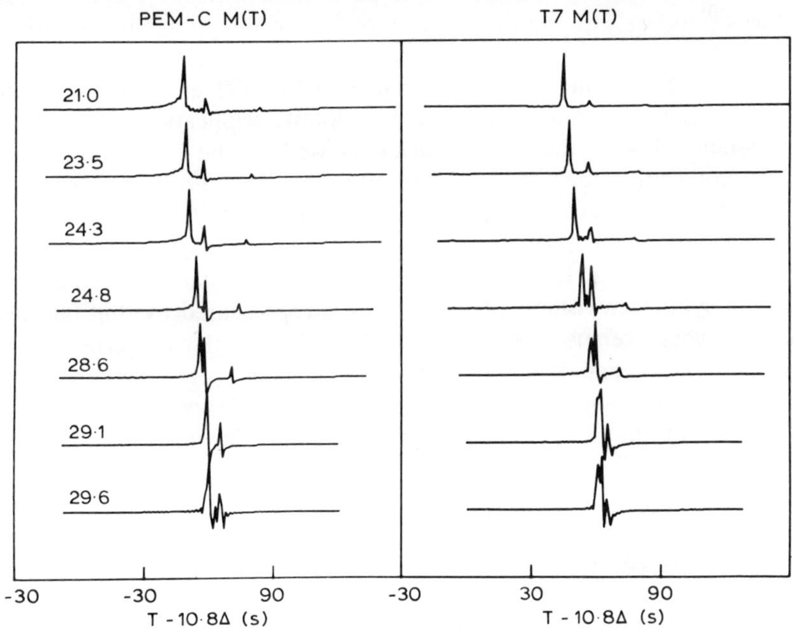

Fig. 12 Displacement responses for an explosive point source for the P-wave arrivals predicted by PEM-C and T7 at distances greater than 20°. Numbers to the left of the traces are distances corresponding to data records studied by Burdick and Helmberger (1978). Note the strong secondary pulse of PEM-C at 29.6°. This is due to the extension of cusp D1 to longer distances in PEM-C due to the larger velocity jump of PEM-C at 670 km depth. (After Cormier and Choy (1981).)

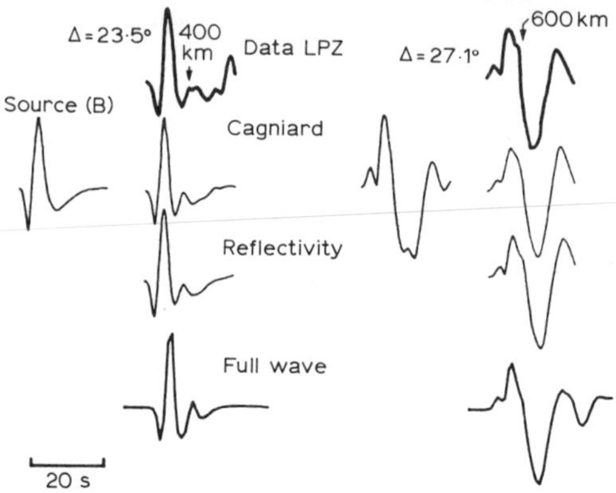

Fig. 13 A comparison of synthetics computed in the T7 model by different methods and model parametrizations. A source function and LP-WWSSN instrument response have been convolved. (After Burdick and Orcutt (1979) and Cormier and Choy (1981).)

vertical wavefunctions on the underside of the 400 and 670 km discontinuities. The integrands described the double triplication (Figs 11 and 12) associated with these boundaries as well as the interference head waves along the underside of the boundaries. The results compare well with those obtained by the reflectivity and the Cagniard methods (Burdick and Orcutt, 1979) in flattened homogeneously layered representation of T7 (Fig. 13). In general, close comparisons are usually found between synthetics computed using models represented by thin homogeneous layers versus models represented by thick vertically inhomogeneous layers. This is true if the displacements are convolved with an instrument response and/or a source pulse longer than the time separation of interlayer multiples in the thinly layered representation (Choy *et al.*, 1980).

3.5 Turning-point calculations

Turning-point radii are needed for the evaluation of the delay times τ used in the Langer approximation and in the construction of the phase factor of turning rays above boundaries. The turning radii are found by the Newton–Raphson method (see e.g. Acton, 1970). An initial guess is made for r_p at a given ray parameter p and r_p is found iteratively to within an accuracy that will depend on the desired accuracy of the phase calculations. The iteration consists in evaluating

$$r_{\mathrm{p}}^{\mathrm{new}} = r_{\mathrm{p}}^{\mathrm{old}} - \frac{pv(r_{\mathrm{p}}^{\mathrm{old}}) - r_{\mathrm{p}}^{\mathrm{old}}}{pv'(r_{\mathrm{p}}^{\mathrm{old}}) - 1}, \tag{19}$$

where v' denotes the radial derivative of velocity returned by CALP or CBETP. Starting along a segment of integration, r_{p} is initially set equal to one of the boundaries of the layer. For a sequence of r_{p} needed in a p integrand, the previous r_{p} obtained at an old p can be saved and then used as the first guess for r_{p} at the new p. This procedure is accomplished by updating an array of turning points stored in a common block labelled RTP. The updating of turning points makes the algorithm converge in one or two iterations for each p along an integration contour Γ. This algorithm makes it simple to incorporate almost any analytic function to describe the P velocity profile in a given layer.

3.6 Delay time τ

τ can be calculated by several methods. The most general method is by numerical integration. Subroutine RAYLEG calculates $\tau = \int_{r_{\mathrm{p}}}^{r_{\mathrm{b}}} (\cos i)/v \, \mathrm{d}r$ by a combination of a Romberg integration (see e.g. Acton, 1970) and a three-point quadrature formula described by Jeffreys and Jeffreys (1956). The accuracy of the Romberg integration can be controlled by input parameters, but the accuracy of the three-point formula decreases the greater the distance between r and r_{p}. For these reasons, Romberg integration is used to within 100 km of the turning point and three-point quadrature is used for the remainder of the path. A linear path in complex r is assumed between r_{b} and r_{p}. The three-point formula for functions that behave like $x^{1/2}$ or $x^{-1/2}$ at one of the integration limits has been found to be sufficiently accurate for synthesis of seismograms having frequencies up to 4 Hz. The transition from Romberg to three-point integration at 100 km can be modified by changing a parameter in a common statement of subroutine ROMTAU. Greater accuracy can be obtained by changing parameters in the convergence test in the Romberg integration.

Alternative τ calculations based on the $v = ar^b$ and the $v = a + br^2$ interpolations have been described by Frazer (1983) and Chapman (1978). Parametrization in terms of polynomials in radius or depth over large depth ranges described in this chapter, while limited in describing extremely complex models, offers some advantages in the regularity of the integrand in (1). Specifically, it has the advantage of avoiding discontinuities in generalized reflection–transmission coefficients due to second-order discontinuities in velocity. Whenever a second-order discontinuity occurs in a model, its effect can be included by calculating a generalized reflection–transmission coefficient or a reflectivity response as described by Doornbos (1981).

Yet another method of calculating τ is to parametrize the thick layers by making the flattened depth coordinate a polynomial in reciprocal flattened velocity, i.e. $\ln r = \text{poly}\,[r/v(r)]$. The integral for τ can then be calculated analytically with formulae given by Červený and Jansky (1983). The turning-point radius need not be known, since a change in the variable of integration from r to r/v makes the lower limit of the integral equal the ray parameter p. A subroutine for this method of calculating τ is provided (see TAUFST) with $\ln r = a + b(r^2/v^2)$.

3.7 Langer-approximated vertical slownesses and wavefunctions

Subroutine LANGER calculates $\pi^{(1),(2)}$, $\sigma^{(1),(2)}$, $\acute{\eta}$, $\grave{\eta}$, $\acute{\xi}$ and $\grave{\xi}$ by formulae from Abramowitz and Stegun (1964). The basic philosophy, which is common to many subroutines that calculate special functions, is to patch together an asymptotic formula appropriate for large arguments $\omega\tau$ with a series solution that converges rapidly for small $\omega\tau$. The two series are patched together at a value of $\omega\tau$ that requires only a small number of terms for either series solution to be highly accurate at the patching point. $\omega\tau = 3.5$ was found to be optimal in this sense.

LANGER also returns the generalized slownesses $\acute{\eta}$, $\grave{\eta}$, $\acute{\xi}$ and $\grave{\xi}$. An option is given in the calling program to calculate the generalized vertical slownesses by their WKBJ limits or by the Langer approximation. If the WKBJ limits are selected, subroutine VALUES places branch cuts in $\acute{\eta}$, $\grave{\eta}$, $\acute{\xi}$ and $\grave{\xi}$ at 60° angles with respect to the real ray-parameter axis. These branch cuts are selected to mimic the high-frequency behaviour of the Langer-approximated $\acute{\eta}$, $\grave{\eta}$, $\acute{\xi}$ and $\grave{\xi}$, which have zeros in the complex p plane aligned at 60° angles with respect to the real p axis for small values of $\omega\tau$ (Figs 6 and 7). To reproduce these WKBJ limits, the generalized vertical slownesses for up and downgoing P waves $\acute{\xi}$ and $\grave{\xi}$ are calculated by

$$\acute{\xi} = e^{-i\pi/3}\left[\left(\frac{1}{\alpha} + \frac{p}{r}\right)e^{i\pi/3}\right]^{1/2}\left[\left(\frac{1}{\alpha} - \frac{p}{r}\right)e^{i\pi/3}\right]^{1/2}, \tag{20a}$$

$$\grave{\xi} = e^{i\pi/3}\left[\left(\frac{1}{\alpha} + \frac{p}{r}\right)e^{-i\pi/3}\right]^{1/2}\left[\left(\frac{1}{\alpha} - \frac{p}{r}\right)e^{-i\pi/3}\right]^{1/2}. \tag{20b}$$

The vertical wavefunctions $\pi^{(1),(2)}$ and $\sigma^{(1),(2)}$ are returned in the form of a complex constant having a magnitude of the order of 1 and an exponential, for example

$$\pi^{(1)}(r) = \hat{C}e^{m}. \tag{21}$$

When wavefunctions are subsequently combined by multiplication and/or division, the values of $\hat{C}$ and m are manipulated to avoid underflow and overflow. In domains of ray parameter p that correspond to travelling waves, which turn below radius r, m is set equal to zero and $\hat{C}$ incorporates the phase $\exp[i\omega \int_{r_p}^{r} (\cos i)/v \, dr]$. In domains corresponding to waves turning above r, m is not equal to zero and $\pi^{(1),(2)}$ can be exponentially large or small.

4 EXTENSIONS AND SPECIAL APPLICATIONS

4.1 Attenuation

The time lag of strain due to an applied stress, which is a feature of all linear mechanisms of intrinsic attenuation, is expressed by a complex modulus $\hat{M}(\omega)$ in the relation between Fourier-transformed stress and strain, $\tilde{e}(\omega) = [1/\hat{M}(\omega)]\tilde{\sigma}(\omega)$. Complex velocities $\hat{v}(\omega)$ can then be defined by $\hat{v}(\omega) = [\hat{M}(\omega)/\rho]^{1/2}$. The effect of attenuation on body waveforms is then simply given by the analytic continuation of all formulae to complex values of velocity (Kennett, 1975; Cormier and Richards, 1976).

The frequency dispersion of complex velocities is an expression of the time lag of the stress–strain hysteresis. High frequencies travel faster than low frequencies. The relations between real and imaginary parts of complex velocity are equivalent to the Kramers–Krönig relations, which were originally formulated for the complex index of refraction for the propagation of electromagnetic waves. The dispersion of complex velocities in the earth is often taken to be that for an attenuation model in which Q is nearly constant over several decades of frequency. Liu *et al.* (1976) give a formula for the complex modulus of such a solid in which Q is nearly constant over a band of radian frequencies from $\omega = 1/\tau_1$ to $\omega = 1/\tau_2$. From this modulus, complex velocities can be constructed. This is the modulus used to define complex, frequency dependent velocities in subroutines CALPHA and CBETA when the flag IAD is set equal to 2. For example, $\hat{\alpha}$ is calculated by

$$\hat{\alpha}(\omega) = \alpha_0(r) \frac{\left[1 + \dfrac{2}{\pi} Q^{-1}(r) D(\omega)\right]^{1/2}}{\left[1 + \dfrac{2}{\pi} Q^{-1}(r) D(2\pi)\right]^{1/2}}, \tag{22}$$

where

$$D(\omega) = \ln\left[\left(\omega i + \frac{1}{\tau_1}\right)\Big/\left(\omega i + \frac{1}{\tau_2}\right)\right]. \tag{23}$$

Here $\alpha_0(r)$ is a real velocity function referenced at 1 Hz. In the subroutines given with this chapter, $\alpha_0(r)$ and $\beta_0(r)$ are taken from the velocity polynomials for the isotropic PREM at 1 Hz. Figure 14 was constructed from output of CALPHA for different values of τ_1 and τ_2.

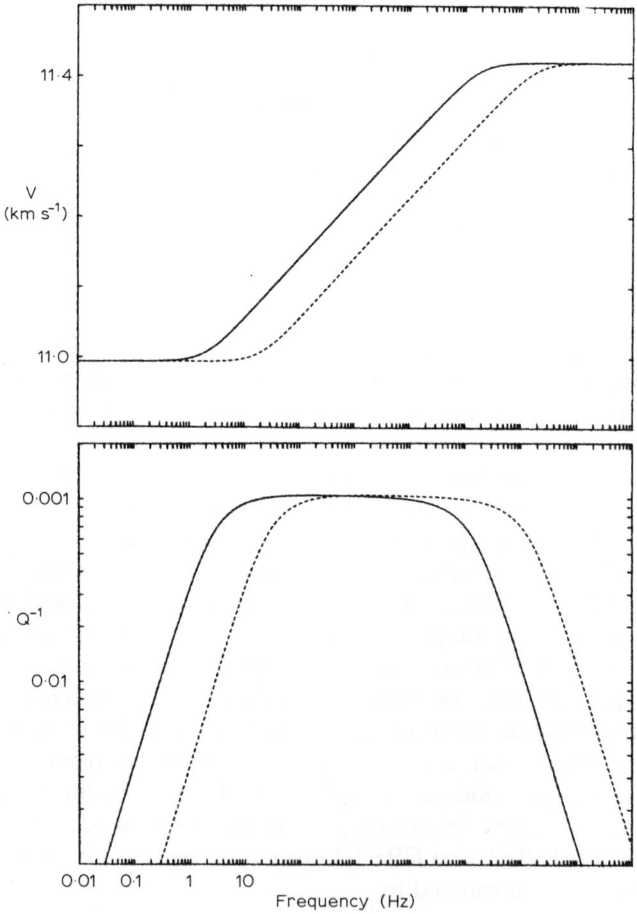

Fig. 14 Velocity dispersion and attenuation for two possible (solid and dashed) absorption-band models of Q_β for the Earth's inner core. The velocity V is that for P waves at the top of the inner core. This figure was prepared using the complex frequency-dependent velocity routines of the programs provided with this chapter.

Re $[v(\omega)]$ and $Q^{-1} = \text{Im}\,[\hat{M}(\omega)]/\text{Re}\,(\hat{M}(\omega))$ are plotted as functions of frequency.

An option is also provided for attenuation, in which dispersion is neglected. This is invoked by IAD = 1. Complex velocities are defined by

$$\hat{\alpha}(r) = \alpha_0(r)[1 - \tfrac{1}{2}\mathrm{i}Q^{-1}(r)]. \tag{24}$$

Velocity dispersion may be neglected for sufficiently narrowband calculations and/or for applications in which details of a pulse shape and rise time are not needed. $\alpha_0(r)$ should ideally be chosen to be the real part of velocity referenced at a frequency in the centre of the narrow band of frequencies.

Calculations with dispersive frequency-dependent velocities are computationally expensive since calculation of quantities such as τ must now be included in a frequency loop. Examples of such calculations are given in Lundquist and Cormier (1980). In that paper τ_2 was tied to a temperature and pressure profile of the Earth's mantle by an expression for a rate activated process:

$$\tau_2 = \tau_0 \exp\,[E^* + PV^*/RT], \tag{25}$$

where E^* and V^* are the activation energy and volume respectively required for the movement of Avogadro's number of lattice defects out of a potential well. The method used for the calculation of τ and turning point r_p easily allows inclusion in CALPHA and CBETA of almost any radius- and frequency-dependent function describing velocity. The algorithm for finding the turning point, however, requires a calculation of the radial derivatives of complex velocities. For complicated parametrizations of velocity, these are found in CALP and CBETP by difference derivatives.

An alternative procedure for handling complex dispersive velocities has been described by Choy and Cormier (1986). In this method, a seismogram is first synthesized in a model having real velocities by either ray theory or by a spectral or slowness technique of ray-parameter integration. A minimum-phase operator is then calculated for a prescribed frequency dependence of path-integrated attenuation $t^* = \int Q^{-1}/v\,\mathrm{d}s$. The problem amounts to finding the phase response of a minimum-phase filter from its amplitude response. This minimum-phase operator is then convolved with the seismogram calculated in the non-attenuating model.

Other options (e.g. Q proportional to a power of frequency) can be incorporated with a code that accommodates higher values of the IAD flag (e.g. IAD ~ = ~3). All modifications can be concentrated in the subroutines for complex velocities.

4.2 Transverse isotropy

A transversely isotropic layer having a vertical axis of symmetry and vertically varying elastic constants A, C, F, L and N can be handled by simple modifications of the τ and turning-point calculations together with simple substitutions in the calculation of reflection–transmission coefficients and fundamental matrix elements (Cormier, 1986). The delay time τ is calculated by numerical integration after finding the turning point r_p by a Newton–Raphson iteration. However, τ is now given by

$$\tau = \int_{r_p}^{r} \lambda \, dr. \tag{26}$$

For SH waves

$$\lambda = \left(\frac{N}{L}\right)^{1/2} \left[\left(\frac{1}{v} + \frac{p}{r}\right)\left(\frac{1}{v} - \frac{p}{r}\right)\right]^{1/2}, \tag{27a}$$

where $v = (N/\rho)^{1/2}$. For P and SV waves

$$\lambda = s_3^{1/2} \left[\left(\frac{1}{v} + \frac{p}{r}\right)\left(\frac{1}{v} - \frac{p}{r}\right)\right]^{1/2} \tag{27b}$$

where $v = [s_3/(s_1 \pm R)]^{1/2}$. s_1, s_2, s_3 and R are functions of the Love (1944) constants A, C, F, L and N and the density; they are defined by Woodhouse (1981). The $\pm R$ corresponds to P waves for $-R$ and SV waves for $+R$. Turning points are found as in the purely isotropic case by an iterative search for zeros of $pv - r$, where v is now one of the specially defined velocities given above. Vertical wavefunctions $\pi^{(1),(2)}$ and $\sigma^{(1),(2)}$ and generalized vertical slownesses $\acute{\xi}$, $\grave{\xi}$, $\acute{\eta}$ and $\grave{\eta}$, which appear in the fundamental matrix of a vertically varying layer, are modified in their WKBJ or Langer forms by simple substitutions of λ for $(1/\alpha^2 - p^2/r^2)^{1/2}$ and $(1/\beta^2 - p^2/r^2)^{1/2}$ and substitution of the modified τ functions.

The modifications of the matrix $\boldsymbol{A}$ in (9) are given by Woodhouse (1978), from which a modified fundamental matrix $\boldsymbol{F}$ can be found by solving an eigenvector problem for each eigenvalue λ. The modified fundamental matrix $\boldsymbol{F}$ is given by Doornbos *et al.* (1986). Reflection/transmission coefficients are determined as in (6b) using this modified fundamental matrix.

These modifications are given in the subroutines associated with this chapter in a section on transverse isotropy. Input for the examples calculated in Cormier (1986) is also provided. This example is for a transversely isotropic layer in the lowermost mantle. Synthetics and particle motion of the transverse component of S + ScS and the radial component of S + SKS + SKKS . . . + ScS for this example are shown in Figs 15(a) and (b).

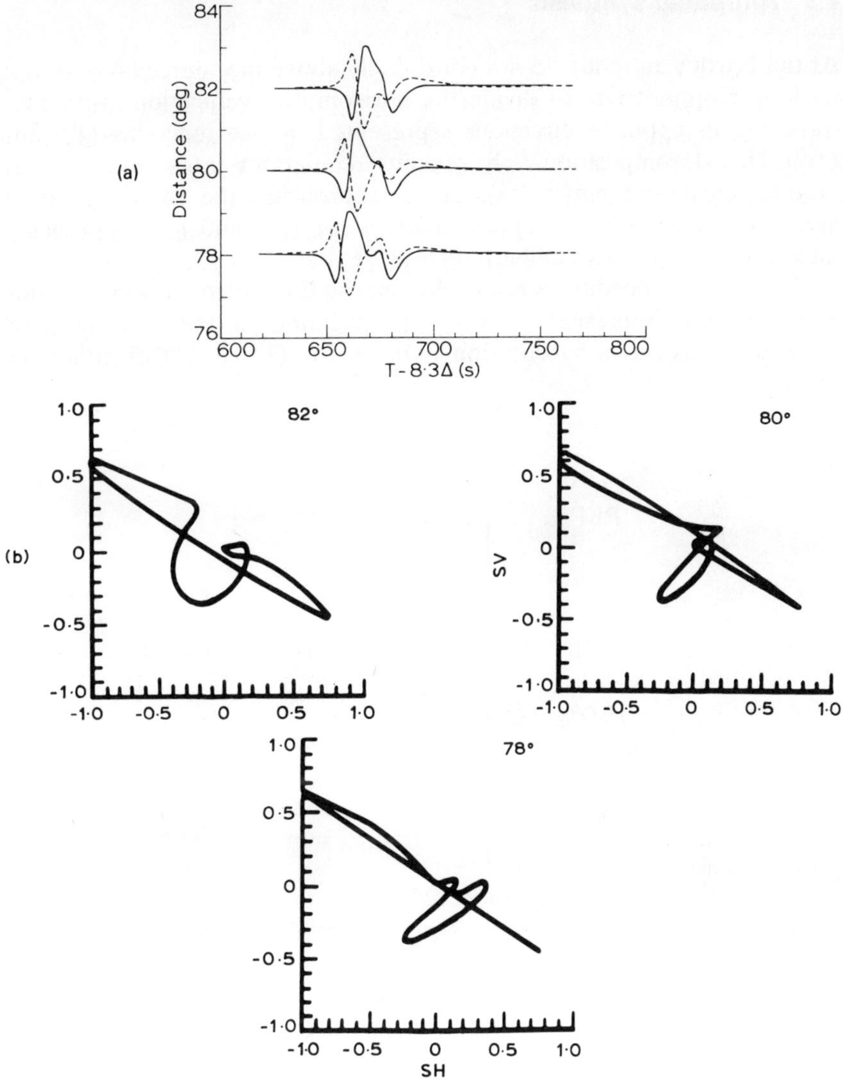

Fig. 15 Synthetic seismograms for SH (dotted) and SV (solid) waves interacting with a transversely isotropic layer at the base of the Earth's mantle. The source radiation pattern and receiver azimuths were chosen such that the direct S wave contains roughly equal amounts of SH and SV energy. (b) Particle motions of the synthetic seismograms shown in (a). The orthogonal rotation of the particle motion is associated with the ScS wave, which reverses polarity for SV motion not SH. The effects of the transversely isotropic layer increase as the S wave approaches more grazing incidence on the core–mantle boundary at greater distances, ScS spending a longer time in the transversely isotropic layer. (After Cormier (1986).)

4.3 Antipodal synthesis

At the Earth's antipode ($\Delta = \pi$) or directly above the source ($\Delta = 0$), it is no longer appropriate to divide the horizontal wavefunction into waves travelling in opposite directions represented by the functions $Q^{(1)}$ and $Q^{(2)}$. This decomposition only remains regular for large values of the product $\omega p \Delta$ and $\omega p(\pi - \Delta)$. As Δ approaches the source point or antipode, the asymptotic approximations for $Q^{(1)}$ and $Q^{(2)}$ approach a logarithmic singularity (Gilbert, 1976).

The correct procedure is not to decompose the horizontal wavefunction into functions representing waves propagating in the $+\Delta$ and $-\Delta$ directions. Instead a substitution is made for $Q^{(2)}$ in (1) of either the

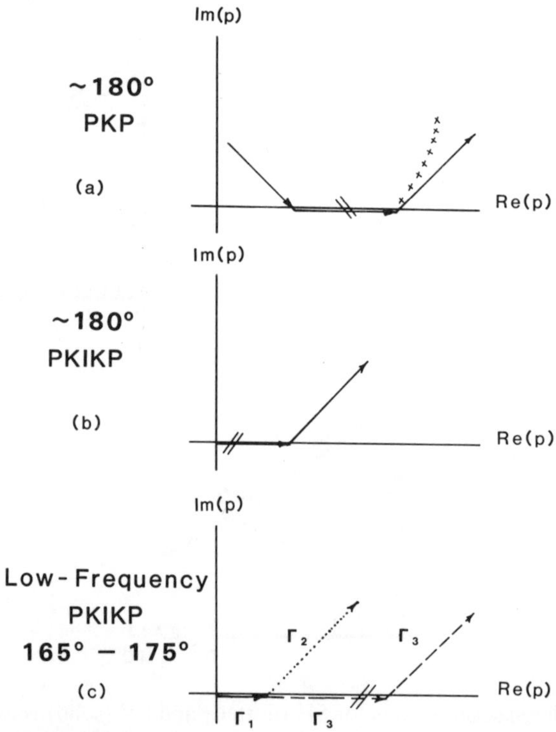

Fig. 16 Integration contours in the complex p plane for antipodal synthesis of PKP (a) and PKIKP (b, c). The Szego asymptotic expansion is used for $P_l(\pi - \Delta)$ near $\Delta = \pi$. At 175° and low frequencies for the PKIKP phase (c) it is convenient to split the integrand along separate contours for $Q_l^{(1)}$ and $Q_l^{(2)}$ when $\omega p(\pi - \Delta)$ is greater than 10. Along Γ_1 $\omega p(\pi - \Delta)$ is less than 10 and the Legendre function P_l is taken as the horizontal wavefunction in the integrand; along Γ_2, $Q_l^{(2)}$ is taken; and along Γ_3, $Q_l^{(1)}$ is taken.

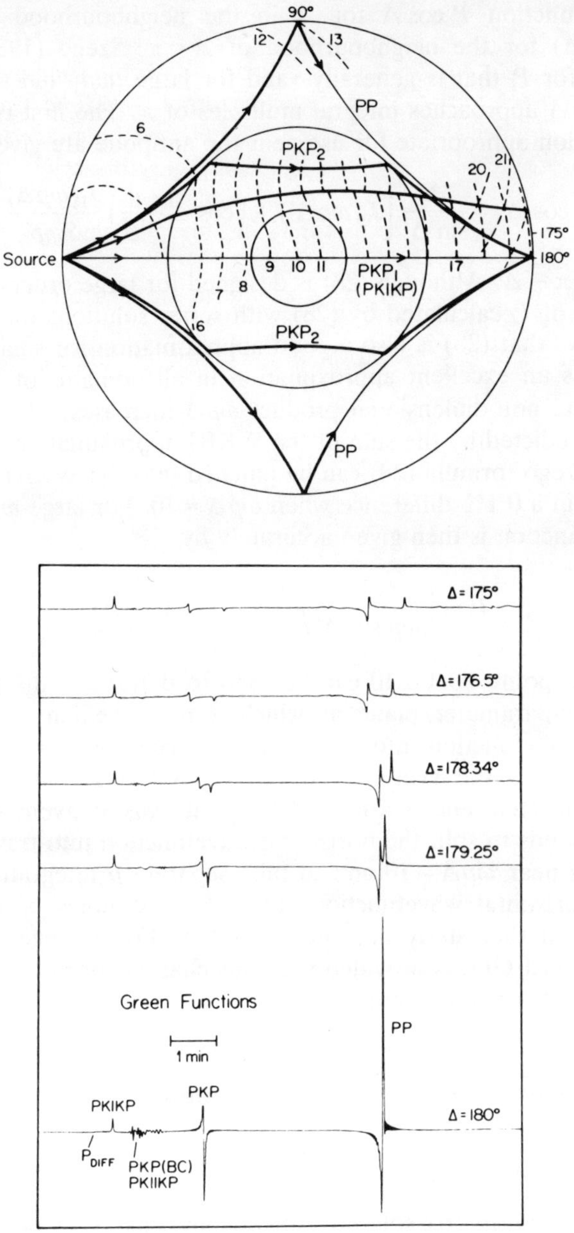

Fig. 17 Ray paths and displacement responses of PKP and PP waves observed at antipodal distances. An explosive point source at zero depth is assumed. (After Rial and Cormier (1980).)

Legendre function $P_l \cos \Delta$ for Δ in the neighbourhood of 0 or of $P_l \cos(\pi - \Delta)$ for the neighbourhood of $\Delta = \pi$. Szego (1934) gave an expression for P_l that is generally valid for large $|\omega p|$ and that remains accurate as Δ approaches integral multiples of π. The first two terms of this expression appropriate for use near the antipode are given by

$$P_l \cos \Delta' = \frac{\Delta'}{\sin \Delta'} \left[J_0(\omega p \Delta') + \left(\cos \Delta - \frac{1}{\Delta} \right) \frac{J_1(\omega p \Delta)}{8 \omega p} \right], \tag{28}$$

where $\Delta' = \pi - \Delta$. Although (28) is designed for large orders $l = \omega p - \frac{1}{2}$, comparison of P_l calculated by (28) with series solutions for P_l valid for small l shows that (28) is also a good approximation for small l. In fact, (28) remains an excellent approximation in all domains of the product $\omega p \Delta$. As the non-dimensional product $\omega p \Delta$ increases, (28) approaches the value predicted by the sum of the WKBJ approximations to $Q^{(1)}$ and $Q^{(2)}$. The Szego formula (28) can be patched into the WKBJ approximation to within a 0.1% difference when $\omega p \Delta = 10$. For large $\omega p \Delta > 10$ the Legendre function is then given accurately by

$$P_l = \left(\frac{2}{\omega p \sin \Delta'} \right)^{1/2} \cos(\omega p \Delta - \tfrac{1}{4}\pi). \tag{29}$$

The patch point $\omega p \Delta = 10$ can be used to determine the point in the complex ray-parameter plane at which it is convenient to divide the horizontal wavefunction into travelling-wave components $Q^{(1)}$ and $Q^{(2)}$. For example, in the synthesis of the PKIKP phase at the antipode described in Rial and Cormier (1980), it was convenient in some frequency bands to split the horizontal wavefunction into travelling wave components near $\omega p \Delta = 10$ and to take separate p integration contours for each horizontal wavefunction (Fig. 16). Examples of PKP phases synthesized in that study are shown in Fig. 17. A Legendre-function subroutine, SZEGO, is provided with this chapter for use with problems at the antipode or source point.

4.4 Reflectivity of vertically inhomogeneous layers

4.4.1 The *n*-layered problem

The fundamental matrix solutions for inhomogeneous layers can be used to find generalized reflectivity and reflection coefficients for a single boundary or a sandwich of several inhomogeneous layers. Equation (6b) expresses the problem of a P wave reflected by a single boundary. The notational extension of (6b) to an n-layered problem is quite simple using propagator matrices. All the features of the n-layered problem can be illustrated by a three-layered problem. Consider a sequence of layers having indices $n-1$, n and $n+1$, the index number increasing with increasing depth. A downgoing P wave is assumed to be incident in layer $n-1$ on the boundary between $n-1$ and n. The stress–displacement vector on the upper side of the boundary r_n between $n-1$ and n is related to the stress–displacement vector on the lower side of the boundary r_{n+1} by

$$\mathbf{F}_{n-1}(r_n)\begin{bmatrix} \grave{P}\acute{P} \\ 1 \\ \grave{P}\acute{S} \\ 0 \end{bmatrix} = \mathbf{P}(r_n, r_{n+1})\mathbf{F}_{n+1}\begin{bmatrix} b_1 \\ b_1 \\ b_2 \\ b_2 \end{bmatrix}, \tag{30}$$

where $\mathbf{P}(r_n, r_{n+1}) = \mathbf{F}_n(r_n)\mathbf{F}_n^{-1}(r_{n+1})$ is the propagator matrix. The vector $\boldsymbol{b}^{\mathrm{T}} = [b_1 \quad b_1 \quad b_2 \quad b_2]$ is chosen to describe turning P and S waves in the lowermost layer. The vertical wavefunctions $\pi^{(3)} = \pi^{(1)} + \pi^{(2)}$ and $\sigma^{(3)} = \sigma^{(1)} + \sigma^{(2)}$ given by the product of $\mathbf{F}_{n+1}(r_{n+1})\boldsymbol{b}^{\mathrm{T}}$ decay exponentially in the ray-parameter domain corresponding to rays bottoming above r_{n+1}.

The vector $\boldsymbol{b}$ may alternatively be taken as $\boldsymbol{b}^{\mathrm{T}} = [0 \quad b_2 \quad 0 \quad b_1]$ to describe downgoing waves in the lowermost layer. This choice corresponds to the boundary condition normally used in calculations with homogeneous boundaries. If this choice is used then the lowermost layer is taken to have arbitrarily high P and S velocity. This capping layer is chosen low enough such that the P and S waves totally reflected from it arrive outside of the time window of interest (Harvey, 1981).

The reflectivity functions $\grave{P}\acute{P}$ etc. can be found by inverting either matrices on the left- or right-hand side of (30) and solving the linear equation by Cramer's rule. This procedure works for the 2×2 fundamental and propagator matrices of SH and acoustic waves but leads to numerical instabilities for P and SV waves in solids. The problem consists in attempting to cancel exponentially large terms, which algebraically cancel, by arithmetic performed on computers having a finite word length. The problem can be crudely handled by dividing the model into thinner layers at high frequencies, but two different algorithms are available, which solve the problem and also minimize the number of arithmetic operations. In order to briefly outline these algorithms it is best to rewrite (30) as

$$[b_1 \quad b_1 \quad b_2 \quad b_2]^{\mathrm{T}} = \mathbf{S}[\grave{P}\acute{P} \quad 1 \quad \grave{P}\acute{S} \quad 0]^{\mathrm{T}}, \tag{31}$$

where $\mathbf{S} = F_{n+1}^{-1}(r_{n+1})\mathbf{P}(r_{n+1}, r_n)\mathbf{F}_{n-1}(r_n)$.

4.4.2 The Kennett algorithm

The algorithm developed by Kennett (1974) and described in Chapter III.1 consists in grouping fundamental matrices in the form of products associated with layer boundaries, i.e.

$$\mathbf{S} = \mathbf{Q}_{n+1}\mathbf{Q}_n, \tag{32}$$

where

$$\mathbf{Q}_{n+1} = \mathbf{F}_{n+1}^{-1}(r_{n+1})\mathbf{F}_n(r_{n+1}),$$
$$\mathbf{Q}_n = \mathbf{F}_n^{-1}(r_n)\mathbf{F}_{n-1}(r_n).$$

From the equation (6b) relating the stress–displacement vector across a single boundary, it is obvious that the elements of the matrix $\mathbf{Q}$ will consist of simple products of generalized reflection–transmission coefficients associated with a single boundary. These coefficients will contain factors specifying phase information through ratios of the type $\pi_{n+1}^{(1)}/\pi_n^{(2)}$. Products of matrices $\mathbf{Q}$ can be built up as a progressive ray expansion, consisting of reverberations between layers. The elements of the $\mathbf{Q}$ products and details on how the reverberatory sequence is constructed are given by Kennett (1974, 1983, and Chapter III.1). Subroutine COEFF included with this chapter can be used to construct a reflectivity function with these techniques, provided that care is taken to include the phase information included in the ratios of vertical wavefunctions. This phase information is returned by calls to the LANGER subroutine for calculation of vertical wavefunctions $\pi^{(1)}$ etc.

4.4.3 The Abo-Zena algorithm

Another algorithm that eliminates the numerical problems associated with P–SV reflectivity propagates an antisymmetric matrix of five independent quantities. The derivation of the algorithm, which is closely related to the delta-matrix method of Kind (1976), is described by Abo-Zena (1979) and Harvey (1981). Applicatons to vertically inhomogeneous layers are described in Cormier (1980) and Doornbos (1981). In this scheme the reflectivity function $\grave{P}\acute{P}$ can be written as

$$\grave{P}\acute{P} = \frac{[0 \;\; 1 \;\; 0 \;\; 0]\mathbf{F}_{n-1}^{\mathrm{T}}\mathbf{P}(r_{n+1}, r_n)^{\mathrm{T}}\mathbf{Y}\mathbf{P}(r_{n+1}, r_n)\mathbf{F}_{n-1}[0 \;\; 0 \;\; 1 \;\; 0]^{\mathrm{T}}}{-[1 \;\; 0 \;\; 0 \;\; 0]\mathbf{F}_{n-1}^{\mathrm{T}}\mathbf{P}(r_{n+1}, r_n)^{\mathrm{T}}\mathbf{Y}\mathbf{P}(r_{n+1}, r_n)\mathbf{F}_{n-1}[0 \;\; 0 \;\; 1 \;\; 0]^{\mathrm{T}}}. \tag{33}$$

Y is an antisymmetric matrix of the form

$$\begin{bmatrix} 0 & d_1 & d_2 & d_3 \\ -d_1 & 0 & d_4 & -d_2 \\ -d_2 & -d_4 & 0 & d_5 \\ -d_3 & d_2 & -d_5 & 0 \end{bmatrix},$$

$$d_1 = -(A^2 + B^2 \xi \check{\eta}), \qquad d_2 = r^{-1} p A + B \xi \check{\eta},$$

$$d_3 = i\rho \check{\eta}, \qquad d_4 = -i\rho \check{\xi}, \qquad d_5 = \check{\xi}\eta + r^{-2} p^{-2}.$$

A subroutine for P–SV reflectivity based on (33) is provided with this chapter.

Although the Kennett and Abo-Zena algorithms are equally efficient and numerically stable, the Kennett algorithm is the more popular. This is because it offers the opportunity of controlling the number of interlayer multiples included in a synthetic calculation. A ray expansion can be easily truncated whenever the amplitude of the next multiple is vanishingly small or if it arrives outside the time window of interest. In algorithms that include all multiples, such as (33), special modifications are required at the stage of transformation into the time domain in order to handle the late-arriving reverberations that fold back into the finite time window defined by the number of frequency points and interval of frequency sampling. The basic modification consists in introducing a small complex part to the frequency, which acts to attenuate late-arriving energy when the complex spectrum is transformed into the time domain (Rosenbaum, 1974).

5 CONCLUSIONS

This chapter has discussed the synthesis of body waves in a radially symmetric model specified by a small number of velocity functions. The velocity functions are analytic in radius over thick regions of the model. The analyticity of the velocity functions offers an opportunity for working with models directly connected to properties related to the state variables of temperature and pressure.

In the vertically varying layers, the radially separated part of the wave equations for P and S waves is solved by a Langer or WKBJ approximation. The Langer approximation is invoked for grazing-incidence problems to correctly account for tunnelling and frequency-dependent reflection and transmission. The computational cost of thick layers and the

Langer approximation is no more and usually less than that associated with the calculation of the reflectivity of a flattened model broken into thin homogeneous layers.

The WKBJ and Langer approximations require the inhomogeneous layers to have a single turning point in the slowness domain of interest. Complicated velocity profiles having low-velocity zones and rapid changes in gradient can be handled by dividing the profile into layers consisting of simple monotonically increasing or decreasing velocity with depth. A reflectivity response can then be calculated to describe the interaction of body waves with the complete velocity profile. The gradients in each layer are also assumed to be sufficiently mild that it is valid to assume the decoupling of the potential equations for P and S waves. The decoupling of potentials is discussed by Richards (1974). That paper gives inequalities in terms of frequency and first and second radial derivatives of density and elastic moduli that must be satisfied to guarantee approximate separation of potentials. If these inequalities are not satisfied in a layer for a frequency band of interest then that layer should be divided into thinner layers and a reflectivity calculation performed.

Intrinsic attenuation can be included in the synthesis method by allowing velocities to be complex. The vertical wavefunctions and fundamental matrices can also be easily generalized to a transversely isotropic layer having a vertical axis of symmetry.

The synthesis method presented is of the spectral type, in which a complex spectrum is calculated by integration over ray parameter or horizontal slowness and is then inverted to the time domain by fast Fourier transform. A subroutine is provided for trapezoidal integration over contours in complex ray parameter as well as subroutines for Fourier transforming to the time domain.

Trapezoidal integration is best used along short segments of the real p axis in which the integrand is not too oscillatory. Once the domain of saddle points or stationary phases has been included along the real p axis, the contour can then be deformed off the real p axis into regions in which the integrand decays exponentially. Deformation of the integration contour into regions of exponential decay insures that the spurious phases are not introduced at slownesses at which the numerical p integration is terminated. These truncation phases are generally present when the p integration is restricted to the real axis. Thomson and Chapman (1986) give some approximate corrections for the truncation phases in a slowness technique of synthesis. Similar corrections can be found for a spectral implementation.

Care was taken in all subroutines to allow the integration path to be deformed off the real axis without encountering any singularities other than those given by the poles of response functions associated with physically diffracted waves. Integration contours can be quite complicated if different contours are taken for different terms in partial-ray expansions of the integrand. Use of the subroutine for trapezoidal integration will entail careful testing of integration contours. Checks should be made for adequacy of the sampling of the integrand along the real p axis as well as whether exponential decay of the integrand has been reached in regions off the real p axis. The p integration can be easily modified, if desired, to remain on the real p axis. In this case, truncation corrections can be applied on or off the real axis and/or alternative integration quadratures can be applied on the p contours. Some of these modifications can be obtained from subroutines provided in other chapters.

ACKNOWLEDGEMENTS

We appreciate the critical reviews of the editor, Bill Menke and Dave Simpson. Cormier and Richards' research in wave propagation, reported here, was supported principally by the National Science Foundation. Lamont-Doherty Geological Observatory Contribution No. 4274.

REFERENCES

Abo-Zena, A. M. (1979). Dispersion function computations for unlimited frequency values. *Geophys. J. R. Astron. Soc.* **58,** 91–105.

Abramowitz, M. and Stegun, I. A. (1964). *Handbook of Mathematical Functions.* Dover, New York.

Acton, F. S. (1970). *Numerical Methods That Work.* Harper and Row, New York.

Aki, K. and Richards, P. G. (1980). *Quantitative Seismology: Theory and Methods.* Freeman, San Francisco.

Burdick, L. J. and Helmberger, D. V. (1978). Upper mantle P velocity structure of the western United States. *J. Geophys. Res.* **83,** 1699–1712.

Burdick, L. J. and Orcutt, J. A. (1979). A comparison of the generalized ray and reflectivity methods of waveform synthesis. *Geophys. J. R. Astron. Soc.* **58,** 261–278.

Červený, V. and Janský, J. (1983). Ray amplitudes of seismic body waves in inhomogeneous, radially symmetric media, *Stud. Geophys. Geod.* **27,** 9–18.

Chapman, C. H. (1973). The Earth flattening transformation in body wave theory. *Geophys. J. R. Astron. Soc.* **35,** 55–70.

Chapman, C. H. (1974). The turning point of elastodynamic waves. *Geophys. J. R. Astron. Soc.* **39,** 613–622.

Chapman, C. H. (1978). A new method for computing synthetic seismograms. *Geophys. J. R. Astron. Soc.* **54,** 481–518.

Choy, G. L. (1977). Theoretical seismograms of core phases calculated by frequency-dependent full wave theory, and their interpretation. *Geophys. J. R. Astron. Soc.* **51,** 275–312.

Choy, G. L. and Cormier, V. F. (1983). The structure of the inner core inferred from short-period and broad band data. *Geophys. J. R. Astron. Soc.* **72,** 1–21.

Choy, G. L. and Cormier, V. F. (1986). Direct measurement of the mantle attenuation operator from broadband P and S waves. *J. Geophys. Res.* **91,** 7326–7342.

Choy, G. L., Cormier, V. F., Kind, K., Müller, G. and Richards, P. G. (1980). A comparison of synthetic seismograms of core phases generated by the full wave theory and by the reflectivity method. *Geophys. J. R. Astron. Soc.* **61,** 21–39.

Cormier, V. F. (1980). The synthesis of complete seismograms in an earth model specified by radially inhomogeneous layers. *Bull Seismol. Soc. Am.* **70,** 691–716.

Cormier, V. F. (1986). Synthesis of body waves in transversely isotropic earth models. *Bull. Seismol. Soc. Am.* **76,** 231–240.

Cormier, V. F. and Choy, G. L. (1981). Theoretical body wave interactions with upper mantle structure. *J. Geophys. Res.* **86,** 1673–1678.

Cormier, V. F. and Richards, P. G. (1976). Comments on "The damping of core waves" by Anthony Qamar and Alfredo Eisenberg. *J. Geophys. Res.* **81,** 3066–3068.

Cormier, V. F. and Richards, P. G. (1977). Full wave theory applied to a discontinuous velocity increase: the inner core boundary. *J. Geophys.* **43,** 3–31.

Doornbos, D. J. (1981). The effect of second-order velocity discontinuities on elastic waves near their turning point. *Geophys. J. R. Astron. Soc.* **64,** 499–511.

Doornbos, D. J., Spiliopoulos, S. and Stacey, F. D. (1986). Seismological properties of D″ and the structure of a thermal boundary layer. *Phys. Earth Planet. Int.* **41,** 225–239.

Dziewonski, A. M. and Anderson, D. L. (1981). Preliminary reference Earth model (PREM). *Phys. Earth Planet. Int.* **25,** 297–356.

Dziewonski, A. M., Hales, A. L. and Lapwood, E. R. (1975). Parametrically simple earth models consistent with geophysical data. *Phys. Earth Planet. Int.* **10,** 12–48.

Frazer, L. N. (1977). Synthesis of shear-coupled PL, Ph.D. thesis, Princeton University.

Frazer, L. N. (1983). Two problems in WKBJ theory: the interpolation of sampled velocity profiles and the use of frequency-dependent, complex velocities. *Geophys. J. R. Astron. Soc.* **75,** 193–211.

Frazer, L. N. and Gettrust, J. F. (1984). On a generalization of Filon's method and the computation of the oscillatory integrals of seismology. *Geophys. J. R. Astron. Soc.* **76,** 461–481.

Gilbert, F. (1976). The representation of seismic displacements in terms of travelling waves. *Geophys. J. R. Astron. Soc.* **44,** 275–280.

Harvey, D. J. (1981). Seismogram synthesis using normal mode superposition: the locked mode approximation. *Geophys. J. R. Astron. Soc.* **66,** 37–69.

Jeffreys, H. and Jeffreys, B. S. (1965). *Methods of Mathematical Physics,* 3rd edn. Cambridge University Press.

Jeffreys, H and Bullen, K. E. (1958). *Seismological Tables.* Brit. Assn, Gray-Milne Trust.

Kennett, B. L. N. (1974). Reflections, rays and reverberations. *Bull. Seismol. Soc. Am.* **64,** 1685–1696.

Kennett, B. L. N. (1975). The effects of attenuation on seismograms. *Bull. Seismol. Soc. Am.* **65,** 1643–1652.

Kennett, B. L. N. (1983). *Seismic Wave Propagation in Stratified Media.* Cambridge University Press.

Kind, R. (1976). Computation of reflection coefficients for layered media. *J. Geophys.* **42,** 191–200.

Langer, R. E. (1932). On the asymptotic solutions of differential equations, with an application to the Bessel functions of large complex order. *Trans. Am. Math. Soc.* **34,** 447–480.

Langer, R. E. (1949). The asymptotic solutions of ordinary differential equations to the second order, with special reference to a turning point. *Trans. Am. Math. Soc.* **67,** 461–490.

Liu, H.-P., Anderson, D. L. and Kanamori, H. (1976). Velocity dispersion due to anelasticity: implications for seismology and mantle composition. *Geophys. J. R. Astron. Soc.* **47,** 41–58.

Love, A. E. H. (1944). *A Treatise on the Mathematical Theory of Elasticity.* Dover, New York.

Ludwig, D. (1970). Diffraction by a circular cavity. *J. Math. Phys.* **11,** 1617–1629.

Lundquist, G. M. and Cormier, V. F. (1980). Constraints on the absorption band model of Q. *J. Geophys. Res.* **85,** 5244–5256,

Menke, W. H. and Richards, P. G. (1980). Crust–mantle whispering gallery phases: a deterministic model of teleseismic P_n wave propagation. *J. Geophys. Res.* **85,** 5416–5422.

Nussenzweig, H. M. (1965). High frequency scattering by an impenetrable sphere. *Ann. Phys.* (*NY*) **34,** 23–95.

Olver, F. W. J. (1954). The asymptotic solution of linear differential equations of the second order for large values of a parameter. *Phil. Trans. R. Soc. Lond.* **A247,** 307–327.

Rial, J. A. and Cormier, V. F. (1980). Seismic waves at the epicenter's antipode. *J. Geophys. Res.* **85,** 2661–2668.

Richards, P. G. (1973). Calculations of body waves for caustics and tunneling in core phases. *Geophys. J. R. Astron. Soc.* **35,** 243–264.

Richards, P. G. (1974). Weakly coupled potentials for high-frequency elastic waves in continuously stratified media. *Bull. Seismol. Soc. Am.* **64,** 1575–1588.

Richards, P. G. (1976). On the adequacy of plane wave reflection/transmission coefficients in the analysis of seismic body waves. *Bull. Seismol. Soc. Am.* **66,** 701–717.

Rosenbaum, J. H. (1974). Synthetic microseisms: logging in porous formations. *Geophysics* **39,** 14–32.

Scholte, J. G. J. (1956). On seismic waves in a spherical Earth. *Koninkl. Ned. Meteorol. Inst. Publ.* **65,** 1–55.

Szego, von G. (1934). Über einege asymptotische enwicklungen der Legendreschen funktionen. *Proc. Land. Math. Soc. Ser. 2* **36,** 427–470.

Thomson, C. J. and Chapman, C. H. (1986). End-point contributions to synthetic seismograms. *Geophys. J. R. Astron. Soc.* **87,** 285–294.

Woodhouse, J. H. (1978). Asymptotic results for elastodynamic propagator matrices in plane stratified and spherically stratified earth models, *Geophys. J. R. Astron. Soc.* **54,** 263–281.

Woodhouse, J. H. (1981). A note on the calculation of travel times in a transversely isotropic Earth model. *Phys. Earth Planet. Int.* **25,** 357–359.

I.2

The WKBJ Seismogram Algorithm

C. H. CHAPMAN, CHU JEN-YI† and D. G. LYNESS

Department of Earth Sciences, Bullard Laboratories, University of Cambridge, UK

1 INTRODUCTION

In many circumstances, seismic body waves can be described using geometrical ray theory. Most seismic body waves propagate with little dispersion or distortion, and non-geometrical effects are only important in limited regions. Nevertheless, non-geometrical effects are of interest as, by their very nature, significant inhomogeneities in the Earth cause non-geometrical signals. In this chapter we describe the simplest extension of geometrical ray theory, the so-called "WKBJ seismogram", which is useful for modelling some non-geometrical signals. At high frequencies it agrees with geometrical ray theory for direct and turning rays, and for partial and total reflections and transmissions at interfaces. At finite frequencies, it remains valid at critical points, for head waves on plane interfaces and other regions where the amplitude function varies discontinuously or rapidly, for example nodes of the source radiation pattern. It also describes signals at and near Airy caustics and in other regions where the geometrical spreading function varies rapidly or is anomalously large or small. At shadows caused by interfaces, it describes the Fresnel part of the shadow, i.e. the discontinuous wavefront, but does not describe the frequency-dependent interaction of the wavefield with the interface, edge or point diffractor nor wave coupling in high gradient regions. A significant advantage of the method is that the anomalous conditions are not treated as special cases and can coexist and overlap. In fact, for instance, the user need not be aware of the existence of small triplications

† Permanent address: Department of Geophysics, Peking University, China.

SEISMOLOGICAL ALGORITHMS
ISBN 0-12-220770-X

and caustics (small in terms of the time separation versus the digitization interval) and will still obtain accurate seismograms.

The method is called the "WKBJ seismogram" algorithm because it depends on using the WKBJ asymptotic approximation for both the horizontal and vertical wavefields (Chapman, 1978). Restricting the solution to the travelling WKBJ approximation and ignoring evanescent WKBJ terms, no further approximation is made and the inverse transforms are evaluated exactly. The theoretical development of the technique has been extensively reviewed elsewhere (Chapman and Orcutt, 1985a), and in the next section we only briefly outline the results of this theory. References for all the theoretical methods can be found in Chapman and Orcutt (1985a) and will not be included here.

Assuming the zeroth-order WKBJ approximation to be valid, the Earth-flattening transformation can be used to obtain results in a spherically symmetric model using the same algorithm as for a plane stratified model. Again assuming the zeroth-order WKBJ approximation at the source, receiver and interfaces and the zeroth-order Langer approximation in layers, the complete vertical wavefunction can be written using the ray expansion. Including only real not evanescent rays, each term can be solved exactly. In the WKBJ seismogram algorithm, the horizontal slowness and frequency transforms effectively cancel and no transform integrals are necessary. Following the theoretical review, the other sections in this chapter describe five subroutines needed to compute WKBJ seismograms in horizontally stratified media: PXINT, COMPON, COEFF, THETAC and CCSQRT. Together these routines form the essence of the WKBJ program—the other requirements are the I/O of the model, travel-time curves and seismograms.

The first routine, PXINT, evaluates the ray integrals in horizontally stratified media. COMPON provides the source radiation pattern and receiver components. COEFF computes reflection and transmission coefficients for plane waves at plane interfaces. These three routines are standard implementations of geometrical ray theory in plane stratified media. They are included for completeness and could be replaced easily by other programs. In particular, the WKBJ seismogram algorithm can be extended to laterally inhomogeneous media using Maslov asymptotic theory (Chapman and Drummond, 1982). Then the results of two- or three-dimensional ray tracing must replace the output of subroutine PXINT, but the crucial routines implementing the inverse slowness and frequency transforms, THETAC and CCSQRT, remain the same. Ray-tracing algorithms are described elsewhere in this book and can be used in conjunction with routines THETAC and CCSQRT to implement Maslov asymptotic theory.

An essential feature of the WKBJ algorithm that leads to its efficiency is the analytic evaluation of the slowness and frequency transform integrals. This in turn depends on the phase being a linear function of frequency and the amplitude being independent of frequency. Unfortunately, for anelastic materials these features are lost. In many circumstances, however, attenuation varies sufficiently slowly that its effect can be factored out of the transform integrals and treated separately. Modifications that can be used to implement this approximation are described in the Appendix.

2 WKBJ SEISMOGRAM THEORY

The WKBJ seismogram method was introduced by Chapman (1976, 1978) and Dey-Sarkar and Chapman (1978) and extensively reviewed by Chapman and Orcutt (1985a). We shall only outline the results here.

We consider a plane stratified, perfectly elastic, isotropic medium. A point source exists on the vertical axis of a cylindrical coordinate system. To simplify the notation, we rotate the coordinate system so that the receiver lies at zero azimuth ($\phi = 0$). This coordinate transformation is applied to the source and receiver components, and the horizontal radiation pattern (the ϕ dependence) enters through this transformation. The azimuthal variable ϕ does not enter explicitly in the following results but is contained implicitly in the source and receiver components. For brevity we only consider the source and receiver components in the plane of propagation ($\phi = 0$), i.e. the P–SV waves (Fig. 1). The transverse

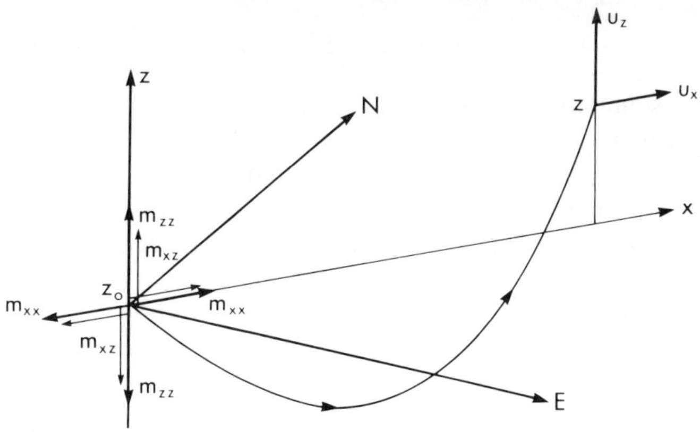

Fig. 1 The cylindrical coordinate system and source and receiver components. Note that the z axis is vertically upwards and the receiver lies in the (x, z) plane.

components, the SH waves, are by comparison trivial and are not included in this chapter or the programs.

Following the standard procedure of expanding the displacement, force and traction vectors in vector cylindrical harmonics (Takeuchi and Saito, 1972), we obtain an ordinary differential system with the independent variable z, the vertical coordinate, positive upwards. The dependent variables are the displacement and traction harmonic coefficients, which are continuous as functions of depth, even at interfaces. The differential system is solved with appropriate boundary and source conditions to find the harmonic coefficients at the receiver. Then the inverse frequency and horizontal slowness transforms are evaluated to obtain the displacement at the receiver. If we approximate the Bessel functions by the first term of their asymptotic expansions (the far-field approximation), we obtain

$$\boldsymbol{u}(t, x, z) \simeq -\frac{1}{2^{5/2}|x|^{1/2}\pi^3}\int_B\int_{-\infty}^{\infty}\left(\frac{\pi}{|\omega|}\right)^{1/2} \times \exp\left[\mathrm{i}\omega(px-t)+\mathrm{i}\,\mathrm{sgn}\,(\omega)\frac{\pi}{4}\right]p^{1/2}\boldsymbol{R}(\omega, p, z)\,\mathrm{d}p\,\mathrm{d}\omega, \quad (1)$$

where $\boldsymbol{R}(\omega, p, z)$ is obtained from the solution of the differential system. Provided that the zeroth-order WKBJ approximation (Richards, 1971) is valid at every interface and at the source and receiver, and the zeroth-order Langer approximation (Chapman, 1974) is valid in every layer, then the propagator matrix (Gilbert and Backus, 1966) can be expanded in terms representing rays (Kennett, 1974). $\boldsymbol{R}(\omega, p, z)$ can be written as

$$\boldsymbol{R}(\omega, p, z) = \sum_{\text{rays}}(-\omega^2\hat{\boldsymbol{m}}(\omega):\boldsymbol{M}_{\text{ray}}(p) - \mathrm{i}\omega\hat{\boldsymbol{f}}(\omega)\cdot\boldsymbol{F}_{\text{ray}}(p))\boldsymbol{U}_{\text{ray}}(p, z)R_{\text{ray}}(p)\mathrm{e}^{\mathrm{i}\omega\tau_{\text{ray}}(p)}, \quad (2)$$

where the summation is over all possible rays. Unfortunately no straightforward method exists to enumerate all possible P–SV rays in a multilayered model, so we have simply indicated the summation by "rays" and a particular ray by the subscript "ray". In practice the user must define each ray by its ray segments and reflection/transmission coefficients (see below) and only a few rays are included.

Before proceeding further, it is useful to consider how the model and rays are defined. Various restrictions have been introduced to simplify the subroutine codes. First we have the model definition (Fig. 2).

I. *Model points d_i.* We consider models with P and S velocities and density, $\alpha(z)$, $\beta(z)$ and $\rho(z)$ respectively, piecewise-continuous

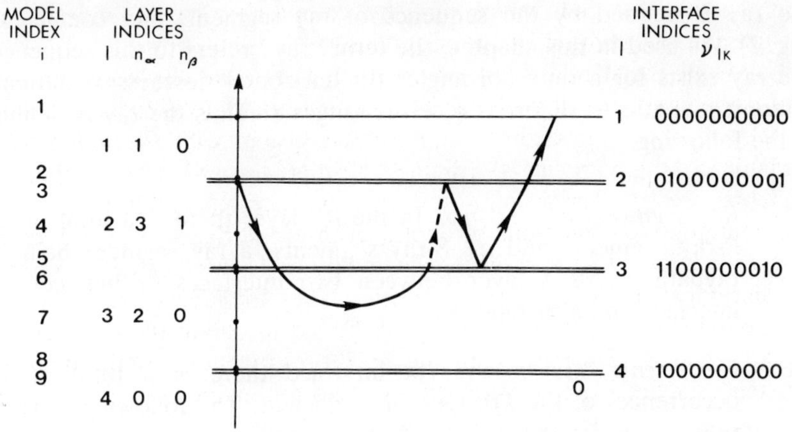

Fig. 2 An example model and ray definition. The solid lines are P rays and the dashed line an SV ray. $I = 8$ and $L = 4$.

functions. The model is defined at discrete depths d_i such that $\alpha_i = \alpha(-d_i)$ etc. and $d_{i+1} \geqslant d_i$ with $i = 1, \ldots, I$. Several methods can be used to interpolate the velocity between model points (see below).

II. *Interfaces* z_l. Interfaces (first-order discontinuities) are defined by two equal depths $d_i = d_{i-1} = -z_l$, say, with $l = 1, \ldots, L$ and $z_{l+1} < z_l$. Note that there is no fixed relationship between the indices of the model points d_i and interfaces z_l. To indicate the model point corresponding to an interface index, we write $i(l)$.

III. *Free surface* z_1. By convention, the first model point is a free surface, $z_1 = -d_1$, and $i(1) = 1$.

IV. *Layers*. The interfaces divide the model into layers. By convention, the lth layer is bounded above by the lth interface and below by the $(l+1)$th, i.e. $z_l \geqslant z \geqslant z_{l+1}$. A layer may contain two or more model points. The model contains $L - 1$ layers.

V. *Model base* z_L. By convention, the model should include a dummy interface at its base, i.e. $d_{I+1} = d_I = -z_L$, $\alpha_{I+1} = \alpha_I$, etc. This permits turning rays in the $(L-1)$th layer.

VI. *Source and receiver coordinates* z_0 and z. By convention, the source and receiver must lie on interfaces. Normally it will be necessary to introduce a dummy interface at the source, but the receiver will be at the free surface.

The ray is defined by the sequence of ray segments and coefficients (Fig. 2). As used in this chapter, the term "ray" refers to this sequence. The ray exists for a range of angles (or horizontal slownesses) defining different raypaths to different receiver ranges x. Thus the ray is defined by the following.

A. *Ray segments* $n_{\alpha l}$ and $n_{\beta l}$. In the lth layer there is a total of $n_{\alpha l}$ P-ray segments and $n_{\beta l}$ S-ray segments, a ray segment being a raypath across a layer between two interfaces or between an interface and a turning point.

B. *Coefficients* ν_{lk}. At the lth interface there is a total of ν_{lk} occurrences of the kth type of coefficient. The following convention is used for the P–SV reflection/transmission coefficients:

$$
\begin{aligned}
k = 1 \quad &\text{for} \quad R_{\mathrm{D}}^{\mathrm{PP}} \\
2 \quad & \qquad T_{\mathrm{D}}^{\mathrm{PP}} \text{ and } T_{\mathrm{U}}^{\mathrm{PP}} \\
3 \quad & \qquad R_{\mathrm{U}}^{\mathrm{PP}} \\
4 \quad & \qquad R_{\mathrm{D}}^{\mathrm{VV}} \\
5 \quad & \qquad T_{\mathrm{D}}^{\mathrm{VV}} \text{ and } T_{\mathrm{U}}^{\mathrm{VV}} \\
6 \quad & \qquad R_{\mathrm{U}}^{\mathrm{VV}} \\
7 \quad & \qquad R_{\mathrm{D}}^{\mathrm{PV}} \text{ and } R_{\mathrm{D}}^{\mathrm{VP}} \\
8 \quad & \qquad T_{\mathrm{D}}^{\mathrm{PV}} \text{ and } T_{\mathrm{U}}^{\mathrm{VP}} \\
9 \quad & \qquad T_{\mathrm{U}}^{\mathrm{PV}} \text{ and } T_{\mathrm{D}}^{\mathrm{VP}} \\
10 \quad & \qquad R_{\mathrm{U}}^{\mathrm{PV}} \text{ and } R_{\mathrm{U}}^{\mathrm{VP}}
\end{aligned}
$$

(see Fig. 3) (the notation follows Chapman and Orcutt, 1985a). Note that we have exploited the symmetry of the coefficients to reduce the number of types. The order of coefficients on the raypath is not important. All values of $n_{\alpha l}$, $n_{\beta l}$ and ν_{lk} are not possible—they must be compatible and describe a real sequence of ray segments joining the source and receiver interfaces in a continuous fashion. The user must specify the numbers $n_{\alpha l}$, $n_{\beta l}$ and ν_{lk}, and the code provided makes no checks of their validity.

C. *Turning points*. Turning rays (see e.g. Fig. 2) are defined as if they were reflected from the base of the layer. Whether the ray is a reflection or turning ray depends on the value of the horizontal slowness p, which is not part of the ray definition. Thus a turning P ray in the lth layer contributes $n_{\alpha l} = 2$ and $\nu_{l+1\,1} = 1$, and a turning SV ray, $n_{\beta l} = 2$ and $\nu_{l+1\,4} = 1$.

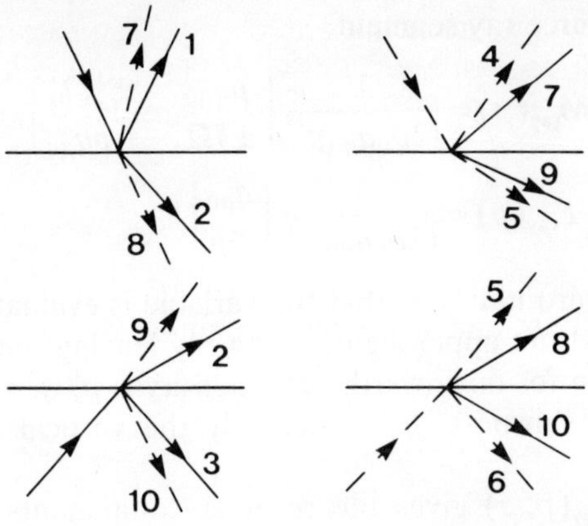

Fig. 3 Illustration of the coefficient index k. The incident, reflected and transmitted rays at an interface are shown. The P rays are solid lines and the SV rays dashed lines.

It is also necessary to specify the model indices, type and directions of the source and receiver ray segments (see COMPON).

Several different rays may have common properties. Any rays with identical $n_{\alpha l}$ and $n_{\beta l}$ form a kinematic group. They have the same function $\tau_{\text{ray}}(p)$. Rays that also have the same v_{lk} and source and receiver specifications form a dynamic group. All rays in a dynamic group have the same $\boldsymbol{M}_{\text{ray}}(p)$, $\boldsymbol{F}_{\text{ray}}(p)$, $\boldsymbol{U}_{\text{ray}}(p)$ and $R_{\text{ray}}(p)$.

In (2) the tensor $\hat{\boldsymbol{m}}(\omega)$ and vector $\hat{\boldsymbol{f}}(\omega)$ contain the spectra of the source components, i.e.

$$\hat{\boldsymbol{m}}(\omega) = \begin{bmatrix} \hat{m}_{xx}(\omega) & \hat{m}_{xz}(\omega) \\ \hat{m}_{xz}(\omega) & \hat{m}_{zz}(\omega) \end{bmatrix}, \tag{3a}$$

$$\hat{\boldsymbol{f}}(\omega) = \begin{bmatrix} \hat{f}_x(\omega) \\ \hat{f}_z(\omega) \end{bmatrix}. \tag{3b}$$

$\hat{m}_{xx}(\omega)$ etc. are the components of the source-moment tensor, i.e. the volume integral of the source glut (Gilbert, 1971; Backus and Mulcahy, 1976), and $\hat{f}_x(\omega)$ etc. are the components of the volume integral of the body-force sources. Remember that these components are in the coordinate system rotated so the receiver is at $\phi = 0$ (Fig. 1). The source components are combined using the tensor $\boldsymbol{M}_{\text{ray}}(p)$ and the vector $\boldsymbol{F}_{\text{ray}}(p)$. For a P-source ray segment

$$\boldsymbol{M}_{\text{ray}}(p) = \frac{1}{(2\rho_0 q_{\alpha 0})^{1/2}} \begin{bmatrix} p^2 & \pm p q_{\alpha 0} \\ \pm p q_{\alpha 0} & q_{\alpha 0}^2 \end{bmatrix}, \tag{4a}$$

$$\boldsymbol{F}_{\text{ray}}(p) = \frac{1}{(2\rho_0 q_{\alpha 0})^{1/2}} \begin{bmatrix} p \\ \pm q_{\alpha 0} \end{bmatrix}, \tag{4b}$$

and for an SV-source ray segment

$$\boldsymbol{M}_{\text{ray}}(p)=\frac{1}{(2\rho_0 q_{\beta 0})^{1/2}}\begin{bmatrix} pq_{\beta 0} & \pm\frac{1}{2}\Omega_0 \\ \pm\frac{1}{2}\Omega_0 & -pq_{\beta 0}\end{bmatrix}, \tag{4c}$$

$$\boldsymbol{F}_{\text{ray}}(p)=\frac{1}{(2\rho_0 q_{\beta 0})^{1/2}}\begin{bmatrix} q_{\beta 0} \\ \mp p\end{bmatrix}, \tag{4d}$$

where subscript zero indicates that the variable is evaluated at the source depth, $z=z_0$, and the upper sign is for a ray leaving the source upwards and the lower sign for downwards. The density is ρ, $q_\alpha=[1/\alpha^2(z)-p^2]^{1/2}$ is the vertical P slowness, q_β is similarly the vertical S slowness, and $\Omega=q_\beta^2-p^2$.

The vector $\boldsymbol{U}_{\text{ray}}(p, z)$ gives the receiver components $[u_x, u_z]^{\text{T}}$. For a P-receiver ray segment

$$\boldsymbol{U}_{\text{ray}}(p, z)=\frac{1}{(2\rho q_\alpha)^{1/2}}\begin{bmatrix} p \\ \pm q_\alpha\end{bmatrix}, \tag{5a}$$

where ρ and q_α are evaluated at the receiver. The upper sign refers to an upgoing ray at the receiver and the lower sign to a downgoing ray. Similarly, for an SV-receiver ray segment

$$\boldsymbol{U}_{\text{ray}}(p, z)=\frac{1}{(2\rho q_\beta)^{1/2}}\begin{bmatrix} q_\beta \\ \mp p\end{bmatrix} \tag{5b}$$

(the factors $(2\rho q)^{-1/2}$ arise from the normalization of the WKBJ solution—see Chapman and Orcutt, 1985a). These receiver vectors refer to a receiver *in* the medium. Normally, seismometers are placed *on* a free surface and the receiver records the combined incident and reflected rays. Then the vectors are

$$\boldsymbol{U}_{\text{ray}}(p, z)=\frac{(2q_\alpha/\rho)^{1/2}}{\beta^2(4p^2 q_\alpha q_\beta+\Omega^2)}\begin{bmatrix} 2pq_\beta \\ \Omega\end{bmatrix} \tag{5c}$$

for an incident P ray, and

$$\boldsymbol{U}_{\text{ray}}(p, z)=\frac{(2q_\beta/\rho)^{1/2}}{\beta^2(4p^2 q_\alpha q_\beta+\Omega^2)}\begin{bmatrix} \Omega \\ -2pq_\alpha\end{bmatrix} \tag{5d}$$

for an incident SV ray.

The subroutine COMPON provides the combinations $\boldsymbol{M}_{\text{ray}}(p)\boldsymbol{U}_{\text{ray}}(p, z)$ and $\boldsymbol{F}_{\text{ray}}(p)\boldsymbol{U}_{\text{ray}}(p, z)$.

The function $R_{\text{ray}}(p)$ is the product of the reflection and transmission coefficients:

$$R_{\text{ray}}(p)=\prod_{l=1}^{L}\left(\prod_{k=1}^{10} T_{lk}^{\nu_{lk}}\right), \tag{6}$$

where T_{lk} are the coefficients at the lth interface, i.e. $T_{l1} = R_D^{PP}$ etc. The coefficients must be normalized with respect to the WKBJ approximation. They are given by

$$R_D^{PP} = (A_{\alpha-}A_{\beta+} + C_{1-}C_{2+} - D)/\Delta^{PV}, \tag{7a}$$
$$R_U^{PP} = (-A_{\alpha-}A_{\beta+} + C_{1+}C_{2-} - D)/\Delta^{PV}, \tag{7b}$$
$$R_D^{VV} = (-A_{\alpha+}A_{\beta-} - C_{1-}C_{2+} + D)/\Delta^{PV}, \tag{7c}$$
$$R_U^{VV} = (A_{\alpha+}A_{\beta-} - C_{1+}C_{2-} + D)/\Delta^{PV}, \tag{7d}$$
$$T_U^{PP} = T_D^{PP} = F_{\alpha1}F_{\alpha2}(q_{\beta1}E_2 + q_{\beta2}E_1)/\Delta^{PV}, \tag{7e}$$
$$T_U^{VV} = T_D^{VV} = F_{\beta1}F_{\beta2}(q_{\alpha1}E_2 + q_{\alpha2}E_1)/\Delta^{PV}, \tag{7f}$$
$$R_D^{PV} = R_D^{VP} = -pF_{\alpha1}F_{\beta1}[2q_{\alpha2}q_{\beta2}E_1B_2 + E_2(E_2 - \rho_1)]/\rho_1\Delta^{PV}, \tag{7g}$$
$$R_U^{PV} = R_U^{VP} = -pF_{\alpha2}F_{\beta2}[2q_{\alpha1}q_{\beta1}E_2B_1 + E_1(E_1 - \rho_2)]/\rho_2\Delta^{PV}, \tag{7h}$$
$$T_D^{PV} = T_U^{VP} = -pF_{\alpha1}F_{\beta2}(2B_2q_{\beta1}q_{\alpha2} + E_1 - \rho_2)/\Delta^{PV}, \tag{7i}$$
$$T_D^{VP} = T_U^{PV} = -pF_{\alpha2}F_{\beta1}(2B_1q_{\beta2}q_{\alpha1} + E_2 - \rho_1)/\Delta^{PV}, \tag{7j}$$

where

$$\Delta^{PV} = A_{\alpha+}A_{\beta+} - C_{1+}C_{2+} + D \tag{7k}$$

and

$$A_{\substack{\alpha+\\ \beta-}} = \rho_2 q_{\substack{\alpha1\\ \beta}} \pm \rho_1 q_{\substack{\alpha2\\ \beta}} \tag{7l}$$
$$B_{\substack{1\\ 2}} = \mu_{\substack{1\\ 2}} - \mu_{\substack{2\\ 1}} \tag{7m}$$
$$C_{\substack{1+\\ 2-}} = 2p[B_{\substack{1\\ 2}}(\pm q_{\alpha1}q_{\beta1} + p^2) - \rho_{\substack{1\\ 2}}] \tag{7n}$$
$$D = p^2(\rho_1 + \rho_2)^2 \tag{7o}$$
$$E_{\substack{1\\ 2}} = \rho_{\substack{1\\ 2}} - 2p^2 B_{\substack{1\\ 2}} \tag{7p}$$
$$F_{\substack{\alpha1\\ \beta2}} = (2\rho_{\substack{1\\ 2}} q_{\substack{\alpha1\\ \beta2}})^{1/2}, \tag{7q}$$

where the alternative subscripts are applied in the obvious fashion. The subscripts 1 and 2 refer to the media just above and just below the interface, for example $\rho_1 = \rho(z_l + 0) = \rho_{i(l)-1}$ and $\rho_2 = \rho(z_l - 0) = \rho_{i(l)}$. When a turning point exists we need effectively $R_D^{PP} = -\mathrm{i}\,\mathrm{sgn}\,(\omega)$ and $R_D^{VV} = \mathrm{i}\,\mathrm{sgn}\,(\omega)$, the "reflection" coefficient at the turning point from the Langer approximation. (The opposite signs occur for P and SV waves because the horizontal displacement changes sign for an SV wave through the turning point and this is not contained in (5b). This and other factors of $\pm\mathrm{i}$ are different between results (7) and Chapman and Orcutt (1985a) because the basis vectors (5) differ.) The routine COEFF described below provides the function $R_{\mathrm{ray}}(p)$.

Finally, in (2), $\tau_{\text{ray}}(p)$ is the delay function for the ray:

$$\tau_{\text{ray}}(p) = \sum_{l=1}^{L-1} \left(n_{\alpha l} \int_{z_{l+1}}^{z_l} q_\alpha(p, z)\, \mathrm{d}z + n_{\beta l} \int_{z_{l+1}}^{z_l} q_\beta(p, z)\, \mathrm{d}z \right). \tag{8}$$

The subroutine PXINT computes this function.

With the WKBJ approximation (2) substituted in (1) and the p range restricted so that the delay function (8) is real, the inverse transforms can be evaluated exactly. We obtain

$$\begin{aligned} \boldsymbol{u}(t, x, z) \simeq & -\frac{1}{2^{3/2}\, |x|^{1/2}\, \pi^2} \\ & \times \left\{ \boldsymbol{m}''(t) :* \operatorname{Im}\left[\Lambda(t) * \sum_{\text{rays}} \sum_{t=\theta_{\text{ray}}} \frac{p^{1/2} \boldsymbol{M}_{\text{ray}}(p) \boldsymbol{U}_{\text{ray}}(p, z) R_{\text{ray}}(p)}{|\partial_p \theta_{\text{ray}}|} \right] \right. \\ & \left. + \boldsymbol{f}'(t) \cdot * \operatorname{Im}\left[\Lambda(t) * \sum_{\text{rays}} \sum_{t=\theta_{\text{ray}}} \frac{p^{1/2} \boldsymbol{F}_{\text{ray}}(p) \boldsymbol{U}_{\text{ray}}(p, z) R_{\text{ray}}(p)}{|\partial_p \theta_{\text{ray}}|} \right] \right\}, \end{aligned} \tag{9}$$

where

$$\theta_{\text{ray}}(p, x) = \tau_{\text{ray}}(p) + px. \tag{10}$$

The operators $:*$ and $\cdot*$ denote convolution in the time domain and summation over source components. The order of convolutions and summations is, in principle, arbitrary and the most convenient or efficient method depends on the form of the source components and the method of convolution. If all the source components have the same time history (e.g. $\boldsymbol{m}(t) = m(t)\boldsymbol{m}_0$, the usual assumption) then the summation over components can be performed inside the innermost sum first (e.g. $\boldsymbol{m}_0 : \boldsymbol{M}_{\text{ray}}(p)$), and convolution performed for a scalar function (e.g. $m(t)$).

The innermost summations in (9) are over p values that satisfy

$$t = \theta_{\text{ray}}(p, x) \tag{11}$$

and are evaluated by the subroutine THETAC. The analytic time series $\Lambda(t)$ is

$$\Lambda(t) = \lambda(t) + i\bar{\lambda}(t), \tag{12}$$

where $\lambda(t) = H(t)t^{-1/2}$ and $\bar{\lambda}(t) = H(-t)(-t)^{-1/2}$, and the convolution operator $\partial_t \operatorname{Im}[\Lambda(t) *$ is evaluated by subroutine CCSQRT. The functions (4)–(6) only depend on the sign of frequency, and in the innermost summations in (9) should be taken for positive frequency, i.e. $\operatorname{Im}(q_\alpha) \geqslant 0$ etc.

3 DESCRIPTION OF SUBROUTINES

3.1 Subroutine PXINT

The subroutine PXINT computes the delay function (8) and its derivative, which gives the range function $X_{\text{ray}}(p) = -\partial_p \tau_{\text{ray}}(p)$, for models as specified above. Three alternative methods of interpolating the velocity between model points are provided. These are

$$\alpha(z) = (a - 2bz)^{-1/2}, \tag{13a}$$

$$\alpha(z) = a - bz \tag{13b}$$

or

$$\alpha(z) = a\mathrm{e}^{-bz}, \tag{13c}$$

and similarly for $\beta(z)$. The constants a and b are determined for each model interval by fitting the function to the two model points, i.e.

$$b_i = \frac{\alpha_i^{-2} - \alpha_{i+1}^{-2}}{2(d_i - d_{i+1})}, \tag{14a}$$

$$b_i = \frac{\alpha_i - \alpha_{i+1}}{d_i - d_{i+1}} \tag{14b}$$

or

$$b_i = \frac{\ln(\alpha_i/\alpha_{i+1})}{d_i - d_{i+1}}. \tag{14c}$$

The required integrals are then

$$\int q_\alpha(p, z)\,\mathrm{d}z = -\frac{q_\alpha^3}{3b}\bigg|, \tag{15a}$$

$$= \frac{1}{b}\left[\ln\left(\frac{1 + \alpha q_\alpha}{\alpha p}\right) - \alpha q_\alpha\right]\bigg| \tag{15b}$$

or

$$= \frac{1}{b}\left[q_\alpha - p\tan^{-1}\left(\frac{q_\alpha}{p}\right)\right]\bigg|, \tag{15c}$$

where the limits are model points or a turning point (note that the right-hand sides of (15a–c) are zero at a turning point). The advantage of interpolating using method (a) is efficiency—only q_α is required—so the method is suitable for finely layered models. Method (b) is a simple interpolation formula (13b) that leads to circular raypaths and is often preferred in plane models. Method (c) is preferred for spherical models as the Earth-flattening transformation maps constant velocities into the function (13c).

The code in subroutine PXINT is straightforward and simple. Redundant calculations of special functions etc. are avoided. The code remains valid for the special cases $b = 0$ or $p = 0$.

The input arguments of the subroutine PXINT are

P	—	p (horizontal slowness)
VEL(IVEL, $*$)	—	VEL(1, i) $= d_i$ (depths) VEL(2, i) $= 1/\alpha_i$ (P slowness) VEL(3, i) $= 1/\beta_i$ (S slowness) VEL(4, i) $= \rho_i$ (density) VEL(5, i) $= b_i$ for $\alpha(z)$ VEL(6, i) $= b_i$ for $\beta(z)$
IVEL	—	multiple array dimension of VEL ($\geqslant 6$)
IL($*$)	—	IL(l) $= i(l)$ (interface indices)
NPS(2, $*$)	—	NPS(1, l) $= n_{\alpha l}$ (P-ray segments) NPS(2, l) $= n_{\beta l}$ (S-ray segments)
L	—	maximum layer with ray segments
IMTH	—	IMTH = 1 for method (a) interpolation IMTH = 2 for method (b) interpolation IMTH = 3 for method (c) interpolation

The output arguments are

TAU	—	$\tau_{\text{ray}}(p)$ (delay function)
X	—	$X_{\text{ray}}(p) = -\partial_p \tau_{\text{ray}}(p)$ (range function)
IRTR	—	IRTR = 1 normal return code IRTR = 2 error return code if $\tau_{\text{ray}}(p)$ complex

The subroutine PXINT should be called for each p_j ($j = 1, \ldots, J$) to define a delay-time table for each kinematic group (THETAC requires the p_j to be in increasing order). The possible range of horizontal slownesses is restricted by the ray type and the model. For a reflection the minimum horizontal slowness is zero (corresponding to a vertical reflection) and the maximum is normally the slowness at the reflecting interface (corresponding to the grazing reflection). For a turning ray the range is normally given by the range of slowness in the turning layer. Sometimes the range of horizontal slowness is further restricted if any ray segments propagate through parts of the model with lower slowness. Often the range function or its derivative is singular at the upper limit of horizontal slowness. Then it is advisable to increase the density of points in the table near this limit.

3.2 Subroutine COMPON

The subroutine COMPON computes the source and receiver terms (4) and (5). It also includes the factor $-p^{1/2}/2^{3/2}\pi^2$ in (9). It is necessary to specify the source and receiver ray types and directions. These are illustrated in Fig. 4. The code in subroutine COMPON is straightforward and simple.

The input arguments of subroutine COMPON are

P — p (horizontal slowness)

VELS(*) — VELS(1) = $1/\alpha_0$ (source-P slowness)
VELS(2) = $1/\beta_0$ (source-S slowness)
VELS(3) = ρ_0 (source density)

ISS — source ray-type index (Fig. 4a)
ISS = 1 (downgoing P-source ray)
ISS = 2 (downgoing SV-source ray)
ISS = 5 (upgoing P-source ray)
ISS = 6 (upgoing SV-source ray)

VELR(*) — VELR(1) = $1/\alpha$ (receiver-P slowness)
VELR(2) = $1/\beta$ (receiver-S slowness)
VELR(3) = ρ (receiver density)

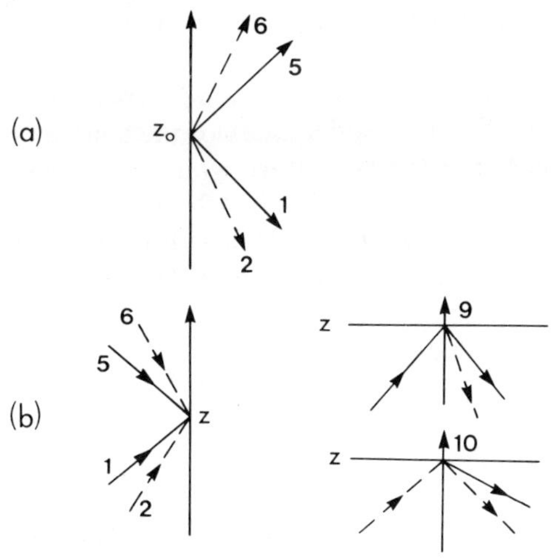

Fig. 4 Source- and receiver-type indices. The solid lines are P rays and the dashed lines SV rays. (a) Indices for upgoing and downgoing source rays. (b) Indices for upgoing and downgoing receiver rays in the medium and rays incident on a free surface.

IRR — receiver ray-type index (Fig. 4b)
IRR = 1 (upgoing P-receiver ray)
IRR = 2 (upgoing SV-receiver ray)
IRR = 5 (downgoing P-receiver ray)
IRR = 6 (downgoing SV-receiver ray)
IRR = 9 (P ray on free surface)
IRR = 10 (SV ray on free surface)

The output argument is

COMP(*) — components of $\boldsymbol{M}_{\text{ray}}\boldsymbol{U}_{\text{ray}}$ and $\boldsymbol{F}_{\text{ray}}\boldsymbol{U}_{\text{ray}}$
COMP(1) = $M_{zz}U_x$
COMP(2) = $M_{xx}U_x$
COMP(3) = $(M_{xz}+M_{zx})U_x$
COMP(4) = $M_{zz}U_z$
COMP(5) = $M_{xx}U_z$
COMP(6) = $(M_{xz}+M_{zx})U_z$
COMP(7) = F_zU_x
COMP(8) = F_xU_x
COMP(9) = F_zU_z
COMP(10) = F_xU_z

A typical call for a teleseismic P wave would be

CALL COMPON (P, VEL(2, IS), 1, VEL(2, 1), 9, COMP)

where we have used parts of the full model array VEL (see PXINT) for VELS and VELR, IS is the model index of the source, and the receiver is on the free surface. For the ray *p*P we would have ISS = 5, IRR = 9 and for *s*P recorded on the ocean floor ISS = 6, IRR = 1 with the ocean surface reflection being ISS = 6, IRR = 5. For an SV ray incident on a free surface, q_α may be imaginary and $\boldsymbol{U}_{\text{ray}}$ complex. Therefore for generality, COMP is complex. One call of COMPON should be made for each p_j for each dynamic group.

3.3 Function COEFF

The complex function COEFF computes the product of the reflection and transmission coefficients for an interface, i.e. the inner product in (6). The routine computes all the required coefficients (7) with one call as many factors are common. The code is a straightforward implementation of (7). The vertical slownesses are either real or imaginary and the complex arithmetic is expanded to avoid unnecessary complex operations. It is assumed that the incident and generated rays are real. If this is not the case for the reflection coefficients R_D^{PP} and R_D^{VV} then the turning-point "reflection" coefficients ($\mp$i) are assumed. Fluid layers are permitted and are indicated by setting the corresponding S slowness(es) $1/\beta_i$ equal to zero. Reflections from below a free surface are permitted and are indicated by setting $\rho_1 = 0$.

The input arguments of function COEFF are

P	—	p (horizontal slowness)
VEL1(*)	—	VEL1(1) = $1/\alpha_1$ (P slowness above interface) VEL1(2) = $1/\beta_1$ (S slowness above interface) VEL1(3) = ρ_1 (density above interface)
VEL2(*)	—	VEL2(1) = $1/\alpha_2$ (P slowness below interface) VEL2(2) = $1/\beta_2$ (S slowness below interface) VEL2(3) = ρ_2 (density below interface)
ID(*)	—	ID(k) = ν_{lk} (coefficient multiplicity)

The output result is

COEFF — $\prod_{k=1}^{10} T_{lk}^{\nu_{lk}}$

The P–SV reflection/transmission coefficients (7) are sufficiently complicated that algebraic or coding mistakes are easily made. The routine COEFF has been tested numerically by combining various coefficients together with the vectors (5) to confirm continuity of displacement at an interface.

The function COEFF should be called once for each p_j, interface and dynamic group. Reflection/transmission coefficients have branch cuts for any slownesses they contain. Any branch cut within the range of p_j is important and may form a head wave. Because low-order methods are used to evaluate the WKBJ seismogram (see THETAC), it is important to sample the coefficients more densely near the branch point and to distribute the p_j (approximately) symmetrically about the branch point (so that the real and imaginary terms cancel in appropriate circumstances).

3.4 Subroutine THETAC

In order to compute the WKBJ seismograms (9), we require

$$v_{\text{ray}}(t, x) = \sum_{t=\theta_{\text{ray}}} \frac{C_{\text{ray}}(p)}{|\partial_p \theta_{\text{ray}}|}, \tag{16a}$$

where v_{ray} and C_{ray} are complex functions. $C_{\text{ray}}(p)$ is one of the components provided by the routines COMPON and COEFF, and $\theta_{\text{ray}}(p, x)$ is provided by routine PXINT and (10). We cannot evaluate (16a) numerically since the denominator may be zero. We therefore smooth (16a) by a normalized boxcar filter $2\,\Delta t$ long, where Δt is the digitization interval. Defining $B(t) = \frac{1}{2}\{H(t+1) - H(t-1)\}$, we obtain

$$\frac{B(t/\Delta t)}{\Delta t} * v_{\text{ray}}(t, x) = \frac{1}{2\,\Delta t} \int_{\theta_{\text{ray}} = t \pm \Delta t} C_{\text{ray}}(p)\, \mathrm{d}p, \tag{16b}$$

where the integral is over all p intervals defined by the band $\theta_{\text{ray}}(p, x) = t - \Delta t$ to $t + \Delta t$, always taken in the positive direction. This is illustrated in Fig. 5(a) by the p interval Δp. A slight simplification results by only computing

$$w_{\text{ray}}(m\,\Delta t + t_0, x) = \sum \int_{p_a}^{p_b} C_{\text{ray}}(p)\, \mathrm{d}p \tag{16c}$$

in subroutine THETAC, where p_a and p_b are defined by the band $\theta_{\text{ray}}(p, x) = (m-1)\,\Delta t + t_0$ to $m\,\Delta t + t_0$ (see Fig. 5a). The summation is over pairs (p_a, p_b) defined by the band.

Two methods exist for finding the intervals (p_a, p_b). We can either specify $t = m\,\Delta t + t_0$, and search through the entire table of $\theta_{\text{ray}}(p, x)$ for intervals (p_a, p_b), or we can work monotonically through the table, solving for the appropriate m and adding into the time series w_{ray}. The former method is inefficient, since for M time values we have to search J values, i.e. the algorithm will contain

$$aMJ \tag{17a}$$

operations, whereas the latter method requires approximately

$$bM + cJ \tag{17b}$$

operations. The constants a, b and c depend on the features of the θ_{ray} curve, but to lowest order are independent of the density of points, i.e. M and J. They are usually a small number (about 10) of arithmetic operations. Naturally we use the latter procedure.

The intervals (p_a, p_b) are found assuming linear interpolation between the tabulated θ_{ray} values $\theta_j = \theta_{\text{ray}}(p_j, x)$ (Fig. 5b). The integrals (16c) are evaluated using the trapezoidal rule with linear interpolation of the tabulated integrand $C_j = C_{\text{ray}}(p_j)$. To obtain an efficient routine, the code in THETAC is fairly compact and some explanation is in order. Figures 6 and 7 illustrate the relevant variables.

Consider the case where we have just completed an interval at p_b (PB)

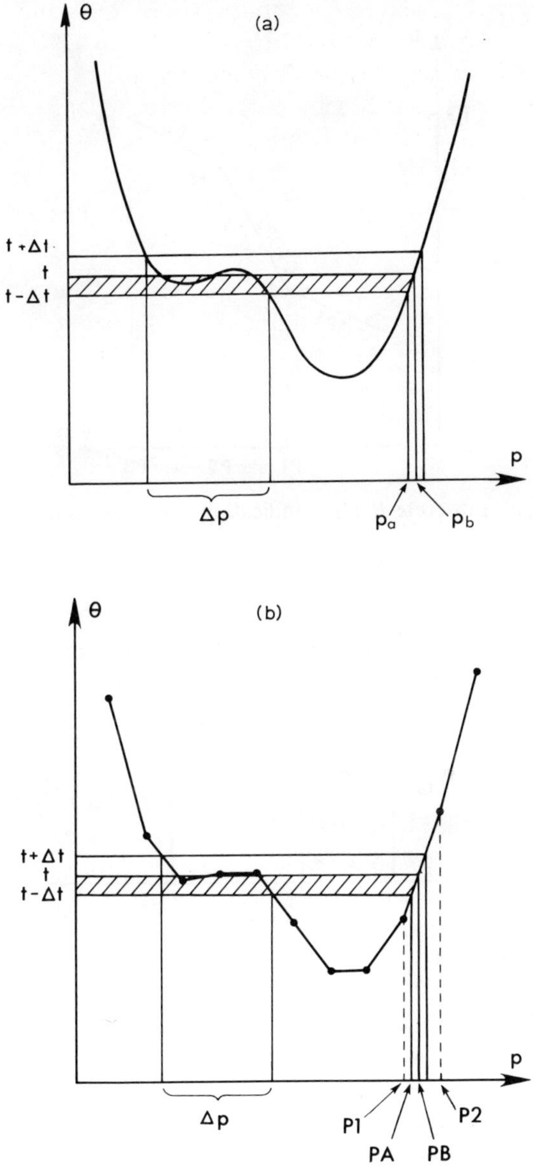

Fig. 5 A typical θ curve illustrating the p values contributing to (16). (a) A typical θ curve with triplication. (b) Discrete version of (a) with variables P1, P2, PA and PB indicated.

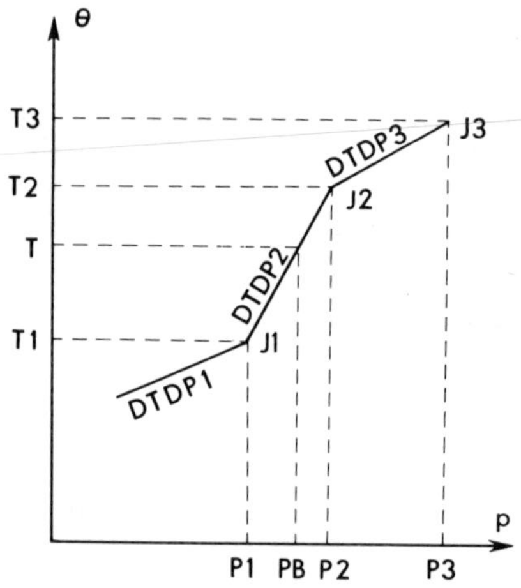

Fig. 6 Part of a discrete θ curve indicating variables used in THETAC.

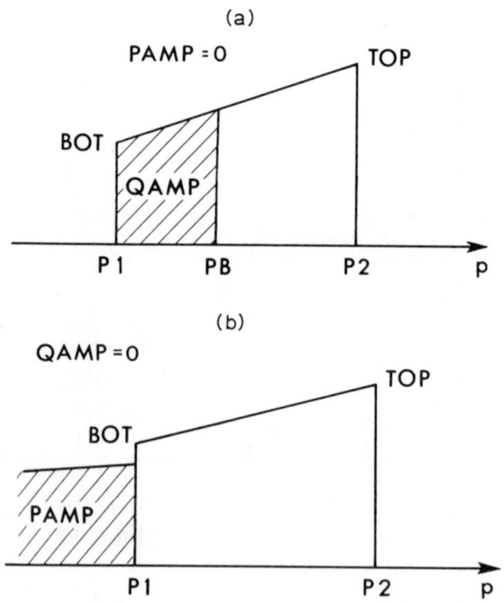

Fig. 7 Part of the integral (16c) indicating variables used in THETAC. (a) Integral to t. (b) Integral to θ_j.

between p_j (P1 = P(J1)) and p_{j+1} (P2 = P(J2)), and the gradient $\theta_j' = (\theta_{j+1} - \theta_j)/(p_{j+1} - p_j)$ (DTDP2) is positive. The integral from p_j to p_b is known (QAMP—see Fig. 7(a)—C_j is BOT and C_{j+1} is TOP). In the next interval, p_b becomes p_a, m (MM) is incremented and we look for the point $t = m\,\Delta t + t_0$ (T). If $t < \theta_{j+1}$ (T2) then p_b is within the same interval and the integral (16c) is calculated by subtracting the old QAMP from its updated value (Fig. 7a). If $t > \theta_{j+1}$ then j is incremented and the search continues. The integral from p_a to the new p_j is stored in PAMP (Fig. 7b).

If the gradient θ_j' is negative then the procedure is similar except that m is decremented and the conditions $t \gtrless \theta_{j+1}$ are reversed. This is controlled by the variables sgn $(1, \theta_j')$ (MDIR) and sgn $(\Delta t, \theta_j')$ (TDIR).

If the sign of the gradient changes when j is incremented then the direction of search reverses, i.e. MDIR and TDIR change sign. If $\theta_j' > 0 > \theta_{j+1}'$ (or $\theta_j' < 0 < \theta_{j+1}'$) then m and t must be decremented (or incremented) before the new search.

A typical seismogram $w(m\,\Delta t + t_0, x)$ is illustrated in Fig. 8. Near the geometrical arrival, θ_j' is small, and typically several tabulated points C_j will contribute to each time point. The user should establish that θ_j and C_j vary sufficiently slowly that the required accuracy is obtained. No automatic accuracy conditions are contained in the code provided. The algorithms are sufficiently simple that it is more efficient to test the

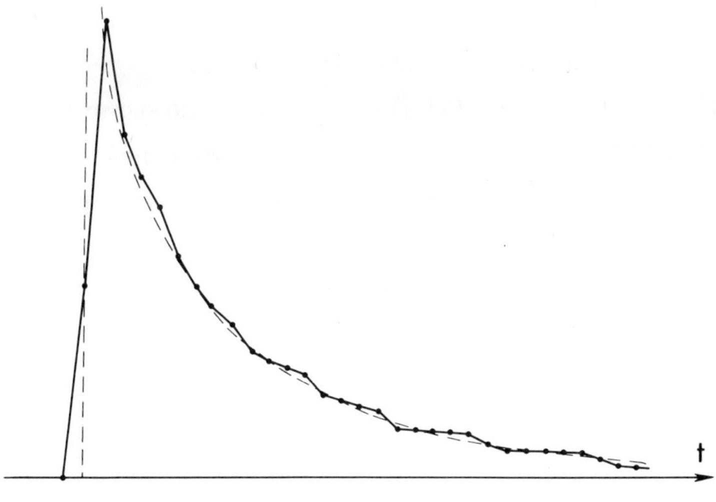

Fig. 8 A typical seismogram for a geometrical arrival. The dashed line corresponds to the analytic expression (16a) and the solid line and dots to the discrete version (16b).

accuracy by increasing the sampling (e.g. M and J) rather than complicate the code with conditions.

For later times, θ_j' is large and often the intervals θ_j to θ_{j+1} are larger than Δt. As $\theta_{\text{ray}}(p, x)$ and $C_{\text{ray}}(p)$ are linearly interpolated, the resultant seismogram is made up of linear segments (Fig. 8). The "discontinuities" between these segments result in higher-order $(O(\omega^{-1/2}))$ arrivals. Normally, these are small compared with the geometrical arrivals and cause no problem. However, the WKBJ seismogram algorithm can also be used to compute differential seismograms (Chapman and Orcutt, 1985b). For differential seismograms the geometrical arrival may not be present and the higher-order arrivals cause a significant error. To reduce this problem, the integrand values are modified if the interval $\Delta\theta_j = \theta_{j+1} - \theta_j$ is large compared with Δt. In the interval (p_j, p_{j+1}) the following modifications are made:

$$\text{BOT} = \frac{2C_j\,|\theta_j'|}{|\theta_j'| + |\theta_{j-1}'|} \quad \text{if } \Delta\theta_j \quad \text{and} \quad \Delta\theta_{j-1} > \Delta t, \tag{18a}$$

$$\text{TOP} = \frac{2C_{j+1}\,|\theta_j'|}{|\theta_j'| + |\theta_{j+1}'|} \quad \text{if } \Delta\theta_j \quad \text{and} \quad \Delta\theta_{j-1} > \Delta t. \tag{18b}$$

It is easily established that when these modifications are made in neighbouring intervals, the integral (16c) is continuous at $t = \theta_j$ and equal to $2C_j\,\Delta t/(|\theta_j'| + |\theta_{j-1}'|)$. The modifications (18) are approximately equivalent to using higher-order interpolation of θ_{ray} without the extra computational difficulty. Again the accuracy is best checked by increasing the sampling J.

The input arguments of subroutine THETAC are

P(ICON, *)	—	$P(1, j) = p_j$ (horizontal slownesses)
TAU(ICON, *)	—	$\text{TAU}(1, j) = \tau_{\text{ray}}(p_j)$ (delay functions)
C(ICON, *)	—	$(C(1, j), C(2, j)) = C_j$ (complex integrands)
J	—	number of tabulated points
ICON	—	multiple array dimension to allow P, TAU and C to be stored in one multidimensional array in calling program (≥2).
X	—	x (range)
T0	—	t_0 (initial time)
DT	—	Δt (time increment)
M	—	number of time points

The seismogram is specified as input and output so different rays can be summed. The series is computed from $t_0 - 2\,\Delta t$ to $t_0 + (M+2)\,\Delta t$ to reduce end errors in the convolution—see below. Thus the input/output argument is

$$W(2,*) \quad - \quad (W(1, m+3),\ W(2, m+3)) = w(m\,\Delta t + t_0, x)$$
$$\text{(complex seismogram, } m = -2, \ldots, M+2\text{)}.$$

The subroutine THETAC should be called once for each kinematic group, source/receiver component and range. Results for different dynamic groups within the kinematic group should be summed before the call.

3.5 Subroutine CCSQRT

To complete the computation of seismograms (9), convolutions must be performed. Often the receiver transfer function would also be included. It is normally more efficient to evaluate a convolution in the frequency domain using the fast Fourier transform. We take the FFT of (16b) summed over all rays, i.e. the FT of

$$v(m\,\Delta t + t_0, x) = \frac{1}{2\,\Delta t}\sum_{\text{rays}} (w_{\text{ray}}(m\,\Delta t + t_0, x) + w_{\text{ray}}(m\,\Delta t + \Delta t + t_0, x)), \tag{19}$$

multiply by a component of the "source" term

$$-\,\omega^2 \hat{\boldsymbol{m}}(\omega)\left(\frac{2\pi}{\omega}\right)^{1/2} H(\omega) \tag{20}$$

(or similarly from $\hat{f}(\omega)$), and invert the FFT. Because of the non-stationary nature of the time series (19), some care is needed taking its FFT. An alternative is to take the inverse FT of (20) (a complex analytic time series) and convolve with (19) in the time domain. If $\hat{\boldsymbol{m}}(\omega)$ is sufficiently band-limited then the discrete time series from (20) will be well behaved. Often this time series can be taken short so the convolution in the time domain is efficient.

An alternative that we have found convenient is to perform the convolution $\partial_t \operatorname{Im}[\Lambda(t) *$ using a rational approximation. The WKBJ seismogram is a high-frequency approximation and the long-period errors introduced by the rational approximation are rarely troublesome. The result of this convolution is to give the impulse response, i.e. simple geometrical arrivals approximate delta functions corresponding to a step-function moment source or a delta-function force source.

First we note that

$$\Lambda(t) = \lambda(t) + \mathrm{i}\bar{\lambda}(t) = \lambda(t) + \mathrm{i}\lambda(-t), \tag{21}$$

so the imaginary part of the operator can be obtained by applying the real part

$$\partial_t \lambda(t) \tag{22}$$

in reverse. Wiggins (1976) first suggested a rational approximation for this operator, but the smoothing applied to (22) was insufficient to reduce aliasing and was asymmetric, producing shifts between the real and imaginary parts (21). We have designed a rational approximation with a more severe, symmetric smoothing of (22). We apply the boxcar $B(t/\Delta t)/\Delta t$ three times to (22) and obtain a discrete time series for (22). The first 100 terms in the time series are fitted by a least-squares procedure to give the coefficients in the rational approximation

$$\frac{1}{6z^2}\,\frac{1 + n_1 z + n_2 z^2 + n_3 z^3 + n_4 z^4 + n_5 z^5}{1 + d_1 z + d_2 z^2 + d_3 z^3 + d_4 z^4} \tag{23}$$

in the z domain. The results are

$$\begin{aligned} &n_1 = 1.265298, && n_2 = -1.508272, && n_3 = -2.047727, \\ &n_4 = 0.5205257, && n_5 = 0.8033268, && \\ &d_1 = -1.563129, && d_2 = 0.7167736, && \\ &d_3 = -0.1569175, && d_4 = 0.03977094. && \end{aligned} \tag{24}$$

The orders of the numerator and denominator in (23) were chosen to obtain a 1% fit to the first 100 terms. Significantly higher-order polynomials were necessary to obtain a significantly better fit. Subroutine CCSQRT implements combining points m and $m+1$ (expression (19)) and the convolution with (23).

The input arguments of subroutine CCSQRT are

W(2, *) — $(W(1, m+3), W(2, m+3)) = w(m\,\Delta t + t_0, x)$ (complex seismogram, $m = -2, \ldots, M+2$—output of THETAC)

M — number of time points

and the output argument is

$U(*)$ — $U(m) = u(m\,\Delta t - \Delta t + t_0, x)$ (real impulse seismogram, $m = 1, \ldots, M$)

Note that factors $(2\,\Delta t)^{-1}$ from (19) and $\frac{1}{6}$ from (23) have been omitted from the code. Overall, the impulse seismogram has been smoothed by a fourth-order boxcar $(B(t/\Delta t)/\Delta t*)^4$. If this is too severe then the simplest alternative as the algorithm is so fast is to decrease Δt and increase M. Alternatively, the other convolution methods can be used.

4 CONCLUSIONS

The five routines described in this chapter provide the crucial elements for the computation of WKBJ seismograms. They can easily be used with two- and three-dimensional ray-tracing routines for the computation of Maslov seismograms (Chapman and Drummond 1982).

A simple program, WKBJ, has also been included in the software library. It uses the five routines to compute impulse WKBJ seismograms. It is intended only as a guide to their use. The user will want to include more versatile input of the model and ray definitions, and output of the travel-time tables and seismograms.

ACKNOWLEDGEMENTS

This chapter was written while one of us (CHC) was a Cecil H. and Ida Green Scholar at the Institute of Geophysics and Planetary Physics, Scripps Institution of Oceanography, University of California, San Diego. We gratefully acknowledge this support and the assistance provided. This is Department of Earth Sciences, Cambridge, publication number 797.

APPENDIX: WKBJ SEISMOGRAMS WITH ATTENUATION

For realistic synthetic seismograms, we must include the anelastic behaviour. As is well known, any solution derived using linear elasticity theory can be transformed into the solution for a linear viscoelastic medium if the elastic moduli are replaced by complex values. This is known as the correspondence principle (Ben-Menahem and Singh, 1981, p. 875). In general the constitutive relationship between stress and strain is expressed as a convolution in the time domain between a creep function and the strain history. In the frequency domain this can be described by complex frequency-dependent elastic moduli. From these we can derive complex velocities. A number of observations suggest that the intrinsic quality factor Q of many Earth materials depends very little upon frequency over the body-wave frequency band (Knopoff, 1964).

Futterman (1962) realized that a constant-Q model for all frequencies was physically impossible as it violated causality. Some dispersion is necessary. Provided that the attenuation is large (terms in Q^{-2} can be ignored), several different models give the same attenuation and dispersion relationship, i.e.

$$\frac{1}{\alpha(\omega)} \simeq \frac{1}{\alpha_{\mathrm{R}}}\left(1 - \frac{1}{\pi Q_\alpha}\ln\left|\frac{\omega}{\omega_\alpha}\right| + \frac{\mathrm{i}\,\mathrm{sgn}\,(\omega)}{2Q_\alpha}\right), \tag{A1}$$

and similarly for $\beta(\omega)$. The velocity α_{R} is the real phase velocity at some reference frequency ω_α, and Q_α is the constant quality factor for P waves. References and more details of the physical models leading to this approximation can be found in Aki and Richards (1980, Section 5.5) and Ben-Menahem and Singh (1981, Chap. 10). The exact form of attenuation used (e.g. the "frequency-independent" model (A1)) is not critical in the following discussion. To obtain an efficient algorithm (e.g. (A6)), it is only necessary that the resultant attenuation vary slowly as a function of horizontal slowness so that it can be factored outside the normal WKBJ algorithm. If this condition is violated, it may still be possible to use the WKBJ algorithm by dividing the p integral into several ranges, and applying the WKBJ algorithm in each range with different attenuation operators.

As already mentioned, the simplicity and efficiency of the WKBJ algorithm depends on the frequency-independence of the amplitude and delay functions. In general, with attenuation included (e.g. (A1)), this property is lost. In this Appendix we only discuss modifications that can be made to the WKBJ seismogram algorithm to include attenuation when several simplifying assumptions can be made. Expression (A1) is only valid for large Q_α and to the same order approximation, the delay function (8) becomes

$$\tau_{\mathrm{ray}}(p) = \tau_{\mathrm{R}}(p) - \frac{1}{\pi}\ln\left|\frac{\omega}{\omega_\alpha}\right| t^*_\alpha(p) - \frac{1}{\pi}\ln\left|\frac{\omega}{\omega_\beta}\right| t^*_\beta(p) + \frac{\mathrm{i}\,\mathrm{sgn}\,(\omega)}{2} t^*(p), \tag{A2}$$

where $\tau_{\mathrm{R}}(p)$ is the delay function (8) evaluated with the real phase velocities α_{R} and β_{R} (for simplicity we drop the subscript "ray" from the right-hand side), $t^*_\alpha(p)$ is defined as

$$t^*_\alpha(p) = \sum_{l=1}^{L-1} n_{\alpha_l}\int_{z_{l+1}}^{z_l} \frac{\mathrm{d}z}{\alpha_{\mathrm{R}}^2 q_{\alpha_{\mathrm{R}}} Q_\alpha}, \tag{A3}$$

and similarly for $t^*_\beta(p)$, and $t^*(p) = t^*_\alpha(p) + t^*_\beta(p)$. We have made the further assumption that ω_α and ω_β are independent of depth but retained the possibility that they differ.

Lockett (1962), Schoenberg (1971), Buchen (1971a,b) and Borcherdt (1973) have applied the general theory to describe the physical characteristics of waves in anelastic media. Borcherdt (1977, 1982) provides a detailed treatment of general plane-wave reflection and transmission. Krebes and Hron (1980a,b) have calculated reflection and transmission coefficients for anelastic SH waves and compared them with the elastic case. Similarly, Kelamis *et al.* (1983) considered the generalized coefficients for SH and P–SV waves. The reflection and transmission coefficients in anelastic media can be calculated using the same algebraic expressions (7) provided that the slownesses are taken to be complex (A1). This leads to frequency-dependent coefficients, but in general the dependence is slight and we can approximate the coefficients by their value at some reference frequency, ω_0 say, in the band of interest. Thus we approximate

$$R_{\text{ray}}(p, \omega) \simeq R_{\text{ray}}(p, \omega_0) \tag{A4}$$

etc. in (2) and (6), and evaluate the coefficients (7) using the complex slownesses (A1) at the reference frequency ω_0.

Finally, to invert the transforms (1), we assume that the t^* functions are slowly varying functions of the horizontal slowness p, and in the transform integrals (1) can be approximated by a constant. Thus

$$t^*_\alpha(p) \simeq t^*_\alpha(p_{\text{ray}}), \tag{A5}$$

where p_{ray} is typically taken as the geometrical value, i.e. $X_{\text{ray}}(p_{\text{ray}}) = x$. With these approximations, the effects of attenuation can be factored outside the integrals and represented in the final result as a convolution, i.e. (9) becomes

$$\begin{aligned}
\boldsymbol{u}(t, x, z) \simeq \frac{1}{2^{3/2}\,|x|^{1/2}\,\pi^2} F^{-1}\Big\{\exp\Big[&-\frac{i\omega}{\pi}\ln\left|\frac{\omega}{\omega_\alpha}\right| t^*_\alpha(p_{\text{ray}}) \\
&-\frac{i\omega}{\pi}\ln\left|\frac{\omega}{\omega_\beta}\right| t^*_\beta(p_{\text{ray}}) - \frac{|\omega|\,t^*(p_{\text{ray}})}{2}\Big]\Big\} \\
\Big\{\boldsymbol{m}''(t) : \operatorname{Im}\Big[\Lambda(t) * \sum_{\text{rays}}\sum_{t=\theta_{\text{ray}}} &\frac{p^{1/2}\boldsymbol{M}_{\text{ray}}(p, \omega_0)\boldsymbol{U}_{\text{ray}}(p, \omega_0)R_{\text{ray}}(p, \omega_0)}{|\partial_p\theta_{\text{ray}}|}\Big] \\
+\boldsymbol{f}'(t) \cdot * \operatorname{Im}\Big[\Lambda(t) * \sum_{\text{rays}}\sum_{t=\theta_{\text{ray}}} &\frac{p^{1/2}\boldsymbol{F}_{\text{ray}}(p, \omega_0)\boldsymbol{U}_{\text{ray}}(p, \omega_0)R_{\text{ray}}(p, \omega_0)}{|\partial_p\theta_{\text{ray}}|}\Big]\Big\},
\end{aligned} \tag{A6}$$

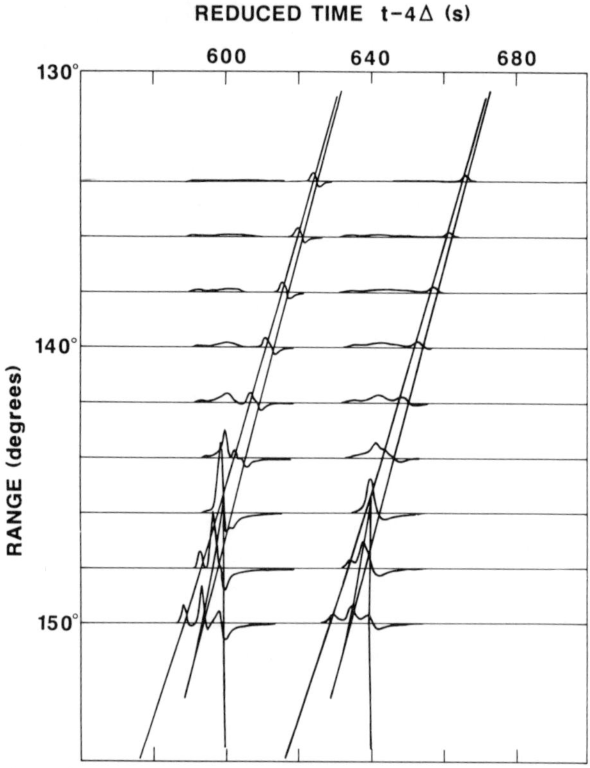

Fig. 9 Impulse seismograms for the vertical component of displacement from an explosion, i.e. $m_{xx}(t) = m_{zz}(t) = m_0H(t)$, for the PREM model with and without attenuation. The attenuated seismograms are offset by a 40 s delay.

where $F^{-1}\{\ \}$ represents the inverse FT. Unfortunately no simple analytic expression is known for this time series, but it is easily evaluated numerically. The convolution is most efficiently evaluated in the frequency domain.

The routines are easily modified to include the effects of attenuation. Extra model parameters $1/Q_{\alpha i}$ and $1/Q_{\beta i}$ are added to the model arrays. Subroutine PXINT returns the functions $t^*_\alpha(p)$ and $t^*_\beta(p)$ in an array, TSTAR. The integrals in (A3) are evaluated, taking the mean value of Q^{-1} between each pair of model points. Routines COMPON and COEFF include the frequencies ω_0, ω_α and ω_β as extra arguments.

Finally in Fig. 9 we illustrate impulse seismograms for the vertical displacement from an explosion for the PREM model (Dziewonski and Anderson, 1981) for the rays PKP, PKiKP and PKIKP with and without attenuation. Note in particular the PKP caustic at about 145°.

REFERENCES

Aki, K. and Richards, P. G. (1980). *Quantitative Seismology: Theory and Methods*. Freeman, San Francisco.

Backus, G. E. and Mulcahy, M. (1976). Moment tensor and other phenomenological descriptions of seismic sources, I. Continuous displacements. *Geophys. J. R. Astron. Soc.* **46,** 341–361.

Ben-Menahem, A. and Singh, S. J. (1981). *Seismic Waves and Sources*. Springer-Verlag, New York.

Borcherdt, R. D. (1973). Energy and plane waves in linear viscoelastic media. *J. Geophys. Res.* **78,** 2442–2453.

Borcherdt, R. D. (1977). Reflection and refraction of type IIS waves in elastic and anelastic media. *Bull. Seismol. Soc. Am.* **67,** 43–67.

Borcherdt, R. D. (1982). Reflection and refraction of general P- and type IS waves in elastic and anelastic solids. *Geophys. J. R. Astron. Soc.* **70,** 621–638.

Buchen, P. W. (1971a). Plane waves in linear viscoelastic media. *Geophys. J. R. Astron. Soc.* **23,** 531–542.

Buchen, P. W. (1971b). Reflection, transmission and diffraction of *SH*-waves in linear viscoelastic solids. *Geophys. J. R. Astron. Soc.* **25,** 87–113.

Chapman, C. H. (1974). The turning point of elastodynamic waves. *Geophys. J. R. Astron. Soc.* **39,** 613–621.

Chapman, C. H. (1976). A first-motion alternative to geometrical ray theory. *Geophys. Res. Lett.* **3,** 153–156.

Chapman, C. H. (1978). A new method for computing synthetic seismograms. *Geophys. J. R. Astron. Soc.* **54,** 481–518.

Chapman, C. H. and Drummond, R. (1982). Body-wave seismograms in inhomogeneous media using Maslov asymptotic theory. *Bull. Seismol. Soc. Am.* **72,** S277–S317.

Chapman, C. H. and Orcutt, J. A. (1985a). The computation of body wave seismograms in laterally homogeneous media. *Rev. Geophys.* **23,** 105–163.

Chapman, C. H. and Orcutt, J. A. (1985b). Least-squares fitting of marine seismic refraction data. *Geophys. J. R. Astron. Soc.* **82,** 339–374.

Dey-Sarkar, S. K. and Chapman, C. H. (1978). A simple method for the computation of body-wave seismograms. *Bull. Seismol. Soc. Am.* **68,** 1577–1593.

Dziewonski, A. M. and Anderson, D. L. (1981). Preliminary reference Earth model. *Phys. Earth Planet. Int.* **25,** 297–356.

Futterman, W. I. (1962). Dispersive body waves. *J. Geophys. Res.* **67,** 5279–5291.

Gilbert, F. (1971). Excitation of normal modes of the Earth by earthquake sources. *Geophys. J. R. Astron. Soc.* **22,** 223–226.

Gilbert, F. and Backus, G. E. (1966). Propagator matrices in elastic wave and vibration problems. *Geophysics* **31,** 326–332.

Kelamis, P. G., Kanasewich, E. R. and Abramovici, F. (1983). Attenuation of seismograms obtained by the Cagniard–Pekeris method. *Geophysics* **48,** 1204–1211.

Kennett, B. L. N. (1974). Reflections, rays and reverberations. *Bull. Seismol. Soc. Am.,* **64,** 1685–1696.

Knopoff, L. (1964). *Q. Rev. Geophys.* **2,** 625–660.

Krebes, E. S. and Hron, F. (1980a). Ray synthetic seismograms for SH waves in anelastic media. *Bull. Seismol. Soc. Am.* **70,** 29–46.

Krebes, E. S. and Hron, F. (1980b). Synthetic seismograms for SH waves in layered anelastic medium by asymptotic ray theory. *Bull. Seismol. Soc. Am.*, **70,** 2005–2020.

Lockett, F. J. (1962). The reflection and refraction of waves at an interface between viscoelastic media. *J. Mech. Phys. Solids* **10,** 53–64.

Richards, P. G. (1971). Elastic waves in stratified media. *Geophysics* **36,** 798–809.

Schoenberg, M. (1971). Transmission and reflection of plane waves at an elasto-viscoelastic media. *J. Acoust. Soc. Am.* **46,** 649–654.

Takeuchi, M. and Saito, M. (1972). Seismic surface waves. *Meth. Comp. Phys.* **11,** 217–295.

Wiggins, R. A. (1976). Body wave amplitude calculations—II. *Geophys. J. R. Astron. Soc.* **46,** 1–10.

I.3

Asphericity and Ellipticity Corrections

D. J. DOORNBOS
Institute of Geophysics, University of Oslo, Norway

1 INTRODUCTION

Recent publications on the aspherical structure of the Earth (e.g. Dziewonski, 1984; Giardini *et al.*, 1987) suggest that there is a need for algorithms to evaluate the travel-time perturbations due to asphericity. This includes the calculation of ellipticity corrections as a special case. The simplest method, valid for slight asphericity, is to evaluate the perturbations to first order. The theoretical basis was given by Dziewonski and Gilbert (1976), and the method can be implemented by modifying conventional one-dimensional ray-tracing algorithms. In this chapter I describe algorithms that can be used to calculate travel times in an aspherical-Earth model. The algorithms can also be used in the inverse problem of inferring the asphericity from travel-time anomalies. The inverse problem has been described in detail by Dziewonski (1984).

The ray-tracing algorithm in this paper accumulates the travel-time changes due to perturbations of the velocity along the ray and/or the velocity discontinuities the ray is crossing. No special form of asphericity is required, but in actual applications, models expanded in spherical harmonics have been used. As pointed out by Dziewonski (1984), it is then an advantage to apply spherical harmonics in a rotated coordinate frame as given by Backus (1964). Another algorithm in this chapter applies this coordinate transformation to give the travel-time perturbation due to each spherical harmonic in the expansion. Some attention will also be given to the implementation of models of asphericity of structure and of ellipticity of the figure of the Earth.

The first-order perturbation method applied in this chapter is based on

SEISMOLOGICAL ALGORITHMS
ISBN 0-12-220770-X

the principle of stationarity of the ray path. The validity of this assumption in current models of asphericity can be tested by a comparison with three-dimensional ray-tracing results. I include an example where the two methods are applied to PcP, PKP and PnKP, $n \geqslant 2$, with an aspherical core–mantle-boundary model.

2 VELOCITY MODELS

The ray-tracing algorithm assumes a spherical reference model where the velocity is interpolated between specified values at radii r_j. For isotropic models there are several interpolation functions, all leading to simple results for the travel time, angular distance and other quantities of the ray through a layer bounded by two subsequent radii, say r_{j-1} and r_j. Elsewhere in this book (Chap. I.2) Chapman *et al.* give three possible interpolation formulae for a flat-Earth model. Note that such simple results can only be obtained for isotropic models; anisotropic layers require numerical integration across the layer. We proceed on the assumption of an isotropic model. The present algorithm uses Bullen's interpolation:

$$v = a(r/r_c)^b, \tag{1}$$

where r_c is a normalizing reference radius. I include a routine MODPS that reads the velocity model $v(r_j)$ and calculates the parameters a_j and b_j for each layer defined by $r_{j-1} \leqslant r \leqslant r_j$. This is done for both P and S velocities. A velocity discontinuity is specified by repeating the radius with two different velocities, i.e. $r_j = r_{j-1}$ and $v(r_j) \neq v(r_{j-1})$.

The aspherical velocity perturbation is given by $\delta v(r, \theta, \varphi)$, where θ is colatitude and φ is longitude. Published models of δv have been given in terms of spherical-harmonics expansion coefficients ${}_kA_l^m$ and ${}_kB_l^m$ (Dziewonski, 1984):

$$\delta v(r, \theta, \varphi) = \sum_{k=0}^{K'} f_k(r) g_k(\theta, \varphi), \tag{2}$$

$$g_k(\theta, \varphi) = \sum_{l=0}^{L} \sum_{m=0}^{l} c_l^m d_l^m ({}_kA_l^m \cos m\varphi + {}_kB_l^m \sin m\varphi) P_l^m(\cos\theta), \tag{3}$$

where

$$c_l^m = [(2l+1)(2-\delta_{m,0})]^{1/2}, \qquad d_l^m = \left[\frac{(l-m)!}{(l+m)!}\right]^{1/2}.$$

Dziewonski (1984) and Woodhouse and Dziewonski (1984) chose to represent the radial dependence $f_k(r)$ by Legendre polynomials in reduced normalized radius. Equations (2) and (3) then represent an expansion in orthogonal basis functions. To evaluate the travel-time perturbation we actually need $\delta v/v$:

$$\frac{\delta v}{v}(r, \theta, \varphi) = w(r) \sum_{k=0}^{K'} \varepsilon_k(r) g_k(\theta, \varphi), \tag{4}$$

from which we specify at the radii of the reference model

$$\varepsilon_k(r_j) = \frac{f_k(r_j)}{v(r_j)}, \quad k = 0, \ldots, K', \tag{5}$$

and interpolate in each layer $r_{j-1} \leq r \leq r_j$. The additional weighting factor $w(r)$ can be used to accommodate possible differences in $\delta v/v$ for P and S velocities. This is especially useful when describing the effect of asphericity of the figure of the Earth's structure. This form of asphericity can be similarly represented:

$$\frac{\delta r}{r}(r, \theta, \varphi) = \sum_{k=K'+1}^{K} \varepsilon_k(r) g_k(\theta, \varphi). \tag{6}$$

The resulting velocity perturbation inside any layer is

$$\frac{\delta v}{v} = -\left(\frac{r}{v}\frac{\mathrm{d}v}{\mathrm{d}r}\right)\frac{\delta r}{r}; \tag{7}$$

hence $w(r) = -(r/v)\,\mathrm{d}v/\mathrm{d}r$, and $w(r) = w_j = -b_j$ in the layer $r_{j-1} \leq r \leq r_j$. Perturbation of a single boundary at $r = r_j$ can be modelled by choosing

$$\varepsilon_k(r) = \varepsilon_k(r_j)\delta(r - r_j).$$

A routine EXMOD specifies $\varepsilon_k(r_j)$, $0 \leq k \leq K'$, and $K' < k \leq K$ for all radii r_j, and w_j for all layers $r_{j-1} \leq r \leq r_j$. The factor w_j is specified for both P and S velocities.

3 RAY-TRACING ALGORITHM

The theoretical basis of one-dimensional ray methods is contained in the works of Bullen (1965), Julian and Anderson (1968), Dziewonski and Gilbert (1976) and others. One-dimensional ray theory is also well covered in more recent textbooks (Aki and Richards, 1980; Bullen and Bolt, 1986). In a spherical Earth the one-way travel time and change in angular distance of a ray between levels r_j and r_0 can be expressed as

$$dT_j = \int_{r_0}^{r_j} \frac{d(\eta^2 - p^2)^{1/2}}{1 - \eta v'}, \qquad d\Delta_j = \int_{r_0}^{r_j} \frac{d\cos^{-1}(p/\eta)}{1 - \eta v'}, \tag{8}$$

where p is the ray parameter, $\eta = r/v$ and $v' = dv/dr$. For a turning ray, r_0 is the turning-point radius. The evaluation of the integrals in (8) is trivial if the velocity model is parametrized in such a way that $1 - \eta v'$ is constant between r_0 and r_j. For the velocity interpolation following (1)

$$1 - \eta v' = 1 - b_j.$$

A velocity perturbation δv gives rise to a travel-time perturbation δT. To first order,

$$\delta T_j = \int_{r_0}^{r_j} -\frac{\delta v}{v}\frac{d(\eta^2 - p^2)^{1/2}}{1 - \eta v'} = \frac{-1}{1 - b_j}\int_{r_0}^{r_j} \frac{\delta v}{v} d(\eta^2 - p^2)^{1/2}. \tag{9}$$

This integral must be evaluated numerically if $\delta v/v$ varies along the ray path. The present algorithm divides $[(\eta^2 - p^2)^{1/2}]_{r_0}^{r_j}$ into equal subintervals, and the simple trapezoidal rule is applied to each subinterval:

$$\delta T_j = \frac{-1}{2N(1 - b_j)}[(\eta^2 - p^2)^{1/2}]_{r_0}^{r_j} \sum_{i=1}^{N} \left[\left(\frac{\delta v}{v}\right)_{i-1} + \left(\frac{\delta v}{v}\right)_i\right]. \tag{10}$$

The number N is reset for each ray in each layer, in such a way that a preset accuracy is obtained. Present models of asphericity are sufficiently smooth that $N = 4$ is usually adequate. The relative velocity perturbation $(\delta v/v)_i$ is a function of r_i and of the angular position, which depends on epicentral distance Δ_i:

$$r_i = r_c\left(\frac{a_j \eta_i}{r_c}\right)^{1/(1-b_j)}, \tag{11}$$

$$\Delta_i = \Delta_j \pm \frac{1}{1 - b_j}\left[\cos^{-1}\left(\frac{p}{\eta}\right)\right]_{r_i}^{r_j}, \tag{12}$$

where Δ_j is the epicentral distance of the ray at r_j, and $+/-$ corresponds to a down/upgoing ray through layer j.

A boundary perturbation $\delta r/r$ gives rise to a travel-time perturbation, to first order,

$$\delta T = \pm\frac{\delta r}{r}(\eta^2 - p^2)^{1/2} + p\,\delta\Delta, \tag{13}$$

where $+/-$ corresponds to a ray at the bottom/topside of the boundary. The second term in (13) is due to migration of the ray–boundary intersection. To first order, this term will be cancelled by the corresponding terms of subsequent ray sections. The first term will also be cancelled, unless r represents a velocity discontinuity or an endpoint of the ray. The algorithm evaluates the first term of (13) at the endpoints of the ray and at the intersection with each interface j:

$$\delta T(r_j^{\mp}) = \pm \frac{\delta r}{r} (\eta^2 - p^2)^{1/2}\big|_{r=r_j}. \tag{14}$$

The ray-racing algorithm is performed by the routines EXRPS, EXTPS and RAYPS. Routine EXRPS accumulates dT_j, $d\Delta_j$ and the perturbations δT_j and $\delta T(r_j^{\mp})$. The routine operates on variations $\delta v/v$ and $\delta r/r$, whose radial and angular dependence is specified in separate subroutines (EXMOD and EXSIG, respectively). Present applications involve models of asphericity in terms of spherical harmonics. The transformation properties of spherical harmonics under rotation of coordinates (next section) ensure that the travel-time perturbations can be reconstructed from perturbations in $\delta v/v$ and $\delta r/r$ of the form $f_k(r)\cos m\Delta$, $f_k(r)\sin m\Delta$, $m = 0, 1, 2, \ldots$, where Δ is the epicentral distance. Consequently, the angular dependence currently specified in EXSIG is just of the form $\cos m\Delta$, $\sin m\Delta$. Routine EXRPS always traces the ray downward through a section of the model, i.e. an actually upgoing ray is traced backward from its endpoint. The sense of direction of the ray is controlled by EXTPS, and the epicentral distance of the endpoint of an upgoing ray is determined by RAYPS, which is a general-purpose subroutine for evaluating several ray-related quantities along the ray.

4 SPHERICAL HARMONICS IN ROTATED COORDINATES

Backus (1964) established the following relation between spherical harmonics under rotation of coordinates:

$$Y_l^m(\theta, \varphi) = e^{im\phi} \sum_{n=-l}^{l} e^{in\psi} Q_l^{mn}(\Theta) Y_l^n(\theta', \varphi'), \tag{15}$$

where (θ, φ) and (θ', φ') are colatitude and longitude in the original and new system, and Θ, ϕ and ψ are the three Euler angles of the rotation (see Fig. 1 of Backus (1964) or Fig. 2 of Dziewonski (1984)). The transformation matrix Q_l^{mn} will be discussed later in this section. Dziewonski (1984) first utilized the relation (15) in the context of ray tracing in an aspherical-Earth model. If we choose $(\theta', \varphi') = (\frac{1}{2}\pi, \Delta)$ i.e. the ray is traced in the equatorial plane of the new coordinate system, then splitting (15) into its real and imaginary parts leads to

$$\left.\begin{aligned}(-1)^m d_l^m P_l^m(\cos\theta)\cos m\varphi &= V_l^m \cos m\phi - W_l^m \sin m\phi,\\(-1)^m d_l^m P_l^m(\cos\theta)\sin m\varphi &= W_l^m \cos m\phi + V_l^m \sin m\phi,\end{aligned}\right\}\tag{16}$$

where d_l^m is defined in (3), and

$$\left.\begin{aligned}V_l^m &= \sum_{n=0}^{l}(1-\tfrac{1}{2}\delta_{n,0})d_l^n P_l^n(0)[(-1)^n Q_l^{mn} + Q_l^{m,-n}]\\&\qquad\times(\cos n\psi\cos n\Delta - \sin n\psi\sin n\Delta),\\W_l^m &= \sum_{n=1}^{l} d_l^n P_l^n(0)[(-1)^n Q_l^{mn} - Q_l^{m,-n}]\\&\qquad\times(\sin n\psi\cos n\Delta + \cos n\psi\sin n\Delta).\end{aligned}\right\}\tag{17}$$

The travel-time perturbation due to asphericity of the form $\cos n\Delta$ or $\sin n\Delta$ is evaluated by the algorithm described in the previous section:

$$\left.\begin{aligned}\varepsilon_k(r)\cos n\Delta &\to {}_k\delta T_c^n,\\\varepsilon_k(r)\sin n\Delta &\to {}_k\delta T_s^n.\end{aligned}\right\}$$

The travel-time perturbation due to asphericity of the form (4) and (6) is then

$$\delta T = \sum_{k=0}^{K}\sum_{l=0}^{L}\sum_{m=0}^{l} {}_kA_l^m({}_k\delta T_l^m)_A + {}_kB_l^m({}_k\delta T_l^m)_B,\tag{18}$$

where

$$\left.\begin{aligned}({}_k\delta T_l^m)_A &= (-1)^m c_l^m[\cos m\phi\,({}_k\delta T_l^m)_V - \sin m\phi\,({}_k\delta T_l^m)_W],\\({}_k\delta T_l^m)_B &= (-1)^m c_l^m[\cos m\phi\,({}_k\delta T_l^m)_W + \sin m\phi\,({}_k\delta T_l^m)_V],\end{aligned}\right\}\tag{19}$$

$$\left.\begin{aligned}({}_k\delta T_l^m)_V &= \sum_{n=0}^{l}(1-\tfrac{1}{2}\delta_{n,0})d_l^n P_l^n(0)[(-1)^n Q_l^{mn} + Q_l^{m,-n}]\\&\qquad\times(\cos n\psi\,{}_k\delta T_c^n - \sin n\psi\,{}_k\delta T_s^n)\\({}_k\delta T_l^m)_W &= \sum_{n=1}^{l} d_l^n P_l^n(0)[(-1)^n Q_l^{mn} - Q_l^{m,-n}]\\&\qquad\times(\sin n\psi\,{}_k\delta T_c^n + \cos n\psi\,{}_k\delta T_s^n).\end{aligned}\right\}\tag{20}$$

Equations (19) and (20) are implemented in routine EXPRO. The Euler angles Θ, ϕ and ψ are determined by EXANG, and the transformation matrix Q_l^{mn} is determined by EXLMN. This routine is based on the following relations given by Backus (1964):

$$Q_l^{mn}(\Theta) = B_l^{mn} P_l^{mn}(\Theta),\tag{21}$$

where

$$B_l^{mn} = (-1)^{l+m}[(l-m)!\,(l+m)!\,(l-n)!\,(l+n)!]^{1/2} \tag{22}$$

and $P_l^{mn}(\Theta)$, $m, n, = -l, \ldots, l$, can be determined by the values for $m = -l$

$$P_l^{-ln} = \frac{(\cos\frac{1}{2}\Theta)^{l-n}(\sin\frac{1}{2}\Theta)^{l+n}}{(l+n)!\,(l-n)!}, \tag{23}$$

the recurrence relation

$$(l+m+1)P_l^{m+1,n} = -(m+n)\tan\tfrac{1}{2}\Theta P_l^{mn} - (l-n+1)P_l^{m,n-1} \tag{24}$$

and the symmetry relations

$$\left.\begin{aligned} P_l^{mn} &= P_l^{nm}, \\ P_l^{mn} &= (-1)^{m+n}P_l^{-m,-n}. \end{aligned}\right\} \tag{25}$$

The program package includes an example program EXASP, which uses the routines described in this and the previous sections to produce, for a given source and receiver location, the terms $({}_k\delta T_l^m)_A$ and $({}_k\delta T_l^m)_B$ from (19), and the final asphericity correction δT from (18).

5 ELLIPTICITY CORRECTIONS

The Earth's ellipticity is given by $\varepsilon(r)$, where

$$\frac{\delta r}{r}(r, \theta, \varphi) = \varepsilon(r)(\tfrac{1}{3} - \cos^2\theta) = -\tfrac{2}{3}\varepsilon(r)P_2^0(\cos\theta), \tag{26}$$

and, following (7),

$$\frac{\delta v}{v}(r, \theta, \varphi) = \frac{2}{3}\left(\frac{r}{v}\frac{dv}{dr}\right)\varepsilon(r)P_2^0(\cos\theta). \tag{27}$$

Thus, in the expansion of (3), (4) and (6),

$$K' = 0, \qquad K = 1, \qquad {}_kA_l^m = -\frac{2}{3\sqrt{5}}\delta_{l2}\delta_{m0}, \qquad {}_kB_l^m = 0, \quad k = 0, 1 \tag{28}$$

at radii r_j of the reference model;

$$\varepsilon_0(r_j) = \varepsilon_1(r_j) = \varepsilon(r_j), \tag{29a}$$

and in layers j, $r_{j-1} \leqslant r \leqslant r_j$

$$w_j = -b_j. \tag{29b}$$

The coefficients $\varepsilon_k(r_j)$ and w_j are determined by the routine ELMOD, which replaces EXMOD. The values adopted for $\varepsilon(r_j)$ have been taken from Bullen (1975).

The advantages of the coordinate rotation can in this case also be realized by applying the addition theorem for Legendre functions:

$$P_l^0(\cos\theta) = \sum_{m=0}^{l} (2-\delta_{m,0}) \frac{(l-m)!}{(l+m)!} P_l^m(\cos\theta_0) P_l^m(\cos\Delta) \cos m\zeta, \quad (30)$$

where θ_0 is the colatitude of the source point and ζ is the azimuth of the ray direction from the source. The algorithm of Dziewonski and Gilbert (1976) is directly based on (30). To implement this algorithm, we need to modify the routine EXSIG specifying the angular dependence. The modified routine ELSIG returns $P_2^m(\cos\Delta)$, $m = 0, 1, 2$. We also need to modify the routine EXPRO returning the travel-time terms from (19): $({}_k\delta T_l^m)_A$ and $({}_k\delta T_l^m)_B$. The modified routine ELPRO uses (30) to evaluate $({}_k\delta T_2^0)_A$, $k = 0, 1$. The ellipticity correction δT then follows directly from (18) and (28). The ray-tracing routines (EXRPS, EXTPS and RAYPS) and the example application program EXASP need not be modified.

6 A TEST CASE: THE CORE–MANTLE BOUNDARY

A significant part of the travel-time correction for asphericity is due to perturbation of internal boundaries. Recent experiments have produced models of the core–mantle boundary characterized by large long-wavelength topography, with peak-to-trough amplitudes of at least 10 km (Morelli and Dziewonski, 1987; Creager and Jordan, 1986). We have applied the algorithms of Sections 3 and 4 to one of the models of Morelli and Dziewonski, which is given in terms of spherical harmonics up to order four, and we have compared the travel-time residuals with those obtained by three-dimensional ray tracing in the same model. The purpose of this exercise is not to validate the model, but rather to test the algorithms in a situation of significant long-wavelength asphericity. Some of the results are shown in Figs 1–4.

In these examples a source was assumed at zero latitude and longitude, and receivers were assumed in four different azimuthal directions from the source (the N, E, S and W directions). Travel-time residuals were calculated for PcP and for the AB and BC branches of PKP and PnKP, $n \geq 2$. The branch of PnKP passing through the inner core was not used. Because the boundary in the model has a mean elevation of −2.5 km with respect to the standard Earth model (PREM), there is a mean positive residual for PcP, and a mean negative residual for PKP and

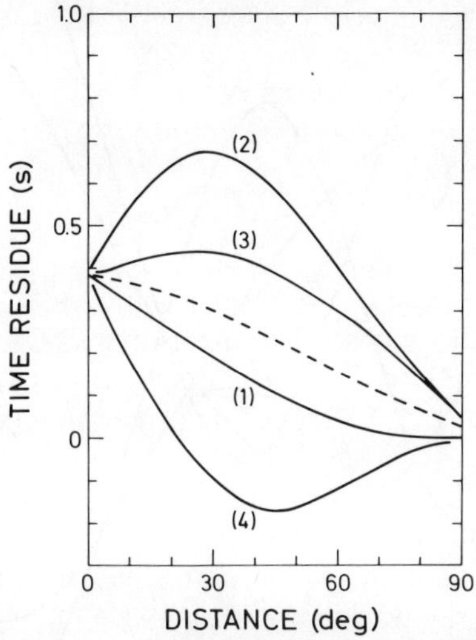

Fig. 1 Travel-time residuals of PcP with aspherical model of the core–mantle boundary from Morelli and Dziewonski (1987). The hypocentre is at zero latitude and longitude, and cases (1), (2), (3) and (4) refer to the N, E, S and W azimuthal directions from the source. The stippled line gives the travel-time residuals due to a decrease in core–mantle boundary radius of 2.5 km.

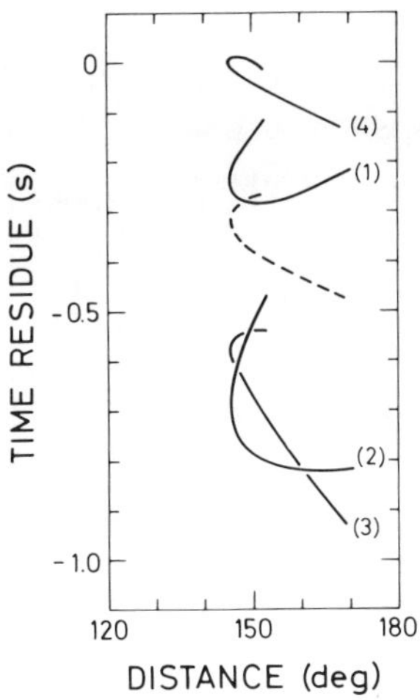

Fig. 2 Travel-time residuals of the AB and BC branches of PKP. Other details are as in Fig. 1.

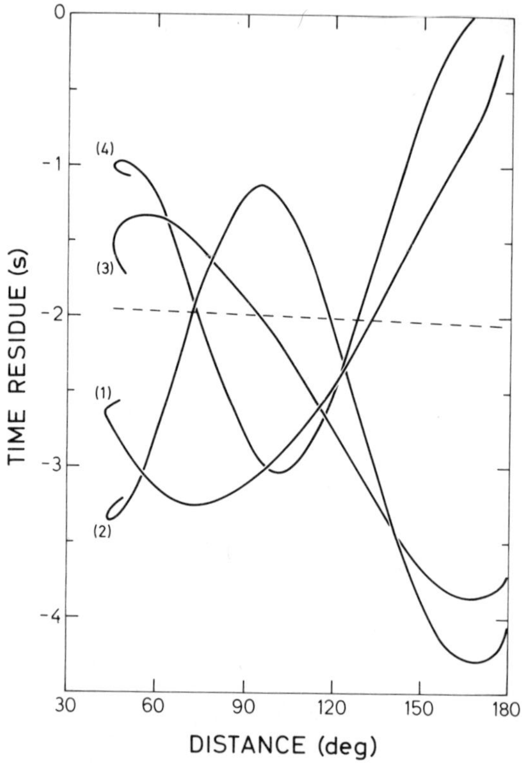

Fig. 3 Travel-time residuals of the AB and BC branches of P4KP. Other details are as in Fig. 1.

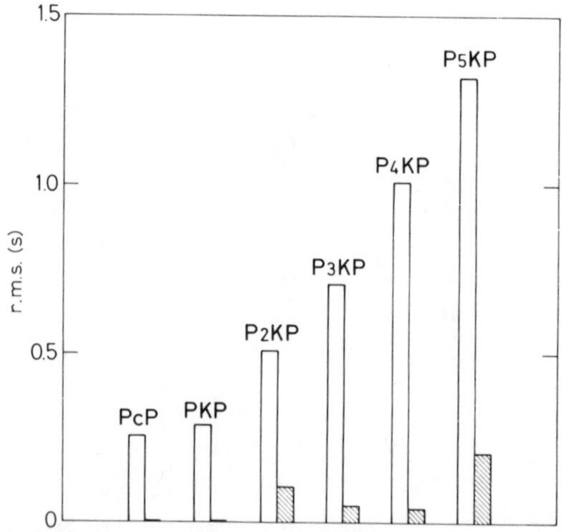

Fig. 4 Root variance (open columns) of travel times of core phases with the aspherical core–mantle boundary model, and r.m.s. error of travel times computed with the perturbation method (shaded columns).

P*n*KP. There is also a significant variation about the mean, and the variation increases with increasing number of reflections, i.e. increasing *n* in P*n*KP. The open columns in Fig. 4 give the root variance of the theoretical data for the different phases, and the shaded columns give the root-mean-square difference between the results by the two methods. The r.m.s. error is negligible for PcP and PKP. The error increases for P*n*KP, but in all cases is less than 20% of the root variance. The largest errors occur near multiples of 180° in epicentral distance (the antipode and the hypocentre). In these regions travel-time analysis would not be reliable anyway. The results suggest that the perturbation method is adequate for calculating travel times in current models of long-wavelength asphericity of the Earth.

REFERENCES

Aki, K. and Richards, P. G. (1980). *Quantitative Seismology,* Vol. 1. Freeman, San Francisco.

Backus, G. E. (1964). Geographical interpretation of measurements of average phase velocities of surface waves over great circular and great semi-circular paths. *Bull. Seismol. Soc. Am.* **54,** 571–610.

Bullen, K. E. (1965). *Introduction to the Theory of Seismology,* 3rd edn. Cambridge University Press.

Bullen, K. E. (1975). *The Earth's Density*. Chapman and Hall, London.

Bullen, K. E. and Bolt, B. A. (1985). *An Introduction to the Theory of Seismology*. Cambridge University Press, New York.

Creager, K. C. and Jordan, T. H. (1986). Aspherical structure of the core–mantle boundary from PKP travel times. *Geophys. Res. Lett.* **13,** 1497–1500.

Dziewonski, A. M. (1984). Mapping the lower mantle: Determination of lateral heterogeneity in P velocity up to degree and order 6. *J. Geophys. Res.* **89,** 5929–5952.

Dziewonski, A. M. and Gilbert, F. (1976). The effect of small, aspherical perturbations on travel times and a re-examination of the corrections for ellipticity. *Geophys. J. R. Astron. Soc.* **44,** 7–17.

Giardini, D., Li, X. D. and Woodhouse, J. H. (1987). Three-dimensional structure of the Earth from splitting in free-oscillation spectra. *Nature* **325,** 405–411.

Julian, B. R. and Anderson, D. L. (1968). Travel times, apparent velocities and amplitudes of body waves. *Bull. Seismol. Soc. Am.* **58,** 339–366.

Morelli, A. and Dziewonski, A. M. (1987). Topography of the core–mantle boundary and lateral homogeneity of the liquid core. *Nature* **325,** 678–683.

Woodhouse, J. H. and Dziewonski, A. M. (1984). Mapping the upper mantle: three dimensional modeling of Earth structure by inversion of seismic waveforms. *J. Geophys. Res.* **89,** 5953–5986.

II

RAYS AND BODY WAVES IN THREE-DIMENSIONAL MEDIA

II.1

Complete Seismic-Ray Tracing in Three-Dimensional Structures

V. ČERVENÝ
Institute of Geophysics, Charles University, Praha, Czechoslovakia

L. KLIMEŠ
Institute of Geology and Geotechnics, Czechoslovak Academy of Sciences, Praha, Czechoslovakia

I. PŠENČÍK
Geophysical Institute, Czechoslovak Academy of Sciences, Praha, Czechoslovakia

1 INTRODUCTION

The ray method and its extensions have recently found many applications in the numerical modelling and interpretation of seismic wavefields in complex two- and three-dimensional structures. The basic and numerically most time-consuming step in the application of the ray method consists in the computation of rays. Ray tracing plays an important role not only in the ray method itself but also in its more sophisticated extensions. In standard ray tracing only the ray trajectory, travel times and slowness vectors along the ray are determined. The components of the slowness vector also represent the first partial derivatives of the travel-time field with respect to spatial coordinates. All these quantities are very important in certain seismological applications, but are not sufficient to solve many other seismological problems.

It is, however, possible to slightly extend the ray-tracing algorithm and to evaluate many other quantities of great seismological importance along the ray (both in the numerical modelling of seismic wavefields and in the solution of inverse problems). Among other things, the ray propagator

SEISMOLOGICAL ALGORITHMS
ISBN 0-12-220770-X

matrix can be computed by means of so-called dynamic ray tracing. If the ray propagator matrix is known then many important seismological problems in the vicinity of the evaluated ray may be solved analytically, without additional numerical ray tracing. It is also not difficult to evaluate several other quantities along the ray; these can then be used for a simple computation of the ray amplitudes.

In this chapter, we propose and describe in detail an algorithm for *complete ray tracing*. Complete ray tracing consists of (a) standard ray tracing, (b) dynamic ray tracing (computation of the ray propagator matrix) and (c) evaluation of the components of the reduced vectorial complex-valued amplitudes. The algorithm for complete ray tracing is, of course, more involved than the algorithm for standard ray tracing, but numerically complete ray tracing is not much more time-consuming than ray tracing itself. Complete ray tracing considerably extends the possibilities and applicability of standard ray tracing and is sure to find an important place in most program packages designed for the numerical modelling of high-frequency (HF) seismic body wavefields in complex structures and for the solution of seismic inverse problems.

In the algorithm for complete ray tracing, a three-dimensional (3D) laterally varying isotropic block structure, specified in an arbitrary curvilinear coordinate system, is considered. The description of the model is rather general, and in principle it follows the proposals of Gjøystdal *et al.* (1985). The proposed algorithms allow us to consider virtually any type of seismic HF body wave propagating in such a medium.

The algorithms described in this chapter are not always simple and straightforward. It would, however, increase the length of the chapter inordinately to try to derive here all the required equations and to explain all steps in the algorithm from the seismological point of view. The most complete and consistent theory of complete ray tracing in general 3D layered media can be found in Červený (1985a). In the latter work Cartesian coordinates are used, but the algorithms proposed here are written for an arbitrary curvilinear coordinate system. For this reason, all the equations required in the algorithms are presented in complete form, even though their derivations are not given.

For specific seismological problems, it would be computationally more efficient to use some simplified algorithms. This chapter considers only the most general case; simplifications for special cases are left to the reader.

The structure of the algorithm for complete ray tracing proposed here is as follows. The full problem of complete ray tracing is divided into several, more or less independent algorithms. This chapter consists of a detailed description of these algorithms. The system of algorithms is full and sufficient for the complete ray tracing of a single ray in 3D structures. The results of complete ray tracing are assumed to be stored in output files, which may be used later in various problems of seismological importance.

In the derivation of the algorithm for complete ray tracing presented here we have profited a great deal from our long experience with numerical modelling of HF seismic wavefields in complex 2D and 3D structures and with 2D and 3D seismic-ray packages. This has allowed us to propose some more general and more efficient algorithms and a more convenient structure of the whole program. This chapter describes algorithms for such a new program.

In the proposed algorithm only a single ray of a selected elementary wave, specified by proper *initial conditions,* is considered. The two-point ray tracing problem is discussed only briefly, and not in full generality, for paraxial rays only. The results of complete ray tracing offer the possibility of solving the two-point ray-tracing problem for paraxial rays analytically (see Section 7.12). The general solution of the two-point ray-tracing problem is not included in the complete ray-tracing algorithm, even though the results of complete ray tracing may be useful in this problem. Likewise, the results of complete ray tracing are of basic importance in the computation of synthetic seismograms, synthetic frequency responses, elastodynamic Green functions etc. Such computations, however, require consideration of rays of various elementary waves. Thus, to perform them the user must supply his or her own routine in which the complete ray-tracing routines will be used successively for individual rays of the elementary waves. For the reader's convenience, the full equations for such computations are given in Sections 7.20–7.22.

In complete ray tracing only standard (zeroth-order) ray approximations are used to evaluate the amplitudes. Complete ray tracing, however, also yields quantities that are very useful and important in the evaluation of HF seismic wavefields in singular regions (vicinity of caustics, critical regions etc.) and in the evaluation of various diffracted waves. Such computations may be performed by programs not described here, but involving the results of complete ray tracing. Similarly, complete ray tracing is a basic routine in the evaluation of Gaussian beams and in the computation of HF seismic body wavefields by summation of Gaussian beams, by the extended WKBJ method, by the Maslov method, and so on. A detailed treatment of this subject is beyond the scope of this chapter.

Most equations in this chapter are written in component notation.

Capital-letter indices ($I, J, K, A, B, \ldots$) take values 1 and 2, lower-case indices ($i, j, k, a, b, \ldots$) take values 1, 2, 3, and Greek lower-case indices ($\alpha, \beta, \gamma, \ldots$) take values 1, 2, 3, 4. The Einstein summation convention is used, with respect to both repeated subscripts and superscripts.

Together with component notation, matrix notation is also used. Matrices are denoted by boldface sans serif letters. In order to distinguish between 2×2 and 3×3 matrices, the latter are denoted with a circumflex above the letter. If the same boldface letter is used for 2×2 and 3×3 matrices with and without the circumflex (e.g. $\mathbf{M}$ and $\hat{\mathbf{M}}$) then the matrix without the circumflex ($\mathbf{M}$) is the 2×2 upper left-hand minor of the 3×3 matrix with a circumflex ($\hat{\mathbf{M}}$). Thus M_{IJ} are components of $\mathbf{M}$, and M_{ij} those of $\hat{\mathbf{M}}$.

2 COORDINATE SYSTEM

2.1 Metric tensor and Christoffel symbols

We consider general right-handed curvilinear coordinates (x^1, x^2, x^3). The local properties of the coordinate system are described by the covariant components $G_{ij} = G_{ij}(x^k)$ or by the contravariant components $G^{ij} = G^{ij}(x^k)$, $i, j, k = 1, 2, 3$, of the metric tensor and by the Christoffel symbols

$$\Gamma^k_{ij} = \tfrac{1}{2} G^{kl}(G_{il,j} + G_{jl,i} - G_{ij,l}). \tag{2.1}$$

The metric tensor is symmetric and generally has six independent components. The covariant G_{ij} and contravariant G^{ij} components of the metric tensor are related by

$$G^{ij} G_{ik} = \delta^j_k,$$

where the mixed co- and contravariant Kronecker delta δ^i_j is equal to 1 for $i = j$ and 0 otherwise. The square of the infinitesimal length $\mathrm{d}s$ can be expressed in terms of the metric tensor as follows:

$$\mathrm{d}s^2 = G_{ij}\, \mathrm{d}x^i\, \mathrm{d}x^j.$$

There are generally 18 independent Christoffel symbols.

The routine designed for the determination of the metric tensor and of Christoffel symbols at any point is here called METRIC.

2.2 Examples of the most important coordinate systems

2.2.1 Cartesian coordinate system

In Cartesian coordinates,

$$G_{ij} = \delta_{ij}, \qquad G^{ij} = \delta^{ij}, \tag{2.2}$$

$$\Gamma^k_{ij} = 0. \tag{2.3}$$

Here the covariant and contravariant forms of the Kronecker delta δ_{ij} and δ^{ij} have the same numerical values as δ^i_j above.

2.2.2 Spherical polar coordinate system

We introduce the spherical polar coordinates $(x^1, x^2, x^3) = (\vartheta, \lambda, r)$ as follows: ϑ is the colatitude (positive southwards), $0 \leqslant \vartheta \leqslant \pi$, λ is the longitude (positive eastwards) $(0 \leqslant \lambda < 2\pi)$ and $r \geqslant 0$ is the radial distance. Then the metric tensor reads

$$G_{ij} = \begin{bmatrix} r^2 & 0 & 0 \\ 0 & r^2 \sin^2\vartheta & 0 \\ 0 & 0 & 1 \end{bmatrix}, \qquad G^{ij} = \begin{bmatrix} r^{-2} & 0 & 0 \\ 0 & r^{-2}\sin^{-2}\vartheta & 0 \\ 0 & 0 & 1 \end{bmatrix}. \tag{2.4}$$

The Christoffel symbols are

$$\left.\begin{aligned} \Gamma^1_{ij} &= \begin{bmatrix} 0 & 0 & r^{-1} \\ 0 & -\sin\vartheta\cos\vartheta & 0 \\ r^{-1} & 0 & 0 \end{bmatrix}, \qquad \Gamma^2_{ij} = \begin{bmatrix} 0 & \cot\vartheta & 0 \\ \cot\vartheta & 0 & r^{-1} \\ 0 & r^{-1} & 0 \end{bmatrix}, \\ \Gamma^3_{ij} &= \begin{bmatrix} -r & 0 & 0 \\ 0 & -r\sin^2\vartheta & 0 \\ 0 & 0 & 0 \end{bmatrix}. \end{aligned}\right\} \tag{2.5}$$

It is more common to use the spherical polar coordinates $(x^1, x^2, x^3) = (r, \vartheta, \lambda)$. Our choice of $(x^1, x^2, x^3) = (\vartheta, \lambda, r)$ is more suitable if we wish to describe interfaces in the form $x^3 = f(x^1, x^2)$.

2.2.3 Geographic spherical coordinate system

We introduce the geographic spherical coordinates $(x^1, x^2, x^3) = (\lambda, \vartheta, r)$ as follows: λ is the longitude (positive eastwards) $(0 \leqslant \lambda < 2\pi)$, ϑ is the latitude (positive northwards) $(-\frac{1}{2}\pi \leqslant \vartheta \leqslant \frac{1}{2}\pi)$ and $r \geqslant 0$ is the radial distance. Then the metric tensor reads

$$G_{ij} = \begin{bmatrix} r^2 \cos^2 \vartheta & 0 & 0 \\ 0 & r^2 & 0 \\ 0 & 0 & 1 \end{bmatrix}, \qquad G^{ij} = \begin{bmatrix} r^{-2} \cos^{-2} \vartheta & 0 & 0 \\ 0 & r^{-2} & 0 \\ 0 & 0 & 1 \end{bmatrix}. \tag{2.6}$$

The Christoffel symbols are

$$\left. \begin{aligned} &\Gamma^1_{ij} = \begin{bmatrix} 0 & -\tan \vartheta & r^{-1} \\ -\tan \vartheta & 0 & 0 \\ r^{-1} & 0 & 0 \end{bmatrix}, \qquad \Gamma^2_{ij} = \begin{bmatrix} \sin \vartheta \cos \vartheta & 0 & 0 \\ 0 & 0 & r^{-1} \\ 0 & r^{-1} & 0 \end{bmatrix}, \\ &\Gamma^3_{ij} = \begin{bmatrix} -r \cos^2 \vartheta & 0 & 0 \\ 0 & -r & 0 \\ 0 & 0 & 0 \end{bmatrix}. \end{aligned} \right\} \tag{2.7}$$

2.2.4 *Other coordinate systems*

Instead of the above coordinate systems, any other system may be used. For instance, if we consider the ellipticity of the Earth then ellipsoidal coordinates may be used. It is usually straightforward to write the metric tensor for other coordinate systems.

3 MODEL OF THE MEDIUM

3.1 The model

The model M is defined inside a volume of space,

$$M: \quad x^i_{\min} \leqslant x^i \leqslant x^i_{\max} \quad (i = 1, 2, 3), \tag{3.1}$$

by functions specifying the distribution of the parameters of the medium. The coordinate surfaces $x^i = x^i_{\min}$ and $x^i = x^i_{\max}$ are *boundaries of the model.* The parameters of the medium are, for example, P- and S-wave velocities, v_{P}, v_{S}, density ρ, loss factors Q_{P}^{-1}, Q_{S}^{-1}, or some powers of these quantities. In this way, the velocity distribution may be specified by velocity values, by values of the square of the velocity, by values of slowness, by values of quadratic slowness, etc. Likewise, instead of loss factors, their reciprocal values Q_{P} and Q_{S}, called quality factors, may be considered. The parameters of the medium must be smooth functions of the coordinates inside the model, where "smooth" means that the parameters and their first and second partial derivatives must be continuous. The smoothness of their distribution may be violated at a finite set of interfaces across which the parameters or their first or second partial derivatives may be discontinuous.

We consider models in which every such interface may be covered by a finite number of *smooth surfaces* $\Sigma: f(x^i) = 0$ with the following property: the functions $f(x^i)$ are defined and are smooth in the whole M i.e. $f(x^i)$ and their first and second derivatives are continuous everywhere in M. The surfaces may intersect each other. They may cross the whole model and intersect the boundaries of the model; they may also form closed surfaces (e.g. of ellipsoidal form) inside the model M. Each surface divides the whole model into two parts, the *positive part,* in which $f(x^i) > 0$, and the *negative part,* in which $f(x^i) < 0$. Accordingly, the side of the surface that faces the positive part of the model is called the *positive side of the surface* Σ, the other one being the *negative side of the surface* Σ. The smooth surfaces Σ may be indexed in ascending order by integers, starting from 1.

As in Gjøystdal *et al.* (1985), we construct the model from two types of blocks formed by smooth surfaces Σ: simple blocks (SB), which are the "building bricks" of our model and have no physical meaning, and complex blocks (CB), which represent the physical units of the model. We distinguish between *material blocks,* in which the density is always non-zero and positive, and *free-space blocks,* in which the density is identically zero.

A *simple block* (*SB*) is defined by two finite sets F^+ and F^- of surfaces $f(x^i) = 0$. A point x^i lies within the block if and only if

$$\left.\begin{array}{ll} f(x^i) > 0 & \text{for any } f \in F^+ \\ f(x^i) < 0 & \text{for any } f \in F^-. \end{array}\right\} \tag{3.2}$$

The simple block is an intersection of the positive parts of the model corresponding to the surfaces from F^+ and the negative parts of the model corresponding to F^- (see Gjøystdal *et al.*, 1985). An example is shown in Fig. 1 (see Section 3.4). Only parts of the surfaces Σ may be boundaries of a block—the remaining parts being only fictitious extensions of the boundary. Boundaries of some simple blocks may also be partially formed by parts of the boundaries of the model. Note that a simple block may be formed by several separated regions, not only by one connected region.

The division of the model into simple blocks is not unique. The SBs need not form a disjoint system, i.e. a point of the model may be situated in several SBs. SBs may be indexed in ascending order by positive integers, starting from 1. It is reasonable to index only the material blocks, and leave the free-space blocks without indices. In this way, the blocks without indices may be immediately identified as free spaces without checking whether the density is zero or non-zero.

The simple block concept does not allow consideration of blocks in which the fictitious extensions of boundaries of the block intrude into the block (see Fig. 2 in Section 3.4). This means that one "physical" block of the model may be simulated by several simple blocks. In order to avoid this artificial division of the model, which would also complicate coding of waves (see Section 4), we introduce complex blocks.

A *complex block* (*CB*) is formed as a union of several simple blocks (it may, of course, also be formed of just one simple block). It may be defined by a table giving the indices of simple blocks that should be included in the complex block. While the SBs need not form a disjoint system, the CBs must form one. Any point of the model must be situated in just one CB. The boundaries of CBs, which may also be called *interfaces,* are formed by boundaries of SBs. The CBs may be indexed by positive indices, starting from 1. The system of indexing the blocks may be arbitrary, but some systems may render computations more effective. The material complex blocks must contain only material SBs. The complex block formed of free-space SBs is called the free-space complex block.

Note that in the following sections we shall introduce additional surfaces, which have the same properties and may be specified in the same way as the smooth surfaces Σ described above. However, they have a different meaning. The purpose of one set of these surfaces is to store all the quantities obtained by complete ray tracing; otherwise they are fully transparent, i.e. they have no effect on the ray tracing (see Section 5.5.2). The other set contains surfaces at which the complete ray tracing is to terminate (see Section 5.4, case (f)).

3.2 The data and routines specifying the model

The specification of data defined in Section 3.2.1 and routines defining the surfaces $f(x^i) = 0$ and the parameters of the medium (see Sections 3.2.2 and 3.2.3) fully determine the model.

3.2.1 *Data specifying the model*

(a) Indices NEXPV, NEXPQ specifying exponents of the power of velocities (NEXPV) and loss-factors (NEXPQ) in input data. For example, unit indices NEXPV and NEXPQ indicate that parameters of the medium are velocities and loss-factors, indices equal to -1 indicate reciprocal values of these quantities, i.e. slowness and quality factors.

(b) The boundaries of the model $x^1_{\min}$, $x^1_{\max}$, $x^2_{\min}$, $x^2_{\max}$, $x^3_{\min}$, $x^3_{\max}$.

(c) The number NSRFC of smooth surfaces Σ in the model. The surfaces are indexed sequentially by positive integers, from 1 to NSRFC.

(d) The number NSB of material simple blocks in the model. The blocks are indexed sequentially by positive integers ISB, from 1 to NSB. The free-space blocks are not indexed.

(e) For each simple block with index ISB, the indices of the surfaces forming the set F^+_{ISB} and the indices of the surfaces forming the set F^-_{ISB}. The indices of surfaces from F^+_{ISB} are stored with positive signs; the indices of surfaces from F^-_{ISB} are stored with negative signs. The indices may be specified in an arbitrary order.

(f) The number NCB of material complex blocks in the model. The blocks are indexed sequentially by positive integers ICB, from 1 to NCB. The free-space blocks are not indexed.

(g) For each complex block, the indices of simple blocks forming the complex block. The indices may be specified in an arbitrary order.

3.2.2 Specification of smooth surfaces

The functions $f(x^i)$ may be specified with the use of simple analytic expressions, interpolation or approximation of values given at discrete points, and so on. The coefficients of these functions may be prepared in the user-defined routine SRFC1, in which the input data concerning the surfaces $f(x^i)=0$ may also be read in. The functional values as well as the first and second partial derivatives of the functions $f(x^i)$ at a specified point may be obtained by a user-defined routine called SRFC2 here.

3.2.3 Specification of the parameters of the medium

The distribution of the parameters of the medium in each complex block may be specified by simple analytic expressions, by interpolation or approximation of values given at discrete points, etc. The parameters may be either P- and S-wave velocities, density and loss-factors, or their powers. The distribution of any of these parameters may also be expressed in terms of the distribution of another parameter, for example, $v_S = 0.577 v_P$ or $\rho = 1.7 + 0.2 v_P$, where v_P, v_S and ρ are P and S velocities and density. The coefficients of these functions may be prepared in a user-defined routine called PARM1 here, in which the input data concerning the distribution of individual parameters within each complex block may be read in. The functional values as well as the first and second partial derivatives of these functions may be obtained in a user-defined routine PARM2.

3.3 Auxiliary procedures

For complete ray tracing, some auxiliary procedures specifying the position of a point with respect to simple and complex blocks and smooth surfaces, or transforming the specified parameters of medium into velocities, density and loss factors, may be useful. The auxiliary procedures do not specify the model, they are only service procedures for complete ray tracing.

3.3.1 *Determination of the index of a block*

For the determination of the index of a simple and a complex block, in which a specified point (e.g. point source) is situated, a routine, called BLOCK here, may be used. This routine may also be used to determine the index of a block touching a specified block at a point situated on the boundary of the specified block (the situation that may arise when a ray impinges on a boundary of a block). Another function of the routine may be to determine the index of a surface bounding a block and separating the block from the given point (surface $f(x^i)=0$ separates sets A and B if $f(x^i)f(y^i)<0$ for any $x^i \in A$, $y^i \in B$).

3.3.2 *Transformation of the parameters of the medium*

The routine transforming the values of the parameters of the medium into the velocity and loss factor of the corresponding type of wave is called VELOC here.

3.4 Examples

We present several examples of how to build models for complete ray tracing. We present cross-sections of 3D models specified in Cartesian coordinates. Of course, they may also be interpreted as 2D models. Individual figures always contain a picture of the model and a picture (or pictures) of its representation in terms of simple blocks. The parts of the model below smooth surfaces are considered as positive, those above

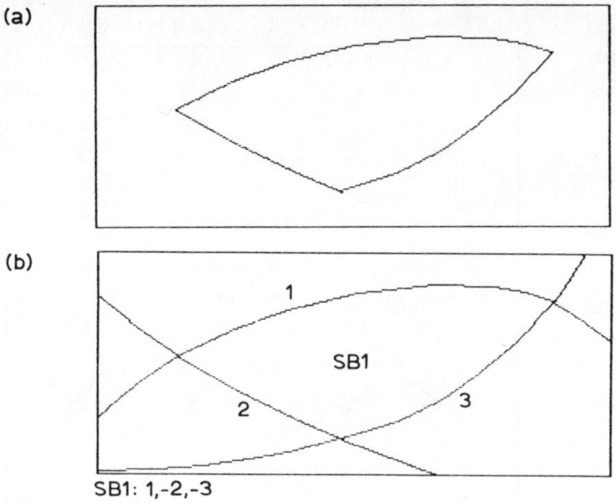

Fig. 1 (a) Simple block. (b) Its representation by three surfaces indexed 1, 2, 3.

them as negative parts of the model. In the case of a closed surface the positive part of the model is situated outside the closed surface. The tables attached to the figures specify the indices of surfaces belonging to the sets F^+_{ISB} (positive numbers) and F^-_{ISB} (absolute values of negative numbers) for each simple block ISB.

To better understand the individual figures of the models shown later in this section, let us first show a schematic picture of a simple block (see Fig. 1). Similarly, Fig. 2 shows that one physical (complex) block may be simulated by several simple blocks in various ways.

Example 1: Layered structure

Figure 3(a) shows a model of a simple layered structure. The layers are separated by non-intersecting smooth interfaces. The representation of such a structure in terms of simple blocks (see Fig. 3b) is very simple. Each interface is represented by one smooth surface Σ, each layer by one simple block. Complex blocks are in this case identical with the simple blocks.

Example 2: Isolated body A

Figure 4(a) shows a layered structure containing an isolated body. The representation of the model in terms of simple blocks for two different systems of indexing surfaces and SBs is shown in Figs 4(b,c). It is again

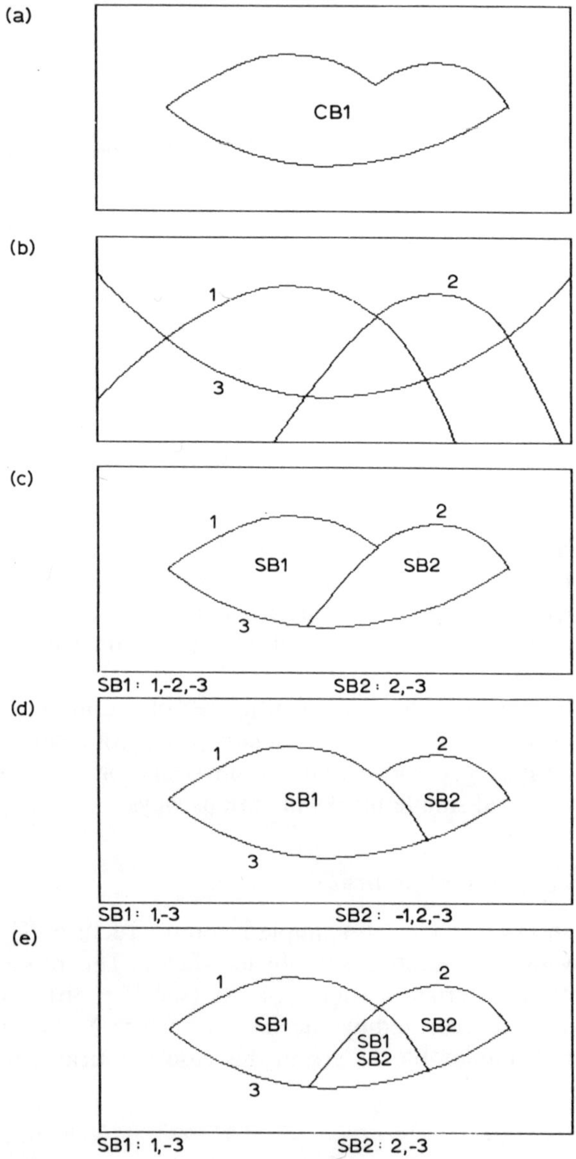

Fig. 2 (a) Complex block. (b) Three surfaces 1, 2, 3 covering the boundary of (a). (c, d, e) Construction of the complex block (a) using two simple blocks in several ways.

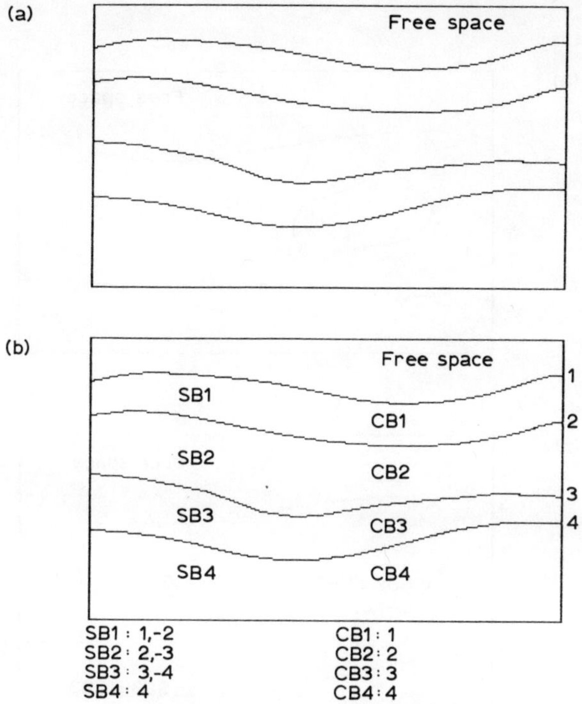

Fig. 3 (a) Model of a simple layered structure. (b) Its computer representation.

very simple. Each interface is represented by one smooth surface, surfaces 2 in Fig. 4(b) and 1 in Fig. 4(c) being closed surfaces. An isolated body represents one simple block, the remaining part of the layer another one. Complex blocks are again identical with simple blocks. In both pictures, (b) and (c), complex blocks are indexed from the top to the bottom.

Example 3: Isolated body B

Figure 5(a) shows a model of a layered structure containing an isolated body with edges. Possible representations of the model in terms of simple blocks are shown in Figs 5(b,c). This example shows that: (i) the division of models into simple blocks may be non-unique; (ii) although the region represented by simple blocks 1 and 3 is one physical unit (see Fig. 5a), it cannot be represented by a single simple block (the region contains points that are situated both in positive and negative parts of the model

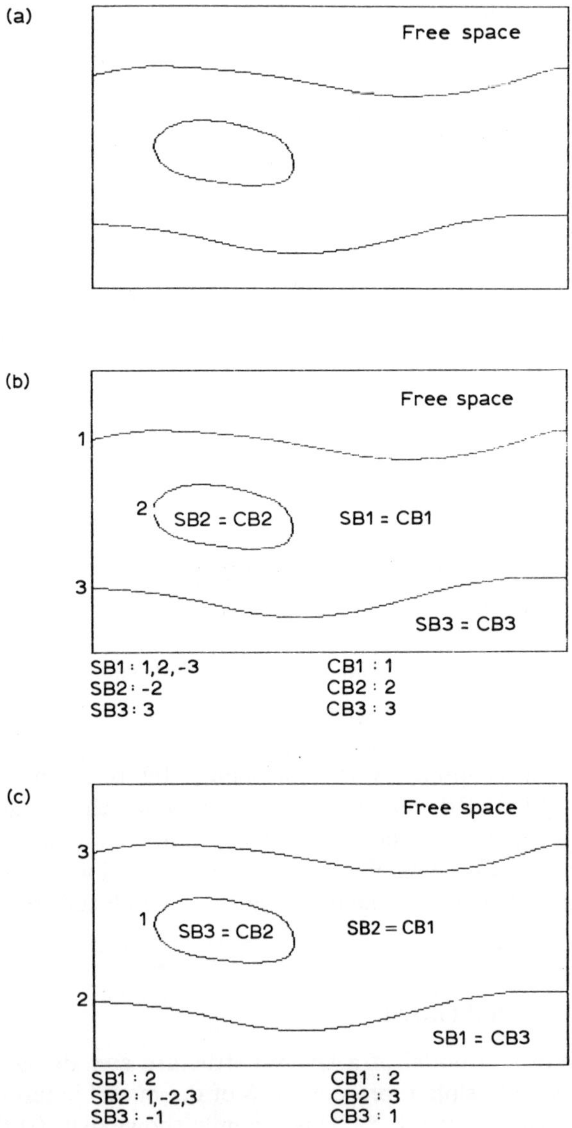

Fig. 4 (a) Model of a smooth isolated body. (b, c) Its two computer representations.

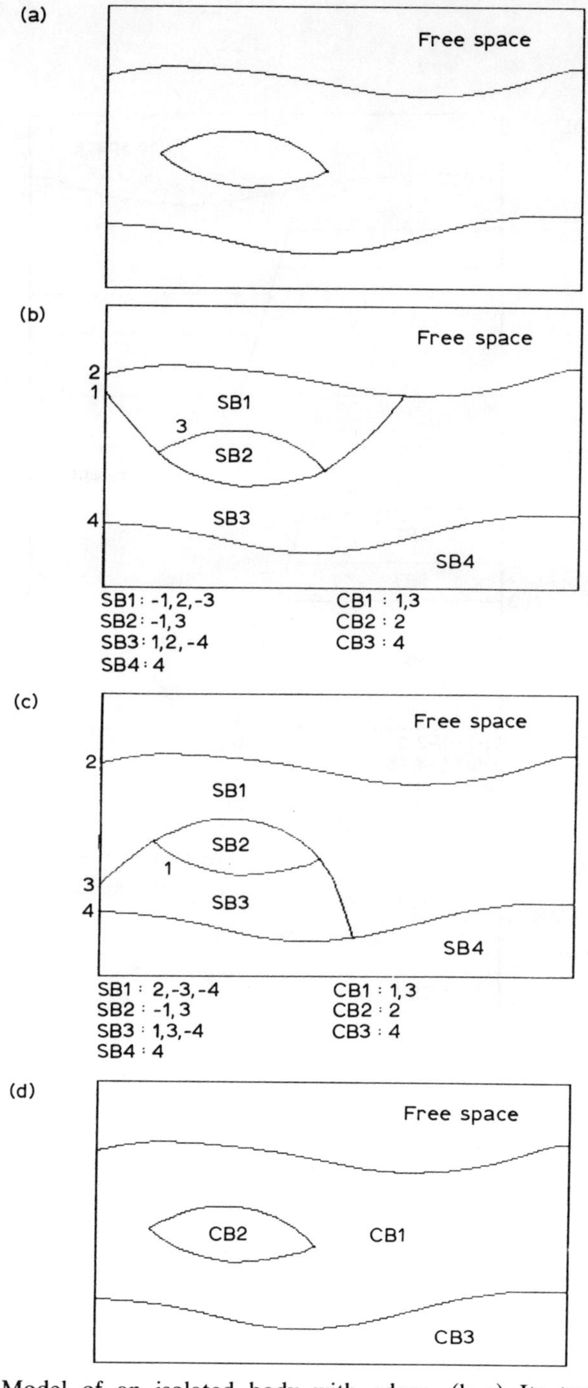

Fig. 5 (a) Model of an isolated body with edges. (b, c) Its two computer representations in terms of simple blocks. The complex blocks in the two representations are equivalent—see (d).

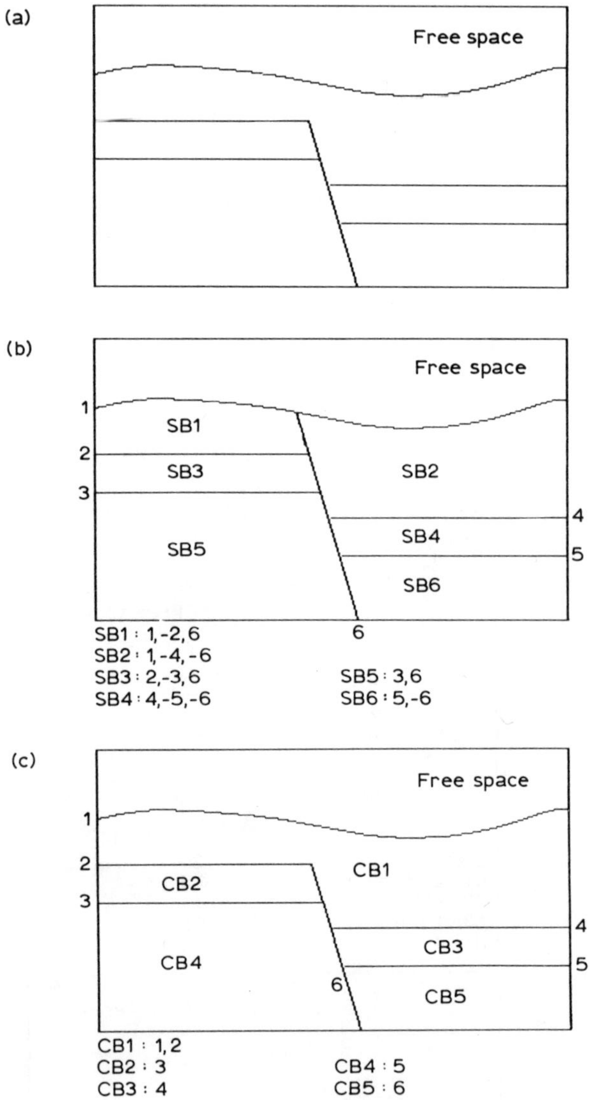

Fig. 6 (a) Model of a fracture structure with a sedimentary cover. (b) Its computer representation in terms of simple blocks. (c) Its computer representation in terms of complex blocks.

with respect to the smooth surface 1 or 3—see also Fig. 2). The simple blocks 1 and 3, however, may be united to form a complex block of the same physical properties (see Fig. 5d). Complex blocks 2 and 3 are identical with the simple blocks.

Example 4: Fracture structure

Figure 6(a) shows a model of a simple fracture structure with a sedimentary cover. Its representation in terms of simple blocks is shown in Fig. 6(b). Simple blocks 1 and 2 may be united to form a complex block, the other complex blocks are identical with simple blocks.

4 CODES FOR ELEMENTARY WAVES

We consider ordinary seismic body waves, such as refracted, primarily or multiply reflected, possibly converted waves. In general, incidence of a wave at an interface (boundary of a complex block) produces four waves, reflected P and S, and transmitted P and S waves. When performing complete ray tracing, we must know a priori which of the four generated waves to follow. This decision must be made at each interface. The alphanumeric string specifying the behaviour of a ray from its initial point to its endpoint is a *code.*

The term *elementary wave* does not have unique meaning in the literature. Here we apply the term to that part of the wavefield that is described by one specific code. Since there may be various types of codes, there is also a variety of divisions of the wavefield into the elementary waves.

We introduce the term *element of a ray,* which has an important meaning in the construction of codes. By an element of a ray, we denote that part of the ray that is situated in one complex block between two successive points of reflection/transmission, or between the initial point or endpoint of the ray and the closest point of reflection/transmission or between the initial point and the endpoint of the ray, if the ray is entirely situated in one complex block.

There are many possible types of codes of elementary waves. Usually they are given by a finite sequence of non-zero integers. For the transformation of the code into instructions specifying the behaviour of the considered elementary wave at the initial points of rays and at all points of incidence at interfaces (boundaries of complex blocks), a routine, called CODE here, should be prepared.

5 COMPLETE RAY TRACING

5.1 Theory

Complete ray tracing consists of: (a) ray tracing and computation of the travel time along the ray; (b) determination of the polarization vectors along the ray; (c) dynamic ray tracing (i.e. computation of the ray propagator matrix); (d) determination of the vectorial complex-valued reduced amplitudes.

5.1.1 Ray tracing and travel-time computation

Ray tracing consists in the computation of a ray, i.e. in a step-by-step evaluation of coordinates of points along the ray. The ray is parametrized by an independent variable

$$\sigma = \sigma_0 + \int_{\tau_0}^{\tau} v^{\mathrm{NEXPS}}\,\mathrm{d}\tau = \sigma_0 + \int_{s_0}^{s} v^{\mathrm{NEXPS}-1}\,\mathrm{d}s, \tag{5.1}$$

where τ is the travel time, s is the arclength along the ray and v is the velocity of the corresponding type of wave. The integer exponent NEXPS may be specified by a user, $\mathrm{NEXPS} = 0, \pm 1, \pm 2, \ldots$. We consider the ray-tracing system in the form of a system of six first-order ordinary differential equations:

$$\left.\begin{aligned} \frac{\mathrm{d}x^i}{\mathrm{d}\sigma} &= v^{2-\mathrm{NEXPS}} G^{ij} p_j, \\ \frac{\mathrm{d}p_i}{\mathrm{d}\sigma} &= v^{2-\mathrm{NEXPS}}\left(-v^{-3}\frac{\partial v}{\partial x^i} + \Gamma^k_{ij} G^{jl} p_k p_l\right). \end{aligned}\right\} \tag{5.2}$$

Here x^i are the coordinates of a point along the ray, p_i are the covariant components of the slowness vector at the point x^i, and $p_i = \partial\tau/\partial x^i$. For the metric tensor G^{ij} and the Christoffel symbols Γ^k_{ij} see Section 2. At an interface, the ray is transformed using the Snell law.

The equation

$$\tau = \tau_0 + \int_{\sigma_0}^{\sigma} v^{-\mathrm{NEXPS}}\,\mathrm{d}\sigma \tag{5.3}$$

for the real-valued travel time follows from (5.1). The imaginary part of the complex-valued travel time is defined as

$$\tau^{\mathrm{IM}} = \tfrac{1}{2}t^* = \int_{\tau_0}^{\tau} (2Q)^{-1}\,\mathrm{d}\tau = \int_{\sigma_0}^{\sigma} (2Q)^{-1} v^{-\mathrm{NEXPS}}\,\mathrm{d}\sigma, \tag{5.4}$$

where Q is the relevant quality factor.

5.1.2 Polarization vectors

Let us consider an arbitrarily selected ray and denote it by Ω. The determination of the polarization vectors along the ray Ω plays an important role in complete ray tracing for several reasons. The polarization vectors must be known if we wish to determine the orientation of the displacement vector of the wave propagating along Ω, especially in the case of S waves. Moreover, the polarization vectors form the vector basis of the *ray-centred coordinate system* connected with Ω. The ray-centred coordinate system is very useful in many applications, especially in dynamic ray tracing (see Section 5.1.3).

The orthogonal ray-centred coordinate system was introduced into seismology by Popov and Pšenčík (1978a,b). We shall denote the ray-centred coordinates by $(q^1, q^2, q^3 = s)$, where s is the arclength along the ray Ω, and q^K are the Cartesian coordinates in the plane tangent to the wavefront at point $s = q^3$ on Ω, with the origin on the ray Ω. The vector basis of the ray-centred coordinate system consists of three unit vectors $\boldsymbol{e}_1$, $\boldsymbol{e}_2$, $\boldsymbol{e}_3$ (polarization vectors), where $\boldsymbol{e}_3$ is tangent to the ray Ω, and $\boldsymbol{e}_1$ and $\boldsymbol{e}_2$ are perpendicular to Ω. They are introduced in such a way as to render the ray-centred coordinate system orthogonal. The equation

$$\frac{\mathrm{d}H_{ij}}{\mathrm{d}\sigma} = v^{1-\mathrm{NEXPS}}\left[\frac{\partial v}{\partial x^k} G^{kl} H_{lj} p_i - p_k G^{kl} H_{lj} \frac{\partial v}{\partial x^i} + v \Gamma^m_{ik} G^{kl} p_l H_{mj}\right] \tag{5.5}$$

holds for the covariant components H_{ij} of the ray-centred basis vector $\boldsymbol{e}_j$, $(e_j)_i = H_{ij}$. This equation simplifies to

$$\frac{\mathrm{d}H_{iJ}}{\mathrm{d}\sigma} = v^{1-\mathrm{NEXPS}}\left[\frac{\partial v}{\partial x^k} G^{kl} H_{lJ} p_i + v \Gamma^m_{ik} G^{kl} p_l H_{mJ}\right] \tag{5.6}$$

for vectors $\boldsymbol{e}_J$ perpendicular to the ray Ω. For the unit vector $\boldsymbol{e}_3$ tangent to the ray Ω ($j = 3$ in (5.5)),

$$H_{i3} = v p_i = G_{ij} \frac{\mathrm{d}x^j}{\mathrm{d}s}, \tag{5.7}$$

and (5.5) is equivalent to the second equation of the ray-tracing system (5.2). It is sufficient to compute only the covariant components H_{i1} ($i = 1, 2, 3$) of the vector $\boldsymbol{e}_1$, because $\boldsymbol{e}_3$ is determined by (5.7) when the ray tracing (5.2) is performed, and because $\boldsymbol{e}_2$ is always perpendicular to both $\boldsymbol{e}_1$ and $\boldsymbol{e}_3$:

$$H_{i2} = \varepsilon_{ijk} G^{jm} H_{m3} G^{kn} H_{n1} [\det (G^{rs})]^{-1/2}. \tag{5.8}$$

Here $\varepsilon_{123} = \varepsilon_{231} = \varepsilon_{312} = 1$, $\varepsilon_{132} = \varepsilon_{321} = \varepsilon_{213} = -1$, $\varepsilon_{ijk} = 0$ for all other combinations of ijk. At an interface, the ray-centred coordinate system is rotated around the vector perpendicular to the plane of incidence in such a way that, after the rotation, the vector $\boldsymbol{e}_3$ coincides with the tangent to the ray of the generated wave under consideration.

5.1.3 Dynamic ray tracing. Ray propagator matrix

Dynamic ray tracing consists in the solution of the system of four first-order linear differential equations along the ray Ω, for the components of a 4×1 column matrix $\boldsymbol{W}$,

$$\frac{\mathrm{d}\boldsymbol{W}}{\mathrm{d}\sigma} = v^{2-\mathrm{NEXPS}}\boldsymbol{S}\boldsymbol{W}, \tag{5.9}$$

where

$$\boldsymbol{S} = \begin{bmatrix} \boldsymbol{0} & \boldsymbol{I} \\ -v^{-3}\boldsymbol{V} & \boldsymbol{0} \end{bmatrix},$$

$\boldsymbol{0}$ and $\boldsymbol{I}$ being zero and identity 2×2 matrices, and $\boldsymbol{V}$ a 2×2 matrix with components

$$V_{IJ} = \left(\frac{\partial^2 v}{\partial x^k\,\partial x^l} - \Gamma^m_{kl}\frac{\partial v}{\partial x^m}\right)G^{kr}G^{ls}H_{rI}H_{sJ}. \tag{5.10}$$

The first two components of $\boldsymbol{W}$ are ray-centred coordinates (q^1, q^2) of a paraxial ray (up to first order in Taylor expansion), and the other two components of $\boldsymbol{W}$ are the corresponding ray-centred components of the slowness vector of the paraxial ray.

The dynamic ray-tracing system (5.9) has four linearly independent solutions. We denote by $\boldsymbol{\Pi}(\sigma, \sigma_0)$ the fundamental 4×4 matrix of linearly independent solutions of (5.9) with the initial condition

$$\boldsymbol{\Pi}(\sigma_0, \sigma_0) = \boldsymbol{I}$$

at the initial point $\sigma = \sigma_0$ of the ray. Here $\boldsymbol{I}$ is the 4×4 identity matrix. The fundamental matrix $\boldsymbol{\Pi}(\sigma, \sigma_0)$ is also called the ray propagator matrix, or the propagator matrix of the dynamic ray-tracing system. It satisfies the following relation along the ray:

$$\boldsymbol{\Pi}^{\mathrm{T}}\boldsymbol{\Sigma}\boldsymbol{\Pi} = \boldsymbol{\Sigma}, \quad \text{with } \boldsymbol{\Sigma} = \begin{bmatrix} 0 & 0 & 1 & 0 \\ 0 & 0 & 0 & 1 \\ -1 & 0 & 0 & 0 \\ 0 & -1 & 0 & 0 \end{bmatrix}. \tag{5.11}$$

This is the so-called symplectic property of $\boldsymbol{\Pi}$. Any solution $\boldsymbol{W}(\sigma)$ of the dynamic ray-tracing system (5.9) can be expressed in the form

$$\boldsymbol{W}(\sigma) = \boldsymbol{\Pi}(\sigma, \sigma_0)\boldsymbol{W}(\sigma_0). \tag{5.12}$$

Solutions of the dynamic ray-tracing system (5.9) have found many important applications in seismology. Let us assume that the computed ray belongs to a two-parameter system of rays, parametrized by ray parameters γ^1, γ^2. Then, the dynamic ray-tracing system can be used to determine the 2×2 matrices $\boldsymbol{Q}^{\mathrm{R}}$ and $\boldsymbol{P}^{\mathrm{R}}$ along the ray, with elements

$$\left.\begin{aligned} Q_{IJ}^{\mathrm{R}} &= \left[\frac{\partial q^I}{\partial \gamma^J}\right]_{q^1=q^2=0}, \\ P_{IJ}^{\mathrm{R}} &= \left[\frac{\partial^2 \tau}{\partial q^I\, \partial \gamma^J}\right]_{q^1=q^2=0} = \left[\frac{\partial p_I^{(q)}}{\partial \gamma^J}\right]_{q^1=q^2=0}. \end{aligned}\right\} \tag{5.13}$$

Here q^I are the ray-centred coordinates of a paraxial ray specified by ray parameters γ^1, γ^2; $p_I^{(q)}$ are the relevant ray-centred components of the slowness vector of the paraxial ray, $p_I^{(q)} = \partial\tau/\partial q^I$. The matrix $\boldsymbol{Q}^{\mathrm{R}}$ is also called the matrix of geometrical spreading, $|\det \boldsymbol{Q}^{\mathrm{R}}|$ being the geometrical spreading. It measures the expansion and contraction of the ray tube. The geometrical spreading is dependent on the parametrization of the rays.

We introduce the ray coordinates $(\gamma^1, \gamma^2, \gamma^3)$, where γ^1 and γ^2 are the ray parameters and γ^3 is a parameter along the ray (e.g. the arclength s). Then the matrix $\boldsymbol{Q}^{\mathrm{R}}$ may be also interpreted as the transformation matrix from ray coordinates (γ^1, γ^2) to the ray-centred coordinates (q^1, q^2) along the ray Ω. Similarly, the matrix $\boldsymbol{P}^{\mathrm{R}}$ represents the transformation matrix from ray coordinates (γ^1, γ^2) to the "phase-space" coordinates $p_1^{(q)}$, $p_2^{(q)}$. The 4×2 matrix $\boldsymbol{X}^{\mathrm{R}}$, defined as

$$\boldsymbol{X}^{\mathrm{R}} = \begin{bmatrix} \boldsymbol{Q}^{\mathrm{R}} \\ \boldsymbol{P}^{\mathrm{R}} \end{bmatrix}, \tag{5.14}$$

satisfies the dynamic ray-tracing system

$$\frac{\mathrm{d}\boldsymbol{X}^{\mathrm{R}}}{\mathrm{d}\sigma} = v^{2-\mathrm{NEXPS}}\boldsymbol{S}\boldsymbol{X}^{\mathrm{R}} \tag{5.15}$$

along the ray Ω. The solution of (5.15) may be expressed in the form

$$\boldsymbol{X}^{\mathrm{R}}(\sigma) = \boldsymbol{\Pi}(\sigma, \sigma_0)\boldsymbol{X}^{\mathrm{R}}(\sigma_0). \tag{5.16}$$

If $\boldsymbol{Q}^{\mathrm{R}}$ and $\boldsymbol{P}^{\mathrm{R}}$ are known then we can also determine the 2×2 matrix of the second derivatives of the travel-time field $\boldsymbol{M}^{\mathrm{R}}(\sigma)$, with elements

$$M_{IJ}^{\mathrm{R}} = \left[\frac{\partial^2 \tau}{\partial q^I\, \partial q^J}\right]_{q^1=q^2=0}. \tag{5.17}$$

$\boldsymbol{Q}^{\mathrm{R}}$, $\boldsymbol{P}^{\mathrm{R}}$ and $\boldsymbol{M}^{\mathrm{R}}$ are related by

$$\boldsymbol{M}^{\mathrm{R}} = \boldsymbol{P}^{\mathrm{R}}(\boldsymbol{Q}^{\mathrm{R}})^{-1}. \tag{5.18}$$

The transformation of solutions of the dynamic ray-tracing system across interfaces is determined by Snell's law for paraxial rays.

5.1.4 Vectorial reduced amplitudes

Vectorial reduced amplitudes

$$A_i = A_i^{\mathrm{R}}(\rho v \, |\det \boldsymbol{Q}^{\mathrm{R}}|)^{1/2} \tag{5.19}$$

are the complex-valued vectorial displacement ray amplitudes A_i^{R} (including a phase shift due to caustics) expressed in the ray-centred coordinate system, multiplied by the square root of the impedance (product of density ρ and velocity v) and of the geometrical spreading $|\det \boldsymbol{Q}^{\mathrm{R}}|$. The vectorial reduced amplitudes A_i are assumed to be unity at the initial point of a ray. If the initial point is situated at the free surface or at any internal interface then the conversion coefficients may be included in A_i.

We define the 3×3 matrix A_{ij} composed of the vectorial reduced amplitude A_{i1} corresponding to the S wave polarized in the direction of the vector $\boldsymbol{e}_1$ at the initial point of the ray, of the vectorial reduced amplitude A_{i2} corresponding to the S wave polarized in the direction of $\boldsymbol{e}_2$ at the initial point, and of the vectorial reduced amplitude A_{i3} corresponding to the P wave at the initial point. The components of the vectorial reduced amplitudes A_{ij} are expressed in the ray-centred coordinate system. They are constant along the elements of a ray inside individual blocks, except at caustics, where they must be multiplied by $-\mathrm{i}$ or -1, according to the type of caustic. At interfaces, the vectorial reduced amplitudes are transformed by means of reduced reflection/transmission (R/T) coefficients

$$R^{\mathrm{E}} = R\left[\frac{\tilde{\rho}\tilde{v} \, |\cos \tilde{\alpha}|}{\rho v \, |\cos \alpha|}\right]^{1/2}, \tag{5.20}$$

where R are plane-wave displacement reflection/transmission coefficients. Here the quantities with a tilde correspond to the relevant R/T wave, those without tilde to the incident wave; in both cases at the point of incidence. α and $\tilde{\alpha}$ are the angles of incidence and of R/T respectively.

5.2 Quantities computed along a ray

5.2.1 Basic quantities computed along a ray

Complete ray tracing is performed to compute the following basic quantities at selected points along a ray:

$\mathrm{Y}(1) = \tau$ travel time;

$\mathrm{Y}(2) = \tau^{\mathrm{IM}} = \frac{1}{2}t^*$ imaginary part of the complex-valued travel time;

$$\left.\begin{array}{l}\mathrm{Y}(3) = x^1\\ \mathrm{Y}(4) = x^2\\ \mathrm{Y}(5) = x^3\end{array}\right\}$$ coordinates of points along the ray;

$$\left.\begin{array}{l}\mathrm{Y}(6) = p_1\\ \mathrm{Y}(7) = p_2\\ \mathrm{Y}(8) = p_3\end{array}\right\}$$ covariant components of the slowness vector;

$$\left.\begin{array}{l}\mathrm{Y}(9) = H_{11}\\ \mathrm{Y}(10) = H_{21}\\ \mathrm{Y}(11) = H_{31}\end{array}\right\}$$ covariant components of the polarization vector $\boldsymbol{e}_1$, see (5.6) for $J = 1$;

$$\begin{bmatrix} \mathrm{Y}(12) & \mathrm{Y}(16) & \mathrm{Y}(20) & \mathrm{Y}(24)\\ \mathrm{Y}(13) & \mathrm{Y}(17) & \mathrm{Y}(21) & \mathrm{Y}(25)\\ \mathrm{Y}(14) & \mathrm{Y}(18) & \mathrm{Y}(22) & \mathrm{Y}(26)\\ \mathrm{Y}(15) & \mathrm{Y}(19) & \mathrm{Y}(23) & \mathrm{Y}(27)\end{bmatrix} = \begin{bmatrix} \Pi_{11} & \Pi_{12} & \Pi_{13} & \Pi_{14}\\ \Pi_{21} & \Pi_{22} & \Pi_{23} & \Pi_{24}\\ \Pi_{31} & \Pi_{32} & \Pi_{33} & \Pi_{34}\\ \Pi_{41} & \Pi_{42} & \Pi_{43} & \Pi_{44}\end{bmatrix},$$

the matrix of fundamental solutions of the dynamic ray tracing system (5.9) (ray propagator matrix);

$\mathrm{Y}(28), \ldots, \mathrm{Y}(\mathrm{NY})$, where $\mathrm{NY} = 27 + \mathrm{NAMPL}$.

There are NAMPL real quantities representing complex-valued vectorial reduced amplitudes. The vectorial reduced amplitudes are specified in the ray-centred coordinate system. Here NAMPL may be either 2, 4 or 8, in the following alternative cases:

(a_1) Case of a P wave at the initial point of the ray, and of a P wave at the point under consideration on the ray. Then NAMPL = 2, and

$$\mathrm{Y}(28) = \mathrm{Re}\, A_{33}, \qquad \mathrm{Y}(29) = \mathrm{Im}\, A_{33}.$$

(a_2) Case of a P wave at the initial point of the ray, and of an S wave at the point under consideration on the ray. Then NAMPL = 4, and

$$Y(28) = \text{Re}\, A_{13}, \qquad Y(29) = \text{Im}\, A_{13},$$
$$Y(30) = \text{Re}\, A_{23}, \qquad Y(31) = \text{Im}\, A_{23}.$$

(b_1) Case of an S wave at the initial point of the ray, and of a P wave at the point under consideration on the ray. Then NAMPL = 4, and

$$Y(28) = \text{Re}\, A_{31}, \qquad Y(29) = \text{Im}\, A_{31},$$
$$Y(30) = \text{Re}\, A_{32}, \qquad Y(31) = \text{Im}\, A_{32}.$$

(b_2) Case of an S wave at the initial point of the ray, and of an S wave at the point under consideration on the ray. Then NAMPL = 8, and

$$Y(28) = \text{Re}\, A_{11}, \qquad Y(29) = \text{Im}\, A_{11},$$
$$Y(30) = \text{Re}\, A_{21}, \qquad Y(31) = \text{Im}\, A_{21},$$
$$Y(32) = \text{Re}\, A_{12}, \qquad Y(33) = \text{Im}\, A_{12},$$
$$Y(34) = \text{Re}\, A_{22}, \qquad Y(35) = \text{Im}\, A_{22}.$$

5.2.2 Auxiliary quantities computed along a ray

Knowledge of the basic quantities, described above, at a point of the ray Ω is not sufficient for continuation of complete ray tracing from the point. In this section we introduce some quantities that are necessary or useful for complete ray tracing, together with some quantities containing some information about ray tracing.

YY(1) = σ	independent variable along the ray.
YY(2) = UEBRAY	upper error bound for ray tracing, which is equal to the input value UEB at the initial point of the ray (see Section 5.6j). It is always doubled when the numerical integration requires more than NHLF bisections of the initial step STEP (see Section 5.6g). UEBRAY > UEB at the end-point of the ray indicates a decreased accuracy of computation.

YY(3) = ERRPP YY(4) = ERRPH YY(5) = ERRHH	deviations (in absolute values) of the two computed basis vectors of the ray-centred coordinate system from the conditions of orthonormality (see Sections 5.8.2d and 5.8.3i), accumulated along the ray. Any of them may be compared with the corresponding specified limit UEBPP, UEBPH or UEBHH (see Section 5.6k) at the endpoint of the ray.
IY(1) = NY = 27 + NAMPL	number of basic quantities describing the point of a ray (see Section 5.2.1).
IY(2) = KODIND	position in the code corresponding to the considered element of a ray; its value is determined by subroutine CODE (see Section 4).
IY(3) = ICB0	index of the complex block from which the ray entered the complex block in which the computed element of the ray is situated; IY(3) = 0 before leaving the complex block in which the initial point of the ray is situated.
IY(4) = ISB1	index of the simple block containing the computed element of the ray.
IY(5) = ICB1	index of the complex block containing the computed element of the ray, supplemented by sign + for a P wave and sign − for an S wave.
IY(6) = ISRF	index of the surface at which the endpoint of the computed element of the ray is situated; undefined inside the complex block, defined only at the endpoint of the element of the ray.
IY(7) = ISB2	index of the simple block touching the complex block ICB1 from the other

	side of the surface ISRF at the endpoint of the computed element of the ray; ISB2 = 0 for a free space on the other side of ISRF; undefined inside the complex block, defined only at the endpoint of the element of the ray.
IY(8) = ICB2	index of the complex block touching the complex block ICB1 from the other side of the surface ISRF at the endpoint of the computed element of the ray; ICB2 = 0 for a free space on the other side of ISRF; undefined inside the complex block, defined only at the endpoint of the element of the ray.
IY(9) = IFCT	number of invocations of subroutine FCT evaluating the right-hand sides of the ordinary differential equations along the computed part of the ray.
IY(10) = IOUTP	number of successful steps of the numerical integration along the ray.
IY(11) = ITRANS	number of transformations at an interface.
IY(12) = KMAH	number of caustic points along the ray (the index of the ray trajectory).

5.2.3 Quantities for the identification of caustics

Computation of the reduced vectorial amplitudes requires knowledge of the position of caustics on the computed ray. The position of a caustic is not a property of a single ray, but a property of the rayfield. It depends on the mutual position of paraxial rays with respect to the computed ray. The mutual position can be determined if the 2×2 matrix

$$\boldsymbol{M}^{\mathrm{INIT}} = \boldsymbol{M}^{\mathrm{R}}(\sigma_0)$$

of the second derivatives of the travel-time field at the initial point $\sigma = \sigma_0$ of the computed ray is known. The matrix $\boldsymbol{M}^{\mathrm{INIT}}$ may be specified in terms of the matrices $\boldsymbol{Q}^{\mathrm{INIT}} = \boldsymbol{Q}^{\mathrm{R}}(\sigma_0)$ and $\boldsymbol{P}^{\mathrm{INIT}} = \boldsymbol{P}^{\mathrm{R}}(\sigma_0)$ (see Section 5.1.3) as

$$\boldsymbol{M}^{\mathrm{INIT}} = \boldsymbol{P}^{\mathrm{INIT}}(\boldsymbol{Q}^{\mathrm{INIT}})^{-1}.$$

5.3 Auxiliary surfaces

In addition to surfaces 1, . . . , NSRFC covering structural interfaces and described in Section 3, some other (auxiliary) surfaces may be useful when we perform complete ray tracing. The surfaces may be used for the limitation of a computational volume for complete ray tracing (see Section 5.4) or for storing the computed quantities (see Section 5.5).

Note that the computational volume may be introduced as a subvolume of the model M under consideration. The actual computations are then performed only in the computational volume, not in the whole of M, and are terminated at the boundaries of the computational volume. This considerably increases the efficiency of computations performed in various regions of the model M.

The auxiliary surfaces are indexed by integers from NSRFC + 1 to NSRFC + NSRFCA, where NSRFCA is the number of auxiliary surfaces. The auxiliary surfaces may be described in the same way as the surfaces covering structural interfaces, the same routine SRFC1 may be used for reading input data and for the preparation of the necessary coefficients, and the same routine SRFC2 may be used for their evaluation.

The auxiliary surfaces are specified by their number NSRFCA, by the input data (read in by the subroutine SRFC1) and by the routine SRFC2.

5.4 Termination of tracing a ray

Computation of the ray Ω is terminated in the following cases:

(a) The computed ray satisfies the whole code, i.e. the last computed element of the ray corresponds to the last element in the code.

(b) At the point of incidence at an interface the ray cannot continue to satisfy the code. For example (i) the point of incidence may be situated at a different surface than that specified by the code; (ii) the next element of the ray may be required by the code to be situated in a block that does not touch the point of incidence (for example, the code may require the transmission into a block that is not in contact with the block containing an incident ray at the point of incidence); (iii) transmission may be required by the code at a free surface; (iv) the ray of the required reflected or transmitted wave may not be real-valued (e.g. overcritical transmission); or (v) an S wave in a liquid block may be required by the code.

(c) Reflection or type conversion at a fictitious part of the interface is required by the code. The amplitudes of reflected or converted waves are zero in this case. In principle, each zero reflection/transmission coefficient could be a reason for the termination of the computation of the ray.

(d) The travel time Y(1) is greater than a specified time limit.

(e) The ray intersects any of the six coordinate planes ($x^i = x^i_{\mathrm{MIN}}$, $x^i = x^i_{\mathrm{MAX}}$) limiting the computational volume for complete ray tracing. The computational volume should not exceed the volume M in which the model is defined (see Section 3.1). Note that in addition to the six coordinate planes, auxiliary surfaces or model surfaces may also be used to limit the computational volume; see (f).

(f) The ray intersects any surface of the given set F_{END} of surfaces limiting the computational volume for complete ray tracing. The surfaces may be either surfaces that cover structural interfaces (see Section 3) or the auxiliary surfaces introduced in Section 5.3. The set F_{END} of surfaces is determined if the indices of the surfaces from F_{END} are specified and if the subroutine SRFC2 is activated.

5.5 Storage of computed quantities

The quantities computed along the rays and described in Section 5.2 may be stored in some files.

5.5.1 Storage of quantities along a ray

The quantities are stored along the whole computed ray with constant step STORE of the independent variable. The quantities at the points of intersection of the ray with interfaces are also stored. They may be, for instance, of use in plotting rays.

5.5.2 Storage of quantities at specified surfaces

The quantities are stored at the points of intersection of a ray with any surface from a given set F_{STORE}. The surfaces may be as follows.

(a) Surfaces 1, . . . , NSRFC covering structural interfaces. The quantities are only stored along those parts of the surfaces that form the boundaries of complex blocks. Since the quantities are discontinuous across interfaces, we must specify, for each surface, its positive or negative side on which the quantities are to be stored. The sign denoting the positive or negative side may be included in the index of the surface.

(b) Auxiliary surfaces NSRFC + 1, . . . , NSRFC + NSRFCA. The quantities are stored along the whole of the surfaces. There is no reason to distinguish between their positive and negative sides.

(c) Boundaries of the computational volume. The coordinate planes (surfaces) forming the boundaries of the computational volume may be indexed as follows: 101 for $x^1 - x^1_{\mathrm{MIN}} = 0$, 102 for $x^1 - x^1_{\mathrm{MAX}} = 0$, 103 for $x^2 - x^2_{\mathrm{MIN}} = 0$, 104 for $x^2 - x^2_{\mathrm{MAX}} = 0$, 105 for $x^3 - x^3_{\mathrm{MIN}} = 0$, and 106 for $x^3 - x^3_{\mathrm{MAX}} = 0$.

(d) Isochrone $\tau - \tau_{\mathrm{MAX}} = 0$. The isochrone may be indexed by 107.

For determination of the surfaces for storage, the indices of the surfaces forming the set F_{STORE} must be specified and a subroutine SRFC2 must be available to evaluate the functions $f(x^i)$ that describe the proper surfaces. Remember that the indices of the surfaces covering interfaces must be supplemented by a $\pm$ sign denoting the side of the surface where the quantities are to be stored.

Each surface used for storage should have its own file with stored quantities.

5.5.3 Storage of quantities at the endpoints of rays of elementary waves

Each elementary wave is defined by means of its code (see Section 4). Left-hand parts of the code usually define some simpler elementary waves, with a smaller number of elements of their rays. It may be useful to store quantities corresponding to these simpler elementary waves at the endpoints of their rays. These quantities may be used in the computation of other elementary waves, with longer chains of elements, but with the same initial elements. Their computation may start from the stored quantities, without repeating the computation of the initial elements.

5.5.4 List of stored quantities

In addition to quantities computed along a ray and described in Section 5.2, some quantities describing local properties of the model at points under consideration along the ray may be useful in further processing of the results of complete ray tracing. These quantities, which will already have been evaluated during complete ray tracing, can, in principle, be recomputed when required. This may, however, be rather time-consuming. It is therefore reasonable to store these quantities together with quantities computed along the ray. The additional (local) quantities to be stored are:

ICB1	index of the complex block in which the point is situated, including the sign + for a P wave or − for an S wave;
$\mathrm{YL}(1) = v_\mathrm{P}$	velocity of P waves at the point;
$\mathrm{YL}(2) = v_\mathrm{S}$	velocity of S waves at the point;
$\mathrm{YL}(3) = \rho$	density at the point;

$$\left.\begin{aligned} \mathrm{YL}(4) &= v_1 = \partial v/\partial x^1 \\ \mathrm{YL}(5) &= v_2 = \partial v/\partial x^2 \\ \mathrm{YL}(6) &= v_3 = \partial v/\partial x^3 \end{aligned}\right\} \text{velocity derivatives in general coordinates.}$$

After these local quantities, the quantities Y(1), . . . , Y(NY) described in Section 5.2 are to be stored.

If the quantities are stored at a point of intersection of the ray with a specified surface, and if the surface coincides with an interface between blocks (i.e. the surface belongs to the set of NSRFC surfaces covering interfaces) then the quantities ICB1, YL and Y should be stored not only once, but either for three waves (incident, reflected P, reflected S) or for two waves (transmitted P, transmitted S), depending on the side of the surface where they are stored.

In certain applications, it may also be sufficient to replace the above three or two sets of quantities by a single set of quantities corresponding to the incident wave. However, in this case the reduced amplitudes Y(28), . . . , Y(NY) in the ray-centred coordinate system should be replaced by reduced amplitudes involving appropriate conversion coefficients, expressed in ray-centred coordinates:

(a) P wave at the initial point (6 reals):

$$\begin{aligned} YC(1) &= \operatorname{Re} A_{13}, & YC(2) &= \operatorname{Im} A_{13}, \\ YC(3) &= \operatorname{Re} A_{23}, & YC(4) &= \operatorname{Im} A_{23}, \\ YC(5) &= \operatorname{Re} A_{33}, & YC(6) &= \operatorname{Im} A_{33}. \end{aligned}$$

(b) S wave at the initial point (12 reals):

$$\begin{aligned} YC(1) &= \operatorname{Re} A_{11}, & YC(2) &= \operatorname{Im} A_{11}, \\ YC(3) &= \operatorname{Re} A_{21}, & YC(4) &= \operatorname{Im} A_{21}, \\ YC(5) &= \operatorname{Re} A_{31}, & YC(6) &= \operatorname{Im} A_{31}, \\ YC(7) &= \operatorname{Re} A_{12}, & YC(8) &= \operatorname{Im} A_{12}, \\ YC(9) &= \operatorname{Re} A_{22}, & YC(10) &= \operatorname{Im} A_{22}, \\ YC(11) &= \operatorname{Re} A_{32}, & YC(12) &= \operatorname{Im} A_{32}. \end{aligned}$$

5.6 Data for complete ray tracing

In this section we summarize the input data necessary for the complete ray tracing (in addition to those discussed in Sections 2, 3, 4 and 6).

(a) The coordinates x^1_{MIN}, x^1_{MAX}, x^2_{MIN}, x^2_{MAX}, x^3_{MIN}, x^3_{MAX}, which specify the coordinate planes bounding the computational volume, and the maximum travel time τ_{MAX}. The coordinate planes may be indexed 101, 102, 103, 104, 105, 106. Boundaries of the model (see Section 3.2.1) may be used as default surfaces for the coordinate planes bounding the computational volume.

(b) The number NSRFCA of auxiliary surfaces. The surfaces are indexed sequentially by positive integers, from NSRFC + 1 to NSRFC + NSRFCA.

(c) The indices of end surfaces (model or auxiliary surfaces bounding the computational volume—see Section 5.4f). The surfaces described by data (a) need not be specified in this set of indices. The indices may be specified in an arbitrary order.

(d) The indices of surfaces for storing computed quantities (see Section 5.5.2). The indices may be specified in an arbitrary order. The indices of surfaces 1, . . . , NSRFC covering structural interfaces should include a sign + or − (see Section 5.5.2a).

(e) Integer KSTORE specifying whether conversion coefficients are to be considered.

(f) Exponent NEXPS specifying independent variable along rays—see (5.1).

(g) Maximum allowed number NHLF of halvings (bisections) of initial increment of independent variable during numerical integration.

(h) Step STORE of independent variables for storing the computed quantities along a ray (see Section 5.5.1). For STORE = 0 the quantities are not stored along rays.

(i) Initial increment $\Delta\sigma = \text{STEP}$ of independent variable for numerical integration.

(j) Upper error bound UEB of travel time per step of numerical integration. Errors in the coordinates of points along the ray are approximately transformed to units of travel time and are also bounded by UEB. The error per step of numerical integration is automatically kept within the limit UEB if this does not require more than NHLF bisections of the initial increment STEP. In the opposite case the upper error bound is 2, 4, 8, . . . times greater for the computation of the rest of the ray. Thus the computation of each ray is completed.

(k) The maximum allowed accumulated deviations UEBPP, UEBPH, UEBHH of the two computed polarization vectors from the conditions of orthonormality (see Sections 5.8.2d and 5.8.3i). The accumulated deviations (quantities YY(3), YY(4), YY(5) defined in Section 5.2.2) may be compared with the specified limits UEBPP, UEBPH, UEBHH. The maximum accumulated deviations UEBPP, UEBPH, UEBHH are expressed in time units.

(l) The maximum allowed deviation UEBDRT of the matrix $\boldsymbol{\Pi}^{\mathrm{T}}\boldsymbol{\Sigma}\boldsymbol{\Pi}$ from the matrix

$$\boldsymbol{\Sigma} = \begin{bmatrix} 0 & 0 & 1 & 0 \\ 0 & 0 & 0 & 1 \\ -1 & 0 & 0 & 0 \\ 0 & -1 & 0 & 0 \end{bmatrix};$$

see (5.11). A deviation of any component may be compared with the specified limit UEBDRT.

5.7 Complete ray tracing

We assume the model (Section 3), the input data for complete ray tracing (Section 5.6) and the code of the elementary wave (Section 4) to be specified. If the quantities Y(1), . . . , Y(NY), YY(1), . . . , YY(5), IY(1), . . . , IY(12), Q_{AB}^{INIT} and P_{AB}^{INIT} (see Section 5.2) are given then the complete ray tracing of the ray of the specified elementary wave may continue from the given point Y(3), Y(4), Y(5).

(a) If the point is situated inside a complex block then computation of the ray continues as described in Section 5.8, until the endpoint of the computed element of the ray is reached (the endpoint of the element is either the endpoint of the ray (see Section 5.4) or the point of intersection of the ray with the boundary of the complex block).

(b) If the point is the endpoint of the element of the ray, situated at the boundary of a complex block, then: if the index IY(6) = ISRF of the surface is specified in the input data for storing the computed quantities (see Section 5.6d), the proper quantities (see Section 5.5.4) are stored in the file corresponding to the surface. If the surface is the end surface (see Section 5.6c) then the computation of the ray is terminated (see Sections 5.4d–f), or else the quantities Y(1), . . . , Y(NY), YY(1), . . . , YY(5), IY(1), . . . , IY(12) must be transformed across the interface. The code for the elementary wave specifies whether the ray is to be transmitted or reflected and whether it has to continue as a P or an S wave. Transformation of the quantities across the interface is described in Section 5.9. Note that also the subroutine CODE (see Section 4) for the interpretation of the code may indicate the endpoint of the ray (see Sections 5.4a, b(i)–b(iii)). Similarly, the subroutine for the transformation at an interface (Section 5.9) may indicate the endpoint of the ray (see Sections 5.4b(iv), b(v), c). After transformation, the computation of the ray continues as described in Section 5.8, until the endpoint of the computed element of the ray is reached.

To start complete ray tracing at the initial point of a ray, the quantities Y(1), . . . , Y(NY), YY(1), . . . , YY(5), IY(1), . . . , IY(12), Q_{AB}^{INIT} and P_{AB}^{INIT} must be specified at the initial point of the ray. The specification of the above quantities at the initial point of the ray is discussed in Section 6.

5.8 Complete ray tracing through a complex block

The element of a ray situated inside a complex block ICB1 is computed by means of numerical integration of the system of 27 ordinary differential equations (5.2)–(5.4) and (5.6) for $J = 1$ and (5.9). The standard subroutines HPCG or RKGS from the IBM Scientific Subroutine Package may be used for the integration. In HPCG the solution of ordinary differential equations is evaluated by means of Hamming's modified predictor–corrector method, which is a fourth-order method. RKGS uses the Runge–Kutta method (also a fourth-order method). Routine HPCG needs two evaluations of the right-hand sides of differential equations per step of the integration, but its starting takes approximately 18 additional evaluations. Routine RKGS needs 11 evaluations of right-hand sides of differential equations per two steps of the integration. Thus HPCG should be usually more effective than RKGS, but for short elements of a ray (up to four steps of numerical integration) RKGS may be more effective.

The numerical computation of the element of a ray situated inside a complex block is organized by a routine called RAYCB here (see Section 5.8.1).

The right-hand sides of the ordinary differential equations are evaluated by SUBROUTINE FCT(X, Y, D) (see Section 5.8.2).

The necessary tests and auxiliary computations performed after each successful step of numerical integration are carried out in SUBROUTINE OUTP(X, Y, D, IHLF, NDIM, PRMT) (see Section 5.8.3).

5.8.1 A short description of complete ray tracing through one block

The subroutine RAYCB transfers the quantities given at the initial point of the element of a ray into the corresponding quantities at the endpoint of the element.

The quantities are Y(1), . . . , Y(NY), described in Section 5.2.1, and the auxiliary quantities YY(1), . . . , YY(5), IY(1), . . . , IY(12), described in Section 5.2.2. The auxiliary quantities IY(2) = KODIND and IY(3) = ICB0 are not required by the routine RAYCB, the quantities IY(6) = ISRF, IY(7) = ISB2 and IY(8) = ICB2 are only output parameters.

The endpoint of the element is the next point of intersection of the ray with a boundary of the computational volume, with a boundary of the complex block ICB1, or with an end surface (see Section 5.4). The index of the crossed surface is stored in IY(6) = ISRF. If the interface between two complex blocks is crossed then the indices of the simple and complex blocks situated on the other side of the interface are stored in IY(7) = ISB2 and IY(8) = ICB2.

The quantities Y(1), . . . , Y(27) are computed by

```
SUBROUTINE HPCG (PRMT, Y, DERY, NDIM, IHLF, FCT, OUTP, AUX)
REAL PRMT(5), Y(NDIM), DERY(NDIM), AUX(16, NDIM)
INTEGER NDIM, IHLF
```

or by

```
SUBROUTINE RKGS (PRMT, Y, DERY, NDIM, IHLF, FCT, OUTP, AUX)
REAL PRMT(5), Y(NDIM), DERY(NDIM), AUX(8, NDIM)
INTEGER NDIM, IHLF
```

respectively, FCT and OUTP being external subroutines in both cases.

The meaning of individual parameters is described in detail in the IBM Scientific Subroutine Package, we shall not repeat it here. We shall only give the specification of these parameters in our case of complete ray tracing.

The right-hand sides of the ordinary differential equations are evaluated by the subroutine FCT. The routine OUTP, called by HPCG (or RKGS) after every successful step, looks for the endpoint of the computed element, controls the phase shift of the reduced amplitudes Y(28), . . . , Y(NY) due to caustics, and stores the computed quantities inside the complex block in proper files. The communication among the subroutines RAYCB, FCT and OUTP is performed by the local common block /RAYC/.

Structure and function of the routine RAYCB

(a) The input auxiliary quantities (see Section 5.2.2) are stored in the common block /RAYC/.

(b) The interval of the independent variable S for the numerical integration extends from PRMT(1) = YY(1) to a sufficiently large value (e.g. PRMT(2) = PRMT(1) + 999 999). The initial increment of the independent variable is PRMT(3) = $\Delta\sigma$ = STEP (see Section 5.6). The upper error bound is PRMT(4) = 13.444 × UEBRAY for HPCG or PRMT(4) = UEBRAY for RKGS. The parameter PRMT(5) controls the termination of numerical integration and need not be specified. Error weights may be selected for example as follows:

for real and imaginary parts of a complex-valued travel time

$$\text{DERY}(1) = 1, \qquad \text{DERY}(2) = 1;$$

for coordinates

$$\text{DERY}(3) = \frac{(G_{11})^{1/2}}{v}, \qquad \text{DERY}(4) = \frac{(G_{22})^{1/2}}{v},$$

$$\text{DERY}(5) = \frac{(G_{33})^{1/2}}{v};$$

for covariant components of the slowness vector

$$\text{DERY}(6) = (G^{11})^{1/2}\,\Delta\sigma\, v^{1-\text{NEXPS}},$$
$$\text{DERY}(7) = (G^{22})^{1/2}\,\Delta\sigma\, v^{1-\text{NEXPS}},$$
$$\text{DERY}(8) = (G^{33})^{1/2}\,\Delta\sigma\, v^{1-\text{NEXPS}};$$

for other computed quantities

$$\text{DERY}(I) = 0 \quad \text{for } I = 9, \ldots, 27.$$

The velocity v of the corresponding type of wave and the components G^{ij} and G_{ij} of the metric tensor are taken at the point where the numerical integration starts, they represent approximately the values of the velocity and the metric tensor along the computed part of the ray. Note that only approximate values of these quantities are needed at this place; they control only the error weights, not the actual computations.

(c) The routine HPCG (or RKGS) is called.

(d) The value of the independent variable at the endpoint of the computed element of the ray is copied from the common block /RAYC/ to the variable YY(1). If the numerical integration is interrupted for a great number of bisections of the initial increment (greater than NHLF, i.e. PRMT(5) < 0, see Section 5.8.3j) then the upper error bound UEBRAY is doubled and the algorithm continues again from (b).

(e) The auxiliary quantities (see Section 5.2.2) are recalled from the common block /RAYC/ to proper arrays.

5.8.2 Right-hand sides of the differential equations (subroutine FCT)

In this section we shall describe the algorithm for evaluating the right-hand sides of the system of 27 ordinary differential equations for Y(1), . . . , Y(27); see (5.2)–(5.4) and (5.6) for $J = 1$ and (5.9). We assume that the algorithm is realized by the routine FCT, by the following steps (a)–(g).

(a) The number IFCT of invocations of FCT (see Section 5.2.2) is increased by one.

(b) The components G^{ij} of the metric tensor and Christoffel symbols Γ^k_{ij} are determined using the routine METRIC (see Section 2.1). The velocity v, its derivatives $\partial v/\partial x^i$ and $\partial^2 v/\partial x^i\,\partial x^j$, and the loss factor Q^{-1} are determined using the routine VELOC (see Section 3.3.2).

(c) The basis of the ray-centred coordinate system and some related quantities are determined as follows:

(i) non-unit vectors

$$p_i = \mathrm{Y}(i+5), \qquad H_{i1} = \mathrm{Y}(i+8);$$

(ii) contravariant components

$$p^i = G^{ij}p_j, \qquad H^{i1} = G^{ij}H_{j1};$$

(iii) norms

$$d_{11} = (H^{i1}H_{i1})^{1/2}, \qquad d_{33} = (p^i p_i)^{1/2}, \qquad d_{13} = \frac{p^i H_{i1}}{d_{11}d_{33}};$$

(iv) orthonormal vectors

$$H_{i3} = \frac{p_i}{d_{33}}, \qquad H^{i3} = \frac{p^i}{d_{33}},$$

$$H_{i1} = \frac{H_{i1}}{d_{11}} - H_{i3}d_{13}, \qquad H^{i1} = \frac{H^{i1}}{d_{11}} - H^{i3}d_{13},$$

$$H_{i2} = (\det G^{rs})^{-1/2}\varepsilon_{ijk}H^{j3}H^{k1}, \qquad H^{i2} = G^{ij}H_{j2}.$$

(d) Quantities useful to test the accuracy of computations: for exact solutions, the norms d_{11}, d_{33}, d_{13} introduced above should be as follows: $d_{11} = 1$, $d_{33} = v^{-1}$, $d_{13} = 0$. The deviations

$$\Delta_{11} = |d_{11} - 1|, \qquad \Delta_{33} = |vd_{33} - 1|, \qquad \Delta_{13} = |d_{13}|$$

of the basis of the ray-centred coordinate system from the orthonormality conditions are stored in the common block /RAYC/ in order to be accumulated along the whole ray by means of routine OUTP. The accumulated values are useful to check the accuracy of the numerical integration.

(e) First and second derivatives of the velocity in the ray-centred coordinate system:

$$V_1 = \frac{\partial v}{\partial x^i}H^{i1}, \qquad V_{AB} = \left(\frac{\partial^2 v}{\partial x^i\,\partial x^j} - \Gamma^k_{ij}\frac{\partial v}{\partial x^k}\right)H^{iA}H^{jB}.$$

(f) Correction of the computed quantities (i.e. renormalization of the slowness vector and of the first polarization vector):

$$\mathrm{Y}(i+5) = H_{i3}/v, \qquad \mathrm{Y}(i+8) = H_{i1}.$$

(g) Right-hand sides of the differential equations

$$\frac{\mathrm{d}\tau}{\mathrm{d}\sigma} = v^{-\mathrm{NEXPS}}, \qquad \frac{\mathrm{d}\tau^{\mathrm{IM}}}{\mathrm{d}\sigma} = \tfrac{1}{2}Q^{-1}v^{-\mathrm{NEXPS}},$$

$$\frac{\mathrm{d}x^i}{\mathrm{d}\sigma} = H^{i3}v^{1-\mathrm{NEXPS}},$$

$$\frac{\mathrm{d}p_i}{\mathrm{d}\sigma} = \left(-v^{-1}\frac{\partial v}{\partial x^i} + \Gamma^k_{ij}H^{j3}H_{k3}\right)v^{-\mathrm{NEXPS}},$$

$$\frac{\mathrm{d}H_{i1}}{\mathrm{d}\sigma} = (V_1 H_{i3} + v\Gamma^k_{ij}H^{j3}H_{k1})v^{-\mathrm{NEXPS}},$$

$$\frac{\mathrm{d}\Pi_{\alpha\beta}}{\mathrm{d}\sigma} = v^{-1-\mathrm{NEXPS}}\begin{bmatrix} 0 & 0 & v^3 & 0 \\ 0 & 0 & 0 & v^3 \\ -V_{11} & -V_{12} & 0 & 0 \\ -V_{12} & -V_{22} & 0 & 0 \end{bmatrix} \times \begin{bmatrix} \mathrm{Y}(12) & \mathrm{Y}(16) & \mathrm{Y}(20) & \mathrm{Y}(24) \\ \mathrm{Y}(13) & \mathrm{Y}(17) & \mathrm{Y}(21) & \mathrm{Y}(25) \\ \mathrm{Y}(14) & \mathrm{Y}(18) & \mathrm{Y}(22) & \mathrm{Y}(26) \\ \mathrm{Y}(15) & \mathrm{Y}(19) & \mathrm{Y}(23) & \mathrm{Y}(27) \end{bmatrix}.$$

5.8.3 Subroutine OUTP

The routine OUTP is called by HPCG (or RKGS) after any successful step of the numerical integration. It includes various tests of the position of the newly computed point of the ray, tests for possible caustic points, etc. It also stores the computed quantities in proper files if required. The detailed step-to-step description of the algorithm is as follows.

(a) The values of the independent variable X, dependent variables Y and their derivatives D, stored one step before in the variable X2 and in arrays Y2(NY), D2(NY), are moved into X1 and into arrays Y1 and D1. The input values of the parameters X, Y, D of the subroutine OUTP are copied to X2 and to arrays Y2, D2. The variable X2 and arrays Y2, D2 are located in the common block /RAYC/, since their values must be saved until the next invocation of OUTP.

At the beginning of the numerical integration (i.e. if the independent variable X equals PRMT(1)), return to HPCG (or to RKGS).

(b) The number IY(10) = IOUTP (see Section 5.2.2) of steps of numerical integration of the ray is increased by 1. The index IY(6) = ISRF of a surface is set equal to zero.

(c) *Check for crossing the coordinate boundaries of the computational volume.* The boundaries are described by seven reals (see Section 5.6). If any one of the seven limits (indexed from 101 to 107) is exceeded, i.e. if any one of $f_{2i+99} = x^i - x^i_{\text{MIN}}$ is less or equal to zero, or if any one of $f_{2i+100} = x^i - x^i_{\text{MAX}}$ is greater or equal to zero, or if $f_{107} = \tau - \tau_{\text{MAX}}$ is greater or equal to zero, then: the index ISRF is set equal to the relevant index from 101 to 107 and the point of intersection is sought. The point of intersection is situated on the ray Ω between X1, Y1 and X, Y. The independent variable X and dependent variables Y are replaced by the values at the point of intersection. The point of intersection may be found by means of the procedure described in Section 5.8.4. Note that the gradients of the functions $f_{101}, \ldots, f_{107}$ are as follows:

$$\frac{\partial f_{2k+99}}{\partial x^j} = \frac{\partial f_{2k+100}}{\partial x^j} = \begin{cases} 1 & \text{for } k = 1, 2, 3, j = k, \\ 0 & \text{for } k = 1, 2, 3, j \neq k, \end{cases}$$

$$\frac{\partial f_{107}}{\partial x^j} = p_j.$$

(d) *Check for crossing the boundary of the complex block.* The indices of the simple and complex blocks in which the point X, Y is situated may be obtained by invocation of the subroutine BLOCK (see Section 3.3.1). If the ray has not left the complex block IY(5) = ICB1 (i.e. if the index of the complex block is equal to ICB1) then the value of IY(4) = ISB1 is replaced by the index of the current simple block. Otherwise (if the ray has left the complex block ICB1), the point of intersection of the ray with the boundary of the complex block ICB1 must be found as follows.

(d_1) The independent variable X is copied to an auxiliary variable XAUX and the coordinates Y(3), . . . , Y(5) are copied to an auxiliary array YAUX.

(d_2) IY(6) = ISRF is set equal to the index of the surface that separates the point X, Y from the simple block ISB1.

(d$_3$) The point X, Y of intersection of the ray with the surface ISRF is found by means of the procedure described in Section 5.8.4. The subroutine BLOCK is called to find the simple and complex blocks IY(7) = ISB2 and IY(8) = ICB2 touching at the point X, Y of intersection the simple block ISB1 from the other side of the surface ISRF. Then there are three possibilities.

(i) The point X, Y of intersection is separated from the simple block ISB1 by a surface bounding ISB1. This means that the point X, Y is not situated at the boundary of the simple block ISB1. The algorithm must be repeated starting from (d$_2$).

(ii) Otherwise, if the neighbouring complex block ICB2 is equal to ICB1, the point X, Y is situated at the boundary between two simple blocks, but not at the boundary of the complex block ICB1. The value ISB1 must be replaced by ISB2 and the subroutine BLOCK must be called to find the new surface ISRF separating the point XAUX, YAUX from the simple block ISB1. The algorithm must be repeated starting from (d$_3$). Note that the point of intersection will be situated between X1, Y1 and X = XAUX, Y = YAUX.

(iii) Otherwise, the point X, Y of intersection is successfully found.

(e) *Check for crossing the end surfaces.* All surfaces on which complete ray tracing should be terminated and that are different from the surfaces 1, . . . , NSRFC covering structural interfaces must be checked for crossing. If the point X, Y is situated on the other side from an end surface than the initial point of the ray, then IY(6) = ISRF is set equal to the index of the surface and the point of intersection is found. The values at the point are stored in variable X and array Y.

(f) *Phase shift due to caustics.* The matrix Q^{R}_{AB} of geometrical spreading is dependent on the matrix Q^{INIT}_{AB} of geometrical spreading and its derivative P^{INIT}_{AB} at the initial point of the ray (see Sections 5.1.3 and 5.2.3):

$$\begin{bmatrix} Q^{R}_{11} & Q^{R}_{12} \\ Q^{R}_{21} & Q^{R}_{22} \end{bmatrix} = \begin{bmatrix} \Pi_{11} & \Pi_{12} & \Pi_{13} & \Pi_{14} \\ \Pi_{21} & \Pi_{22} & \Pi_{23} & \Pi_{24} \end{bmatrix} \begin{bmatrix} Q^{INIT}_{11} & Q^{INIT}_{12} \\ Q^{INIT}_{21} & Q^{INIT}_{22} \\ P^{INIT}_{11} & P^{INIT}_{12} \\ P^{INIT}_{21} & P^{INIT}_{22} \end{bmatrix}.$$

The matrices Q_{AB}^{INIT} and P_{AB}^{INIT} are stored in the array YI (see Section 6.1). We denote

$$\begin{bmatrix} Q_{11}^1 & Q_{12}^1 \\ Q_{21}^1 & Q_{22}^1 \end{bmatrix} = \begin{bmatrix} \mathrm{Y1(12)} & \mathrm{Y1(16)} & \mathrm{Y1(20)} & \mathrm{Y1(24)} \\ \mathrm{Y1(13)} & \mathrm{Y1(17)} & \mathrm{Y1(21)} & \mathrm{Y1(25)} \end{bmatrix} \times \begin{bmatrix} \mathrm{YI(12)} & \mathrm{YI(16)} \\ \mathrm{YI(13)} & \mathrm{YI(17)} \\ \mathrm{YI(14)} & \mathrm{YI(18)} \\ \mathrm{YI(15)} & \mathrm{YI(19)} \end{bmatrix},$$

the matrix of geometrical spreading at the point X1, Y1. Similarly we denote

$$\begin{bmatrix} Q_{11}^2 & Q_{12}^2 \\ Q_{21}^2 & Q_{22}^2 \end{bmatrix} = \begin{bmatrix} \mathrm{Y(12)} & \mathrm{Y(16)} & \mathrm{Y(20)} & \mathrm{Y(24)} \\ \mathrm{Y(13)} & \mathrm{Y(17)} & \mathrm{Y(21)} & \mathrm{Y(25)} \end{bmatrix} \times \begin{bmatrix} \mathrm{YI(12)} & \mathrm{YI(16)} \\ \mathrm{YI(13)} & \mathrm{YI(17)} \\ \mathrm{YI(14)} & \mathrm{YI(18)} \\ \mathrm{YI(15)} & \mathrm{YI(19)} \end{bmatrix},$$

the matrix of geometrical spreading at the point X, Y.

If

$$\det \boldsymbol{Q}^1 \det \boldsymbol{Q}^2 < 0,$$

where $\det \boldsymbol{Q}^1 = Q_{11}^1 Q_{22}^1 - Q_{12}^1 Q_{21}^1$ and $\det \boldsymbol{Q}^2 = Q_{11}^2 Q_{22}^2 - Q_{12}^2 Q_{21}^2$, then there is a caustic point of first order (line caustic) between the points X1, Y1 and X, Y. Then the reduced amplitudes Y(28), . . . , Y(NY) must be multiplied by −i, and IY(12) = KMAH is increased by one.

Otherwise, if

$$(Q_{11}^1 Q_{22}^2 - Q_{12}^1 Q_{21}^2 + Q_{22}^1 Q_{11}^2 - Q_{21}^1 Q_{12}^2) \det \boldsymbol{Q}^1 < 0$$

then there is a caustic point of second order (point caustic) between points X1, Y1 and X, Y. Then the reduced amplitudes Y(28), . . . , Y(NY) must change their signs, and IY(12) = KMAH is increased by two.

(g) *Storage of the computed quantities at given surfaces.* The loop over all surfaces specified for storage of the computed quantities, and different from the surfaces 1, . . . , NSRFC covering interfaces, must be performed. If a surface is identical with the surface ISRF on which the endpoint X, Y of the element is situated then the quantities at this point are stored in the corresponding file (see Section 5.5.2). Otherwise, if the point X, Y is situated on the other side from the surface from the point X1, Y1 then the point of intersection of the ray with the surface is found and the independent variable and the dependent variables at the point are temporarily stored in the auxiliary variable XAUX and auxiliary array YAUX. The reduced amplitudes YAUX(28), . . . , YAUX(NY) are obtained from the amplitudes Y(28), . . . , Y(NY) by adding the phase shift due to the caustics between the points X, Y and XAUX, YAUX. The phase shift may be determined in the same way as described in (f). We must remember, however, that the point XAUX, YAUX is situated against the direction of propagation from the point X, Y. Then a factor i must be used instead of −i in (f).

(h) *Storage of computed quantities along the ray.* The quantities are stored along the ray with the given step STORE (see Section 5.6h) in the independent variable if STORE $\neq 0$. We denote

$$k_1 = \text{int}\,(\text{X1/STORE}) + 1, \qquad k_2 = \text{int}\,(\text{X/STORE}).$$

The quantities are stored at the points given by the following values of the independent variable:

$$\text{XAUX} = k \times \text{STORE} \quad (k = k_1, \ldots, k_2).$$

For $k_1 > k_2$ the data are not stored. The dependent variables YAUX(1), . . . , YAUX(27) at the point XAUX may be obtained by interpolation from the points X1, Y1 and X2, Y2. The interpolation method is described in Section 5.8.4. The reduced amplitudes YAUX(28), . . . , YAUX(NY) may be obtained in the same way as in (g).

(i) *Accumulation of the renormalization errors for a test of accuracy.* The quantities Δ_{33}, Δ_{13} and Δ_{11} (see Section 5.8.2d), stored in the common block /RAYC/, are multiplied by $\frac{1}{2}(\text{X} - \text{X1})v^{-\text{NEXPS}}$ and are added to the quantities YY(3) = ERRPP, YY(4) = ERRPH, YY(5) = ERRHH (see Section 5.2.2), temporarily stored in the common block /RAYC/.

(j) *Large number of bisections of the initial step.* If the number IHLF of bisections is greater than the given limit NHLF then the numerical integration is interrupted by setting PRMT(5) negative (numerical integration is interrupted for PRMT(5) non-zero).

(k) *Final operations.* The value X of the independent variable is stored in the common block /RAYC/. If ISRF$\neq$0 (i.e. the endpoint of the element of the ray is reached) then the numerical integration is terminated by setting PRMT(5) positive.

5.8.4 Auxiliary procedures

In the routine OUTP it is necessary to perform the interpolation of certain quantities along the ray, and to find the points of intersection of the ray with certain given surfaces. The algorithms for the solution of these two problems are described in this section.

(a) Interpolation along the ray

The part of the ray between two consequent points obtained by numerical integration may be interpolated by a third-order method using the functional values and their first derivatives at the two points. We denote by x_1, y_1, y_1' (x_2, y_2, y_2') the independent variable, the dependent variables and their derivatives in the first (second) of the two given points. We introduce four functions of the independent variable x:

$$\begin{aligned}
a_1(x) &= [2a(x)+3]a(x)a(x), \\
a_2(x) &= 1-a_1(x), \\
b_1(x) &= [a(x)+1]a(x)(x-x_2), \\
b_2(x) &= [a(x)+1]a(x)(x-x_1),
\end{aligned}$$

where

$$a(x) = (x-x_2)/(x_2-x_1).$$

They have the following properties:

$$\begin{bmatrix} a_1(x_1) & a_1(x_2) & a_1'(x_1) & a_1'(x_2) \\ a_2(x_1) & a_2(x_2) & a_2'(x_1) & a_2'(x_2) \\ b_1(x_1) & b_1(x_2) & b_1'(x_1) & b_1'(x_2) \\ b_2(x_1) & b_2(x_2) & b_2'(x_1) & b_2'(x_2) \end{bmatrix} = \begin{bmatrix} 1 & 0 & 0 & 0 \\ 0 & 1 & 0 & 0 \\ 0 & 0 & 1 & 0 \\ 0 & 0 & 0 & 1 \end{bmatrix}.$$

Here the functions

$$\begin{aligned}
a_1'(x) &= 6[a(x)+1]a(x)/(x_2-x_1), \\
a_2'(x) &= -a_1'(x), \\
b_1'(x) &= [3a(x)+2]a(x), \\
b_2'(x) &= [3a(x)+1][a(x)+1],
\end{aligned}$$

are the first derivatives of the functions $a_1(x)$, $a_2(x)$, $b_1(x)$ and $b_2(x)$. The values of the dependent variables and their derivatives at x are then determined by

$$y = a_1 y_1 + a_2 y_2 + b_1 y_1' + b_2 y_2',$$
$$y' = a_1'(y_1 - y_2) + b_1' y_1' + b_2' y_2'.$$

(b) The search for the intersection of the ray with a given surface

The search for the point of intersection of a ray parametrized by an independent variable X with a surface $f(x^i) = 0$ is started if

$$f(x^i(\mathrm{XA}))f(x^i(\mathrm{XB})) < 0$$

for two points XA, XB of the ray. The point of intersection must be situated between XA and XB.

The ray is interpolated between its two last points obtained by numerical integration using the third-order method described in (a). The function $f(x^i)$ is evaluated by means of the subroutine SRFC2 if the index ISRF of the surface is less than 100. For ISRF > 100 the functions $f_{101}, \ldots, f_{107}$ are used.

The point of intersection can be found iteratively by a combined two-point method. The *regula falsi* method is used in each odd iteration, the Newton–Raphson method in each even iteration. Since the Newton–Raphson method need not converge, it is replaced by the method of bisections in the case of emergency. The last approximation of X is taken for a new XB. That one of the old XA, XB at which the function $f(x^i)$ has the opposite sign to $f(x^i(X))$ is taken for a new XA. The iterations are terminated if the new interval $\langle \mathrm{XA}, \mathrm{XB} \rangle$ of the independent variable is sufficiently small:

$$|\mathrm{XA} - \mathrm{XB}| \leq \mathrm{ERR},$$

with

$$\mathrm{ERR} = \mathrm{UEB}\ v^{-\mathrm{NEXPS}},$$

where v is the velocity of propagation of the corresponding type of wave.

The *regula falsi* iteration (odd iteration) is as follows:

$$\mathrm{X} = (\mathrm{FA} \times \mathrm{XB} - \mathrm{FB} \times \mathrm{XA})/(\mathrm{FA} - \mathrm{FB}),$$

where

$$\mathrm{FA} = f(x^i(\mathrm{XA})), \qquad \mathrm{FB} = f(x^i(\mathrm{XB})).$$

The modified Newton–Raphson iteration (even iteration) is as follows.

We denote

$$\begin{aligned} \mathrm{XC} &= (\mathrm{XA}+\mathrm{XB})/2, \\ \mathrm{XCA} &= \mathrm{XA}-\mathrm{FA}/\mathrm{DA}+(\mathrm{ERR}/50)\,\mathrm{sign}\,(\mathrm{XA}-\mathrm{XB}), \\ \mathrm{XCB} &= \mathrm{XB}-\mathrm{FB}/\mathrm{DB}+(\mathrm{ERR}/50)\,\mathrm{sign}\,(\mathrm{XA}-\mathrm{XB}), \end{aligned}$$

where

$$\mathrm{DA}=\frac{\partial f}{\partial x^k}(x^i(\mathrm{XA}))\frac{\partial x^k}{\partial X}(\mathrm{XA}), \qquad \mathrm{DB}=\frac{\partial f}{\partial x^k}(x^i(\mathrm{XB}))\frac{\partial x^k}{\partial X}(\mathrm{XB}).$$

If XCA and XCB are situated between points XB and XC then

$$\mathrm{X}=\begin{cases}\mathrm{XCA} & \text{for } |\mathrm{XCA}-\mathrm{XB}|<|\mathrm{XCB}-\mathrm{XB}|, \\ \mathrm{XCB} & \text{for } |\mathrm{XCB}-\mathrm{XB}|\leqslant|\mathrm{XCA}-\mathrm{XB}|.\end{cases}$$

If only XCA is situated between XB and XC then

$$\mathrm{X}=\mathrm{XCA}.$$

If only XCB is situated between XB and XC then

$$\mathrm{X}=\mathrm{XCB}.$$

If neither XCA nor XCB is situated between XB and XC then

$$\mathrm{X}=\mathrm{XC}.$$

5.9 Complete ray tracing across a curved interface

The quantities Y(1), . . . , Y(NY), YY(1), . . . , YY(5) and IY(1), . . . , IY(12) computed along a ray and defined in Section 5.2 must be transformed at an interface. We denote the quantities Y, YY, IY (described in Section 5.2) corresponding to the incident wave and the reflected/transmitted wave at the point of incidence by Y1, YY1, IY1 and Y2, YY2, IY2 respectively. We also denote by KODNEW the new position in the code, and by |ICBNEW| the index of the complex block in which the generated wave is to propagate. The sign of ICBNEW specifies the type of generated wave (+ for P, − for S).

5.9.1 Transformation of auxiliary quantities, travel time and coordinates

(a) The number of real-valued quantities describing the reduced amplitudes of the incident wave is NAMPL1 = IY1(1) − 27.

(b) The number NAMPL2 of real-valued quantities specifying the reduced amplitudes of the generated wave is
   (b_1) if IY1(5) > 0 and ICBNEW < 0 then NAMPL2 = 2 NAMPL1;
   (b_2) if IY1(5) < 0 and ICBNEW > 0 then NAMPL2 = NAMPL1/2;
   (b_3) otherwise NAMPL2 = NAMPL1.

(c) The output number of basic quantities is IY2(1) = 27 + NAMPL2.

(d) The new position in the code is IY2(2) = KODNEW.

(e) Indices of the simple and complex blocks:
   (e_1) for reflection (|IY1(5)| = |ICBNEW|): IY2(3) = IY1(3), IY2(4) = IY1(4), IY2(5) = ICBNEW;
   (e_2) for transmission: IY2(3) = |IY1(5)|, IY2(4) = IY1(7), IY2(5) = ICBNEW.

(f) IY2(6), IY2(7), IY2(8) are undefined.

(g) IY2(9) = IY1(9), IY2(10) = IY1(10), IY2(11) = IY1(11) + 1, IY2(12) = IY1(12).

(h) YY2(I) = YY1(I) for $I = 1, \ldots, 5$.

(i) The real and imaginary parts of the travel time and the coordinates remain unchanged: Y2(I) = Y1(I) for $I = 1, \ldots, 5$.

5.9.2 Metric tensor and velocities

The components G_{ij} and G^{ij} of the metric tensor and the Christoffel symbols Γ^k_{ij} at the point $x^i = \text{Y1}(i+2)$ are the result of the routine METRIC (see Section 2.1).

The parameters of the complex block |IY1(5)| for the incident wave are obtained by invocation of subroutine PARM2. The parameters of the complex block IY1(8) on the other side of the surface IY1(6) are obtained similarly. The velocities VP1, VS1 in the complex block |IY1(5)|, and the velocity v of the incident wave and its derivatives are obtained by invocation of routine VELOC. Similarly, the velocities VP2 and VS2 in the complex block IY1(8) are also obtained by invocation of routine VELOC. In the case of transmission, i.e. if |ICBNEW| = IY1(8), the same invocation yields the velocity $\tilde{v}$ of the transmitted wave, together with its derivatives. In the case of reflection without conversion, i.e. if ICBNEW = IY1(5) then the velocity v with its derivatives is equal to the velocity v of the incident wave with its derivatives. In the case of reflection with conversion, i.e. if ICBNEW = −IY1(5), the routine VELOC must be called once again to give the velocity $\tilde{v}$ of the reflected wave and its derivatives.

If the velocity $\tilde{v}$ of the reflected/transmitted wave is equal to zero (i.e. an S wave in liquid) then the computation of the ray must be terminated.

5.9.3 Transformation of the slowness vector and the polarization vectors

The polarization vectors (basis vectors of the ray-centred coordinate system) of the incident wave are given by

$$H_{i3} = p_i(p_k G^{kl} p_l)^{-1/2},$$

where

$$p_i = \mathrm{Y1}(i+5)$$

is the slowness vector, and by

$$H_{i1} = \mathrm{Y1}(i+8).$$

We define the basis of the local Cartesian coordinate system at the surface IY1(6) specified by $f(x^i) = 0$. The unit normal to the surface is

$$Z_{i3} = f_i(f_k G^{kl} f_l)^{-1/2}.$$

Here we denote $f_i = \partial f/\partial x^i$, $f_{ij} = \partial^2 f/\partial x^i\, \partial x^j$. The cosine of the angle of incidence α $(0 \leqslant \alpha \leqslant \pi)$ is

$$\cos\alpha = H_{i3} G^{ij} Z_{j3},$$

and the sine is

$$\sin\alpha = (1 - \cos^2\alpha)^{1/2}.$$

The basis vector in the plane of incidence is

$$Z_{i1} = (H_{i3} - Z_{i3}\cos\alpha)/\sin\alpha,$$

and the basis vector perpendicular to the plane of incidence is

$$Z_{i2} = \varepsilon_{ijk} Z^j{}_3 Z^k{}_1 (\det G_{rs})^{1/2},$$

where $Z^j{}_k = G^{jm} Z_{mk}$ are the contravariant components of the basis vectors, and where $(\det G_{rs})^{1/2}$ is an output parameter of the routine METRIC. In the case of normal incidence, i.e. if $\sin\alpha = 0$, the vectors Z_{i1} and Z_{i2} may be chosen arbitrarily (e.g. $Z_{i1} = H_{i1}$).

The sine of the angle $\tilde{\alpha}$ of reflection/transmission may be evaluated using Snell's law

$$\sin \tilde{\alpha} = \frac{\tilde{v}}{v} \sin \alpha.$$

The cosine is

$$\cos \tilde{\alpha} = \pm \varepsilon (1 - \sin^2 \tilde{\alpha})^{1/2}, \qquad \varepsilon = \text{sign} (\cos \alpha),$$

where the plus sign is taken for transmission (|ICBNEW| = IY1(8)) and the minus sign for reflection (|ICBNEW| = |IY1(5)|).

The revolution angle ψ of the vector Z_{i2} with respect to the polarization vector H_{i2} is determined by

$$\cos \psi = H_{i1} Z^i{}_1 / \cos \alpha, \qquad \sin \psi = -H_{i1} Z^i{}_2.$$

The unit vector perpendicular to the reflected/transmitted ray in the plane of incidence is

$$E_i = Z_{i1} \cos \tilde{\alpha} - Z_{i3} \sin \tilde{\alpha}.$$

The unit vector tangent to the reflected/transmitted ray is

$$\tilde{H}_{i3} = Z_{i1} \sin \tilde{\alpha} + Z_{i3} \cos \tilde{\alpha}.$$

We wish the vector $\tilde{H}_{i2}$ to be rotated with respect to the vector Z_{i2} through the same angle ψ as the vector H_{i2}. Then

$$\mathrm{Y2}(i+8) = \tilde{H}_{i1} = E_i \cos \psi - Z_{i2} \sin \psi.$$

The slowness vector of the reflected/transmitted ray is

$$\mathrm{Y2}(i+5) = \tilde{p}_i = \tilde{H}_{i3} / \tilde{v}.$$

5.9.4 The curvature of the interface and the velocity gradients in the local Cartesian coordinate system

The second derivatives of the function $f(x^i)$ describing the surface IY1(6) in local Cartesian coordinates divided by the norm of the gradient of $f(x^i)$ are

$$D_{AB} = (f_{ij} - \Gamma^k_{ij} f_k) Z^i{}_A Z^j{}_B (f_r G^{rs} f_s)^{-1/2}.$$

This equation is obtained by transforming the second covariant derivatives of $f(x^i)$ into local Cartesian coordinates.

The velocity gradients in local Cartesian coordinates are

$$V_i = \frac{\partial v}{\partial x^j} Z^j{}_i, \qquad \tilde{V}_i = \frac{\partial \tilde{v}}{\partial x^j} Z^j{}_i.$$

5.9.5 Dynamic ray tracing across a curved interface

First the ray-centred coordinate system is rotated through an angle ψ:

$$Q_{A\beta} = \Psi_{AC}\Pi_{C\beta} = \Psi_{AC}\mathrm{Y1}(7 + C + 4\beta),$$
$$P_{A\beta} = \Psi_{AC}\Pi_{C+2,\beta} = \Psi_{AC}\mathrm{Y1}(9 + C + 4\beta),$$

where

$$\boldsymbol{\Psi} = \begin{bmatrix} \cos\psi & \sin\psi \\ -\sin\psi & \cos\psi \end{bmatrix}.$$

Then the transformation equations are

$$\tilde{Q}_{A\alpha} = \tilde{C}_{AB}C_{BC}^{-1}Q_{C\alpha},$$
$$\tilde{P}_{A\alpha} = \tilde{C}_{AB}^{-1}C_{BC}P_{C\alpha} + \tilde{C}_{AB}^{-1}F_{BC}C_{CD}^{-1}Q_{D\alpha},$$

where

$$\boldsymbol{C} = \begin{bmatrix} \cos\alpha & 0 \\ 0 & 1 \end{bmatrix}, \qquad \tilde{\boldsymbol{C}} = \begin{bmatrix} \cos\tilde{\alpha} & 0 \\ 0 & 1 \end{bmatrix}.$$

The matrix F_{AB} is given by

$$F_{AB} = (v^{-1}\cos\alpha - \tilde{v}^{-1}\cos\tilde{\alpha})D_{AB} + v^{-1}\sin\alpha\,(E_{AB} - \tilde{E}_{AB}),$$

where

$$\boldsymbol{E} = \begin{bmatrix} -V_1v^{-1}(1 + \cos^2\alpha) + V_3v^{-1}\sin\alpha\cos\alpha & -V_2v^{-1} \\ -V_2v^{-1} & 0 \end{bmatrix},$$

$$\tilde{\boldsymbol{E}} = \begin{bmatrix} -\tilde{V}_1\tilde{v}^{-1}(1 + \cos^2\tilde{\alpha}) + \tilde{V}_3\tilde{v}^{-1}\sin\tilde{\alpha}\cos\tilde{\alpha} & -\tilde{V}_2\tilde{v}^{-1} \\ -\tilde{V}_2\tilde{v}^{-1} & 0 \end{bmatrix}.$$

Finally, the ray-centred coordinate system is rotated through an angle $-\psi$

$$\mathrm{Y2}(7 + A + 4\beta) = \tilde{\Pi}_{A\beta} = \Psi_{CA}Q_{C\beta}, \quad \mathrm{Y2}(9 + A + 4\beta) = \tilde{\Pi}_{A+2,\beta} = \Psi_{CA}P_{C\beta}.$$

5.9.6 Transformation of reduced amplitudes

In the case of an incident S wave the ray-centred coordinate system is rotated through an angle ψ

$$\mathrm{Re}\,A_{Ai} = \Psi_{A1}\mathrm{Y1}(28) + \Psi_{A2}\mathrm{Y1}(30), \quad \mathrm{Im}\,A_{Ai} = \Psi_{A1}\mathrm{Y1}(29) + \Psi_{A2}\mathrm{Y1}(31)$$

for $i = 1$ or 3, and, possibly,

$$\mathrm{Re}\,A_{A2} = \Psi_{A1}\mathrm{Y1}(32) + \Psi_{A2}\mathrm{Y1}(34), \quad \mathrm{Im}\,A_{A2} = \Psi_{A1}\mathrm{Y1}(33) + \Psi_{A2}\mathrm{Y1}(35)$$

if there is an S wave at the initial point of the ray. The reduced amplitudes of the incident P wave are not rotated:

$$\operatorname{Re} A_{3i} = \mathrm{Y1}(28), \qquad \operatorname{Im} A_{3i} = \mathrm{Y1}(29)$$

for $i = 1$ or 3, and, possibly,

$$\operatorname{Re} A_{32} = \mathrm{Y1}(30), \qquad \operatorname{Im} A_{32} = \mathrm{Y1}(31).$$

Then the above reduced amplitudes are transformed by one of the following equations:

$$\begin{bmatrix} \tilde{A}_{1i} \\ \tilde{A}_{2i} \end{bmatrix} = \left(\frac{\tilde{\rho}\tilde{v}\,|\cos\tilde{\alpha}|}{\rho v\,|\cos\alpha|} \right)^{1/2} \begin{bmatrix} R_{\mathrm{SV\to SV}} & 0 \\ 0 & R_{\mathrm{SH\to SH}} \end{bmatrix} \begin{bmatrix} A_{1i} \\ A_{2i} \end{bmatrix},$$

$$\begin{bmatrix} \tilde{A}_{1i} \\ \tilde{A}_{2i} \end{bmatrix} = \left(\frac{\tilde{\rho}\tilde{v}\,|\cos\tilde{\alpha}|}{\rho v\,|\cos\alpha|} \right)^{1/2} \begin{bmatrix} R_{\mathrm{P\to SV}} \\ 0 \end{bmatrix} A_{3i},$$

$$\tilde{A}_{3i} = \left(\frac{\tilde{\rho}\tilde{v}\,|\cos\tilde{\alpha}|}{\rho v\,|\cos\alpha|} \right)^{1/2} [R_{\mathrm{SV\to P}} \quad 0] \begin{bmatrix} A_{1i} \\ A_{2i} \end{bmatrix},$$

$$\tilde{A}_{3i} = \left(\frac{\tilde{\rho}\tilde{v}\,|\cos\tilde{\alpha}|}{\rho v\,|\cos\alpha|} \right)^{1/2} R_{\mathrm{P\to P}} A_{3i},$$

where $i = 1, 2$ for the S wave at the initial point of the ray and $i = 3$ for the P wave at the initial point of the ray.

The reduced amplitudes of generated P waves may be directly stored in the output array Y2,

$$\mathrm{Y2}(28) = \operatorname{Re} \tilde{A}_{3i}, \qquad \mathrm{Y2}(29) = \operatorname{Im} \tilde{A}_{3i}$$

for $i = 1$ or 3, and, for an S wave at the initial point of the ray,

$$\mathrm{Y2}(30) = \operatorname{Re} \tilde{A}_{32}, \qquad \mathrm{Y2}(31) = \operatorname{Im} \tilde{A}_{32}.$$

For S waves the ray-centred coordinate system is rotated through an angle $-\psi$,

$$\mathrm{Y2}(26 + 2A) = \Psi_{BA} \operatorname{Re} \tilde{A}_{Bi}, \qquad \mathrm{Y2}(27 + 2A) = \Psi_{BA} \operatorname{Im} \tilde{A}_{Bi}$$

for $i = 1$ or 3, and, for an S wave at the initial point of the ray,

$$\mathrm{Y2}(30 + 2A) = \Psi_{BA} \operatorname{Re} \tilde{A}_{B2}, \qquad \mathrm{Y2}(31 + 2A) = \Psi_{BA} \operatorname{Im} \tilde{A}_{B2}.$$

5.9.7 The reflection/transmission coefficients

The R/T coefficients are given by the following expressions:

Reflection coefficients

$$\begin{aligned}
R_{\mathrm{SV}\to\mathrm{SV}} &= D^{-1}[q^2p^2P_1P_2P_3P_4 + \rho_1\rho_2(\alpha_1\beta_2P_2P_3 - \beta_1\alpha_2P_1P_4) \\
&\quad - \alpha_1\beta_1P_3P_4Y^2 + \alpha_2\beta_2P_1P_2X^2 - \alpha_1\alpha_2\beta_1\beta_2p^2Z^2], \\
R_{\mathrm{P}\to\mathrm{P}} &= D^{-1}[q^2p^2P_1P_2P_3P_4 + \rho_1\rho_2(\beta_1\alpha_2P_1P_4 - \alpha_1\beta_2P_2P_3) \\
&\quad - \alpha_1\beta_1P_3P_4Y^2 + \alpha_2\beta_2P_1P_2X^2 - \alpha_1\alpha_2\beta_1\beta_2p^2Z^2], \\
R_{\mathrm{SV}\to\mathrm{P}} &= -2\varepsilon\beta_1pP_2D^{-1}(qP_3P_4Y + \alpha_2\beta_2XZ), \\
R_{\mathrm{P}\to\mathrm{SV}} &= 2\varepsilon\alpha_1pP_1D^{-1}(qP_3P_4Y + \alpha_2\beta_2XZ), \\
R_{\mathrm{SH}\to\mathrm{SH}} &= \bar{D}^{-1}(\rho_1\beta_1P_2 - \rho_2\beta_2P_4);
\end{aligned}$$

Transmission coefficients

$$\begin{aligned}
R_{\mathrm{SV}\to\mathrm{SV}} &= 2\beta_1\rho_1P_2D^{-1}(\alpha_1P_3Y + \alpha_2P_1X), \\
R_{\mathrm{P}\to\mathrm{P}} &= 2\alpha_1\rho_1P_1D^{-1}(\beta_2P_2X + \beta_1P_4Y), \\
R_{\mathrm{SV}\to\mathrm{P}} &= 2\varepsilon\beta_1\rho_1pP_2D^{-1}(qP_1P_4 - \alpha_1\beta_2Z), \\
R_{\mathrm{P}\to\mathrm{SV}} &= -2\varepsilon\alpha_1\rho_1pP_1D^{-1}(qP_2P_3 - \beta_1\alpha_2Z), \\
R_{\mathrm{SH}\to\mathrm{SH}} &= 2\rho_1\beta_1P_2\bar{D}^{-1}.
\end{aligned}$$

Here

$$\begin{aligned}
D &= q^2p^2P_1P_2P_3P_4 + \rho_1\rho_2(\beta_1\alpha_2P_1P_4 + \alpha_1\beta_2P_2P_3) \\
&\quad + \alpha_1\beta_1P_3P_4Y^2 + \alpha_2\beta_2P_1P_2X^2 + \alpha_1\alpha_2\beta_1\beta_2p^2Z^2, \\
\bar{D} &= \rho_1\beta_1P_2 + \rho_2\beta_2P_4, \\
q &= 2(\rho_2\beta_2^2 - \rho_1\beta_1^2), \quad X = \rho_2 - qp^2, \quad Y = \rho_1 + qp^2, \quad Z = \rho_2 - \rho_1 - qp^2, \\
P_1 &= (1 - \alpha_1^2p^2)^{1/2}, \quad P_2 = (1 - \beta_1^2p^2)^{1/2}, \\
P_3 &= (1 - \alpha_2^2p^2)^{1/2}, \quad P_4 = (1 - \beta_2^2p^2)^{1/2}.
\end{aligned}$$

The square roots P_κ are positive imaginary for negative argument, e.g. $P_3 = \mathrm{i}(\alpha_2^2p^2 - 1)^{1/2}$ for $p > 1/\alpha_2$. The quantity p is the ray parameter, $p = \sin\alpha/v$, where α is the angle of incidence, $0 \leqslant \alpha \leqslant \pi$ (see Section 5.9.3), and v is the velocity of the incident wave at the point of incidence. The quantities $\alpha_1 = \mathrm{VP1}$, $\beta_1 = \mathrm{VS1}$ and ρ_1 correspond to the point of incidence at that side of the interface containing the incident wave; $\alpha_2 = \mathrm{VP2}$, $\beta_2 = \mathrm{VS2}$ and ρ_2 correspond to the other side of the interface at the same point. The quantity ε is given by the relation $\varepsilon = \mathrm{sign}(\cos\alpha)$ (see Section 5.9.3). For further details see Červený (1985a).

For a liquid–liquid interface we have

$$R_{\mathrm{P}\to\mathrm{P}} = \frac{\rho_2\alpha_2 P_1 - \rho_1\alpha_1 P_3}{\rho_2\alpha_2 P_1 + \rho_1\alpha_1 P_3}$$

for reflection, and

$$R_{\mathrm{P}\to\mathrm{P}} = \frac{2\rho_1\alpha_1 P_1}{\rho_2\alpha_2 P_1 + \rho_1\alpha_1 P_3}$$

for transmission.

For reflections from the Earth's surface we put $\alpha_2 = \beta_2 = \rho_2 = 0$ in the above expressions.

6 INITIAL POINTS OF RAYS

Let us consider a two-parameter system of rays of a specified elementary wave. A ray of the elementary wave is specified by its two parameters γ^1 and γ^2. The computed rays may start from a single initial point common to all rays, from an initial surface along which an initial travel time is defined, from a curve along which an initial travel time is defined, etc.

6.1 Important quantities at the initial point of the ray

At the initial point of the ray, we are interested in particular in the quantities ICB1 and YL(1), . . . , YL(6), defined in Section 5.5.4 and describing the local properties of the model, and in the following quantities describing the properties of the rays and of the travel-time field:

YI(1)	initial travel time;
YI(2)	initial imaginary part of the complex travel time;
YI(3), . . . , YI(5)	coordinates of the initial point of the ray;
YI(6), . . . , YI(8)	covariant components of the initial slowness vector;
YI(9), . . . , YI(11)	covariant components of the first basis vector of the ray-centred coordinate system at the initial point of the ray (perpendicular to the slowness vector YI(6), . . . , YI(8));

$$\begin{bmatrix} \mathrm{YI(12)} & \mathrm{YI(16)} \\ \mathrm{YI(13)} & \mathrm{YI(17)} \\ \mathrm{YI(14)} & \mathrm{YI(18)} \\ \mathrm{YI(15)} & \mathrm{YI(19)} \end{bmatrix} = \begin{bmatrix} Q_{11}^{\mathrm{INIT}} & Q_{12}^{\mathrm{INIT}} \\ Q_{21}^{\mathrm{INIT}} & Q_{22}^{\mathrm{INIT}} \\ P_{11}^{\mathrm{INIT}} & P_{12}^{\mathrm{INIT}} \\ P_{21}^{\mathrm{INIT}} & P_{22}^{\mathrm{INIT}} \end{bmatrix},$$

the elements of the matrix $\boldsymbol{Q}^{\mathrm{R}}$ of geometrical spreading, and the matrix $\boldsymbol{P}^{\mathrm{R}}$, at the initial point of the ray, see (5.13);

$\mathrm{YI}(20) = \tilde{\gamma}^1$, $\mathrm{YI}(21) = \tilde{\gamma}^2$ take-off parameters of the computed ray.

In addition to the above quantities describing the properties defined for a single ray, there are also quantities describing the properties of the discrete system of computed rays in the vicinity of the computed ray. Of these quantities, we are interested in particular in the following:

YI(22) area Γ of element $\delta\Gamma$ of the ray-parameter surface, corresponding to the ray,

$$\Gamma = \int_{\delta\Gamma} \mathrm{d}\gamma^1 \, \mathrm{d}\gamma^2; \tag{6.1}$$

YI(23), YI(24), YI(25) components Γ_{11}, Γ_{12}, Γ_{22} of the symmetric matrix Γ_{KL} inverse to the specific moment

$$\Gamma^{IJ} = \Gamma^{-1} \int_{\delta\Gamma} (\tilde{\gamma}^I - \gamma^I)(\tilde{\gamma}^J - \gamma^J) \, \mathrm{d}\gamma^1 \, \mathrm{d}\gamma^2 \tag{6.2}$$

of the element $\delta\Gamma$ of the ray-parameter surface, evaluated with respect to the ray. Here γ^I are the ray parameters.

The quantities ICB1, YL(1), . . . , YL(6) and YI(1), . . . , YI(21) corresponding to the initial point of the ray, together with the quantities YI(22), . . . , YI(25), are required in processing the results of complete ray tracing; they must be stored in some file. The quantities YI(22), . . . , YI(25) are not generally known during complete ray tracing; they can be determined after it. Thus they cannot be stored when the results of complete ray tracing are stored, but must be stored after the complete tracing of the ray has been finished. For these reasons, the quantities YI(22), . . . , YI(25) should be stored in another file than that where the results of the complete ray tracing are stored. It seems reasonable to store all the quantities ICB1, YL(1), . . . , YL(6) and

YI(1), . . . , YI(25) corresponding to the initial points of rays in a separate file.

The index ISB1 of the simple block containing the initial point of the ray, and abovementioned quantities ICB1, YL(1), . . . , YL(6), YI(1), . . . , YI(21) defined at the initial point of the ray, may be necessary or useful during the complete tracing of the ray. For this reason, they should be stored in a common block.

6.2 Initial values for complete ray tracing

The initial values of the basic quantities Y(1), . . . , Y(NY), where NY = 27 + NAMPL, computed along a ray and defined in Section 5.2.1, are

$$\mathrm{Y}(I) = \mathrm{YI}(I) \quad \text{for } I = 1, \ldots, 11;$$

$$\begin{bmatrix} \mathrm{Y}(12) & \mathrm{Y}(16) & \mathrm{Y}(20) & \mathrm{Y}(24) \\ \mathrm{Y}(13) & \mathrm{Y}(17) & \mathrm{Y}(21) & \mathrm{Y}(25) \\ \mathrm{Y}(14) & \mathrm{Y}(18) & \mathrm{Y}(22) & \mathrm{Y}(26) \\ \mathrm{Y}(15) & \mathrm{Y}(19) & \mathrm{Y}(23) & \mathrm{Y}(27) \end{bmatrix} = \begin{bmatrix} 1 & 0 & 0 & 0 \\ 0 & 1 & 0 & 0 \\ 0 & 0 & 1 & 0 \\ 0 & 0 & 0 & 1 \end{bmatrix};$$

and

(a) for a P wave at the initial point of the ray

$$\mathrm{NAMPL} = 2,$$
$$\mathrm{Y}(28) = 1, \qquad \mathrm{Y}(29) = 0;$$

(b) for an S wave at the initial point of the ray

$$\mathrm{NAMPL} = 8,$$
$$\mathrm{Y}(28) = 1, \qquad \mathrm{Y}(29) = 0,$$
$$\mathrm{Y}(30) = 0, \qquad \mathrm{Y}(31) = 0,$$
$$\mathrm{Y}(32) = 0, \qquad \mathrm{Y}(33) = 0,$$
$$\mathrm{Y}(34) = 1, \qquad \mathrm{Y}(35) = 0.$$

The initial values of the auxiliary quantities defined in Section 5.2.2 are

$$\mathrm{YY}(1) = 0;$$
$$\mathrm{YY}(2) = \mathrm{UEB},$$

which is the value specified in the input data (see Section 5.6);

$$\mathrm{YY}(3) = 0, \qquad \mathrm{YY}(4) = 0, \qquad \mathrm{YY}(5) = 0;$$
$$\mathrm{IY}(1) = 27 + \mathrm{NAMPL},$$

where NAMPL is given above;

$$\text{IY}(2) = 0, \qquad \text{IY}(3) = 0;$$

$$\text{IY}(4) = \text{ISB1},$$

where ISB1 is the index of the simple block in which the initial point of the ray is situated;

$$\text{IY}(5) = \text{ICB1},$$

where ICB1 is the index of the complex block in which the initial point of the ray is situated, together with sign + for a P wave and − for an S wave;

$$\text{IY}(6) \quad \text{undefined} \quad (\text{e.g. IY}(6) = 0);$$

$$\text{IY}(7) \quad \text{undefined} \quad (\text{e.g. IY}(7) = 0);$$

$$\text{IY}(8) \quad \text{undefined} \quad (\text{e.g. IY}(8) = 0);$$

$$\text{IY}(9) = 0, \qquad \text{IY}(10) = 0, \qquad \text{IY}(11) = 0, \qquad \text{IY}(12) = 0.$$

7 APPLICATIONS AND PROCESSING OF THE RESULTS OF COMPLETE RAY TRACING

Using the complete ray-tracing procedure described here, we determine and store various quantities of great seismological importance along the ray and at intersections of the ray with some selected surfaces. Moreover, additional routines can be used to modify and/or considerably generalize the initial conditions used in the procedure and to process the results. In this way, the complete ray-tracing procedure can be used as a basic procedure in many program packages dealing with high-frequency seismic body waves propagating in complex 3D laterally varying layered and block structures. In this section we shall briefly describe several such possibilities. We believe that complete ray tracing will find many other important seismological applications, in addition to those listed in this section, in the near future.

We denote here the computed ray by Ω, the initial point of the ray by 0_0, and another point of the ray Ω by 0_s. We shall also consider a point S situated close to the point 0_s, and a point 0 close to 0_0 (see Fig. 7). The general coordinates of these points are denoted by $x^i(0_0)$, $x^i(0_s)$, $x^i(S)$ and $x^i(0)$ respectively. The coordinates of the initial point 0_0 are specified in the initial conditions, see YI(3), . . . , YI(5) in Section 6.1, and the coordinates of the point 0_s are obtained by the complete ray tracing, see Y(3), . . . , Y(5) in Section 5.2.1.

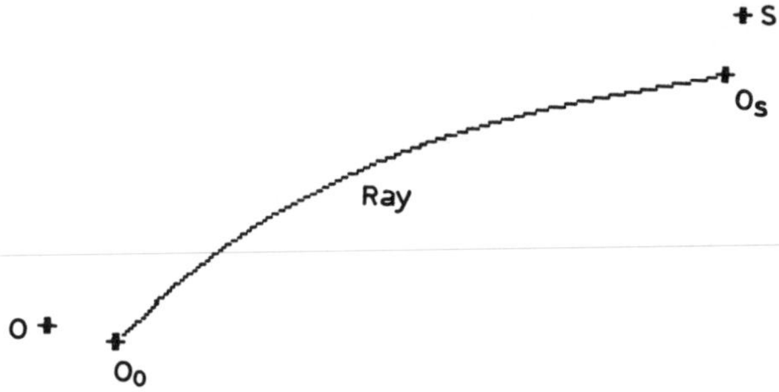

Fig. 7

7.1 Travel time. Imaginary travel time

We denote the initial travel time by $\tau_0 = \tau(0_0)$, and the initial imaginary travel time (i.e. the imaginary part of the complex-valued travel time) by $\tau^{\mathrm{IM}}(0_0)$. They are specified by YI(1) and YI(2) in the initial conditions (see Section 6.1). By complete ray tracing we obtain $\tau(0_s)$, the arrival time of the wave under consideration at 0_s, and $\tau^{\mathrm{IM}}(0_s)$, the imaginary part of the complex-valued arrival time, see Y(1) and Y(2) in Section 5.2.1. We introduce the travel time $\tau(0_s, 0_0)$ and the imaginary travel time $\tau^{\mathrm{IM}}(0_s, 0_0)$ from 0_0 to 0_s along Ω by the relations

$$\left.\begin{aligned} &\tau(0_s, 0_0) = \tau(0_s) - \tau(0_0), \\ &\tau^{\mathrm{IM}}(0_s, 0_0) = \tfrac{1}{2}t^*(0_s, 0_0) = \tau^{\mathrm{IM}}(0_s) - \tau^{\mathrm{IM}}(0_0). \end{aligned}\right\} \tag{7.1}$$

Instead of $\tau^{\mathrm{IM}}(0_s, 0_0)$, we shall also use $\frac{1}{2}t^*(0_s, 0_0)$, and call t^* the "t-star". The importance of the travel time $\tau(0_s, 0_0)$ is obvious; the quantity t^* plays an important role in the evaluation of the amplitudes of waves propagating in dissipative media (see Section 7.18).

7.2 Slowness vector. First partial derivatives of the travel-time field

At the initial point 0_0 the covariant components of the slowness vector $p_i(0_0)$ are given by YI(6), . . . , YI(8) (see Section 6.1). At 0_s they are obtained by complete ray tracing, see Y(6), . . . , Y(8) in Section 5.2.1. In both cases they represent the first partial derivatives of the travel-time field with respect to the general coordinates x^i:

$$p_i(0_0) = \left[\frac{\partial\tau(P)}{\partial x^i}\right]_{P=0_0}, \qquad p_i(0_s) = \left[\frac{\partial\tau(P)}{\partial x^i}\right]_{P=0_s}. \tag{7.2}$$

Thus $p_i(0_0)$ can be used in the hypocentre location procedures, as $-p_i(0_0)$ represent the partial derivatives of the travel time $\tau(0_s, 0_0)$ with respect to the coordinates of the source 0_0. Equations (7.2) can be also used in the calculation of the paraxial travel times (see Section 7.10).

From the covariant components $p_i(0_0)$ and $p_i(0_s)$, we can easily obtain the contravariant components using the relations

$$p^i(0_0) = G^{ik}(0_0)p_k(0_0), \qquad p^i(0_s) = G^{ik}(0_s)p_k(0_s). \tag{7.3}$$

7.3 Vector basis of the ray-centred coordinate system

The basis vectors of the ray-centred coordinate system $\boldsymbol{e}_1$, $\boldsymbol{e}_2$, $\boldsymbol{e}_3 = \boldsymbol{t}$ (also called the polarization vectors) are easily evaluated from the quantities obtained by complete ray tracing.

The unit vector $\boldsymbol{t}$ is tangent to the ray, so it is immediately obtained from the slowness vector:

$$\left.\begin{aligned} t_i(0_0) &= v(0_0)p_i(0_0), \qquad & t_i(0_s) &= v(0_s)p_i(0_s), \\ t^i(0_0) &= v(0_0)p^i(0_0), \qquad & t^i(0_s) &= v(0_s)p^i(0_s). \end{aligned}\right\} \tag{7.4}$$

The triplet $\boldsymbol{e}_1$, $\boldsymbol{e}_2$, $\boldsymbol{e}_3 = \boldsymbol{t}$ is mutually perpendicular at any point of the ray. The initial unit vectors $\boldsymbol{e}_1(0_0)$ and $\boldsymbol{e}_2(0_0)$ satisfying this property may be chosen arbitrarily.

We denote by $H^{\mathrm{U}}_{ij}(0_0)$ the ith covariant component of the initial polarization vector $\boldsymbol{e}_j(0_0)$, and, similarly, by $H^{\mathrm{U}}_{ij}(0_s)$ the ith covariant component of the polarization vector $\boldsymbol{e}_j(0_s)$. We assume that the vector H^{U}_{i3} is perpendicular to the wavefront. In this section we shall call the following choice of initial polarization vectors the *intrinsic choice*:

$$H^{\mathrm{U}}_{ij}(0_0) = H_{ij}(0_0), \tag{7.5}$$

where

$$\left.\begin{aligned} &H_{i3}(0_0) = t_i(0_0), \\ &H_{i1}(0_0) = \mathrm{YI}(i+8), \\ &H_{i2}(0_0) = \varepsilon_{ijk}G^{jm}(0_0)H_{m3}(0_0)G^{kn}(0_0)H_{n1}(0_0)[\det G^{rs}(0_0)]^{-1/2}. \end{aligned}\right\} \tag{7.6}$$

The intrinsic choice of polarization vectors is implicitly used in the procedure of complete ray tracing proposed here. Any general choice $H^{\mathrm{U}}_{ij}(0_0)$ of the initial polarization vectors $\boldsymbol{e}_j(0_0)$ will be called here the *user's choice*. Of course, the matrix $H^{\mathrm{U}}_{ij}(0_0)$ must be specified by the user; we assume here that it is known.

At the point 0_s, the covariant components $H^{\mathrm{U}}_{ij}(0_s)$ of the polarization vectors $\boldsymbol{e}_j(0_s)$ are given by the relation

$$H^{\mathrm{U}}_{ij}(0_s) = H_{ik}(0_s)H_{mk}(0_0)G^{mn}(0_0)H^{\mathrm{U}}_{nj}(0_0), \tag{7.7}$$

where

$$\left.\begin{aligned} &H_{i3}(0_s) = t_i(0_s),\\ &H_{i1}(0_s) = \mathrm{Y}(i+8),\\ &H_{i2}(0_s) = \varepsilon_{ijk}G^{jm}(0_s)H_{m3}(0_s)G^{kn}(0_s)H_{n1}(0_s)[\det G^{rs}(0_s)]^{-1/2}.\end{aligned}\right\} \tag{7.8}$$

Let us conclude that for any user's choice of the initial polarization vectors at 0_0 we can evaluate the polarization vectors at 0_s using Y(6), . . . , Y(11) and the above equations.

For the intrinsic choice (7.5) of initial polarization vectors, (7.7) simplifies to

$$H^{\mathrm{U}}_{ij}(0_s) = H_{ij}(0_s). \tag{7.9}$$

Note that the quantities H^{U}_{ij} represent the elements of the 3×3 transformation matrix from the user's ray-centred to the general coordinate system. Similarly, H_{ij} represent the elements of the 3×3 transformation matrix from the intrinsic ray-centred to the general coordinate system.

The 3×3 transformation matrix $\hat{\boldsymbol{H}}^{\mathrm{UI}}$ from the intrinsic ray-centred to the user's ray-centred coordinate system,

$$H^{\mathrm{UI}}_{jk} = H^{\mathrm{U}}_{mj}(0_0)G^{mn}(0_0)H_{nk}(0_0), \tag{7.10}$$

is constant along the ray.

7.4 Ray propagator matrix

The elements of the 4×4 propagator matrix $\boldsymbol{\Pi}(0_s, 0_0)$ at 0_s are stored in Y(12), . . . , Y(27) (see Section 5.2.1). Remember that $\boldsymbol{\Pi}(0_0, 0_0) = \boldsymbol{I}$, where $\boldsymbol{I}$ is the 4×4 identity matrix (see Section 6.2). The ray propagator matrix is expressed in the intrinsic ray-centred coordinates.

We introduce the 2×2 real-valued matrices $\boldsymbol{Q}_1(0_s, 0_0)$, $\boldsymbol{Q}_2(0_s, 0_0)$, $\boldsymbol{P}_1(0_s, 0_0)$ and $\boldsymbol{P}_2(0_s, 0_0)$ by the relation

$$\boldsymbol{\Pi}(0_s, 0_0) = \begin{bmatrix} \boldsymbol{Q}_1(0_s, 0_0) & \boldsymbol{Q}_2(0_s, 0_0) \\ \boldsymbol{P}_1(0_s, 0_0) & \boldsymbol{P}_2(0_s, 0_0) \end{bmatrix}. \tag{7.11}$$

Here $\boldsymbol{Q}_1(0_s, 0_0)$ and $\boldsymbol{P}_1(0_s, 0_0)$ are the matrix solutions of the dynamic ray-tracing system (5.15), with the telescopic (plane-wavefront) initial conditions at 0_0. Similarly, $\boldsymbol{Q}_2(0_s, 0_0)$ and $\boldsymbol{P}_2(0_s, 0_0)$ correspond to the point-source initial conditions at 0_0.

Note that the inverse of the ray propagator matrix is given by

$$\boldsymbol{\Pi}(0_0, 0_s) = \boldsymbol{\Pi}^{-1}(0_s, 0_0) = \begin{bmatrix} \boldsymbol{P}_2^{\mathrm{T}}(0_s, 0_0) & -\boldsymbol{Q}_2^{\mathrm{T}}(0_s, 0_0) \\ -\boldsymbol{P}_1^{\mathrm{T}}(0_s, 0_0) & \boldsymbol{Q}_1^{\mathrm{T}}(0_s, 0_0) \end{bmatrix}, \tag{7.12}$$

which follows from (5.11). Thus, using the results of complete ray tracing from 0_0 to 0_s, we can also determine the inverse ray propagator matrix from 0_s to 0_0.

7.5 Matrix of geometrical spreading $\boldsymbol{Q}$

The 2×2 matrix of geometrical spreading

$$Q_{AB} = \left[\frac{\partial q^A}{\partial \gamma^B}\right]_{q^1=q^2=0} \tag{7.13}$$

represents the transformation matrix from ray coordinates (γ^1, γ^2) to the user's ray-centred coordinates (q^1, q^2). The general relation for the matrix of geometrical spreading at the point 0_s is

$$\boldsymbol{Q}(0_s) = \boldsymbol{H}^{\mathrm{UI}}\{\boldsymbol{Q}_1(0_s, 0_0)(\boldsymbol{H}^{\mathrm{UI}})^{\mathrm{T}}\boldsymbol{Q}(0_0) + \boldsymbol{Q}_2(0_s, 0_0)(\boldsymbol{H}^{\mathrm{UI}})^{\mathrm{T}}\boldsymbol{P}(0_0)\}. \tag{7.14}$$

The transformation matrix $\boldsymbol{P}$ from the ray coordinates to the user's ray-centred components of the slowness vector is defined in Section 7.6.

Equation (7.14) is valid for any system of rays containing the ray Ω and for any parametrization of the rays, not only for the initial conditions and the ray parametrization specified before the complete ray tracing of the ray Ω (see Section 6). For this reason, we use here the notation $\boldsymbol{Q}$ instead of $\boldsymbol{Q}^{\mathrm{R}}$, see (5.13). Note that the change from the parameters γ^A to the parameters $\bar{\gamma}^A$ (for a fixed system of rays) represents the multiplication of both left- and right-hand sides of (7.14) by a constant 2×2 matrix

$$C_{AB} = \frac{\partial \gamma^A}{\partial \bar{\gamma}^B}. \tag{7.15}$$

Considering the system of rays and its parametrization specified for the complete tracing of the ray Ω (see Section 6), we can evaluate the matrix of geometrical spreading at the initial point of the ray in the user's ray-centred coordinates as

$$\boldsymbol{Q}(0_0) = \boldsymbol{H}^{\mathrm{UI}}\boldsymbol{Q}^{\mathrm{INIT}}, \tag{7.16}$$

where the components of the 2×2 matrix $\boldsymbol{Q}^{\mathrm{INIT}}$ are stored in YI(12), YI(13), YI(16), YI(17). Similarly,

$$\boldsymbol{P}(0_0) = \boldsymbol{H}^{\mathrm{UI}}\boldsymbol{P}^{\mathrm{INIT}}, \tag{7.17}$$

where $\boldsymbol{P}^{\mathrm{INIT}}$ is stored in YI(14), YI(15), YI(18), YI(19) (see Section 6.1). Then (7.14) reads

$$\boldsymbol{Q}(0_s) = \boldsymbol{H}^{\mathrm{UI}}\{\boldsymbol{Q}_1(0_s, 0_0)\boldsymbol{Q}^{\mathrm{INIT}} + \boldsymbol{Q}_2(0_s, 0_0)\boldsymbol{P}^{\mathrm{INIT}}\}. \qquad (7.18)$$

7.6 Transformation matrix *P*

The 2×2 matrix $\boldsymbol{P}$ with components

$$P_{AB} = \left[\frac{\partial p_A^{(q)}}{\partial \gamma^B}\right]_{q^1, q^2 = 0}$$

represents the transformation matrix from ray coordinates (γ^1, γ^2) to the user's ray-centred components $p_1^{(q)} = \partial\tau/\partial q^1$, $p_2^{(q)} = \partial\tau/\partial q^2$ of the slowness vector. The general relation for the matrix $\boldsymbol{P}$ is

$$\boldsymbol{P}(0_s) = \boldsymbol{H}^{\mathrm{UI}}\{\boldsymbol{P}_1(0_s, 0_0)(\boldsymbol{H}^{\mathrm{UI}})^{\mathrm{T}}\boldsymbol{Q}(0_0) + \boldsymbol{P}_2(0_s, 0_0)(\boldsymbol{H}^{\mathrm{UI}})^{\mathrm{T}}\boldsymbol{P}(0_0)\}. \quad (7.19)$$

The discussion of this equation and the meaning of the individual quantities is similar to those in Section 7.5.

For the system of rays and its parametrization specified for complete ray tracing along the ray Ω, $\boldsymbol{Q}(0_0) = \boldsymbol{H}^{\mathrm{UI}}\boldsymbol{Q}^{\mathrm{INIT}}$, $\boldsymbol{P}(0_0) = \boldsymbol{H}^{\mathrm{UI}}\boldsymbol{P}^{\mathrm{INIT}}$ (see Section 7.5), (7.19) yields

$$\boldsymbol{P}(0_s) = \boldsymbol{H}^{\mathrm{UI}}\{\boldsymbol{P}_1(0_s, 0_0)\boldsymbol{Q}^{\mathrm{INIT}} + \boldsymbol{P}_2(0_s, 0_0)\boldsymbol{P}^{\mathrm{INIT}}\}. \qquad (7.20)$$

7.7 Geometrical spreading

By the geometrical spreading $J(0_s)$, we understand here the quantity

$$J(0_s) = |\det \boldsymbol{Q}(0_s)|. \qquad (7.21a)$$

The geometrical spreading can be evaluated if the matrix of geometrical spreading $\boldsymbol{Q}$ is known. The geometrical spreading does not depend on the initial orientation of the polarization vectors $\boldsymbol{e}_1(0_0)$ and $\boldsymbol{e}_2(0_0)$.

For a point source at 0_0, (7.21a) yields

$$J(0_s) = |\det \boldsymbol{Q}_2(0_s, 0_0)|\,|\det \boldsymbol{P}(0_0)|. \qquad (7.21b)$$

The geometrical spreading is of great importance in the evaluation of ray amplitudes: the ray amplitudes are inversely proportional to the square root of geometrical spreading.

7.8 Matrix of second derivatives of the travel-time field

The elements of the 2×2 matrix $\boldsymbol{M}$ of the second derivatives of the travel-time field with respect to the user's ray-centred coordinates (q^1, q^2), $M_{JK} = [\partial^2\tau/\partial q^J\,\partial q^K]_{q^1=q^2=0}$, can be evaluated by $\boldsymbol{M} = \boldsymbol{P}\boldsymbol{Q}^{-1}$ both at the initial point 0_0 and at 0_s. Using (7.14) and (7.19), we obtain the continuation formula for $\boldsymbol{M}$ as

$$\begin{aligned}\boldsymbol{M}(0_s) = \boldsymbol{H}^{\mathrm{UI}}&\{\boldsymbol{P}_1(0_s, 0_0) + \boldsymbol{P}_2(0_s, 0_0)(\boldsymbol{H}^{\mathrm{UI}})^{\mathrm{T}}\boldsymbol{M}(0_0)\boldsymbol{H}^{\mathrm{UI}}\}\\ &\times\{\boldsymbol{Q}_1(0_s, 0_0) + \boldsymbol{Q}_2(0_s, 0_0)(\boldsymbol{H}^{\mathrm{UI}})^{\mathrm{T}}\boldsymbol{M}(0_0)\boldsymbol{H}^{\mathrm{UI}}\}^{-1}(\boldsymbol{H}^{\mathrm{UI}})^{\mathrm{T}}. \end{aligned}\tag{7.22}$$

Let us now consider a local Cartesian coordinate system (y^1, y^2, y^3) at 0_s, with $y^1 = q^1$, $y^2 = q^2$ and with y^3 being the length coordinate measured along the tangent to the ray Ω at 0_s. Then the 3×3 matrix $\hat{\boldsymbol{M}}(0_s)$, with elements $M_{ij} = [\partial^2\tau/\partial y^i\,\partial y^j]_{0_s}$, is given by

$$\hat{\boldsymbol{M}}(0_s) = \begin{bmatrix} M_{11}(0_s) & M_{12}(0_s) & -V_1 v^{-2}\\ M_{12}(0_s) & M_{22}(0_s) & -V_2 v^{-2}\\ -V_1 v^{-2} & -V_2 v^{-2} & -V_3 v^{-2}\end{bmatrix}, \tag{7.23}$$

where

$$V_j = \frac{\partial v}{\partial x^m} G^{mn} H_{nj} \tag{7.24}$$

is the velocity gradient in the local Cartesian coordinates y^j at 0_s.

Finally, we introduce the 3×3 matrix $\hat{\boldsymbol{N}}(0_s)$, with elements $N_{ij}(0_s) = [\partial^2\tau/\partial x^i\,\partial x^j]_{0_s}$ given by

$$N_{jk}(0_s) = H_{jm}(0_s)H_{kn}(0_s)M_{mn}(0_s) + \Gamma^s_{jk}(0_s)p_s(0_s). \tag{7.25}$$

7.9 Curvature of the wavefront

The 2×2 matrix of the curvature of the wavefront $\boldsymbol{K}$ in the user's ray-centred coordinate system can be simply evaluated from the 2×2 matrix $\boldsymbol{M}$ using the relation

$$\boldsymbol{K} = v\boldsymbol{M} = v\boldsymbol{P}\boldsymbol{Q}^{-1}. \tag{7.26}$$

Thus we have

$$\boldsymbol{K}(0_0) = v(0_0)\boldsymbol{M}(0_0), \qquad \boldsymbol{K}(0_s) = v(0_s)\boldsymbol{M}(0_s).$$

Similarly, the 2×2 matrix of the radii of curvature $\boldsymbol{R}$ is given by

$$\boldsymbol{R} = \boldsymbol{K}^{-1} = v^{-1}\boldsymbol{M}^{-1} = v^{-1}\boldsymbol{QP}^{-1}. \tag{7.27}$$

Using $\boldsymbol{K}$ (or $\boldsymbol{R}$), we can determine the main geometrical characteristics of the wavefront (the principal curvatures, the principal directions, the wavefront ellipse or the wavefront hyperbola, etc.). For details see Červený and Pšenčík (1983b).

7.10 Paraxial travel times

By the paraxial travel times, we understand the travel times in the vicinity of the central ray Ω.

Let us consider a single initial point at 0_0 and denote

$$x^k(S, 0_s) = x^k(S) - x^k(0_s), \qquad x^k(0, 0_0) = x^k(0) - x^k(0_0). \tag{7.28}$$

Then the results of complete ray tracing from 0_0 to 0_s can be used to find not only the travel time from 0_0 to 0_s, but also approximate expressions for the travel time from 0_0 to S or even for the travel time from 0 to S if S is close to 0_s and 0 is close to 0_0:

$$\tau(S, 0_0) = \tau(0_s, 0_0) + p_k(0_s)x^k(S, 0_s) + \tfrac{1}{2}x^m(S, 0_s)x^n(S, 0_s)N_{mn}(0_s, 0_0). \tag{7.29}$$

Similarly, we obtain (see Červený *et al.*, 1984)

$$\begin{aligned} \tau(S, 0) = {} & \tau(0_s, 0_0) + x^k(S, 0_s)p_k(0_s) - x^k(0, 0_0)p_k(0_0) \\ & + \tfrac{1}{2}x^m(S, 0_s)N_{mn}(0_s, 0_0)x^n(S, 0_s) \\ & - \tfrac{1}{2}x^m(0, 0_0)N_{mn}(0_0, 0_s)x^n(0, 0_0) \\ & - x^j(0, 0_0)H_{jJ}(0_0)[Q_2^{-1}(0_s, 0_0)]_{JK}H_{kK}(0_s)x^k(S, 0_s). \end{aligned} \tag{7.30}$$

In (7.29) and (7.30) $N_{mn}(0_s, 0_0)$ (respectively $N_{mn}(0_0, 0_s)$) are elements of the 3×3 matrix $\hat{\boldsymbol{N}}(0_s, 0_0)$ (respectively $\hat{\boldsymbol{N}}(0_0, 0_s)$) of the second derivatives of the travel time field with respect to the general coordinates x^i at the point 0_s (respectively 0_0) due to a point source at 0_0 (respectively 0_s). They are given by (7.25) and (7.23), where M_{JK} are replaced by

$$\boldsymbol{M}(0_s, 0_0) = \boldsymbol{P}_2(0_s, 0_0)(\boldsymbol{Q}_2(0_s, 0_0))^{-1} \tag{7.31a}$$

(respectively

$$\boldsymbol{M}(0_0, 0_s) = -(\boldsymbol{Q}_2(0_s, 0_0))^{-1}\boldsymbol{Q}_1(0_s, 0_0). \qquad (7.31b))$$

Equations (7.29) and (7.30) are, of course, only approximate; they are valid for small $x^i(S, 0_s)$ and $x^i(0, 0_0)$.

7.11 Paraxial rays

Paraxial rays are rays situated *close to the central ray Ω.* If the ray propagator matrix is known along Ω then the paraxial rays can be approximately evaluated analytically.

Let us first consider the simplest problem of *initial-value paraxial ray tracing*. We assume that the initial point 0 of the paraxial ray is situated in a plane perpendicular to Ω at 0_0 and denote its user's ray-centred coordinates by $(q^1(0), q^2(0))$. Similarly, we assume that the point S is situated in the plane perpendicular to Ω at 0_s and denote its user's ray-centred coordinates by $(q^1(S), q^2(S))$. We denote the user's ray-centred components of the slowness vector at 0 by $p_1^{(q)}(0)$ and $p_2^{(q)}(0)$, and those at S by $p_1^{(q)}(S)$ and $p_2^{(q)}(S)$. We also denote

$$\left.\begin{aligned} \boldsymbol{W}(0_0) &= [q^1(0) \quad q^2(0) \quad p_1^{(q)}(0) \quad p_2^{(q)}(0)]^{\mathrm{T}}, \\ \boldsymbol{W}(0_s) &= [q^1(S) \quad q^2(S) \quad p_1^{(q)}(S) \quad p_2^{(q)}(S)]^{\mathrm{T}}. \end{aligned}\right\} \tag{7.32}$$

Then the dynamic ray-tracing system yields the following analytical solution for the initial conditions given by $\boldsymbol{W}(0_0)$:

$$\boldsymbol{W}(0_s) = \boldsymbol{U}\boldsymbol{\Pi}(0_s, 0_0)\boldsymbol{U}^{\mathrm{T}}\boldsymbol{W}(0_0), \tag{7.33}$$

where $\boldsymbol{U}$ is the 4×4 matrix given by

$$\boldsymbol{U} = \begin{pmatrix} \boldsymbol{H}^{\mathrm{UI}} & \boldsymbol{0} \\ \boldsymbol{0} & \boldsymbol{H}^{\mathrm{UI}} \end{pmatrix}. \tag{7.34}$$

Thus we determine the user's ray-centred coordinates of the point S situated on the paraxial ray by simple matrix multiplication. In addition, we obtain also the user's ray-centred components of the slowness vector at S.

Let us now consider a more complex situation: the initial point 0 of the paraxial ray is not, in general, situated in the plane perpendicular to Ω at 0_0, but it is close to 0_0. Moreover, its position is specified in the general coordinates $x^i(0)$. The initial slowness vector $\boldsymbol{p}(0)$ is specified by its covariant components $p_i(0)$ in the general coordinate system. We assume that both $x^i(0)$ and $p_i(0)$ are selected in such a way that the whole paraxial ray is situated close to the central ray Ω. We wish to determine the analytic equation for the paraxial ray in the vicinity of the point 0_s. We denote a point situated on the paraxial ray close to 0_s by S, and $x^i(S, 0_s) = x^i(S) - x^i(0_s)$. Then we obtain the following parametric equation for $x^i(S, 0_s)$:

$$x^i(S, 0_s) = G^{ik}(0_s)H_{kn}(0_s)q^n(S), \tag{7.35}$$

where q^3 is a (small) free parameter, and $q^1(S)$ and $q^2(S)$ are given by

$$q^K(S) = (Q_2(0_s, 0_0))_{KL} H_{mL}(0_0) G^{mn}(0_0) \times \{p_n(0) - p_n(0_0) - N_{nj}(0_0, 0_s) x^j(0, 0_0)\}. \qquad (7.36)$$

The parametric equation (7.35) with (7.36) represents a straight-line approximation of the paraxial ray in the vicinity of the point 0_s. All the symbols in (7.36) have the same meaning as in (7.30) in Section 7.10.

It is simple to see that the point S is situated in the plane perpendicular to Ω at 0_s for $q^3 = 0$. For $q^3 > 0$ it is shifted "behind" this plane, and for $q^3 < 0$ "in front of" the plane.

7.12 Two-point ray tracing for paraxial rays

Assume that we know the position of the points 0 and S and we wish to find the ray that passes through these two points. This is the classical two-point ray-tracing problem. For paraxial rays this problem can be solved analytically. This means that we can find analytically approximate expressions for the slowness vector $\boldsymbol{p}(0)$ and/or $\boldsymbol{p}(S)$. The slowness vector determines the initial direction of the ray.

We again denote the general coordinates of S and 0 by $x^i(S)$ and $x^i(0)$ and use the notation (7.28). Differentiating (7.30), we obtain

$$\left.\begin{aligned} p_j(0) &= p_j(0_0) + N_{jk}(0_0, 0_s) x^k(0, 0_0) \\ &\quad + H_{jK}(0_0)(Q_2^{-1}(0_s, 0_0))_{KL} H_{mL}(0_s) x^m(S, 0_s), \\ p_j(S) &= p_j(0_s) + N_{jk}(0_s, 0_0) x^k(S, 0_s) \\ &\quad + H_{jK}(0_s)(Q_2^{-1}(0_s, 0_0))_{LK} H_{mL}(0_0) x^m(0, 0_0). \end{aligned}\right\} \qquad (7.37)$$

All the symbols in (7.37) have the same meanings as in Section 7.10.

The solution (7.37) of the two-point ray-tracing problem is only approximate. The accuracy is, of course, higher for paraxial rays situated closer to the central ray Ω. The solution, however, can be repeated iteratively. An iterative loop in which the initial slowness vector at 0 is changed at each step can be used to find the ray that passes through the point S with the required accuracy.

7.13 Fresnel volumes

Fresnel volumes represent some vicinity of the central ray Ω that actually influences the wavefield at the receiver point 0_s. The Fresnel volumes are sometimes called the "*physical rays*" (see Kravtsov and Orlov, 1980). Let us again assume that a point source is situated at 0_0 and the receiver at 0_s. Then the Fresnel volume is composed of points C for which the following condition holds:

$$|\tau(C, 0_0) + \tau(0_s, C) - \tau(0_s, 0_0)| < \tfrac{1}{2}T. \tag{7.38}$$

Here T is the period of the harmonic wave under consideration.

The Fresnel volume can be simply computed approximately by some algebraic manipulation with the propagator matrices. Let us consider a point 0_F situated on the ray Ω between 0_0 and 0_s. Any point F in the vicinity of the point 0_F is situated in the Fresnel volume if and only if

$$\begin{aligned}|[x^j(F) - x^j(0_F)]H_{jM}(0_F)[M_{MN}(0_F, 0_0)\\ - M_{MN}(0_F, 0_s)]H_{kN}(0_F)[x^k(F) - x^k(0_F)]| < T.\end{aligned} \tag{7.39}$$

Here the matrices

$$\boldsymbol{M}(0_F, 0_0) = \boldsymbol{P}_2(0_F, 0_0)(\boldsymbol{Q}_2(0_F, 0_0))^{-1} \tag{7.40a}$$

and

$$\begin{aligned}\boldsymbol{M}(0_F, 0_s) = [-\boldsymbol{P}_1(0_F, 0_0)\boldsymbol{Q}_2^{\mathrm{T}}(0_s, 0_0) + \boldsymbol{P}_2(0_F, 0_0)\boldsymbol{Q}_1^{\mathrm{T}}(0_s, 0_0)]\\ \times[-\boldsymbol{Q}_1(0_F, 0_0)\boldsymbol{Q}_2^{\mathrm{T}}(0_s, 0_0) + \boldsymbol{Q}_2(0_F, 0_0)\boldsymbol{Q}_1^{\mathrm{T}}(0_s, 0_0)]^{-1},\end{aligned} \tag{7.40b}$$

are composed of the 2×2 submatrices of the ray propagator matrices

$$\boldsymbol{\Pi}(0_F, 0_0) = \begin{bmatrix} \boldsymbol{Q}_1(0_F, 0_0) & \boldsymbol{Q}_2(0_F, 0_0) \\ \boldsymbol{P}_1(0_F, 0_0) & \boldsymbol{P}_2(0_F, 0_0) \end{bmatrix} \tag{7.41}$$

and $\boldsymbol{\Pi}(0_s, 0_0)$ (see (7.11)). Both the matrices $\boldsymbol{\Pi}(0_F, 0_0)$ and $\boldsymbol{\Pi}(0_s, 0_0)$ are obtained by complete ray tracing along Ω from 0_0 to 0_s, passing through 0_F.

Equation (7.39) can be modified in many ways. Similar equations can be found for a point 0_F situated at a curved interface Σ. The intersection of the Fresnel volume with Σ will usually be represented by an ellipse. Only the part of the interface inside the ellipse will influence the wavefield at 0_s considerably.

7.14 Phase shift due to caustics. KMAH index

To evaluate the complex-valued amplitude of the displacement vector of the seismic body wave under consideration, it is necessary to take into account the phase shift due to caustics. In the algorithm for complete ray tracing proposed here the amplitudes are *automatically* modified by the relevant phase shift at any caustic point (see Section 5.8.3f). It may, however, still be of some interest to know the complete phase shift $\delta\tau(0_s, 0_0)$ due to caustic points along Ω between 0_0 and 0_s. The phase shift $\delta\tau(0_s, 0_0)$ can be expressed in terms of the index $k(0_s, 0_0)$ of the ray trajectory Ω between 0_0 and 0_s by

$$\delta\tau(0_s, 0_0) = -\tfrac{1}{2}\pi k(0_s, 0_0). \tag{7.42}$$

The *index of the ray trajectory* $k(0_s, 0_0)$, also known as the *KMAH index*, is defined as the sum of the caustic points along Ω between 0_0 and 0_s; caustic points of second order (point caustics) being considered twice in this sum. We recall that $\det \boldsymbol{Q}^{\mathrm{R}} = 0$, $\boldsymbol{Q}^{\mathrm{R}} \neq \boldsymbol{0}$ at a caustic point of first order (simple caustic), and that $\boldsymbol{Q}^{\mathrm{R}} = \boldsymbol{0}$ at a caustic point of second order (point caustic). Here $\boldsymbol{Q}^{\mathrm{R}}$ is the matrix of geometrical spreading $\boldsymbol{Q}$ corresponding to the initial conditions (7.16) and (7.17). In the proposed procedure the KMAH index is stored in IY(12).

Note that the phase shift due to caustics is the only quantity evaluated in complete ray tracing that depends on the initial matrices $\boldsymbol{Q}^{\mathrm{INIT}}$ and $\boldsymbol{P}^{\mathrm{INIT}}$ (see Section 5.8.3f).

The existence, position, type and number of caustic points, and consequently the KMAH index, are changed if $\boldsymbol{Q}^{\mathrm{INIT}}$ and $\boldsymbol{P}^{\mathrm{INIT}}$ are changed.

7.15 Ray amplitudes

In the algorithm for complete ray tracing the vectorial complex-valued *reduced* amplitudes $A_i(0_s)$ given by (5.19) are evaluated at 0_s. They are expressed in the ray-centred coordinate system at 0_s. For convenience, the 3×3 matrix $\hat{\boldsymbol{A}}(0_s)$ is considered in this section. The element $A_{ij}(0_s)$ represents the ray-centred component in the direction of $\boldsymbol{e}_i(0_s)$ of the vectorial complex-valued *reduced* amplitude at 0_s corresponding to the unit initial vectorial complex-valued reduced amplitude at 0_0, polarized in the direction of $\boldsymbol{e}_j(0_0)$ (intrinsic choice). The non-zero quantities $\mathrm{Re}\,(A_{ij}(0_s))$ and $\mathrm{Im}\,(A_{ij}(0_s))$ are stored in Y(28), . . . , Y(27 + NAMPL), where NAMPL = 2, 4 or 8, depending on the type of the wave at 0_0 and 0_s (see Section 5.2.1).

The matrix A_{ij} is constructed from the stored quantities Y(28), . . . , Y(NY), with NY = NAMPL + 27, in the following way:

(a) P wave at the initial point of the ray 0_0, P wave at the point 0_s (NAMPL = 2):

$$A_{ij} = \begin{bmatrix} 0 & 0 & 0 \\ 0 & 0 & 0 \\ 0 & 0 & \mathrm{Y}(28) + \mathrm{i}\mathrm{Y}(29) \end{bmatrix};$$

(b) P wave at the initial point of the ray 0_0, S wave at the point 0_s (NAMPL = 4):

$$A_{ij} = \begin{bmatrix} 0 & 0 & Y(28) + iY(29) \\ 0 & 0 & Y(30) + iY(31) \\ 0 & 0 & 0 \end{bmatrix};$$

(c) S wave at the initial point of the ray 0_0, P wave at the point 0_s (NAMPL = 4):

$$A_{ij} = \begin{bmatrix} 0 & 0 & 0 \\ 0 & 0 & 0 \\ \mathrm{Y}(28) + \mathrm{iY}(29) & \mathrm{Y}(30) + \mathrm{iY}(31) & 0 \end{bmatrix};$$

(d) S wave at the initial point of the ray 0_0, S wave at the point 0_s (NAMPL = 8):

$$A_{ij} = \begin{bmatrix} \mathrm{Y}(28) + \mathrm{iY}(29) & \mathrm{Y}(32) + \mathrm{iY}(33) & 0 \\ \mathrm{Y}(30) + \mathrm{iY}(31) & \mathrm{Y}(34) + \mathrm{iY}(35) & 0 \\ 0 & 0 & 0 \end{bmatrix}.$$

We denote by $U^j(0_s)$ the contravariant components of the non-reduced complex-valued ray amplitude in the *local recording coordinate system* z^m *at* 0_s. If the local recording coordinate system z^m is Cartesian then we need not distinguish between covariant and contravariant components of the ray amplitudes. The local recording coordinate system may, of course, coincide with the general coordinate system, the user's ray-centred coordinate system, etc. We shall specify the local recording coordinate system z^m in terms of the 3×3 transformation matrix:

$$Z_k{}^j(0_s) = \frac{\partial z^j}{\partial x^k}(0_s), \qquad Z^j{}_k(0_s) = \frac{\partial x^j}{\partial z^k}(0_s). \tag{7.43}$$

Likewise, we introduce the contravariant components $U^j(0_0)$ of the ray amplitude with respect to the local coordinate system z^j at the initial point 0_0 of the ray, and the relevant transformation matrix:

$$Z^k{}_j(0_0) = \frac{\partial x^k}{\partial z^j}(0_0), \qquad Z_k{}^j(0_0) = \frac{\partial z^j}{\partial x^k}(0_0). \tag{7.44}$$

The local coordinates at the initial point 0_0 and the local recording coordinates at 0_s need not be of the same type.

The complex-valued vectorial ray amplitude (including the phase shift due to caustics) at 0_s is given by

$$U^j(0_s) = \left[\frac{v(0_0)\rho(0_0)}{v(0_s)\rho(0_s)}\right]^{1/2} (J(0_s))^{-1/2} C^j{}_n(0_s, 0_0) a^n(0_0), \tag{7.45}$$

where

$$C^j{}_n(0_s, 0_0) = Z_k{}^j(0_s)G^{km}(0_s)H_{mu}(0_s)A_{ur}(0_s, 0_0)H_{sr}(0_0)Z^s{}_n(0_0), \quad (7.46)$$

$$a^n(0_0) = \lim_{0 \to 0_0} \{(J(0))^{1/2}U^n(0)\}. \quad (7.47)$$

The path for the evaluation of the limit is along the ray, against the direction of propagation. The geometrical spreading $J = |\det \boldsymbol{Q}|$ is defined in Section 7.7.

For a *point source* at 0_0, (7.45) can be simplified. We can write

$$(J(0_s))^{-1/2}a^n(0_0) = |\det \boldsymbol{Q}_2(0_s, 0_0)|^{-1/2}v(0_0)g^n(0_0), \quad (7.48)$$

where

$$g^n(0_0) = \lim_{0 \to 0_0} \{(q^3(0) - q^3(0_0))U^n(0)\} \quad (7.49)$$

is the *radiation pattern of the point source*. The path for the evaluation of the limit is the same as in the case of (7.47). Thus, for a point source at 0_0, we can write

$$U^j(0_s) = v(0_0)\left[\frac{v(0_0)\rho(0_0)}{v(0_s)\rho(0_s)}\right]^{1/2} |\det \boldsymbol{Q}_2(0_s, 0_0)|^{-1/2}C^j{}_n(0_s, 0_0)g^n(0_0). \quad (7.50)$$

The radiation pattern $g^n(0_0)$ is, of course, a function of the ray parameters γ^1 and γ^2: $g^n = g^n(0_0, \gamma^1, \gamma^2)$.

7.16 Paraxial ray approximation for the ray amplitudes

Owing to the curvature of the wavefront, the direction of the slowness vector changes in the neighbourhood of the ray Ω. As the displacement vector is parallel to the slowness vector for P waves, and perpendicular to it for S waves, the vectorial complex-valued amplitudes also vary in the vicinity of Ω. Let us consider a point S close to 0_s. Then we can write approximately

$$U^j(S) = Z_k{}^j(S)G^{km}(0_s)H_{mn}(0_s)B_{nr}(S, 0_s)H_{sr}(0_s)Z^s{}_t(0_s)U^t(0_s). \quad (7.51)$$

Here $B_{nr}(S, 0_s)$ are elements of the 3×3 *paraxial approximation matrix* $\hat{\boldsymbol{B}}(S, 0_s)$. This matrix can be evaluated from the quantities obtained by the complete ray tracing. It is given by

$$\hat{\boldsymbol{B}}(S, 0_s) = \begin{bmatrix} 1 & 0 & \theta_1(S, 0_s) \\ 0 & 1 & \theta_2(S, 0_s) \\ -\theta_1(S, 0_s) & -\theta_2(S, 0_s) & 1 \end{bmatrix}. \quad (7.52)$$

Here $\theta_i(S, 0_s)$ are components of a 3×1 column matrix $\hat{\boldsymbol{\theta}}(S, 0_s)$:

$$\theta_i(S, 0_s) = v(0_s)H_{ki}(0_s)G^{kl}(0_s)N_{ln}(0_s)x^n(S, 0_s), \tag{7.53}$$

where $N_{ln}(0_s)$ is given by (7.25). Equation (7.52) may also be written in the form

$$B_{ij}(S, 0_s) = \delta_{ij} + v(0_s)[H_{ki}(0_s)\delta_{j3} - H_{kj}(0_s)\delta_{i3}]G^{kl}(0_s)N_{ln}(0_s)x^n(S, 0_s). \tag{7.54}$$

7.17 Amplitudes at structural interfaces or on the Earth's surface

If the point 0_s is situated at the structural interface then the above expressions for the ray amplitudes must be modified. Together with the incident wave, there are also two reflected waves (reflected P, reflected S) and two transmitted waves (transmitted P, transmitted S). Thus, to compute the amplitudes at 0_s, not only the quantities corresponding to the incident wave, but also those corresponding to the two reflected waves or two transmitted waves, must be stored.

Such a possibility is considered in the algorithm for complete ray tracing proposed here (see Section 5.5.4).

Let us assume that the quantities corresponding to the incident, reflected P and reflected S waves are stored at the point 0_s. We denote the quantities corresponding to the incident wave, to the reflected P wave and to the reflected S wave at 0_s by the superscripts IW, RP and RS respectively. We now denote by $\hat{\boldsymbol{U}}^{\mathrm{IW}}(0_s)$, $\hat{\boldsymbol{U}}^{\mathrm{RP}}(0_s)$ and $\hat{\boldsymbol{U}}^{\mathrm{RS}}(0_s)$ the 3×1 column matrices of the components of the ray amplitudes at 0_s, expressed in local recording or general coordinates, corresponding to the three waves under consideration. Then we can write

$$\hat{\boldsymbol{U}}(0_s) = \hat{\boldsymbol{U}}^{\mathrm{IW}}(0_s) + \hat{\boldsymbol{U}}^{\mathrm{RP}}(0_s) + \hat{\boldsymbol{U}}^{\mathrm{RS}}(0_s). \tag{7.55}$$

On the opposite side of the interface,

$$\hat{\boldsymbol{U}}(0_s) = \hat{\boldsymbol{U}}^{\mathrm{TP}}(0_s) + \hat{\boldsymbol{U}}^{\mathrm{TS}}(0_s), \tag{7.56}$$

where TP and TS correspond to the transmitted P and S waves.

If the point 0_s is situated on the Earth's surface then it is also possible to use, instead of (7.55), standard conversion coefficients (see Červený *et al.*, 1977). Such a possiblility is also optionally considered in the proposed algorithm (see Section 5.5.4). Equations (7.55) or (7.56) are, however, fully general and applicable at any interface, not only on the Earth's surface.

7.18 Ray amplitudes in slightly dissipative media

The quantities computed by complete ray tracing can be used to evaluate the amplitudes of seismic body waves in slightly dissipative media, for both causal and non-causal absorption. For simplicity, we shall only work here in the frequency domain.

In slightly dissipative media we can consider a complex-valued velocity v_C with a small imaginary part, which is formally assumed to be of order ω^{-1} for $\omega \to \infty$ (the so-called Debye procedure). The imaginary part of the velocity is not considered in the evaluation of rays and geometrical spreading and in the computation of the reflection/transmission coefficients, but it leads to an amplitude decay factor,

$$A_d(0_s) = \exp\left[-\omega \int_{0_0}^{0_s} \mathrm{Im}\left(\frac{1}{v_C(s)}\right) \mathrm{d}s\right], \tag{7.57}$$

where the integral is taken along the ray Ω from 0_0 to 0_s.

The amplitude decay factor $A_d(0_s)$ can be simply evaluated analytically for non-causal and various types of causal absorption if the quantity $t^* = 2\mathrm{Y}(2)$ is known. This quantity, however, is evaluated by complete ray tracing. For *non-causal absorption* the amplitude decay factor is

$$A_d(0_s) = \exp(-\pi f t^*). \tag{7.58}$$

Similar equations for the amplitude decay factor for various *causal absorption* models (Futterman, Müller) can be found in Červený (1985a).

7.19 Displacement vector

In a perfectly elastic medium the contravariant components $u^j(0_s)$ of the displacement vector $\boldsymbol{u}(0_s)$ at the point 0_s are given by

$$u^j(0_s, \omega) = U^j(0_s) \exp[\mathrm{i}\omega\tau(0_s, 0_0)] \tag{7.59}$$

in the *frequency domain,* and by

$$u^j(0_s, t) = \mathrm{Re}\{U^j(0_s)\delta^{(A)}(t - \tau(0_s, 0_0))\} \tag{7.60}$$

in the *time domain.* Here ω is the circular frequency, $U^j(0_s)$ are given by (7.45) or (7.50), and $\tau(0_s, 0_0)$ by (7.1). The function $\delta^{(A)}(\xi)$ is the analytic unit impulse corresponding to the Dirac delta function $\delta(\xi): \delta^{(A)}(\xi) = \delta(\xi) - \mathrm{i}/\pi\xi$. The time factor $\exp(-\mathrm{i}\omega t)$ in (7.59) is omitted. The time dependence at 0_0 in (7.60) is assumed here to be of the form $\delta^{(A)}(t)$. The Fourier transform is assumed here to be of the form

$$\left.\begin{aligned} u^j(0_s, t) &= \frac{1}{\pi}\,\mathrm{Re}\int_0^{\infty} u^j(0_s, \omega)\mathrm{e}^{-\mathrm{i}\omega t}\,\mathrm{d}\omega, \\ u^j(0_s, \omega) &= \int_{-\infty}^{\infty} u^j(0_s, t)\mathrm{e}^{\mathrm{i}\omega t}\,\mathrm{d}t. \end{aligned}\right\} \tag{7.61}$$

Both expressions (7.59) and (7.60) can be simply modified to include effects of slightly dissipative media. For a non-causal absorption, this can be done using a complex-valued travel time $\tau(0_s, 0_0) + \frac{1}{2}\mathrm{i}t^*$ instead of the real-valued travel time $\tau(0_s, 0_0)$. The function $\delta^{(A)}(\xi)$ in (7.60) is then given by the relation

$$\delta^{(A)}(\xi) = -\frac{\mathrm{i}}{\pi\xi} = \mathrm{Im}\left(\frac{1}{\pi\xi}\right) - \mathrm{i}\,\mathrm{Re}\left(\frac{1}{\pi\xi}\right). \tag{7.62}$$

Let us now consider a real-valued source-time function $f(t)$ with a Fourier spectrum $F(\omega)$. Then, instead of (7.59) and (7.60), we obtain

$$u^j(0_s, \omega) = F(\omega)U^j(0_s)\exp\,[\mathrm{i}\omega\tau(0_s, 0_0)], \tag{7.63}$$

$$\begin{aligned} u^j(0_s, t) &= f(t) * \mathrm{Re}\,\{U^j(0_s)\delta^{(A)}(t - \tau(0_s, 0_0))\} \\ &= \mathrm{Re}\,\{f^{(A)}(t) * U^j(0_s)\delta(t - \tau(0_s, 0_0))\}. \end{aligned} \tag{7.64}$$

For absorbing media $\delta(\xi)$ is replaced by $\mathrm{Im}\,(1/\pi\xi)$. In (7.64), $f^{(A)}(t)$ denotes the complex-valued analytical signal corresponding to the real-valued function $f(t)$:

$$f^{(A)}(t) = \frac{1}{\pi}\int_0^{\infty} F(\omega)\mathrm{e}^{-\mathrm{i}\omega t}\,\mathrm{d}\omega = f(t) + \mathrm{i}g(t), \tag{7.65}$$

where $g(t)$ is the Hilbert transform of the function $f(t)$.

Expressions (7.59), (7.60), (7.63) and (7.64) correspond to the selected elementary wave propagating along the ray Ω from 0_0 to 0_s. In the time domain these expressions describe the so-called *elementary synthetic seismograms*.

7.20 Ray-synthetic body-wave seismograms

Until now, we have considered only one elementary wave specified by an appropriate alphanumeric code, and its ray passing through 0_s. In a layered medium, however, there may be many body waves that travel from the point source 0_0 or from the initial surface Σ to the point 0_s along various ray trajectories Ω. They correspond to various reflected, refracted, multiply reflected, converted and other seismic body waves. Moreover, even a single elementary wave may travel to 0_s along different ray trajectories (so-called multiple rays). Thus for the complete high-frequency seismic body wavefield $\bar{u}^j$ at 0_s we can write the following *ray expansion*:

$$\bar{u}^j(0_s) = \sum_{(\Omega)} u^j(0_s), \tag{7.66}$$

where the summation runs over all rays Ω starting from 0_0 or from the initial surface Σ and passing through 0_s. The contributions $u^j(0_s)$ correspond to individual rays and are given by the equations of the previous section.

In a laterally varying layered structure the number of rays Ω arriving at 0_s may be infinite so that only a *partial-ray expansion* is actually possible. This, of course, introduces some errors into the ray computations.

In the *frequency domain* we can write

$$\bar{u}^j(0_s, \omega) = F(\omega) \sum_{(\Omega)} U^j(0_s) \exp[\mathrm{i}\omega\tau(0_s, 0_0)]. \tag{7.67}$$

Here $U^j(0_s)$ and $\tau(0_s, 0_0)$ correspond to the previously derived and discussed expressions. Equation (7.67) represents the *synthetic frequency response* for $F(\omega) = 1$. The so-called FFR (fast frequency response) algorithm can be used to evaluate the synthetic frequency response very efficiently (see Červený, 1985c).

In the *time domain* the HF synthetic body-wave seismogram is obtained by the Fourier transform of (7.67) (see (7.61)):

$$\bar{u}^j(0_s, t) = \frac{1}{\pi} \operatorname{Re} \int_0^\infty \bar{u}^j(0_s, \omega) \mathrm{e}^{-\mathrm{i}\omega t} \, \mathrm{d}\omega. \tag{7.68a}$$

Alternatively, we can write for the HF synthetic body-wave seismogram the expression

$$\begin{aligned} \bar{u}^j(0_s, t) &= f(t) * \operatorname{Re} \sum_{(\Omega)} U^j(0_s)\delta^{(A)}(t - \tau(0_s, 0_0)) \\ &= \operatorname{Re}\left\{f^{(A)}(t) * \sum_{(\Omega)} U^j(0_s)\delta(t - \tau(0_s, 0_0))\right\}. \end{aligned} \tag{7.68b}$$

Both of these expressions can be used even for complex-valued $\tau(0_s, 0_0)$ (dissipative media). $\delta^{(A)}(\xi)$ is then given by (7.62); $\delta(\xi)$ should be interpreted as $\operatorname{Im}(1/\pi\xi)$. The sum $\Sigma_{(\Omega)} \ldots$ in the second expression of (7.68b) represents the complex-valued *synthetic impulse response*.

For more details on ray-synthetic seismograms see the review by Červený (1985c).

7.21 Ray-theory elastodynamic Green function

We define the elastodynamic Green function $\bar{G}^m{}_n(0_s, t; 0_0, t_0)$ as follows: $\bar{G}^m{}_n(0_s, t; 0_0, t_0)$ is the mth contravariant component of the displacement vector in the local recording coordinate system at 0_s at time t, due to the application of a *single-force unit impulse* in the direction of the nth axis of the local initial coordinate system at 0_0 and time t_0. We can write approximately

$$\bar{G}^m{}_n(0_s, t; 0_0, t_0) = \sum_{(\Omega)} G^m{}_n(0_s, t; 0_0, t_0). \qquad (7.69)$$

Here $G^m{}_n$ are the contributions corresponding to rays Ω of individual elementary waves (including multiple arrivals).

To determine $G^m{}_n$, we use (7.50) and specify the radiation pattern $g^j(0_0)$ for the single force oriented along the nth axis of the local initial coordinate system. We obtain (see Kennett, 1983),

$$g^j(0_0) = \begin{cases} (4\pi\rho(0_0)v^2(0_0))^{-1} & \text{for } j = n, \\ 0 & \text{for } j \neq n. \end{cases}$$

In the *time domain* this yields

$$G^m{}_n(0_s, t; 0_0, t_0) = A(0_s, 0_0) \operatorname{Re}\{C^m{}_n(0_s, 0_0)\delta^{(A)}(t - t_0 - \tau(0_s, 0_0))\}, \qquad (7.70)$$

where

$$A(0_s, 0_0) = \frac{1}{4\pi}[v(0_0)\rho(0_0)v(0_s)\rho(0_s)\,|\det \boldsymbol{Q}_2(0_s, 0_0)|]^{-1/2}. \qquad (7.71)$$

In the *frequency domain* we have

$$G^m{}_n(0_s, 0_0, \omega) = A(0_s, 0_0)C^m{}_n(0_s, 0_0)\exp(i\omega\tau(0_s, 0_0)), \qquad (7.72)$$

where $G^m{}_n(0_s, 0_0, \omega)$ denotes the Fourier spectrum of $G^m{}_n(0_s, t; 0_0, 0)$.

7.22 Moment-tensor point source

Using the Green function, we can easily write expressions for point sources of more general types in 3D laterally varying layered block structures. Let us denote the contravariant components of the moment tensor by $M^{kl}(t)$. Then the contravariant components $\bar{u}^n(0_s, t)$ of the displacement vector at 0_s are given by

$$\bar{u}^n(0_s, t) = \dot{M}^{ij}(t) * \sum_{(\Omega)} G^n{}_i(0_s, t; 0_0, 0)p_j(0_0). \qquad (7.73)$$

Here $\dot{M}^{ij}(t)$ denotes the time derivatives of the moment tensor and $p_j(0_0)$ are covariant components of the slowness vector at the source point 0_0. This equation can be rewritten in many alternative forms; one of which is

$$\bar{u}^n(0_s, t) = \mathrm{Re}\left\{(\dot{M}^{ij}(t))^{(\mathrm{A})} * \sum_{(\Omega)} A(0_s, 0_0)\right.$$

$$\left. \times\, C^n{}_i(0_s, 0_0)p_j(0_0)\delta(t - t_0 - \tau(0_s, 0_0))\right\}. \quad (7.74)$$

Here $(\dot{M}^{ij}(t))^{(\mathrm{A})}$ denotes the analytical signal corresponding to the time derivative of $M^{ij}(t)$, $A(0_s, 0_0)$ is given by (7.71) and $C^n{}_i(0_s, 0_0)$ by (7.46).

The ray method can only be used to compute the high-frequency part of the seismic wavefield. Thus the resulting seismograms must be high-pass filtered to exclude the low-frequency contributions if such contributions are included in the source-time function.

7.23 Particle-motion diagrams

As the local recording coordinate system z^n at 0_s can be chosen arbitrarily by the user, complete ray tracing offers the possibility of determining and plotting particle-motion diagrams in an arbitrary plane passing through 0_s. The particle-motion diagrams of the complete wavefield may, in general, be rather complicated, mainly in various interference regions of two or more elementary waves, even in the ray approximation. The particle-motion diagrams considerably simplify for individual elementary waves if they are separated from other waves. For simplicity, let us consider harmonic waves, and a point 0_s situated inside the medium. When the wave arrives at 0_s as a P wave it is linearly polarized at that point. When the wave arrives at 0_s as an S wave its polarization is more complicated: as a rule, the S wave is elliptically polarized. The elliptical polarization may be caused by radiation patterns of the source and by overcritical reflections at interfaces. If the point 0_s is situated at an interface then the polarization may be elliptical for both the incident P and S waves, as both waves may impinge at the interface overcritically. Finally, at the Earth's surface, the polarization remains linear for the incident P wave (the P wave always impinges subcritically at the surface). The S wave, however, very often impinges overcritically at the Earth's surface. In this case its polarization is elliptical at 0_s even if the incident S wave is polarized linearly.

7.24 Gaussian beams and Gaussian packets

Expressions for the displacement vector (7.63) and (7.64) are high-frequency solutions of the elastodynamic equation even if the real-valued solutions $\mathbf{Q}$, $\mathbf{P}$ and $\mathbf{M}$ of the dynamic ray-tracing system are replaced by complex-valued solutions of that system. For positive-definite Im $\mathbf{M}$ such a solution represents a *Gaussian beam* concentrated near the ray Ω. The amplitude profile of the Gaussian beam in the plane perpendicular to the ray Ω is Gaussian, with its maximum on the ray Ω. The matrix $\mathbf{L} = (\frac{1}{2}\omega \operatorname{Im} \mathbf{M})^{-1/2}$ represents the matrix of the halfwidth of the Gaussian beam.

If the matrices $\operatorname{Re} \mathbf{M}(0_0)$ and $\operatorname{Im} \mathbf{M}(0_0)$ (or $\mathbf{L}(0_0)$) are given at the initial point 0_0 of the ray Ω, the matrices $\operatorname{Re} \mathbf{M}(0_s)$, $\operatorname{Im} \mathbf{M}(0_s)$ (and $\mathbf{L}(0_s)$) can be simply determined at any point 0_s of the ray Ω, using the results of complete ray tracing (particularly the ray propagator matrix $\boldsymbol{\Pi}(0_s, 0_0)$) and (7.22).

Similarly, if $\mathbf{Q}(0_0)$ and $\mathbf{P}(0_0)$ are given then $\operatorname{Re} \mathbf{Q}(0_s)$ and $\operatorname{Im} \mathbf{Q}(0_s)$ may be determined by means of the propagator matrix $\boldsymbol{\Pi}(0_s, 0_0)$ and (7.14). To apply (7.63) and (7.64) to the computation of Gaussian beams, the geometrical spreading $|\det \mathbf{Q}(0_s)|^{-1/2}$ must be replaced by $[\det \mathbf{Q}(0_s)]^{-1/2}$ and the phase shift due to caustics must be removed from the amplitude factor. As the KMAH index is known, the phase shift due to caustics can be removed simply. The sign of $[\det \mathbf{Q}(0_s)]^{-1/2}$, however, is not determined by complete ray tracing. Thus we can use complete ray tracing to obtain the final expressions for the displacement vector of the single Gaussian beam concentrated near Ω at 0_s, *except for the sign*. If the phase shift due to caustics is not removed then the ambiguous multiplicative factor is i^n ($n = 0, 1, 2, \ldots$).

In some applications of Gaussian beams—for example in the computation of the wavefield by the *summation* of Gaussian beams—the ambiguous multiplicative factor need not be known, it is cancelled by other factors in the summation formulae. Thus complete ray tracing is quite suitable for determining the Gaussian-beam contributions for the summation procedures.

As for Gaussian beams, the results of complete ray tracing can be used to compute the Gaussian packets (also called "quasiphotons"; see Babich and Ulin, 1981) concentrated near points along Ω, again except for the abovementioned ambiguous multiplicative factor. The amplitudes of the Gaussian packet concentrated near a point 0_s of the ray Ω decay exponentially with the square of the distance from 0_s, not only in the direction perpendicular to Ω, but in all directions.

If the results of complete ray tracing are stored in a sufficiently dense system of points along Ω between 0_0 and 0_s then the ambiguity factor can be removed, as $[\det \boldsymbol{Q}]^{-1/2}$ must vary continuously and smoothly along Ω in the case of Gaussian beams, even through caustic points.

For more details regarding Gaussian beams and Gaussian packets see the reviews by Červený (1985a,b) and references cited therein.

7.25 Summation of Gaussian beams or Gaussian packets

A time-harmonic high-frequency seismic body wavefield can be approximately expressed by a summation of Gaussian beams. For each elementary wave composing the wavefield, a sufficiently dense two-parameter system of rays is evaluated by the standard complete ray tracing. The wavefield can be expanded into a two-parameter system of Gaussian beams concentrated near individual rays. The final wavefield at any receiving point is then obtained by summation of all beams passing in some vicinity of the receiver. The Gaussian beams far from the receiver need not be considered in the summation owing to the finite width of the beam. For details and for the expansion formulae see Klimeš (1984a, 1986) and Červený (1985a).

In a similar way, the wavefield can be asymptotically expressed by summation of a three-parameter system of Gaussian packets. For each ray of the abovementioned two-parameter system of rays, a system of Gaussian packets concentrated near regularly distributed points along the ray is constructed. The summation is over all Gaussian packets influencing the wavefield at the receiver. As in the case of Gaussian beams, the Gaussian packets concentrated near points far from the receiver need not be considered. For details see Klimeš (1984b).

In the summation formulae, infinitely broad Gaussian beams (paraxial ray approximations) can be used instead of Gaussian beams of finite width. In this way, the Maslov method (see Chapman and Drummond, 1982; Klimeš, 1984b; Thompson and Chapman, 1985) may be obtained.

In 2D computations it is not necessary to evaluate a two-parameter system of rays—a one-parameter system of rays is sufficient. Two-dimensional (ribbon) Gaussian beams must, however, be used in the summation formulae. See Červený and Pšenčík (1983a) and Červený (1985a).

A similar simplified procedure based on the summation of Gaussian beams can be used even in 3D media if we are interested in the wavefield only along a specified profile (generally curvilinear). In such a case the tracing of only those rays that approximately follow the profile is sufficient. The take-off parameters of these rays form a curve in the 2D set of ray parameters γ^1 and γ^2. Such a computation is called the *profile mode computation* here.

In this case the quantity YI(22) = Γ (see (6.1)) must be replaced by the arclength of that part of the abovementioned curve in the ray-parameter surface that corresponds to the ray under consideration. The matrix Γ_{KL} (see (6.2)) is then singular, with one zero eigenvalue. The eigenvector corresponding to the non-zero eigenvalue is tangential to the curve in the ray-parameter surface; the other one is perpendicular to the curve. However, the algorithm for complete ray tracing remains the same.

Profile-mode computation in 3D media is *numerically very efficient*; it is only slightly more time-consuming than standard 2D computations. It may be used in the numerical modelling of high-frequency seismic wavefields in various source–receiver configurations: for surface profiles (in deep seismic sounding of the lithosphere, etc.), for borehole profiles (vertical seismic profiling), for borehole-to-borehole computations, in normal ray section computations, etc.

7.26 Integrals of the ray propagator matrix along the ray

Some integrals of the type

$$B_\beta = \int_{\sigma_0}^{\sigma} F_\alpha(\sigma)\Pi_{\alpha\beta}(\sigma, \sigma_0)\,\mathrm{d}\sigma \qquad (7.75)$$

or

$$B_{\alpha\beta} = \int_{\sigma_0}^{\sigma} \Pi_{\gamma\alpha}(\sigma, \sigma_0)F_{\gamma\delta}(\sigma)\Pi_{\delta\beta}(\sigma, \sigma_0)\,\mathrm{d}\sigma, \qquad (7.76)$$

where the quadrature is performed along the ray Ω and σ is defined by (5.1), are important in *perturbation theory* (Farra and Madariaga, 1987) and in *optimization of the shape of Gaussian beams and packets* (Klimeš, 1985). Here the weighting functions $F_\alpha(\sigma)$ and $F_{\gamma\delta}(\sigma)$ along the ray may be dependent on the velocities and their derivatives and/or on the velocity perturbations. $\sigma = \sigma_0$ denotes the initial point of the ray Ω.

There are two ways of evaluating the integrals.

(a) The numerical quadratures may be substituted directly into the complete ray-tracing algorithm. This approach has two disadvantages: (i) the complete ray-tracing algorithm is more complicated;

(ii) the integrals are only evaluated for the previously specified weighting functions and the weighting functions cannot be changed without repeating the complete ray-tracing computations.

(b) The results of complete ray tracing are stored along the ray Ω with the step STORE (see Section 5.5.1) of the independent variable σ. Then the integrals may be computed later by means of another program. This approach has two disadvantages: (i) since the quantities are stored with a given step of an independent variable, the points of intersection of the ray with the interfaces must also be stored in a file and taken into account; (ii) since the weighting functions are usually dependent on the velocity distribution, the routine computing the velocity and its derivatives must be called again for each stored point of the ray if some unstored derivatives of velocity are required.

The matrix $B_{\alpha\beta}$ defined above is often ill-conditioned, even in the case of a well-conditioned positive-definite weighting matrix $F_{\alpha\beta}(\sigma)$. For this reason, it is recommended that rather than directly evaluating the matrix $B_{\alpha\beta}$, with

$$\boldsymbol{B} = \begin{bmatrix} \boldsymbol{B}_{11} & \boldsymbol{B}_{12} \\ \boldsymbol{B}_{21} & \boldsymbol{B}_{22} \end{bmatrix}, \tag{7.77}$$

one should evaluate the matrix (Klimeš, 1985)

$$\boldsymbol{C} = \begin{bmatrix} \boldsymbol{C}_{11} & \boldsymbol{C}_{12} \\ \boldsymbol{C}_{21} & \boldsymbol{C}_{22} \end{bmatrix} = \begin{bmatrix} \boldsymbol{B}_{11} - \boldsymbol{B}_{12}\boldsymbol{B}_{22}^{-1}\boldsymbol{B}_{21} & \boldsymbol{B}_{12} \\ \boldsymbol{B}_{21} & \boldsymbol{B}_{22} \end{bmatrix}. \tag{7.78}$$

The differential equations

$$\begin{aligned} \frac{\mathrm{d}}{\mathrm{d}\sigma}\boldsymbol{C}_{11} &= [\boldsymbol{\Pi}_1(\sigma, \sigma_0) - \boldsymbol{\Pi}_2(\sigma, \sigma_0)\boldsymbol{C}_{22}^{-1}\boldsymbol{C}_{21}]^{\mathrm{T}} \\ &\qquad \times \boldsymbol{F}(\sigma)[\boldsymbol{\Pi}_1(\sigma, \sigma_0) - \boldsymbol{\Pi}_2(\sigma, \sigma_0)\boldsymbol{C}_{22}^{-1}\boldsymbol{C}_{21}], \\ \frac{\mathrm{d}}{\mathrm{d}\sigma}\boldsymbol{C}_{21} &= \boldsymbol{\Pi}_2^{\mathrm{T}}(\sigma, \sigma_0)\boldsymbol{F}(\sigma)\boldsymbol{\Pi}_1(\sigma, \sigma_0), \\ \frac{\mathrm{d}}{\mathrm{d}\sigma}\boldsymbol{C}_{12} &= \boldsymbol{\Pi}_1^{\mathrm{T}}(\sigma, \sigma_0)\boldsymbol{F}(\sigma)\boldsymbol{\Pi}_2(\sigma, \sigma_0), \\ \frac{\mathrm{d}}{\mathrm{d}\sigma}\boldsymbol{C}_{22} &= \boldsymbol{\Pi}_2^{\mathrm{T}}(\sigma, \sigma_0)\boldsymbol{F}(\sigma)\boldsymbol{\Pi}_2(\sigma, \sigma_0) \end{aligned} \tag{7.79}$$

for the 2×2 submatrices of $\boldsymbol{C}$ follow directly from (7.76). Here

$$\boldsymbol{\Pi}_1(\sigma, \sigma_0) = \begin{bmatrix} \boldsymbol{Q}_1(\sigma, \sigma_0) \\ \boldsymbol{P}_1(\sigma, \sigma_0) \end{bmatrix}, \qquad \boldsymbol{\Pi}_2(\sigma, \sigma_0) = \begin{bmatrix} \boldsymbol{Q}_2(\sigma, \sigma_0) \\ \boldsymbol{P}_2(\sigma, \sigma_0) \end{bmatrix}.$$

In (7.79), σ is again the variable along the ray Ω defined by (5.1).

7.27 Other applications

In principle, complete ray tracing can be effectively used in any sort of numerical modelling of high-frequency seismic wavefields in complex 2D and 3D structures, including the computation of the high-frequency seismic wavefield generated by finite sources (Červený *et al.*, 1987), the solution of various diffraction problems, the investigation of the effects of local geological conditions on surface motion, and the evaluation of Kirchhoff integrals, as well as those applications mentioned above. Similarly, complete ray tracing is sure to find applications in the solution of inverse seismic problems, both kinematic and dynamic. For example, Klimeš (1987) applied complete ray tracing and the Gaussian-packet approach in an algorithm for the kinematic location of hypocentres in general 3D models.

The number of applications of complete ray tracing is growing rapidly, and complete ray tracing is bound to play an important role in many other problems of seismological importance.

REFERENCES

Babich, V. M. and Ulin, V. V. (1981). Complex space-time ray method and "quasiphotons". *Mathematical Problems of the Theory of Wave Propagation* (ed. V. M. Babich), pp. 5–12. Nauka, Leningrad. (In Russian.)

Červený, V. (1985a). The application of ray tracing to the numerical modelling of seismic wave fields in complex structures. *Seismic Shear Waves,* Part A: *Theory* (ed. G. Dohr), pp. 1–124. Geophysical Press, London.

Červený, V. (1985b). Gaussian beam synthetic seismograms. *J. Geophys.* **58,** 44–72.

Červený, V. (1985c). Ray synthetic seismograms for complex two-dimensional and three-dimensional structures. *J. Geophys.* **58,** 2–26.

Červený, V. and Pšenčík, I. (1983a). Gaussian beams in two-dimensional elastic inhomogeneous media. *Geophys. J. R. Astron. Soc.* **72,** 419–435.

Červený, V. and Pšenčík, I. (1983b). Gaussian beams and paraxial ray approximations in three-dimensional elastic inhomogeneous media. *J. Geophys.* **53,** 1–15.

Červený, V., Molotkov, I. A. and Pšenčík, I. (1977). *Ray Method in Seismology.* Universita Karlova, Praha.

Červený, V., Klimeš, L. and Pšenčík, I. (1984). Paraxial ray approximations in the computation of seismic wave field in inhomogeneous media. *Geophys. J. R. Astron. Soc.* **79,** 89–104.

Červený, V., Klimeš, L., Pšenčík, I. and Pleinerová, J. (1987). High-frequency radiation from earthquake sources in laterally varying layered structures. *Geophys. J. R. Astron. Soc.* **88,** 43–79.

Chapman, C. H. and Drummond, R. (1982). Body wave seismograms in inhomogeneous media using Maslov asymptotic theory. *Bull. Seismol. Soc. Am.* **72,** S277–S317.

Farra, V. and Madariaga, R. (1987). Seismic waveform modelling in heterogeneous media by ray perturbation theory. *J. Geophys. Res.* **92,** 2697–2712.

Gjøystdal, H., Reinhardsen, J. E. and Astebøl, K. (1985). Computer representation of complex 3-D geological structures using a new "solid modeling" technique. *Geophys. Prosp.*, **33,** 1195–1211.

Kennett, B. L. N. (1983). *Seismic Wave Propagation in Stratified Media.* Cambridge University Press.

Klimeš, L. (1984a). Expansion of a high-frequency time-harmonic wave field given on an initial surface into Gaussian beams. *Geophys. J. R. Astron. Soc.* **79,** 105–118.

Klimeš, L. (1984b). The relation between Gaussian beams and Maslov asymptotic theory. *Stud. Geophys. Geod.* **28,** 237–247.

Klimeš, L. (1985). Computation of seismic wavefields in 3-D media by the Gaussian beam method. Program SW84. Research Report 68. Institute of Geophysics, Charles University, Praha. (In Czech.)

Klimeš, L. (1986). Discretization error for the superposition of Gaussian beams. *Geophys. J. R. Astron. Soc.* **86,** 531–551.

Klimeš, L. (1987). Kinematic hypocentre location. *Acta Montana* **75,** 51–64. (In Czech, English abstract.)

Kravtsov, Y. A. and Orlov, Y. I. (1980). *Geometrical Optics of Inhomogeneous Media.* Nauka, Moscow. (In Russian.)

Popov, M. M. and Pšenčík, I. (1978a). Ray amplitudes in inhomogeneous media with curved interfaces. *Travaux Géophysiques,* Vol. 24 (ed. A. Zátopek), pp. 118–129. Academia, Praha.

Popov, M. M. and Pšenčík, I. (1978b). Computation of ray amplitudes in inhomogeneous media with curved interfaces. *Stud. Geophys. Geod.* **22,** 248–258.

Thompson, C. J. and Chapman, C. H. (1985). An introduction to the Maslov asymptotic method. *Geophys. J. R. Astron. Soc.* **83,** 143–168.

II.2

Numerical Methods for Tracing Rays and Wavefronts

ANDRZEJ HANYGA

Institute of Geophysics, Polish Academy of Sciences, Warszawa, Poland

1 BASIC ASSUMPTIONS AND METHODS

1.1 Introduction. The extended asymptotic ray theory (ART)

The algorithms presented here purport to be a logically consistent and reasonably complete collection for the computation of body-wave traces in seismology and in exploration geophysics. In particular, they also allow for edge and caustic diffraction as well as for head waves generated at plane interfaces. The underlying method is an extension of asymptotic ray theory (Hanyga, 1980, 1982, 1983, 1984a–d, 1986, 1988). These algorithms have been implemented in the software package DANUTA.

In this chapter we merely describe the algorithms and quote the equations applied in the computations. Derivations are outlined only if they have not been published elsewhere.

In order to put the method on a firm conceptual basis, we begin with a presentation of the underlying notions. For the associated theoretical developments the reader is referred to Hanyga *et al.* (1984) and Hanyga (1980, 1982, 1983, 1984a–d, 1986).

Quite generally, a seismogram/seismic trace can be represented as a superposition of a finite number of *body wave* arrivals and normal modes and a negligible remainder term (Kamel and Felsen, 1981). The body waves dominate in the earlier part of a seismogram. On account of complicated geometry and small horizontal offset between a seismic source and a recording instrument, a generic seismic trace in exploration geophysics is essentially a superposition of body-wave arrivals.

SEISMOLOGICAL ALGORITHMS
ISBN 0-12-220770-X

In order to give a definition of the body-wave content of a seismogram $\boldsymbol{u}(\boldsymbol{x}, t)$ recorded at a point $\boldsymbol{x}$, it is expedient to deconvolve the source signature $s(t)$:

$$\boldsymbol{u}(\boldsymbol{x}, t) = s(t) * \boldsymbol{R}_\delta(\boldsymbol{x}, t). \tag{1.1}$$

The kernel $\boldsymbol{R}_\delta$ is a distribution representing the wavefield produced by a source whose signature $s(t)$ is the Dirac delta function $\delta(t)$. It exhibits a number of singularities localized at surfaces propagating from the source. On account of its sharp localization each singularity can be assigned a *travel time* $S(\boldsymbol{x}_0, \boldsymbol{x})$ from the source $\boldsymbol{x}_0$ to the receiver $\boldsymbol{x}$.

The wavefield $\boldsymbol{R}_\delta$ admits a decomposition

$$\boldsymbol{R}_\delta(\boldsymbol{x}, t) = \boldsymbol{S}_\delta(t, x) + \boldsymbol{T}_\delta(t, \boldsymbol{x}) \tag{1.2}$$

in terms of its singular component $\boldsymbol{S}_\delta$ and a locally integrable and sufficiently smooth function $\boldsymbol{T}_\delta$. The former represents the *body-wave* content of the wavefield and the latter represents the *tail*. The decomposition (1.2) can be made unique if it is agreed that the singular component is a superposition of a specified family of distributions localized at the associated *wavefronts* $S(\boldsymbol{x}_0, \boldsymbol{x}) = t$. An appropriate family of distributions for $\boldsymbol{R}_\delta$ consists of the Dirac delta function, its indefinite integrals and their Hilbert transforms:

$$\boldsymbol{S}_\delta(\boldsymbol{x}, t) = \sum_{r,\sigma} a_{r,\sigma}(\boldsymbol{x}) g_r(t - S_\sigma(\boldsymbol{x}_0, \boldsymbol{x})) \boldsymbol{r}_\sigma(\boldsymbol{x}, \boldsymbol{\nabla} S_\sigma(\boldsymbol{x}_0, \boldsymbol{x})), \tag{1.3}$$

with $g_0 = \delta$, $g_1 = H$ (Heaviside step function), $g_\lambda := t_+^{\lambda-1} = H(t)t^{\lambda-1}$, $\lambda = 2, \ldots,$ or $g_0 = -1/\pi t$, $g_1 = \ln|t|$ etc. These distributions are rigorously defined by Gel'fand and Shilov (1964). The index σ on S_σ refers to the ray code (conversions and reflections of the signal on its way from $\boldsymbol{x}_0$ to $\boldsymbol{x}$). The polarization vector $\boldsymbol{r}_\sigma(\boldsymbol{x}, \boldsymbol{p})$ is a function of *slowness* $\boldsymbol{p} = \boldsymbol{\nabla} S_\sigma(\boldsymbol{x}_0, \boldsymbol{x}) = \partial S_\sigma / \partial \boldsymbol{x}$. The coefficients $a_{r,\sigma}(\boldsymbol{x})$ are identified as *body-wave* amplitudes.

In contrast with the tail, the body waves obey relatively simple propagation laws. For each family of wavefronts $S_\sigma(\boldsymbol{x}_0, \boldsymbol{x}) = t$ a congruence of rays $\boldsymbol{x} = \boldsymbol{X}(t; u, v)$ can be constructed in such a way that the parameter t coincides with the travel time

$$S_\sigma(\boldsymbol{x}_0, \boldsymbol{X}(t; u, v)) = t. \tag{1.4}$$

The parameters u and v are constant along each ray. They can be chosen to be the takeoff angles of the ray at a point source. The amplitudes $a_{r,\sigma}$ satisfy a system of ordinary differential equations (transport equations) along each ray $\boldsymbol{X}(t; u_0, v_0)$, generically in the form

$$\frac{\mathrm{d}a_{r,\sigma}}{\mathrm{d}t} = F_\sigma(\boldsymbol{x}, \boldsymbol{p}) a_{r,\sigma} + G_{r,\sigma}(\boldsymbol{x}, \boldsymbol{p}; \{a_{s,\sigma}, s < r\}), \tag{1.5}$$

where $\boldsymbol{p}$ denotes the slowness vector along the ray.

For the highest-order discontinuity $r=0$, (1.5) are homogeneous: $G_{0,\sigma}=0$. In this case (1.5) amount to two laws: energy conservation (Hanyga *et al.*, 1984) and the law of rotation of the polarization of S waves. The former asserts that the energy content E of a portion of a ray tube contained between wavefronts $S_\sigma(\boldsymbol{x}_0, \boldsymbol{x})=t$ and $S_\sigma(\boldsymbol{x}_0, \boldsymbol{x})=t+\Delta$ is independent of t as long as the tails are disregarded (the energy associated with $g_0(t-S_\sigma)$ is constrained to follow the ray tube). The energy density is then given by $E\,|J|^{-1}$, where

$$J(x, t)\,\mathrm{d}t\,\mathrm{d}u\,\mathrm{d}v = \frac{\mathrm{d}\boldsymbol{X}}{\mathrm{d}t}\cdot\left(\frac{\mathrm{d}\boldsymbol{X}}{\mathrm{d}u}\times\frac{\mathrm{d}\boldsymbol{X}}{\mathrm{d}v}\right)\mathrm{d}t\,\mathrm{d}u\,\mathrm{d}v \tag{1.6}$$

denotes the volume of an infinitesimal ray tube contained between the wavefronts t and $t+\mathrm{d}t$. Since the energy density associated with a single body wave S_σ is, to lowest order in the asymptotic expansion, proportional to $|a_{0,\sigma}|^2$, with $r=0$ corresponding to the highest-order singularity g_0, we derive the familiar relationship

$$a_{0,\sigma}=A_{0,\sigma}\,|J|^{-1/2} \tag{1.7}$$

between the ray spreading J and the amplitude of the strongest singularity, with $A_{0,\sigma}$ piecewise-constant along the ray.

Every body wave propagates independently of the others until its rays intersect an interface or a layer with high gradient of the propagation speed. The former case is dealt with satisfactorily by asymptotic solution of the associated boundary conditions (Sections 4.12 and 4.13), while the latter case remains outside the scope of asymptotic ray theory. Boundary conditions provide the only coupling between the amplitudes $a_{r,\sigma}$ of various body waves σ.

The main benefit we derive from the above formalism is a method for inexpensive computation of body waves. In addition, inversion of the relationship between propagation speeds and travel times is itself an invaluable source of information about geological structures. A fairly obvious method of computing the amplitudes travel times for a given point source and receiver, followed here, consists of two steps.

Step 1 Solving the two-point boundary-value problem for the rayfield associated with a body wave. The body wave is specified in terms of a raycode listing its successive conversions and reflections at interfaces. The output includes the travel time and the take-off angles of the ray joining the source and receiver.

Step 2 Integration of the transport equation (1.5), with the initial data determined by the take-off angles and source characteristics.

Since we restrict ourselves to the lowest-order asymptotic approximation ($r = 0$), Step 2 boils down to the integration of a system of matrix-valued equations (Sections 4.4, 4.8 and 4.10), which allows computation of the ray-spreading Jacobian J.

In the following the above procedure is referred to as *dynamic ray tracing* (*DRT*).

The ART theory outlined above breaks down in the neighbourhood of *caustics* (defined by the equation $J = 0$), shadow boundaries and critical rays. These failures can, however, be remedied by an appropriate extension of the ART method.

At a caustic of the ray field $X(t; u, v)$, (1.7) implies an infinite amplitude. The non-linear equations exhibit a $0 \cdot \infty$ singularity. All this is a simple consequence of the fact that the cross-section of a ray tube $\boldsymbol{X}(t; u, v)$ shrinks to zero. The energy density, however, remains finite if it is referred to a more natural measure associated with the flow than the volumetric measure in the physical space. For example it is constant if referred to the measure $\mathrm{d}t\, \mathrm{d}u\, \mathrm{d}v$. In practice we choose a measure that is non-singular with respect to $\mathrm{d}t\, \mathrm{d}u\, \mathrm{d}v$ at the caustic and define the absolute value of a new amplitude by the square root of the new energy density. The associated transformation of the DRT equations (Section 4) involves non-linear algebraic transformations that are easily implemented numerically. It also yields the phase shift experienced by the body wave after touching a caustic.

We note here that the phase shift experienced by the signal at a simple caustic need not be the familiar phase delay $\frac{1}{2}\pi$ if the medium is strongly anisotropic.

A body-wave contribution to a seismic trace recorded at a generic receiver $\boldsymbol{x}$ takes the form

$$\boldsymbol{u}_\sigma(t, s) = a_{0,\sigma}\tilde{s}(t)\boldsymbol{r}_\sigma(\boldsymbol{x}, \boldsymbol{p}), \tag{1.8}$$

where

$$\tilde{s}(t) = s(t - S_\sigma(\boldsymbol{x}_0, \boldsymbol{x})) \cos \alpha - \mathrm{H}s(t - S_\sigma(\boldsymbol{x}_0, \boldsymbol{x})) \sin \alpha, \tag{1.9}$$

$$\mathrm{H}s(t) = \frac{1}{\pi} \int_{-\infty}^{\infty} \mathrm{d}t' \frac{1}{t' - t} s(t') \tag{1.10}$$

denotes the Hilbert transform and α is the phase shift experienced by the body wave at the caustics and total reflections. The input data to (1.8) can be obtained from dynamic ray tracing, even if the body wave has touched a cau tic on its way from the source to the receiver.

The formula (1.8) breaks down if the receiver is situated in the vicinity of a caustic, shadow boundary or critically reflected ray. In each of these cases the amplitude $a_{0,\sigma}$ tends to infinity at a critical surface separating two regions with a different coverage by rays. At the same time, travel times of a few body waves coincide at the critical surface. In its vicinity these waves exhibit strong interference, giving rise to a composite signal. The amplitude of the composite signal is bounded at the caustic and can be computed by the following modification of dynamic ray tracing (for definiteness we specialize the discussion to caustics here).

Step 1 Solve the two-point boundary-value problem for the body waves participating in the complex signal. The output involves the travel times $T_1, \ldots, T_K$ of these waves and their take-off angles at the source.

Step 2 Carry out DRT for each constituent body wave. The output involves the amplitudes referred to an appropriate measure that is non-singular at the caustic.

Step 3 The composite signal is an explicitly computable non-linear function of the travel times $T_1, \ldots, T_K$, depending linearly on the amplitudes obtained in Step 2.

The method described here can be considered as an alternative to Gaussian beam tracing (see Chapter II.1 of this book). The latter approach avoids an exact solution of the two-point boundary-value problem and accounts for diffraction phenomena in a different way. Exact two-point ray tracing can also be avoided by use of paraxial approximation (Červený and Pšenčík, 1983).

The more explicit equations presented in this chapter allow for isotropic and transversely isotropic media.

The chapter is organized in the following way. In Sections 2 and 3 we present the basic algorithms and methods for solving the initial-value problem and the two-point boundary-value problem for seismic rays (Step 1). In Section 4 we proceed to discuss the integration of the DRT equations. For the underlying theory we refer to Hanyga *et al.* (1984) and Hanyga (1982, 1984a–c). We rely mainly on non-linear DRT equations, but Section 4.17 contains a brief discussion of the alternative linear DRT equations and the advantages of each method. Finally, in Section 5 we present some methods of computing the composite signal at a caustic or shadow boundary. For details of the theory the reader is referred to Hanyga (1988) and Hanyga and Seredyńska (1988).

The assumptions about the model and the description of the model required by the tracing algorithms are discussed in the next section.

1.2 Representation of the model

The input of a ray-tracing algorithm includes information about the medium supporting the wavefield. This item of the input is referred to as the *model.*

The model is specified by a library of model subprograms called by the tracing programs and perhaps also by files containing numerical data. The model subprograms can define a class of models while the numerical data may single out a specific model within this class. It is assumed that the model library has been supplied by the user. The executing image of a ray-tracing program is obtained by linking the object file and an associated subprogram library with the user's model library. Model data files are accessed by the model subprograms only and therefore they are left at the user's discretion.

The model subprograms must meet two groups of requirements:

(i) the model must be consistent with the tracing algorithms;
(ii) those model subprograms that are called directly by the tracing programs and their subprograms must bear appropriate names and their inputs and outputs must be compatible with the calling programs.

In this section we discuss the first type of requirements. The second group of requirements can be found in an information file accompanying the software package.

The physically relevant part of the model is assumed to be contained in a *model box* $\mathscr{B}: z_{1,k} < x_k < z_{2,k}$, $k = 1, 2, 3$. The coordinate system (x_k) is either Cartesian or curvilinear orthogonal.

The medium is assumed to be linear-elastic. In particular, attenuation and dispersion are not allowed for. The elastic moduli are assumed to be piecewise-continuous and to have piecewise-continuous derivatives up to second order. A *block* $\mathscr{L}$ is defined to be a maximal subset of $\mathscr{B}$ with the property that all the elastic moduli are $\mathscr{C}^2$ smooth in $\mathscr{L}$. In particular, two model blocks do not intersect and the union of all the blocks covers $\mathscr{B}$. For the sake of convenient reference, each block in the model is assigned an integer β and the assignment is one-to-one.

Every block is bounded by a few surfaces called *interfaces.* The set of interfaces is one of the primary notions defining the model. It is assumed that an interface is a surface without boundary, i.e. either it extends to infinity or it is a closed surface topologically equivalent to a sphere. In particular, it is defined beyond the box $\mathscr{B}$ unless it is closed. If there are N interfaces in the model then each interface is assigned a reference number λ between 1 and N and the assignment is one-to-one.

Convergence of iterative ray-tracing algorithms is guaranteed provided that model interfaces have continuous normals and curvatures, although in practice discontinuities of normals and curvatures need not have any significant effect in this respect.

The points of an interface are given by an equation $f(\boldsymbol{x};\lambda)=0$. Some two-point tracing algorithms require parametric representation $x_k = g_k(\xi, \eta, \lambda)$ of interfaces (Section 3.4). For consistency reasons the two representations are connected by the following construction (Gjøystdal *et al.*, 1984). A possibly curvilinear coordinate system (ξ, η, ζ) is associated with each interface λ. It is defined in terms of the coordinate transformations $\boldsymbol{x}=\boldsymbol{H}(\xi, \eta, \zeta;\lambda)$ and $(\xi, \eta, \zeta)=\boldsymbol{G}(\boldsymbol{x};\lambda)$. Interface λ is now defined by an equation $\zeta = h(\xi, \eta;\lambda)$, and its two alternative representations are defined in terms of $f(\boldsymbol{x};\lambda)=\zeta - h(\xi, \eta;\lambda)$ and $\boldsymbol{g}(\xi, \eta;\lambda)=\boldsymbol{H}(\xi, \eta, h(\xi, \eta;\lambda);\lambda)$.

Each interface is endowed with an orientation defined in terms of a continuous unit normal $\boldsymbol{n}(\boldsymbol{x};\lambda)$, where λ is the reference number of the interface. The normal $\boldsymbol{n}(\boldsymbol{x};\lambda)$ points into the region $f(\boldsymbol{x};\lambda)>0$. Two-point ray tracing requires evaluation of $\boldsymbol{n}(\boldsymbol{x};\lambda)$ at points lying in a neighbourhood of interface λ. It is therefore assumed that $\boldsymbol{n}(\boldsymbol{x};\lambda)$ is defined everywhere and satisfies the following conditions: (1) $\boldsymbol{n}(\boldsymbol{x};\lambda)$ is a continuous function of $\boldsymbol{x}$; (2) $\boldsymbol{n}(\boldsymbol{x};\lambda)$ is a unit vector; (3) for $\boldsymbol{x}$ lying on interface λ the vector $\boldsymbol{n}(\boldsymbol{x};\lambda)$ is orthogonal to λ. A natural choice is $\boldsymbol{n}(\boldsymbol{x};\lambda)=|\boldsymbol{\nabla} f(\boldsymbol{x};\lambda)|^{-1}\boldsymbol{\nabla} f(\boldsymbol{x};\lambda)$.

A point $\boldsymbol{x}$ of an interface $\mathscr{I}$ is said to be *non-physical* if the reference numbers of the two blocks adjacent to $\mathscr{I}$ at $\boldsymbol{x}$ are equal. This notion is relevant for initial-value tracing (Section 2), which excludes reflection and conversion of rays at non-physical points of an interface.

The medium filling a block is assumed to be either isotropic or transversely isotropic. In the former case the block is assigned an index $\rho=-1, 0$ or 1. If $\rho=1$ then the block is interpreted as a generic isotropic block and the propagation speeds $c_P=f_1(\boldsymbol{x};\beta)$, $c_S=f_2(\boldsymbol{x};\beta)$ are defined in a neighbourhood of the block. If $\rho=0$ or -1 then the block is assumed to be homogeneous or to have a constant vertical gradient, and the appropriate numerical data must be supplied by a model subprogram.

A transversely isotropic block is assigned $\rho=2$, 3 or 4. If $\rho=2$ then the isotropy axes (Appendix B) are assumed to be everywhere oriented in the direction of the x_3 axis; five elastic moduli must then be defined throughout and in a neighbourhood of the block. For $\rho=3$ and 4 the orientation of the isotropy axes is defined in terms of two angles θ and φ on the unit sphere with $\theta=0$ corresponding to the x_3 direction. In the case $\rho=3$ it is assumed that θ and φ assume constant values throughout the block. Negative values of ρ indicate that the block is homogeneous.

Iterative ray-tracing algorithms require that all the functions representing the variation of elastic properties of a block are smoothly extrapolated beyond the boundary of the block (Section 3).

2 INITIAL-VALUE RAY TRACING

2.1 Basic concepts

We now describe a method of solving the initial-value problem for a piecewise smooth curve $w_k(s)$ in an N-dimensional space $\mathbb{R}^N = \mathbb{R}^3 \times \mathbb{R}^{N-3}$; the curve is assumed to project onto a seismic ray $x_k(t)$, $k = 1, 2, 3$: $w_k(t) = x_k(t)$ for $k \leqslant 3$. The case $N = 6$ is referred to as *kinematic* (initial-value) tracing.

In the most general case the intersection of the curve $w_k(t)$ with a single model block is determined by integrating a system of ordinary differential equations; the first six equations can be expressed in Hamiltonian form

$$\frac{\mathrm{d}x_k}{\mathrm{d}t} = \frac{\partial H}{\partial p_k}(x, p; l, \sigma), \qquad \frac{\mathrm{d}p_k}{\mathrm{d}t} = -\frac{\partial H}{\partial x_k}(x, p; l, \sigma), \qquad k = 1, 2, 3, \tag{2.1}$$

with $p_k(t) = w_{k+3}(t)$ for $k = 1, 2, 3$ (Hanyga *et al.*, 1984). The integer argument l is the reference number of the block containing the point $\boldsymbol{x}$, and σ refers to the wave species. The definition of the *wave Hamiltonian* H and related definitions can be found in Appendices A and B.

At the intersection of the ray with an interface $\mathcal{J}$ the curve $w_k(t)$ satisfies appropriate boundary conditions. The boundary conditions involve a choice of the signal path after crossing interface $\mathcal{J}$. If l_1 is the block entered by the ray after crossing $\mathcal{J}$ and σ_1 is the wave species assigned to the ray segment in block l_1 then the boundary conditions for thc first six components of w_k assume the form

$$\boldsymbol{x}(t_1 + 0) = \boldsymbol{x}(t_1 - 0) \in \mathcal{J}, \tag{2.2}$$

$$\boldsymbol{\nu} \times [\boldsymbol{p}(t_1 + 0) - \boldsymbol{p}(t_1 - 0)] = 0, \tag{2.3}$$

$$H(\boldsymbol{x}(t_1 + 0), \boldsymbol{p}(t_1 + 0); l_1, \sigma_1) = 1. \tag{2.4}$$

In (2.3), $\boldsymbol{\nu}$ denotes a normal vector to the interface $\mathcal{J}$ evaluated at $\boldsymbol{x}(t_1 - 0)$. From (2.3) and (2.4), the outgoing slowness $\boldsymbol{p}_0 = \boldsymbol{p}(t_1 + 0)$ is determined.

The function $H(\boldsymbol{x}, \boldsymbol{p}; l, \sigma)$ is convex if either block l is isotropic or σ is the fastest wave species in l (Courant and Hilbert, 1962). In these cases (2.3) and (2.4) have two real solutions in the subcritical range of incident slowness $\boldsymbol{p}(t_1 - 0)$ and two complex conjugate ones in the supercritical range. In the subcritical range $\boldsymbol{p}_0$ is determined uniquely by the obvious geometrical condition

$$\frac{\partial H}{\partial \boldsymbol{p}_0}(\boldsymbol{x}_1, \boldsymbol{p}_0; l_1, \sigma_1) \cdot \boldsymbol{v} > 0 \quad \text{or} \quad < 0, \tag{2.5}$$

implying that the outgoing ray enters block l_1. In the supercritical range the condition (2.5) is replaced by the radiation condition (evanescence of a surface wave propagating along $\mathscr{I}$)

$$\operatorname{Im}(\boldsymbol{p}_0 \cdot \boldsymbol{v}) \begin{cases} >0 & \text{if } \boldsymbol{v} \text{ points into } l, \\ <0 & \text{otherwise.} \end{cases} \tag{2.6}$$

2.2 Component algorithms of an initial-value tracer

Initial-value ray tracing involves numerical computation of the relevant points of a curve and determination of its intersections with a family of surfaces (*interrupt surfaces*). At the intersection with an interrupt surface the tracing program takes appropriate action, such as application of the boundary conditions at an interface, transformation of dependent variables, recording the intersection point or termination of tracing.

The task of tracing an arc of the curve between two successive interrupt surfaces is achieved by a call of module INTFIND or CHINTR2. Module INTFIND performs numerical integration of a system of ordinary differential equations defined by a generic subroutine RHS(t, $\boldsymbol{Y}$, $\boldsymbol{F}$) computing the right-hand-side vector $\boldsymbol{F}$ for given values t and $\boldsymbol{Y}$ of the curve parameter and the dependent-variable vector. Module CHINTR2 computes points on the curve according to a closed-form expression $\boldsymbol{y} = \boldsymbol{r}(t; \boldsymbol{x})$, $\boldsymbol{r}(0; \boldsymbol{x}) = \boldsymbol{x}$, defined by a generic subroutine RAY($\boldsymbol{X}$, $\boldsymbol{Y}$, t, IFAIL). The names RHS and RAY are arguments of INTFIND and CHINTR2 and the calling program substitutes appropriate subroutines for them.

The interrupt surfaces are defined by a generic Fortran function SUR($\boldsymbol{X}$) whose name is an argument of INTFIND and CHINTR2. An interrupt surface $f(\boldsymbol{x}; k) = 0$ is specified by the index k passed to module SUR through common block /A/, and the value of $f(\boldsymbol{x}; k)$ is returned as the value of SUR($\boldsymbol{X}$). Although the set of interrupt surfaces included in SUR($\boldsymbol{X}$) is fixed at the compile time, only a subset of this set is active during a call of INTFIND or CHINTR2 at the execution time. Elimination of temporarily irrelevant interrupt surfaces is achieved by including an integer array SURIND in the argument of INTFIND and CHINTR2.

An interrupt surface $f(\boldsymbol{x}; k)=0$ is ignored if SURIND$(k)=0$. In particular, this allows eliminating computations of functions $f(\boldsymbol{x}; k)=0$ representing those interfaces that are not contained in the boundary of the current block. For this purpose the model library must include subroutine MODBLB described in the introductory file of the package.

Module INTFIND allows for additional interrupts that are not defined in terms of interrupt surfaces. The interrupt occurs in the course of estimating the next integration step if a subprogram called by RHS has substituted a non-zero value for the integer variable MARK in common block /MK/. If MARK<0 then INTFIND performs a transformation of dependent variables defined by a generic subroutine TRANS and resumes integration; for any other non-zero value of MARK module INTFIND exists. The name TRANS is an argument of INTFIND. The additional interrupts and variable transformations play an important role in dynamic ray tracing (Section 4). The facilities described in this paragraph are not provided by a simplified version INTFINC of INTFIND, applied in kinematic tracing.

In some kinematic tracers CHINTR2 is replaced by module CHINTR1. CHINTR1 allows for one interrupt surface only and is called in a loop over such k that SURIND$(k)\neq 0$. For kinematic ray tracing in homogeneous and constant gradient blocks (Appendix C) CHINTR1 is used in combination with module RAYCUR1 defining the arc.

For a detailed description of INTFIND and CHINTR2 the interested reader is referred to Appendix F.

A module computing the initial data for INTFIND or CHINTR2 after crossing an interface is another important component of an initial-value tracer. It is based on (2.2)–(2.4). For isotropic blocks and for T waves (Appendix B) the solution of (2.2)–(2.4) is available in explicit form.

2.3 Formulation of an initial-value problem

An initial-value tracing problem is defined in terms of initial data and a set of rules specifying the outgoing wave to be followed after crossing an interface. These rules are expressed in terms of a list of events attached to some particular interfaces (reflection with or without conversion, transmission with conversion); at any other interface the default (transmission without conversion) is applied. Successful tracing stops if the list has been exhausted and a termination condition has been satisfied. Otherwise tracing stops if the ray leaves the model box (1.2) or if overcritical incidence excludes the continuation indicated by the rules imposed in the formulation of the problem.

For kinematic tracing the initial data are specified in terms of the initial point $\boldsymbol{x}_0$ of the ray and two take-off angles θ and φ. The initial value of slowness $\boldsymbol{p}$ is then given by the following expressions

$$\left.\begin{aligned} &n_1 = \sin\theta\cos\varphi, \qquad n_2 = \sin\theta\sin\varphi, \qquad n_3 = \cos\theta \\ &H = H(\boldsymbol{x}_0, \boldsymbol{n}), \qquad p_k = \frac{n_k}{H} \quad \text{for } k = 1, 2, 3. \end{aligned}\right\} \tag{2.8}$$

Additional initial conditions for DRT are discussed in Section 4.

The continuation rules are conveniently expressed in terms of the *reduced raycode* RD(k). RD(1) is the number of non-default events required by the user. Each non-default event is specified in terms of three successive entries of RD: the interface reference number i, the incidence side ι and a number specifying the event (reflection, conversion or both); for anisotropic media the outgoing wave species has to be specified. The incidence side $\iota = 1$ if the incident wave arrives from the region $f(\boldsymbol{x}; i) > 0$. A reflection or conversion is not applied if the incidence point lies on a non-physical part of the interface.

Alternatively, the entries i, ι in some reduced raycodes are replaced by *block faces,* defined in terms of ordered pairs (β_1, β_2) of block reference numbers. In this case the event is applied at any interface separating blocks β_1 and β_2, provided that the incident wave arrives from block β_1. We note here that a block face consists of several pieces of smooth interfaces.

The order of events on the list is irrelevant in some tracers. In addition, it can be required that the ray has a given number of turning points.

Initial-value tracers stop if one of the following conditions is satisfied: (1) an over-critical incidence has occurred; (2) the ray has left the model box; (3) an inconsistency of the model has been detected; (4) the reduced raycode has been exhausted and, *subsequently,* a termination condition defined by the user has been satisfied.

The following termination conditions have been made available to the user: (1) the ray intersects a surface $r(\boldsymbol{x}) = 0$ defined by the user ("receiver surface"); (2) the ray reaches the closest point to a given point; (3) the ray reaches the closest point to a straight line (e.g. the axis of a well). All the three conditions are expressed in terms of the six-dimensional curve $w_k(t)$, $k = 1, \ldots, 6$ intersecting a hypersurface.

The most important items in the output of kinematic ray tracing are the raycode, a sequential file containing points of the ray for plotting and an unformatted file containing a description of the ray appropriate for processing by an iterative two-point ray-tracing program (Section 3).

2.4 Error control in initial-value tracing

The error is controlled by two control parameters T_1 and T_2, which limit the local integration error e_i (incurred in one integration step) of w_i, $i=1, \ldots, N$,

$$\sum_{i=1}^{N} \frac{|e_i|}{s_i \min(f_i, |w_i|)} < T_1, \tag{2.9}$$

and the error in the determination of an intersection of the curve $w_i(t)$ with an interrupt surface,

$$|t_2 - t_1| < T_2, \tag{2.10}$$

where $[t_1, t_2]$ is the interval known to contain the intersection point.

The parameter T_1 controls the number of significant digits in the integration. The floors f_i and the scales s_i for $i \leqslant 6$ are fixed at 0.001 and 1 respectively. The remaining floors and scales, as well as the two tolerances T_1 and T_2, are determined by the user.

Another factor affecting the accuracy of tracing is the accuracy of the model subprograms involved.

The number of significant digits in the integration can be checked by an optional computation of the wave Hamiltonian along the ray at equi-spaced time points. The Hamiltonian is expected to be constant along the ray. A constant Hamiltonian not equal to unity indicates an error in the initial data. A significant variation of the wave Hamiltonian indicates an inconsistency between the user-supplied model subprograms MODELV1 and MODELG1 computing wave speeds (or elastic moduli divided by the density if $|\rho| > 1$) and their gradients respectively.

The wave Hamiltonian depends on w_i, $i \leqslant 6$, only. For $i > 6$ more elaborate tests are available (Section 4).

3 TWO-POINT RAY TRACING

3.1 Introduction

The input of a two-point ray-tracing algorithm always includes the following data: (1) the coordinates of the two endpoints $\boldsymbol{x}_0$ and $\boldsymbol{x}_1$ of the ray; (2) the raycode $R(i)$, $i=1, \ldots, 3R(1)+3$.

The *raycode* $R(i)$ specifies the path of the signal through the model and the wave conversions of the signal at interfaces. The raycode has the following form:

$R(1)$	number of interfaces intersected by the ray and listed in the raycode;
$R(2)$	reference number l_0 of the block containing the ray origin $\boldsymbol{x}_0$;
$R(3)$	initial wave species σ_0;
$R(3k+1)$	reference number of the successive interface intersected by the ray;
$R(3k+2)$	reference number of the successive block intersected by the ray;
$R(3k+3)$	wave species of the signal in block $R(3k+2)$;

with $k = 1, \ldots, R(1)$.

Intersections of the ray with non-physical interfaces can be ignored in the raycode. Their inclusion increases array dimensions and in some cases (edge diffraction; Section 3.8) can introduce some errors. The output of two-point ray tracing includes a numerical specification of the ray, travel time and initial data for initial-value tracing along the output ray.

The two-point ray-tracing problem need not have a unique solution. Since two-point ray-tracing algorithms are iterative, the output ray depends on the first-approximation ray. The first-approximation ray is either generated by the tracing algorithm or included in the input data. In the latter case it is referred to as the *input ray*.

The input ray need not satisfy Snell's law. Deviations from Snell's law are gradually reduced in the course of iterations. This fact allows a fairly arbitrary construction of the input ray (3.7).

The two-point ray-tracing problem is equivalent to a two-point boundary-value problem for (2.1).

Iterative methods of solving two-point boundary-value problems for systems of ordinary differential equations fall into two classes. In *bending* (Julian and Gubbins, 1975; Pereyra *et al.*, 1980) boundary conditions are satisfied throughout the iterative process while deviation from the differential equations is gradually reduced in the course of iterations. *Shooting* methods consist in determining the initial direction of the solution curve $\mathscr{J}$ at the point $\boldsymbol{x}_0$ in such a way that the point $\boldsymbol{x}_1$ lies on $\mathscr{J}$. At each iteration step the curve satisfies the differential equations. Our discussion is restricted to some adaptations of the shooting method.

The general two-point boundary-value problem for seismic rays is complicated by the fact that the ray satisfies interface boundary conditions (2.2)–(2.4) in addition to the differential equations (2.1). Another departure from a standard two-point BVP (Roberts and Shipman, 1972) consists in the fact that the interval of the curve parameter (travel time) is not known *a priori*. These facts necessitate a reformulation of the shooting algorithm.

Two variants of shooting are presented in Sections 3.2 and 3.4. We then proceed to discuss further adaptations of these basic algorithms for dealing with edge-diffracted rays, head wave rays and NIP rays. Some methods of generating the first-approximation ray are also discussed.

3.2 Parallel ray tracing. Basic concepts

The method of parallel ray tracing takes a middle road between shooting and bending by treating the interface boundary conditions on an equal footing with the two-point boundary condition. The ray is allowed to bend at interface intersections but is constrained to satisfy the appropriate differential equations in each block.

At the same time, parallel ray tracing circumvents the other difficulty mentioned in Section 3.1. In two-point ray tracing the travel time of the signal is not known *a priori,* while the initial data are constrained by the condition $H(\boldsymbol{x}(0), \boldsymbol{p}(0); l_0, \sigma_0) = 1$. On account of the homogeneity property

$$H(\boldsymbol{x}, \theta\boldsymbol{p}) = \theta H(\boldsymbol{x}, \boldsymbol{p}) \quad \text{for } \theta > 0 \tag{3.1}$$

of wave Hamiltonians (Hanyga *et al.*, 1984; Hanyga, 1984b), we are free to manipulate the two direction cosines of the initial slowness in order to reach $\boldsymbol{x}_1$ from a fixed ray origin $\boldsymbol{x}_0$. The third degree of freedom necessary to reach the point $\boldsymbol{x}_1$ in a three-dimensional model space is provided by the travel time. In a standard two-point BVP (Roberts and Shipman, 1972) the three components of the initial slowness would be manipulated in order to reach $\boldsymbol{x}_1$ in a preassigned time.

All the above difficulties are circumvented by a transformation of the two-point ray-tracing problem that reduces it to the standard two-point BVP for (2.1). The ray is split into a finite collection of segments $\mathcal{S}_1, \ldots, \mathcal{S}_N$. Each segment $\mathcal{S}_A$ represents the intersection of the ray with one block. In accordance with (2.1), each segment $\mathcal{S}_A$ is treated as a curve $\boldsymbol{y}^A(t)$ in the six-dimensional space $\boldsymbol{y} = (\boldsymbol{x}, \boldsymbol{p})$. After a shift and rescaling of the parametrization of each segment, the ray is represented by a curve $\boldsymbol{z}(s) = \{\tilde{\boldsymbol{y}}^1(s), \ldots, \tilde{\boldsymbol{y}}^N(s)\}$ in a $6N$-dimensional space, where $\tilde{\boldsymbol{y}}^A(s)$, $s \in [0, U]$, is the new parametric representation of $\mathcal{S}_A$.

Rescaling of the parametrization is necessitated by the fact the curves $\boldsymbol{y}^A(t)$ are defined over intervals of varying length T_A. However, since the travel times T_A are *a priori* unknown, rescaling has to be compensated by lifting the constraint $H(\boldsymbol{x}, \boldsymbol{p}) = 1$ and varying the length of the slowness vector of each segment. On account of (3.1), (2.1) is insensitive to the dilatations $\boldsymbol{p} \to \theta \boldsymbol{p}$, and therefore (2.1) are replaced by the Hamiltonian system

$$\frac{\mathrm{d}\boldsymbol{x}}{\mathrm{d}s} = \frac{1}{2}\frac{\partial H^2}{\partial \boldsymbol{p}}, \qquad \frac{\mathrm{d}\boldsymbol{p}}{\mathrm{d}s} = -\frac{1}{2}\frac{\partial H^2}{\partial \boldsymbol{x}} \quad (0 < s < U). \tag{3.2}$$

Since H^2 is constant along any solution curve of (3.2), (3.2) are equivalent to (2.1) if $t = \text{const} + Hs$. In addition to the possibility of varying the initial data instead of interval lengths, we benefit from the fact that computation of H^2 involves one square root less than computation of H.

The original two-point ray-tracing problem turns out to be equivalent to a standard two-point BVP with non-separated boundary conditions (Roberts and Shipman, 1972) for the $6N$-dimensional curve $z(s) = [y^1(s) \quad \ldots \quad y^N(s)]$:

$$\boldsymbol{x}^1(0) = \boldsymbol{x}_0, \tag{3.3}$$

$$\boldsymbol{x}^N(U) = \boldsymbol{x}_1, \tag{3.4}$$

$$\boldsymbol{x}^A(U) - \boldsymbol{x}^{A+1}(0) = 0, \tag{3.5}$$

$$f(\boldsymbol{x}^A(U); R(3A+1)) = 0, \tag{3.6}$$

$$\boldsymbol{v} \times [H_A(\boldsymbol{x}^A(U), \boldsymbol{p}^A(U))^{-1}\boldsymbol{p}^A(U) - H_{A+1}(\boldsymbol{x}^{A+1}(0), \boldsymbol{p}^{A+1}(0))^{-1}\boldsymbol{p}^{A+1}(0)] = 0, \tag{3.7}$$

where $(\boldsymbol{x}(s), \boldsymbol{p}(s)) = \boldsymbol{y}(s)$, $H_A(\boldsymbol{x}, \boldsymbol{p}) = H(\boldsymbol{x}, \boldsymbol{p}; R(3A-1), R(3A))$, $\boldsymbol{v}$ is the unit normal to the interface $f(\boldsymbol{x}; R(3A+1)) = 0$ evaluated at $\boldsymbol{x}^A(0)$ and $A = 1, \ldots, N-1$. It is assumed that each wave Hamiltonian $H(\boldsymbol{x}, \boldsymbol{p}; l, \sigma)$ is defined outside the boundaries of block l and the normal $\boldsymbol{v}$ is defined in a neighbourhood of the interface $R(3A+1)$ (Section 1.2).

Since (3.2) define $\boldsymbol{x}^A(U)$ and $\boldsymbol{p}^A(U)$ as functions of $\boldsymbol{x}^A(0)$ and $\boldsymbol{p}^A(0)$, we end up with a system of $6N + 6$ non-linear equations

$$\boldsymbol{F}(\boldsymbol{V}) = 0, \tag{3.8}$$

where

$$\boldsymbol{V} = [\boldsymbol{x}^1(0) \quad \boldsymbol{p}^1(0) \quad \ldots \quad \boldsymbol{x}^N(0) \quad \boldsymbol{p}^N(0)]^{\mathrm{T}}$$

is the numerical representation of the ray. In order to reduce the array dimensions, (3.3) are usually considered as a constraint and $\boldsymbol{x}^1(0)$ is deleted from $\boldsymbol{V}$. In this case we deal with $6N + 3$ equations and $6N + 3$ unknowns $\boldsymbol{V}$.

The Jacobian $\mathrm{D}\boldsymbol{F}(\boldsymbol{V})$ of $\boldsymbol{F}$ is invertible unless for some $A \leq N$, $\boldsymbol{x}^A(U)$ lies at a caustic of the source $\boldsymbol{x}^A(0)$.

3.3 Parallel ray tracing. Iterative solution of the problem (3.8)

Equations (3.8) are solved iteratively by an algorithm of Polak (1974) combining the method of secants (Ortega and Rheinboldt, 1970) with local variations. The method of secants is a modified Newton search

$$V_{r+1} = V_r + \beta^\kappa G_r^{-1} F(V_r), \quad r = 0, 1, 2, \ldots, \tag{3.9}$$

where G_r approximates $DF(V_r)$, β is a relaxation parameter satisfying the inequality $0 < \beta < 0.5$ and κ runs over the set of integers $0, \ldots, \kappa_0 \geqslant 0$ until the Armijo criterion (Bertsekas, 1982)

$$|F(V_{r+1})|^2 < (1 - 2\alpha\beta^{\kappa_0})\,|F(V_r)|^2 \tag{3.10}$$

is met. α is a fixed positive real number satisfying the inequality $\alpha < \frac{1}{6}$. V_{r+1} is accepted as a new iterate if (3.10) is satisfied. If (3.10) fails to be satisfied for $\kappa = 0, \ldots, \kappa_0$ then the last perturbation V_r' (3.13) of V_r is accepted as a new iterate provided that it satisfies the inequality

$$|F(V_r')| < |F(V_r)| \tag{3.11}$$

(*local variations*). Iterations are stopped if $|F(V_r)| < \gamma$, where γ is a numerical tolerance. On account of their different geometrical meaning, the components of $F(V)$ differ by orders of magnitude. In order to give the same weight to the errors (3.5)–(3.7), it is recommended that the corresponding components of $F(V)$ be multiplied by appropriate scale factors.

In the algorithm of Polak (1974) G_r is generated from an initial matrix $G_0 = 0$ by updating one column of G_r at every step of the algorithm:

$$(G_r')_{AB} = \pm[F_A(V_r') - F_A(V_r)]/\varepsilon, \quad A = 1, \ldots, \dim V, \tag{3.12}$$

with a fixed $B \leqslant \dim V$ and V_r' defined componentwise by

$$(V_r')^A = (V_r)^A \quad \text{for } A \neq B, \qquad (V_r')^B = (V_r')^B \pm \varepsilon. \tag{3.13}$$

The value of B and the sign $\pm$ of the perturbation vary cyclically.

Unless a segment $\mathscr{S}_A$ joins the source $x^A(0)$ to its caustic, the matrix G_r is invertible as soon as all its columns are different from zero. The first inversion is iterative and it takes advantage of the banded structure of G_r. At the following steps of the algorithm the following formula is applied:

$$G_r'^{-1} = G_r^{-1} - (1 - e_B \cdot G_r^{-1}\Delta)^{-1} G_r^{-1}\Delta \otimes (G_r^{\mathrm{T}})^{-1} e_B, \tag{3.14}$$

where $(e_B)_A = \delta_{AB}$ componentwise and advantage is taken of the fact that $G_r' - G_r = \Delta \otimes e_B$ for some vector Δ. Equation (3.14) is readily verified by multiplying both sides by $G_r' = G_r + \Delta \otimes e_B$. The vector Δ has at most 12 non-zero components, which allows an additional simplification of the computations. In the algorithm of Section 3.4, $\boldsymbol{\Delta}$ has at most six non-zero components.

The combination of the method of secants with local variations ensures that the algorithm is globally convergent. In practical ray-tracing problems significant progress is achieved only by the secant steps without relaxation ($\kappa = 0$). In typical successful applications of the algorithm local variations are invoked until all the columns of $\mathbf{G}_r$ are different from zero; the following steps generate new iterates by the secant method. Persistent failure of the secant method indicates that the solution does not exist, although in some cases it can be due to the fact that the perturbation ε does not match the accuracy of the computation of $\boldsymbol{F}(\boldsymbol{V})$.

In many cases the secant iterate reduces the error by a factor of 10–100. The main expenditure of CPU time is then due to the necessity of generating the secant matrix *ex nihilo*. For a ray intersecting 10 interfaces dim $\boldsymbol{V} = 63$ and it takes 63 algorithm steps before the secant method can be applied. The CPU expenditure per ray tracing is, however, considerably reduced in tracing large families of rays of similar shape if for every tracing the initial secant matrix $\mathbf{G}_0$ is set equal to the final secant matrix of the previous tracing. In such cases the first-approximation ray $\boldsymbol{V}_0$ is taken to be the final ray of the previous tracing. In the case of poor progress the secant matrix is refreshed by setting it equal to zero.

Computation of the function $\boldsymbol{F}(\boldsymbol{V})$ in general involves integration of the differential equations (3.2). For every block with $\rho = 0$ or -1 closed-form solutions (Appendix C) replace integration of (3.2). Use of closed-form formulae combines high accuracy with a considerable speed-up of the algorithm.

3.4 Two-point ray tracing in piecewise-homogeneous models

In a model consisting of homogeneous blocks a more efficient algorithm replaces parallel ray tracing. It reduces the order of the secant matrix $\mathbf{G}_r$ from $6N + 3$ to $2N + 2$. The ray is represented by a connected broken line. The broken line is specified by listing the points $\boldsymbol{x}^A$, $A = 1, \ldots, N$, joined successively by straight segments. The list starts with the ray origin $\boldsymbol{x}_0$, which is either fixed (as in ITH and ITHGEN) or allowed to move on a constraint surface $\mathcal{S}$ (as in ITHEX). The next points $\boldsymbol{x}^A$ represent the successive intersections of the ray with (physical) interfaces i_A. The point $\boldsymbol{x}^A$ is *assumed* to lie on interface i_A throughout the iterations. The end $\boldsymbol{x}^R$ of the ray is either fixed (as in ITHGEN) or allowed to move on a constraint surface $\mathcal{R}$.

Constraining surfaces allow significant reduction of the dimension of the vector $\boldsymbol{V}$ representing the ray. Each constraining surface (including interfaces; Section 1.2) is given in the parametric form $x_k = g_k(\xi, \eta)$, $k = 1, 2, 3$, and the corresponding point $\boldsymbol{x}^A$ is represented in terms of two parameters ξ and η. The parameters ξ and η of all the movable points $\boldsymbol{x}^A$ constitute the components of $\boldsymbol{V}$. The functions g_k representing the constraint surfaces $\mathscr{R}$ and $\mathscr{S}$ are defined by user-supplier Fortran functions.

Snell's law is expressed by the equations

$$(\boldsymbol{p}_{\text{out}} - \boldsymbol{p}_{\text{in}}) \cdot \boldsymbol{t}_r = 0, \quad r = 1, 2, \tag{3.15}$$

where $\boldsymbol{t}_1$ and $\boldsymbol{t}_2$ denote two linearly independent tangent vectors on the interface. The incident slowness $\boldsymbol{p}_{\text{in}}$ and the outgoing slowness $\boldsymbol{p}_{\text{out}}$ are given by the equations

$$\boldsymbol{p}_{\text{in}} = C_{\text{in}}(\boldsymbol{x}^A - \boldsymbol{x}^{A-1}), \qquad \boldsymbol{p}_{\text{out}} = C_{\text{out}}(\boldsymbol{x}^{A+1} - \boldsymbol{x}^A), \tag{3.16}$$

where

$$C_{\text{in}} = H_{\text{in}}(\boldsymbol{x}^A, \boldsymbol{x}^A - \boldsymbol{x}^{A-1})^{-1}, \qquad C_{\text{out}} = H_{\text{out}}(\boldsymbol{x}^A, \boldsymbol{x}^{A+1} - \boldsymbol{x}^A)^{-1}$$

and H_{in} and H_{out} denote the wave Hamiltonians appropriate for the incident and outgoing ray segment according to the raycode.

The requirement that the endpoint $\boldsymbol{x}^R$ of the ray should coincide with its target $\boldsymbol{x}_1$ is expressed in terms of two equations:

$$\xi^R - \xi_1 = 0, \qquad \eta^R - \eta_1 = 0. \tag{3.17}$$

Equations (3.17) are applied to every movable endpoint of the ray.

Equations (3.15) and (3.17) constitute a non-linear system of equations $\boldsymbol{F}(\boldsymbol{V}) = 0$. It is easy to see that the number of equations is always equal to the dimension of $\boldsymbol{V}$ (= twice the number of movable points of the ray).

The non-linear system $\boldsymbol{F}(\boldsymbol{V}) = 0$ is solved iteratively by Polak's algorithm. The Jacobian matrix $\mathrm{D}\boldsymbol{F}(\boldsymbol{V})$ is banded and invertible provided that $\boldsymbol{x}^R$ does not lie at a caustic of $\boldsymbol{x}^1$. Updating and inversion of the secant matrix is handled in the way described in Section 3.3. The algorithm based on (3.15) and (3.17) is more efficient and robust than parallel ray tracing. It has an extra advantage if both ends of the ray are fixed at $\boldsymbol{x}_0$ and $\boldsymbol{x}_1$ (Section 3.7). In this case all the components of the error vector $\boldsymbol{F}(\boldsymbol{V})$ are given by the same kind of expression (3.15) and no scaling is needed to balance their orders of magnitude in the stopping criterion.

The above algorithm allows tracing *head-wave rays* along plane interfaces. Indeed, Snell's law (3.15) implies critical incidence if the outgoing ray segment is tangent to the interface. However, construction of the first-approximation ray for head-wave tracing requires some comments. We describe here how a head-wave ray is generated from a reflected ray $\mathcal{R}_1$ obtained in a previous tracing. The algorithm described in Section 3.7 does not require tracing a reflected ray.

Let $\boldsymbol{x}_1$ denote an end of the ray $\mathcal{R}_1$. The target endpoint of the head-wave ray is shifted from $\boldsymbol{x}_1$ to a point $\boldsymbol{x}_2 = \boldsymbol{x}_1 + \boldsymbol{v}$. The algorithm replaces the reflection point $\boldsymbol{x}_2$ of $\mathcal{R}_1$ at an interface i_0 by a straight-line segment $[\boldsymbol{x}_2, \boldsymbol{x}_2 + \boldsymbol{v}']$. The vector $\boldsymbol{v}'$ is defined by a projection of $\boldsymbol{v}$ onto i_0. The ray code is supplemented with three entries: the reference number of the block on the transmission side of i_0, the wave species of $[\boldsymbol{x}_2, \boldsymbol{x}_2 + \boldsymbol{v}']$ and another occurrence of i_0.

3.5 Methods of generating the first-approximation ray

A fairly obvious method of generating the input ray V_0 for a two-point ray tracer consists in initial-value tracing of a ray with the required raycode and a guessed take-off direction. Initial-value ray tracers optionally express the result of tracing in terms of the vector V_0 appropriate for a two-point tracer.

The input ray V_0 for two-point ray tracing in a complex inhomogeneous or anisotropic model $\mathcal{M}_1$ can also be generated by solving an analogous two-point BVP in a simplified model $\mathcal{M}_0$. The relevant interfaces of the two models must be related by a continuous deformation. A two-point ray tracer based on the algorithm described in Section 3.4 optionally expresses the result of tracing in the form appropriate for parallel ray tracing. In most cases such an input ray ensures satisfactory convergence. In some cases the discrepancy between the input ray and the final ray is too large and the transition from model $\mathcal{M}_0$ to model $\mathcal{M}_1$ must involve intermediate models (Section 3.6).

A user-friendly two-point ray-tracing algorithm should not require an input ray. In Section 3.7 we describe such an algorithm for ray tracing in piecewise-homogeneous models.

In many applications large families of rays are traced. The rays can be grouped in such a way that the rays within one group are *kinematically analogous,* i.e. all the entries in their raycodes, except perhaps for the wave species, are identical. An appropriate ordering of rays within each group allows very efficient tracing requiring one or two secant iterations to achieve high accuracy. The final ray and the final secant matrix in each tracing provide the input ray and the initial secant matrix for the next tracing. The following input data can be varied for successive ray tracings: (1) coordinates of one or both endpoints; (2) wave species in the raycode; (3) numerical model parameters. The first case finds numerous applications in modelling DSS data, VSP profiles and constant-offset tracing. In the second case kinematic data for successive signal arrivals between a fixed pair of points are obtained. Kinematic inversion involves (3).

3.6 Continuation

In many applications the input ray $\mathcal{R}_0$ for parallel ray tracing of a ray $\mathcal{R}_1$ in a complex inhomogeneous model $\mathcal{M}_1$ is a solution of an analogous two-point ray-tracing problem in a geometrically related simpler model $\mathcal{M}_0$. Most often $\mathcal{M}_0$ is a piecewise-homogeneous approximation of $\mathcal{M}_1$ with the same or slightly shifted and deformed interfaces.

It is assumed that the interfaces i_A, $A=1, \ldots, N$, of $\mathcal{M}_0$ intersected by $\mathcal{R}_0$ can be put in a one-to-one correspondence with the interfaces j_A, $A=1, \ldots, N$, of $\mathcal{M}_1$ intersected by $\mathcal{R}_1$. The set i_A, $A=1, \ldots, N$, can be obtained by a continuous deformation of j_A, $A=1, \ldots, N$ in such a way that no interface intersections are created or annihilated.

In view of these assumptions it is possible to consider $\mathcal{R}_0$ as an approximation of $\mathcal{R}_1$. Usually the input ray $\mathcal{R}_0$ ensures fairly good convergence. Occasionally the effect of inhomogeneities in $\mathcal{M}_1$ is so strong that $\mathcal{R}_0$ is a very crude approximation of $\mathcal{R}_1$ and convergence is poor. In the hard cases the powerful *continuation method* (Roberts and Shipman, 1972) allows considerable speed-up of tracing.

The continuation method introduces a homotopy of models $\mathcal{M}_\theta$, $0 \leq \theta \leq 1$. The numerical data for $\mathcal{M}_\theta$ depend continuously on θ and for $\theta=0$ and 1 the models $\mathcal{M}_\theta$ coincide with $\mathcal{M}_0$ and $\mathcal{M}_1$. More specifically, the models $\mathcal{M}_\theta$ are defined in terms of the wave Hamiltonians

$$H(\boldsymbol{x}, \boldsymbol{p}; l, \sigma; \theta) = [\theta H_1(\boldsymbol{x}, \boldsymbol{p}; l, \sigma)^2 + (1-\theta) H_0(\boldsymbol{x}, \boldsymbol{p}; l, \sigma)^2]^{1/2} \quad (3.18)$$

and interfaces

$$f(\boldsymbol{x}; i; \theta) = \theta f_1(\boldsymbol{x}; i) + (1-\theta) f_0(\boldsymbol{x}; i) = 0 \quad (3.19)$$

where the subscripts 0 and 1 refer to the models $\mathcal{M}_0$, $\mathcal{M}_1$.

In the first step of ray tracing by continuation the ray $\mathcal{R}_0$ is used as the input ray for tracing in $\mathcal{M}_{\theta_1}$, where $0 < \theta_1 < 1$. The result of tracing in $\mathcal{M}_{\theta_1}$ provides an input ray for tracing in a model $\mathcal{M}_{\theta_2}$ with $\theta_2 > \theta_1$. A typical sequence of θ-values is $0, \frac{1}{2}, 1$ or $0, \frac{1}{3}, \frac{2}{3}, 1$.

The possibility of continuation has been provided for in two-point tracers by including the parameter θ in the subprograms computing wave

Hamiltonians and their derivatives. Continuation with respect to interfaces (3.19) was seldom used since two-point ray traces deal very efficiently with complicated interfaces involving strong dips, curvatures and dense corrugations. It is assumed that the model subprograms called by the tracing program involve a common block $/B/$ IMOD. If IMOD $=0$ then information about model $\mathcal{M}_0$ is required; if IMOD $=1$ then information about model $\mathcal{M}_1$ should be returned on exit.

A powerful and elegant application of continuation in two-dimensional ray tracing is described by Keller and Perozzi (1983). It does not, however, extend to three-dimensional models.

3.7 Autonomous ray tracers

In a relatively simple piecewise-homogeneous model the first-approximation ray can be generated by a largely arbitrary construction. This method allows tracing several solutions of the same two-point ray-tracing problem provided that the construction depends on some input parameters (Example 3 in Appendix G).

We describe here the algorithm applied in the ray tracer ITHGEN. The algorithm allows tracing reflected/transmitted and head-wave rays (edge-diffracted rays can easily be included by redefining the function $\boldsymbol{F}(\boldsymbol{V})$; cf. Section 3.8). The input data include the two endpoints $\boldsymbol{x}_0, \boldsymbol{x}_1$ and a sequence $i_1, \ldots, i_N$ of interface reference numbers. Each interface i_A is assigned three Boolean constants $\rho_A, \sigma_A, \lambda_A$. If $\rho_A=1$ then the ray is reflected at i_A; otherwise it is transmitted. $\sigma_A=1$ entails conversion at i_A. If $i_A=i_{A+1}$ then $\lambda_A=1$ indicates that i_A is a plane interface and the two successive intersection points of the ray with i_A should be joined by a straight segment lying in the plane i_A. We denote by $j_1, \ldots, j_M$ the sequence of such i_A that $\rho_A=1$.

The construction depends on a three-component vector $\boldsymbol{N}'$. The algorithm replaces $\boldsymbol{N}'$ by a unit vector $\boldsymbol{N}=c\boldsymbol{N}'$, $c>0$. The segment $[\boldsymbol{x}_0, \boldsymbol{x}_1]$ is subdivided into $M+1$ subsegments $[\boldsymbol{y}_A, \boldsymbol{y}_{A+1}]$ of equal length, with $\boldsymbol{y}_0=\boldsymbol{x}_0$ and $\boldsymbol{y}_{M+1}=\boldsymbol{x}_1$. The reflection points $\boldsymbol{x}_A$ are defined by the following rule. If the oriented segment $\boldsymbol{y}_A+s\boldsymbol{N}$, $0\leqslant s\leqslant S$, intersects interface i_A then $\boldsymbol{x}_A$ is the first intersection of $\boldsymbol{y}_A+s\boldsymbol{N}$ with i_A. Otherwise, if the oriented segment $\boldsymbol{y}_A-s\boldsymbol{N}$, $0\leqslant s\leqslant S$, intersects i_A then the first intersection defines $\boldsymbol{x}_A$. Failure stop follows if neither segment intersects i_A.

In the next step module SUBDIV determines successive intersections of each segment $[\boldsymbol{y}_A, \boldsymbol{y}_{A+1}]$ such that $\lambda_A\neq 1$ with model interfaces. On exit, SUBDIV returns the raycode and the intersection points of the input ray with physical interfaces.

In the third step the intersection points are expressed in terms of the parameters ξ and η and the algorithm of Section 3.4 is applied.

In most cases the algorithm converges very quickly. Rapid convergence (5–10 secant steps for a travel-time error of order 10^{-8}) has been observed in the case of strongly dipping plane interfaces (Fig. 1) as well as for curved interfaces defined in terms of splines. In case of difficulties the algorithm offers the possibility of applying the continuation method.

In order to allow for continuation, module SUBDIV defines a *local model* $\mathcal{M}_0$ in such a way that the input ray satisfies the law of reflection in $\mathcal{M}_0$. Model $\mathcal{M}_0$ consists of locally defined almost plane mirrors; transmitting interfaces are ignored. The adjective "local" refers to the fact that $\mathcal{M}_0$ is in general inconsistent on a global scale on account of interface intersections, which have no counterparts in $\mathcal{M}_1$.

The intermediate models $\mathcal{M}_\theta$, $0 \leqslant \theta \leqslant 1$, are defined by the homotopy

$$\boldsymbol{x} = \boldsymbol{g}_A(\xi, \eta; \theta) = \boldsymbol{g}(\xi, \eta) - (1-\theta)[(\boldsymbol{g}(\xi, \eta) - \boldsymbol{x}_A) \cdot \boldsymbol{v}_A]\boldsymbol{v}_A \quad (3.20)$$

where $\boldsymbol{x} = \boldsymbol{g}(\xi, \eta)$ is a parametric representation of the interface containing $\boldsymbol{x}_A$ in model $\mathcal{M}_1$ and $\boldsymbol{v}_A$ is a unit vector parallel to the bisector of the angle subtended by the two legs of the input ray at $\boldsymbol{x}_A$. For $\theta = 1$, $\boldsymbol{g}_A = \boldsymbol{g}$, while for $\theta = 0$ the plane tangent to $\boldsymbol{g}_A$ is orthogonal to $\boldsymbol{v}_A$.

3.8 Two-point tracing of edge-diffracted rays

Snell's law of edge diffraction

$$(\boldsymbol{p}_{\text{out}} - \boldsymbol{p}_{\text{in}}) \cdot \boldsymbol{t} = 0 \quad (3.21)$$

follows from the condition that the eikonal functions of the incident and diffracted wave coincide along the edge: $S_{\text{in}}(\boldsymbol{x}(s)) = S_{\text{out}}(\boldsymbol{x}(s))$, with $\boldsymbol{t} = \mathrm{d}\boldsymbol{x}/\mathrm{d}s$. In piecewise-homogeneous media $\boldsymbol{p}_{\text{in}}$ and $\boldsymbol{p}_{\text{out}}$ are defined by (3.16), while in parallel ray tracing

$$\boldsymbol{p}_{\text{out}} = \frac{\boldsymbol{p}^{A+1}(0)}{H_{A+1}(\boldsymbol{x}^{A+1}(0), \boldsymbol{p}^{A+1}(0))}, \qquad \boldsymbol{p}_{\text{in}} = \frac{\boldsymbol{p}^{A}(U)}{H_A(\boldsymbol{x}^A(U), \boldsymbol{p}^A(U))}.$$

Equation (3.21) is supplemented by the requirement that the ray hits the edge, defined by the intersection of an interface $f(\boldsymbol{x}; i_0) = 0$ with an auxiliary user-defined surface

$$g(\boldsymbol{x}) = 0 \quad (3.22)$$

which can be another interface. It is assumed that the two surfaces intersect transversally:

$$\boldsymbol{v}_1 \times \boldsymbol{v}_2 \neq 0 \tag{3.23}$$

where $\boldsymbol{v}_1$ and $\boldsymbol{v}_2$ denote the respective normals.

At the diffracting interface i_0, (3.21) and (3.22) replace (3.7) for parallel ray tracing and (3.15) for two-point tracing in piecewise-homogeneous media.

We describe below the output of two modules tracing edge-diffracted rays: ITID (parallel tracing) and ITHDIF (tracing in piecewise-homogeneous models).

The output data include the following input data for initial-value tracing along the final ray: the ray origin $\boldsymbol{x}_0$; the initial slowness; the reference number of the block entered by the diffracted ray and the initial slowness $\boldsymbol{p}_\mathrm{d}$ of the diffracted ray (cf. Section 4.16 concerning the use of these data in DRT).

Tracing edge-diffracted rays requires some caution. Owing to the error inherent in the solution of the two-point problem, the actual ray retraced by an initial-value tracer does not hit the edge. In particular, it can additionally intersect one of the two surfaces meeting at the edge, which in turn might entail a misinterpretation of the input data by the initial-value tracer. DYNTRAC avoids such a misinterpretation by excluding the auxiliary surface $g(\boldsymbol{x}) = 0$ from the list of interrupt surfaces if $g(\boldsymbol{x}) = 0$ coincides with an interface.

The value of $\boldsymbol{p}_\mathrm{d}$ is not reliable if the diffracted ray intersects an interface at a point lying in the vicinity of the diffraction point. For this reason ITHDIF estimates the error E inherent in the determination of the intersection points. A warning is issued if the distance of the first interface intersection of the diffracted ray from the diffraction point does not exceed $10E$.

In many practical applications a diffracting edge is detected and located before the physical nature of the surface $g(\boldsymbol{x}) = 0$ defining the edge is known. The surface $g(\boldsymbol{x}) = 0$ can also represent a fault plane rather than a discontinuity surface of model parameters. In such cases one can avoid the difficulties mentioned in the two preceding paragraphs by defining the edge in terms of an auxiliary surface $g(\boldsymbol{x}) = 0$ that is *not* an interface.

3.9 Edge tracing

Model geometry is primarily defined in terms of interfaces determined from reflection data and expressed in terms of spline interpolation of grid data. In such a model diffracting edges are implicitly defined by interface intersections, although their precise location is unknown. In order to relate reflection and diffraction data, it is desirable to visualize diffracting edges by means of computer graphics. It is also desirable to have a parametric representation $\boldsymbol{x}=\boldsymbol{x}(s)$ of the edge, for example for the sake of tracing a family of edge-diffracted rays by two-point tracing. It is possible to achieve these goals by using an algorithm related to initial-value tracing.

The edge $\boldsymbol{x}(s)$ of $f(\boldsymbol{x}; i)=0$ and $g(\boldsymbol{x})=0$ satisfies the differential equations

$$\frac{\mathrm{d}\boldsymbol{x}}{\mathrm{d}s}=|\boldsymbol{\nu}_1\times\boldsymbol{\nu}_2|^{-1}\,\boldsymbol{\nu}_1\times\boldsymbol{\nu}_2, \tag{3.24}$$

where $\boldsymbol{\nu}_1$ and $\boldsymbol{\nu}_2$ denote the normals to the two surfaces and transversality is assumed.

The initial data for (3.24) are found by defining a start surface $\boldsymbol{x}=\boldsymbol{s}(\xi, \eta)$, which intersects the edge transversally. The initial point $\boldsymbol{x}(0)$ is defined by the intersection of the start surface with the edge. It is determined by applying Polak's algorithm (3.3) to the equations $f(\boldsymbol{x}; i)=0$ and $g(\boldsymbol{x})=0$ restricted to $\boldsymbol{x}=\boldsymbol{s}(\xi, \eta)$. The point $\boldsymbol{x}(0)$ is found with a precision limited only by the errors of computing f and g in two or three secant steps.

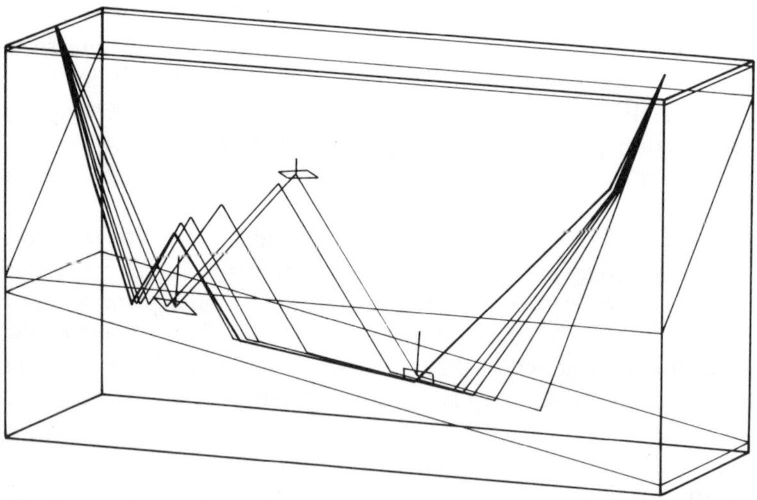

Fig. 1 Two-point tracing of a head-wave ray in a model with strongly dipping plane interfaces. Local model interfaces are indicated by small rectangles and by their normals at the reflection points of the input ray. The iterates are indicated by thin solid lines while the final solution is indicated by a heavy solid line, the wire diagram of the model interfaces was obtained by the method of Section 3.9.

Integration of (3.24) is carried out by a version of INTFINC. The interrupt surfaces include all the interfaces and model box faces that can intersect the edge. Edge points $\boldsymbol{x}(s)$ are stored on a sequential file at equispaced arclength values.

A somewhat more elaborate version of this algorithm allows tracing a contour of an interface i_0 enclosed in a box $\mathcal{B}'$. $\mathcal{B}'$ can coincide with the model box $\mathcal{B}$ (Section 1.2) or be bounded by some interfaces (e.g. fault planes). The algorithm determines the point $\boldsymbol{x}(0)$ and INTFINC traces the intersection of i_0 with another interface or model box face. INTFINC exits at an interface or model box face i_1, returning the number i_1 and the intersection point $\boldsymbol{x}_1$. The algorithm then proceeds to trace the intersection of i_0 and i_1 with the initial data $\boldsymbol{x}_1$. Contour tracing continues until the loop is closed. Repeated application of this algorithm to a collection of interfaces produces a complete wire diagram of the model, displaying all the diffracting edges (Fig. 1).

3.10 Tracing NIP rays (normal ray tracing)

The results of CMP stacking are commonly interpreted in terms of signals that travel to a reflecting interface i_0 and retrace their path backwards to the source. It follows immediately that the ray is orthogonal to i_0 at the reflection point. Modelling CMP stacking data leads to the problem of finding a ray that is orthogonal to a given reflector i_0 at one end and terminates at a given shotpoint $\boldsymbol{x}_1$ at the other end. The raycode is also specified. Such a ray represents the path of the signal after reflection.

Our discussion of NIP rays is restricted to piecewise-homogeneous models, although a generalization to inhomogeneous models does not present any difficulty. The algorithm for tracing *NIP rays* is a modification of the algorithm described in Section 3.4. At every interface except for i_0, (3.15) are applied. The following equations are imposed at i_0:

$$(\boldsymbol{x}^2 - \boldsymbol{x}^1) \cdot \boldsymbol{t}_r = 0, \quad r = 1, 2, \tag{3.25}$$

where $\boldsymbol{t}_1$ and $\boldsymbol{t}_2$ denote two linearly independent tangent vectors at i_0 and $[\boldsymbol{x}^1, \boldsymbol{x}^2]$ denotes the ray segment adjacent to i_0.

Interpretation of CMP stacking data involving curved and irregular reflectors in marine seismics often indicates presence of NIP rays reflected from a diffracting edge of i_0 (J. Pajchel, private communication). *Edge-diffracted NIP rays* satisfy the boundary-value problem defined by (3.15), except at i_0, where the equations

$$g(\boldsymbol{x}^1) = 0, \tag{3.26}$$

$$(\boldsymbol{x}^2 - \boldsymbol{x}^1) \cdot \boldsymbol{t} = 0 \tag{3.27}$$

are applied. The edge is defined by the intersection of $f(\boldsymbol{x}; i_0) = 0$ with an auxiliary user-defined surface $g(\boldsymbol{x}) = 0$. The tangent $\boldsymbol{t}$ to the edge is defined by the vector product of the normals of the two interfaces. The normal to $g(\boldsymbol{x}) = 0$ is defined by a user-supplied subroutine NORG.

The ray code of an NIP ray assumes the following form:

RC(1)	number of physical interfaces intersected by the ray, including i_0;
RC(2) = RC(3) = 0;	
RC(3k + 1)	interface reference number (RC(4) = i_0);
RC(3k + 2)	block reference number;
RC(3k + 3)	wave species.

Both NIP ray-tracing algorithms are very efficient and robust. They have found application in modelling data from marine seismics.

3.11 Travel time and travel-time error in two-point tracing

In parallel ray tracing the travel time is given by

$$T = U \sum_{A=1}^{N} H(\boldsymbol{x}^A(0), \boldsymbol{p}^A(0)), \tag{3.28}$$

while in ray tracing in an isotropic piecewise-homogeneous model

$$T = \sum_{A=1}^{N} |\boldsymbol{x}^{A+1} - \boldsymbol{x}^A|/c_A, \tag{3.29}$$

where c_A denote the appropriate propagation speeds.

Travel-time error is computed according to the formula

$$\Delta T = \sum_k \sum_A \sum_i \frac{\mathrm{d}T}{\mathrm{d}x_i^A} \frac{\mathrm{d}x_i^A}{\mathrm{d}V_k} (\boldsymbol{GF}(\boldsymbol{V}))_k, \tag{3.30}$$

where $\boldsymbol{G}$ denotes the last update of the secant matrix and k refers to the components of the vectors $\boldsymbol{V}$ and $\boldsymbol{GF}(\boldsymbol{V})$.

3.12 Warnings

Two-point ray tracers work on the assumption that the raycode is fixed throughout the iterations. It follows that each ray segment is assigned a fixed block reference number. The actual signal path defined by the geometry of the final ray can intersect a different sequence of blocks. In this case the solution is wrong and the program issues a warning.

Persistent failure in the secants usually indicates lack of solutions, but it can also result from poor accuracy of the model programs or from inappropriate values of control parameters. Ray tracing in piecewise-homogeneous models is very robust even if the model geometry is complicated. On the other hand, success and efficiency of parallel ray tracing in general inhomogeneous blocks may be strongly affected by the choice of control parameters. It is therefore recommended that optimal values of control parameters for a given class of models be determined by running tests. The key parameters are the perturbation step and the tolerance for integration of Hamiltonian equations. The remaining parameters can affect the number of computations and the speed.

4 DYNAMIC RAY TRACING

4.1 Introduction

The following quantities are traced in DRT: (1) the ray $\boldsymbol{x}(t)$ and the slowness vector $\boldsymbol{p}(t)$; (2) a ray-spreading Jacobian $J(t)$; (3) the amplitude A; (4) the phase shift α. The pulse recorded at the point $\boldsymbol{x}(t)$ is given by (1.8) and (1.7) provided that $\boldsymbol{x}(t)$ is not situated in the vicinity of a caustic, a shadow boundary or a critical ray. A composite pulse recorded in the vicinity of a caustic or a shadow boundary is expressed in terms of the same quantities traced along a few contributing rays (Section 5).

Ray spreading Jacobians and S-wave polarization are traced by integration of appropriate differential equations accompanied by solving boundary conditions at interfaces. The amplitude A is piecewise-constant, and its value varies at the intersection of interfaces only. The *total phase shift*

$$\alpha = \alpha_1 + \tfrac{1}{4}\pi\alpha_2 \tag{4.1}$$

consists of a real component α_1 generated at intersections of the ray with interfaces and an integer multiple of $\frac{1}{4}\pi$ generated at the contact points of the ray with caustics. The variables A and α_1 are manipulated exclusively by the module TRNSFM6 responsible for solving the boundary conditions at interfaces, while the value of the integer α_2 is updated at every transition from regular to singular DRT and vice versa.

It follows easily from equations (56) of Hanyga (1988a) that in a regular region $\alpha_2 = 2M$ is an even number. In this case M is the well-known Maslov (or KMAH) index. Unlike the KMAH index, the integer phase shift α_2 in a singular region depends on the choice of the canonical transformation underlying singular DRT (Section 4.8).

As it stands, the program DYNTRAC allows for a general point source. Minor modifications (Section 4.10) would allow for a distributed source.

In order to avoid more specific assumptions about the directivity of the point source, DYNTRAC assumes the initial data $A = 1$ and $\alpha = 0$ at the source. For the same reason the initial polarization of an S wave generated at the source is defined by a user-supplied subprogram INPOL.

On exit, DYNTRAC records the final values of position $\boldsymbol{x}$ and slowness $\boldsymbol{p}$. This allows continuation of DRT and is recommended in DRT along branching ray paths. For a path consisting of two parts traced separately the amplitude A, travel time T and phase shift α are given by $A = A_1 A_2$, $T = T_1 + T_2$, $\alpha = \alpha_1 + \alpha_2$.

4.2 DRT of continuously varying quantities

In general, ray spreading and S-wave polarization are obtained by integration of appropriate systems of differential equations. The equation governing the variation of S-wave polarization is discussed in Section 4.11. On the other hand, ray spreading requires special handling in the vicinity of caustics (including the point source). The differential equations governing ray spreading fall into three classes: regular, singular and linear DRT, discussed separately. The appropriate regions of validity are defined in terms of inequalities involving the ray spreading and auxiliary quantities traced. Each class is handled by a separate submodule. The transformations of the traced quantities at transitions from one class to another are carried out by appropriate transformation modules.

Singular and linear DRT are well defined in caustic regions. We concentrate here on a method involving singular DRT in the vicinity of generic caustics and linear DRT in the vicinity of a point source; regular DRT is applied in the remaining regions. An alternative approach based exclusively on linear DRT is discussed in Section 4.18.

4.3 Regular DRT. Ray-centred coordinates

At a distance from caustics the most economical way of tracing ray spreading is *regular DRT*. The fundamental system of 13 equations (2.3.7), (2.5.41) and (2.5.42) of Hanyga *et al.* (1984) is reduced by recourse to a system of *ray-centred coordinates* (τ, y_1, y_2) (Hanyga, 1982, 1986) defined in a neighbourhood of the ray $\boldsymbol{x}(t)$, $\boldsymbol{p}(t)$.

The ray-centred coordinate system of index $I = 1, 2$ or 3 (Fig. 2) is

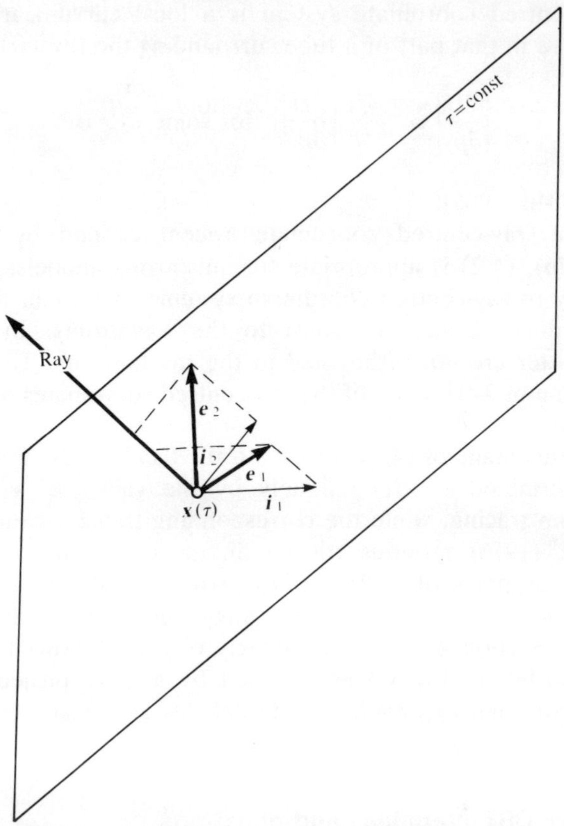

Fig. 2 Ray-centred coordinates. The planes $\tau = \text{const}$ are tangent to the wavefronts at the central ray. For $I = 3$ the coordinate vectors $\boldsymbol{e}_1$, $\boldsymbol{e}_2$ of the y coordinates are projections of the x_1, x_2 coordinate vectors $\boldsymbol{i}_1$, $\boldsymbol{i}_2$ onto the plane $\tau = \text{const}$:

$$\boldsymbol{e}_a = \textstyle\sum_k P_{ka}\boldsymbol{i}_k = \boldsymbol{i}_a - p_a(\partial H/\partial \boldsymbol{p})/H, \quad a = 1, 2.$$

defined by the formulae

$$x_k = \sum_{\lambda=1}^{3}{}' P_{k\lambda}(\tau) y_{\pi(\lambda)} \quad \text{for } k = 1, 2, 3, \tag{4.2}$$

where the prime indicates that $\lambda = I$ should be left out. If λ is the smaller of the numbers 1, 2, 3 not equal to I then $\pi(\lambda) = 1$, otherwise $\pi(\lambda) = 2$. The matrix P_{kl} is defined in terms of the projection operator

$$P_{kl}(\boldsymbol{x}, \boldsymbol{p}) = \delta_{kl} - \frac{1}{H(\boldsymbol{x}, \boldsymbol{p})} p_l \frac{\partial H}{\partial p_k}(\boldsymbol{x}, \boldsymbol{p}) \tag{4.3}$$

evaluated at $(\boldsymbol{x}(\tau), \boldsymbol{p}(\tau))$.

The ray-centred coordinate system is a local curvilinear coordinate system defined in that part of a tube surrounding the ray $x(t)$ for which

$$\left|\frac{\partial H}{\partial p_I}\right| > \varepsilon_2 \sum_i^3 \left|\frac{\partial H}{\partial p_i}\right| \quad \text{for some } \varepsilon_2 > 0 \tag{4.4}$$

(Hanyga, 1984d, 1986).

Unlike the ray-centred coordinate system defined by Popov and Pšenčík (1976), (4.2) is appropriate for anisotropic models. Indeed, the key property of ray-centred coordinate systems consists in the tangency of the coordinate planes $\tau = \text{const}$ to the wavefronts. In anisotropic media the latter are not orthogonal to the ray tangents. This, however, contradicts the orthogonality of the ray-centred coordinates of Popov and Pšenčík (1976).

Another advantage of (4.2) consists in the fact that the transformation (4.2) is determined at $\boldsymbol{x}(t)$ uniquely by the values of $\boldsymbol{x}(t)$ and $\boldsymbol{p}(t)$ obtained from tracing, while the corresponding transformation in Popov and Pšenčík (1976) requires tracing an angle φ. Except for S-wave tracing, this property of (4.2) implies saving one differential equation. For S waves φ is essentially the angle ψ of rotation of S-wave polarization (Section 4.11). This property of (4.2) follows from the fact that the coordinates (y_1, y_2) are defined by a mere projection of two Cartesian coordinates x_i, $i \neq I$, onto the planes $\tau = \text{const}$.

4.4 Regular DRT. Variables and equations

The ray-centred coordinate system I underlying the definition of the equations and variables is initially chosen in such a way that

$$\left|\frac{\partial H}{\partial p_I}\right| \geqslant \left|\frac{\partial H}{\partial p_k}\right| \quad \text{for } k = 1, 2, 3 \tag{4.5}$$

and is updated as soon as $|\partial H/\partial p_I|$ has fallen below some $\varepsilon_2 > 0$.

We define the *wavefront curvature matrix* $C_{\alpha\beta}$, with α and β ranging over $1, 2$, by the formula

$$C_{\alpha\beta} := \frac{\partial^2 \tilde{S}(\tau, y_1, y_2)}{\partial y_\alpha \, \partial y_\beta} = \sum_{k,l} \frac{\partial^2 S}{\partial x_k \, \partial x_l} P_{ka}(\tau) P_{kb}(\tau), \tag{4.6}$$

with $\alpha = \pi(a)$ and $\beta = \pi(b)$, where $S(\boldsymbol{x})$ denotes the eikonal and $\tilde{S}(\tau, y)$ is an expression for the eikonal in terms of ray-centred coordinates.

The ray-tracing equations assume the form

$$\frac{\mathrm{d}\boldsymbol{C}}{\mathrm{d}t} = -\boldsymbol{CGC} - \boldsymbol{W}^{\mathrm{T}}\boldsymbol{C} - \boldsymbol{CW} - \boldsymbol{K}, \tag{4.7}$$

$$\frac{\mathrm{d}J}{\mathrm{d}t} = J\left(\sum_i \frac{\partial^2 H}{\partial x_i\, \partial p_i} + \sum_{\alpha,\beta=1}^{2} G^{\alpha\beta} C_{\alpha\beta}\right), \tag{4.8}$$

with the 2×2 matrices $G^{\alpha\beta}$, $K_{\alpha\beta}$ and

$$W^{\alpha}{}_{\beta} = U^{\alpha}{}_{\beta} + \sum_{\gamma} G^{\alpha\gamma} H_{\gamma} p_{\beta} \tag{4.9}$$

defined in Hanyga (1982, 1984c, 1986), $H_\gamma = \partial H/\partial y_\gamma$ and $\boldsymbol{C} = [C_{\alpha\beta}]$.

Specific expressions for $\boldsymbol{G}$, $\boldsymbol{K}$ and $\boldsymbol{U}$ for isotropic and transversely isotropic media can be found in the Appendices to Hanyga (1986).

The vector w_k, $k = 1, \ldots, 10$, is defined by the following assignments:

$$w_k = x_k, \qquad w_{k+3} = p_k \qquad \text{for } k = 1, 2, 3 \tag{4.10a}$$

$$w_{4+\alpha+\beta} = C_{\alpha\beta}, \qquad w_{10} = J. \tag{4.10b}$$

For large values of $|J|$ the variable J is expressed in terms of a new dependent variable $L: J = \operatorname{sign}(J)\, \mathrm{e}^{L}$. The transformation from J to L and vice versa occurs at an interrupt surface $|J| = B$, where B is a large number.

4.5 Retrieving the second derivatives $\partial^2 S/\partial x_i\, \partial x_j$ from $C_{\alpha\beta}$

Equations (68)–(71) of Hanyga (1984c), expressing $S_{ij} = \partial^2 S/\partial x_i\, \partial x_j$ in terms of $C_{\alpha\beta}$ contain an error. The correct formulae are as follows:

$$S_{ab} = \Gamma_{ab} + A p_a p_b + D_a p_b + D_b p_a, \tag{4.11}$$

with

$$\Gamma_{ab} = C_{\pi(a),\pi(b)} \quad \text{for} \quad a \neq I \text{ (Section 4.3)}, \tag{4.12}$$

$$A = -\sum_i^3 \frac{\partial H}{\partial x_i} \frac{\partial H}{\partial p_i} \bigg/ H^2, \tag{4.13}$$

$$D_a = -\left(\frac{\partial H}{\partial x_a} \bigg/ H + A p_a\right) \quad \text{for } a \neq I, \tag{4.14}$$

$$D_I = -\sum_{a(\neq I)} D_a E_a, \tag{4.15}$$

$$S_{Ia} = S_{aI} = A p_a p_I + D_I p_a + p_I D_a - \sum_{b(\neq I)} \Gamma_{ab} E_b p_I, \tag{4.16}$$

$$E_a = \sum_{b(\neq I)} Q_{ab} \frac{\partial H}{\partial p_b} \Big/ H, \tag{4.17}$$

$$S_{II} = Ap_I^2 + 2D_I p_I + \sum_{\substack{a,b \\ (\neq I)}} \Gamma_{ab} E_a E_b p_I^2, \tag{4.18}$$

$$[Q_{ab}] = \text{inverse of } [P_{ab}, a, b \neq I], \tag{4.19}$$

where I denotes the index of the ray-centred coordinate system and $a \neq I$.

The transformation from w_k to $\partial^2 S / \partial x_k \, \partial x_l$ is performed by the subroutine SECDER.

4.6 Transformations of w between two ray-centred coordinate systems

Transformation of the dependent-variable vector $\boldsymbol{w}$ from a ray-centred coordinate system I to a ray-centred coordinate system I' in regular DRT is a composition of (4.11)–(4.19) and (4.6) with I replaced by I'. Transformation of dependent variables in linear DRT is derived in Hanyga (1987).

4.7 Singular DRT. General remarks

The main objective of singular DRT consists in avoiding the singularities $J \cdot C_{\alpha\beta} \to 0 \cdot \infty$ in (4.8) and $J \to 0$ in (1.7). These singularities appear at the intersection points of the ray with the caustics. For this reason in singular DRT the variables J and $C_{\alpha\beta}$ are replaced by their well-behaved equivalents. Transition from regular to singular DRT occurs whenever $|J|$ falls below a small number ε_1.

Singular DRT is also recommended whenever excessively large values of wavefront curvature, $|C_{\alpha\beta}| > B$, in regular DRT cause a considerable slowdown of integration. Unlike ε_1, the value of the threshold B should be related to the tolerance parameters (Appendix F). Although large values of $C_{\alpha\beta}$ typically accompany small values of J, no *a priori* relationship between their magnitudes can be established.

These considerations suggest the following definition of the *singular region*

$$\text{either } |J| < \varepsilon_1 \quad \text{or} \quad \exists(\alpha, \beta) \;\; |C_{\alpha\beta}| > B. \tag{4.20}$$

In regular DRT the boundary of the singular region (4.20) is defined in terms of four interrupt surfaces $J = \varepsilon_1 \operatorname{sign} J$ and $|C_{\alpha\beta}| = B$, $\alpha, \beta = 1, 2$, $\alpha \geqslant \beta$. The sign of J is stored in a variable and updated after every crossing of a caustic or an interface. An apparently simpler interrupt surface $|J| = \varepsilon_1$ is often crossed twice in one scanning step of CHINTR2 (Section 2.1) if the ray intersects the caustic $J = 0$. In this case the caustic and the transition to singular DRT are overlooked and the final value of phase shift is wrong. This difficulty is due to the extremely quick variation of J in the vicinity of a caustic.

Monitoring intersections with the boundary of (4.20) in singular DRT would be computationally expensive, and therefore the following compromise solution has been applied.

If the transition to singular DRT has been caused by an intersection with $|J| = \varepsilon_1$ then $|J| = \varepsilon_1$ is treated as an interrupt surface in singular DRT. Otherwise, the transition to singular DRT is due to the second disjunct of (4.20), and the validity of (4.20) is checked by the subroutine OUTPUT at time intervals much larger than the integration step. Exit from singular DRT occurs as soon as (4.20) turns out to be false.

For a sufficiently high threshold B the singular region is effectively defined by the first disjunct of (4.20), since a blow-up of wavefront curvature is possible at caustics only. A decrease of B can, however, reduce the number of step-size reductions by eliminating excessively high values of $C_{\alpha\beta}$ in the non-linear equation (4.7). Although this entails a more frequent recourse to the larger system (4.23), (4.24) of equations, the CPU expenditure can be reduced in some cases.

Singular DRT involves three additional equations in comparison with regular DRT. Computation of the functions $f(\boldsymbol{x}; k)$ associated with the interrupt surfaces is more expensive in singular DRT. It is therefore recommended that the singular region be restricted as much as possible, provided only that its objectives are satisfied. This precaution also reduces the possibility of a failure exit caused by the appearance of an interface intersection in the singular region.

4.8 Singular DRT. Equations

The specific equations applied in singular DRT are determined by an index ISING referring to the underlying canonical transformation (Hanyga, 1984c).

For ISING $= 0$ the transformation is an identity, and singular DRT is identical to the regular DRT without the reduction associated with a recourse to ray-centred coordinates.

If $0 < \text{ISING} = \iota \leqslant 3$ then the canonical transformation becomes

$$\left.\begin{aligned} \tilde{x}_k &= x_k, & \tilde{p}_k &= p_k & \text{for } k \neq \iota, \\ \tilde{x}_\iota &= p_\iota, & \tilde{p}_\iota &= -x_\iota. \end{aligned}\right\} \tag{4.21}$$

For $3 < \text{ISING} = \iota < 7$ the canonical transformation is defined by

$$\tilde{x}_k = p_k, \qquad \tilde{p}_k = -x_k \qquad \text{for } k \neq \iota,$$

$$\tilde{x}_\iota = x_\iota, \qquad \tilde{p}_\iota = p_\iota. \tag{4.22}$$

For ISING = 7 we assume that $\tilde{x}_k = p_k$ and $\tilde{p}_k = -x_k$ for $k = 1, 2, 3$.

We now define the sign function $\sigma(K)$ and the permutation $\rho(K)$ in such a way that the canonical transformation can be expressed in the form $\tilde{y}_K = \sigma(K) y_{\rho(K)}$, with $y_k = x_k$ and $y_{k+3} = p_k$ for $k = 1, 2, 3$; $K = 1, \ldots, 6$. Equations (65) and (66) of Hanyga (1984c) assume the forms

$$\frac{\mathrm{d}\tilde{C}_{ij}}{\mathrm{d}t} = -\sum_{K,L=1}^{6} R_{Ki} H_{KL} R_{Lj}, \tag{4.23}$$

$$\frac{\mathrm{d}\tilde{J}}{\mathrm{d}t} = \tilde{J} \sum_{k}^{3} \sum_{L}^{6} R_{LK} H_{\rho(k+3),L}\, \sigma(k+3), \tag{4.24}$$

where R_{Ki} is defined by

$$\tilde{R}_{K,i} = \begin{cases} \delta_{Ki} & (K \leqslant 3), \\ \tilde{C}_{K-3,i} & (K > 3), \end{cases} \tag{4.25}$$

$$R_{\rho(K),i} = \sigma(K) \tilde{R}_{Ki} \quad \text{for } K = 1, \ldots, 6,\ i = 1, \ldots, 3. \tag{4.26}$$

$$H_{KL} = \frac{\partial^2 H}{\partial y_K\, \partial y_L}, \quad K, L = 1, \ldots, 6.$$

The symmetric matrix $\tilde{C}_{ij} = \partial \tilde{y}_{i+3} / \partial \tilde{y}_j$ and the *generalized ray-spreading Jacobian* $\tilde{J}$ define the components $7, \ldots, 13$ of the dependent-variable vector w_k in the following linear order: $\tilde{C}_{11}$, $\tilde{C}_{21}$, $\tilde{C}_{31}$, $\tilde{C}_{22}$, $\tilde{C}_{32}$, $\tilde{C}_{33}$, $\tilde{J}$. If S waves are being traced then w_{14} is the polarization angle ψ.

Whenever a stop in singular DRT is admitted, singular DRT is restricted to ISING $\leqslant 3$ by setting the argument IOPT = 3 in the module carrying out the transformation.

4.9 Transformation from regular to singular DRT

The transformation is based on the transformation $\boldsymbol{w} \rightarrow \boldsymbol{S}$ described in Section 4.5 followed by equations (72) of Hanyga (1984c). The index ISING is chosen in such a way that the associated generalized ray-spreading Jacobian J assumes a maximum absolute value under the constraint ISING $\leqslant$ IOPT, with IOPT $= 0$, 3, 6 or 7. The Jacobian is computed according to equation (73) of Hanyga (1984c).

The inverse transformation, carried out by TRNSFM3, is based on equations (75)–(77) of Hanyga (1984c) followed by (4.6); the value of the ray-centred coordinate index I in (4.6) is determined by recourse to (4.4).

The modules TRNSFM2 and TRNSFM3 carrying out the transformation also compute integer phase shift according to equation (56) of Hanyga (1984a). In the case of termination in a singular region the value of the total phase shift *before* the ray entered the singular region is returned on exit. Relative phase shifts of the signals that have or have not yet travelled through the caustic are properly accounted for by the composite pulse expressions mentioned in Section 5.

In the case of termination in a singular region the value of the total phase shift just before entering the singular region is returned on exit. The relative phase shifts of the signals that have or have not yet reached the caustic are accounted for by the composite pulse expressions (Section 5).

4.10 Linear DRT

Although the point source is a caustic, it is not convenient to apply singular DRT in its vicinity on account of particular geometry. A point caustic is an exceptional case that is unlikely to occur unless it has been imposed by a symmetry of the problem. Therefore it is not covered by the theory of Hanyga and Seredyńska (1988).

Following Červený and Pšenčík (1983), the linear DRT equations of Hanyga (1983) are applied in the vicinity of the source:

$$\frac{\mathrm{d}\boldsymbol{Q}}{\mathrm{d}t} = \boldsymbol{WQ} + \boldsymbol{G\Pi}, \tag{4.27}$$

$$\frac{\mathrm{d}\boldsymbol{\Pi}}{\mathrm{d}t} = -\boldsymbol{W}^{\mathrm{T}}\boldsymbol{\Pi} - \boldsymbol{KQ} \tag{4.28}$$

(note that the sign of $\boldsymbol{K}$ in equation (80) of Hanyga (1984c) is misprinted).

The matrices $\boldsymbol{G}$, $\boldsymbol{W}$ and $\boldsymbol{K}$ are identical with those appearing in Section 4.4. They depend on the index I of the underlying ray-centred coordinate system determined by SETIND and passed to the subroutine RHSI computing the right-hand sides of (4.27) and (4.28) through common block /I/.

The 2×2 matrices $\boldsymbol{Q}$ and $\boldsymbol{\Pi}$ define the dependent-value vector $\boldsymbol{w}$ for linear ray-tracing:

$$w_{k+2j+4} = Q_{kj}, \qquad w_{k+2j+8} = \Pi_{kj}. \tag{4.29}$$

The initial data for (4.27) and (4.28) are

$$\boldsymbol{p} = |\boldsymbol{p}|\,[\sin\theta\cos\varphi, \sin\theta\sin\varphi, \cos\theta]^{\mathrm{T}} = |\boldsymbol{p}|\,\boldsymbol{n}, \tag{4.30}$$

$$\boldsymbol{Q} = 0,$$

$$\Pi_{\alpha A} = \sum_{k}^{3} P_{k,\pi^{-1}(\alpha)} T_{kA}, \quad \alpha, A = 1, 2, \tag{4.31}$$

where

$$T_{k1} = \frac{\partial n_k}{\partial\theta}\Big/ H(\boldsymbol{x}, \boldsymbol{n}), \qquad T_{k2} = \frac{\partial n_k}{\partial\varphi}\Big/ (H(\boldsymbol{x}, \boldsymbol{n})\sin\theta) \tag{4.32}$$

(Hanyga, 1984d) are generated by the submodule INDAT.

The transition from linear to regular or singular ray tracing is defined by the interrupt surface det $\boldsymbol{Q} = \varepsilon_2$.

The transformation from linear to regular DRT is carried out by the submodule TRNSFM1 according to the formulae

$$\boldsymbol{C} = \boldsymbol{\Pi}\boldsymbol{Q}^{-1}, \tag{4.33}$$

$$J = (\det \boldsymbol{Q})\,\frac{\partial H}{\partial p_I}. \tag{4.34}$$

The matrix $\boldsymbol{C}$ is symmetric on account of (4.27), (4.28) and (4.30). Indeed, (4.27), (4.28) and (4.30) imply that $\boldsymbol{T} = \boldsymbol{Q}\boldsymbol{\Pi}^{-1}$ satisfies the initial-value problem $\boldsymbol{T} = 0$, $\mathrm{d}\boldsymbol{T}/\mathrm{d}t = \boldsymbol{TKT} + \boldsymbol{TW}^{\mathrm{T}} + \boldsymbol{WT} + \boldsymbol{G}$, with *exactly* symmetric $\boldsymbol{K}$ and $\boldsymbol{G}$. Therefore $\boldsymbol{C} = \boldsymbol{T}^{-1}$ is symmetric within the accuracy of integration in linear in DRT.

Equations (4.31), (4.32) and (4.34) imply that the initial ray-spreading Jacobian is given by

$$J = \frac{\partial H}{\partial \boldsymbol{p}} \cdot \left(\frac{\partial \boldsymbol{x}}{\partial\theta} \times \frac{\partial \boldsymbol{x}}{\partial\varphi}\right)\Big/ \sin\theta,$$

with

$$\boldsymbol{x} \approx \boldsymbol{x}_0 + \frac{\partial H}{\partial \boldsymbol{p}}(\boldsymbol{x}_0, \boldsymbol{n}(\theta, \varphi))\tau$$

in the vicinity of the source. In an isotropic medium

$$J \approx c(\boldsymbol{x}_0)\left|\frac{\partial \boldsymbol{x}}{\partial\theta} \times \frac{\partial \boldsymbol{x}}{\partial\varphi}\right|\Big/ \sin\theta = c(\boldsymbol{x}_0)^3 t^2$$

for $t \approx 0$, and the amplitude a returned by DYNTRAC on exit corresponds to an initial amplitude $a_0 = [c(\boldsymbol{x}_0)]^{-1/2}$ on the unit sphere centred at $\boldsymbol{x}_0$.

For a distributed source the sequence

call INDAT(. . .), call DRTINI(. . .), call TRNSFM1(. . .)

in DYNTRAC should be replaced by calling a subprogram generating appropriate initial conditions for regular DRT.

A DRT module based exclusively on linear DRT is described in Section 4.17.

4.11 Tracing S-wave polarization

The S-wave polarization vector $\boldsymbol{v}$ is defined in terms of two basis polarization vectors

$$\boldsymbol{r}_1 = |(\boldsymbol{e} \cdot \boldsymbol{p})\boldsymbol{p} - \boldsymbol{p}^2\boldsymbol{e}|^{-1}((\boldsymbol{e} \cdot \boldsymbol{p})\boldsymbol{p} - \boldsymbol{p}^2\boldsymbol{e}) \tag{4.35a}$$

$$\boldsymbol{r}_2 = |\boldsymbol{p} \times \boldsymbol{e}|^{-1}\boldsymbol{p} \times \boldsymbol{e} \tag{4.35b}$$

(cf. equation (F6–7) of Hanyga, 1986). The auxiliary vector $\boldsymbol{e}$ cannot be (approximately) parallel to the slowness vector $\boldsymbol{p}$. It is initially computed by POLSET according to the formula

$$\tilde{e}_k = P_{kJ}, \quad k = 1, 2, 3, \qquad \boldsymbol{e} = |\tilde{\boldsymbol{e}}|^{-1}\tilde{\boldsymbol{e}}, \tag{4.36}$$

where $1 \leqslant J \leqslant 3$ and J is not equal to the index I of the initial ray-centered coordinate system. Indeed, $\tilde{\boldsymbol{e}} \cdot \boldsymbol{p} = 0$ on account of (4.3), while $\tilde{\boldsymbol{e}} = 0$ contradicts (4.4). Orthogonality of $\boldsymbol{e}$ and $\boldsymbol{p}$ is restored in the course of tracing by a rotation of $\boldsymbol{e}$ in the plane of $\boldsymbol{e}$ and $\boldsymbol{p}$ whenever the value of $|\boldsymbol{e} \times \boldsymbol{p}|/|\boldsymbol{p}|$ falls below a threshold value ε_3. In tracing through a homogeneous block declared $\rho = 0$ the vector $\boldsymbol{e}$ is updated after an application of Snell's law only.

The polarization angle ψ defines the rotation of the basis vectors in the course of propagation through a single block:

$$\left.\begin{aligned} \boldsymbol{r}_1 \to \boldsymbol{r}_1' &= \cos\psi\, \boldsymbol{r}_1 + \sin\psi\, \boldsymbol{r}_2, \\ \boldsymbol{r}_2 \to \boldsymbol{r}_2' &= -\sin\psi\, \boldsymbol{r}_1 + \cos\psi\, \boldsymbol{r}_2. \end{aligned}\right\} \tag{4.37}$$

The angle ψ satisfies

$$\frac{\mathrm{d}\psi}{\mathrm{d}t} = -\frac{\boldsymbol{e} \cdot \boldsymbol{p}}{|\boldsymbol{e} \times \boldsymbol{p}|^2}(\boldsymbol{e} \times \boldsymbol{p}) \cdot \boldsymbol{\nabla} c_{\mathrm{S}} \tag{4.38}$$

(Hanyga, 1986), where c_{S} denotes the S-wave speed. At the beginning of tracing through a block ψ is set equal to zero.

Adjustments of the auxiliary vector $\boldsymbol{e}$ do not affect the momentary vectors $\boldsymbol{r}_1$ and $\boldsymbol{r}_2$, but they do affect the rates of rotation of these vectors induced by the variation of slowness and entail a discontinuity in the right-hand side of (4.38). Since the current step size should be estimated for the transformed right-hand side, an adjustment of $\boldsymbol{e}$ necessitates a momentary interruption of integration. Technically this is achieved by treating an adjustment of $\boldsymbol{e}$ on a par with ray-centred coordinate transformations. (See Appendix F for the treatment of dependent variable transformations.)

The S-wave polarization vector $\boldsymbol{v}$ is given by

$$\boldsymbol{v} = g_1\boldsymbol{r}_1' + g_2\boldsymbol{r}_2'\mathrm{e}^{\mathrm{i}\delta}, \tag{4.39}$$

where the real numbers g_1, g_2 and δ are calculated at the source (by the user-supplied subroutine INPOL) or at an interface (by SNELLX; Section 4.13). They remain fixed throughout a block and are stored in the common block /SPOL/.

The right-hand side of (4.38) is returned as the value of Fortran function ROTRAT. The polarization angle ψ defines the last component of the dependent-value vector whenever an S wave is being traced through a block. The subroutine RHSI, RHSR or RHSS computing the right-hand-side vector of the differential equations calls ROTRAT for the computation of the last component.

If the ray terminates with an S arrival then on exit the S-wave arrival is specified in terms of two unit polarization vectors $\boldsymbol{v}_1$, $\boldsymbol{v}_2$, amplitude a, total phase shift α and a relative phase shift δ. The total phase shift of the $\boldsymbol{v}_1$ component is α, while the total phase shift of the $\boldsymbol{v}_2$ component is $\alpha + \delta$.

4.12 Boundary conditions at interfaces. Kinematical variables

In the first step the slowness vectors $\boldsymbol{p}_B$ of all six outgoing rays are determined from the equation

$$H(\boldsymbol{x}_\mathrm{I}, \boldsymbol{p}_\mathrm{T} + \lambda_B\boldsymbol{\nu}; l_B, \sigma_B) = 1, \tag{4.40}$$

where $\boldsymbol{p}_\mathrm{T}$ denotes the component of the incident slowness vector $\boldsymbol{p}^{\mathrm{in}}$ tangent to the interface at the point of incidence $\boldsymbol{x}_\mathrm{I}$, $\boldsymbol{\nu}$ is a normal to the interface at $\boldsymbol{x}_\mathrm{I}$ and $\boldsymbol{p}_B = \boldsymbol{p}_\mathrm{T} + \lambda_B\boldsymbol{\nu}$ is in general complex-valued. The block reference number l_B in (4.40) assumes two alternative values—one for reflected waves and another for transmitted waves. For either value of l_B the wave species σ_B ranges over all the three possible wave species.

Since for a fixed value of l_B, (4.40) with $\sigma_B = 1, 2, 3$ are equivalent to a sextic equation (Hanyga *et al.*, 1984, equation (2.4.26)), the total number of solutions of (4.40) is $6 + 6 = 12$. The *outgoing* waves $\boldsymbol{p}^{\mathrm{out}}$ are defined by the condition

$$\operatorname{sign}\left(\sum_i^3 \frac{\partial H^{\text{out}}}{\partial p_i}(\boldsymbol{x}_I, \boldsymbol{p}^{\text{out}})\nu_i\right) = \pm\operatorname{sign}\left(\sum_i^3 \frac{\partial H^{\text{in}}}{\partial p_i}(\boldsymbol{x}_I, \boldsymbol{p}^{\text{in}})\nu_i\right), \quad (4.41)$$

with the upper sign for transmitted waves and the lower sign for reflected waves, provided that $\boldsymbol{p}^{\text{out}}$ is real; otherwise it is required that

$$\operatorname{sign}\left[\operatorname{Im}\left(\boldsymbol{p}^{\text{out}} \cdot \boldsymbol{\nu}\right)\right] = \iota, \quad (4.42)$$

with $\iota = 1$ if the wave propagates in the block into which the normal $\boldsymbol{\nu}$ points and $\iota = -1$ otherwise.

Equation (4.41) has an obvious geometrical meaning. Equation (4.42) expresses the condition that the outgoing wave decays exponentially in a local sense. A simple geometrical argument (Hanyga, 1980; Hanyga *et al.*, 1984, equation (2.4.3)) shows that (4.41) reduces the number of the real solutions $\boldsymbol{p}_{\mathrm{T}} + \lambda\boldsymbol{\nu}$ of (4.40) by a factor of 2; (4.42) eliminates half of the complex-valued solutions, since complex solutions λ of (4.40) occur in complex-conjugate pairs.

The continuation of the ray is determined in terms of the values of l_B and σ_B indicated by the ray code, with account taken of (4.41) and (4.42).

The surfaces $H = 1$ in $\boldsymbol{p}$ space are convex for P, S, T and QL waves. Indeed, the first two are spheres and the third is an ellipsoid; the slowness surface of the fastest wave species is convex by a general theorem (Courant and Hilbert, 1962). For a convex surface $H = 1$, (4.40) has either two real or two complex-conjugate solutions λ_B. Exactly one of them satisfies (4.41) or (4.42).

For the wave species QT there can be 0, 2 or (for $F + 2L$ significantly different from A and C; cf. (B7)) 4 real solutions λ_B. In the last case there are two outgoing waves of the same wave species $\sigma_B = 2$ (Fig. 3), and consequently there are two different continuations of the ray starting with QT segments. In practical terms this entails two different rays for the same reduced raycode (Section 4.18 explains how the problem is resolved).

4.13 Boundary conditions at interfaces. Dynamical variables

The boundary condition for the ray-spreading Jacobian J is given by equation (105) Hanyga (1984b):

$$\frac{J^{\text{out}}}{\boldsymbol{\nu} \cdot (\partial H/\partial \boldsymbol{p})^{\text{out}}} = \frac{J^{\text{in}}}{\boldsymbol{\nu} \cdot (\partial H/\partial \boldsymbol{p})^{\text{in}}} \quad (4.43)$$

(Appendix D). The superscripts "in" and "out" on $\partial H/\partial \boldsymbol{p}$ indicate that this vector is evaluated at $\boldsymbol{p}^{\text{in}}$ and $\boldsymbol{p}^{\text{out}}$ respectively.

The incoming wavefront-curvature matrix $C^{\text{in}}_{\alpha\beta}$ is transformed into the 3×3 matrix S^{in}_{ij} according to Section 4.5, and the outgoing S^{out}_{ij} is then calculated from equations (100)–(102) of Hanyga (1984b):

$$\sum_{i,j}^{3} S^{\text{out}}_{ij} h_{ia} h_{jb} = \sum_{i,j} S^{\text{in}}_{ij} h_{ia} h_{jb} + \sum_{i}^{3} (p^{\text{in}}_i - p^{\text{out}}_i) h_{iab} \tag{4.44}$$

$$\sum_{j}^{3} \frac{\partial H}{\partial p_j} S^{\text{out}}_{ij} + \frac{\partial H}{\partial x_i} = 0, \quad i = 1, 2, 3, \; a, b = 1, 2, \tag{4.45}$$

where $\boldsymbol{x} = \boldsymbol{h}(s_1, s_2)$ is a parametric representation of the interface, $h_{ia} = \partial h_i / \partial s_a$ and $h_{iab} = \partial^2 h_i / \partial s_a\, \partial s_b$; the derivatives of the wave Hamiltonian $H = H(\boldsymbol{x}, \boldsymbol{p}; l_{\text{out}}, \sigma_{\text{out}})$ are evaluated at $\boldsymbol{p} = \boldsymbol{p}^{\text{out}}$.

In DRT interfaces are preferably represented by equations $f(\boldsymbol{x}) = 0$. We therefore note that $\boldsymbol{p}^{\text{in}} - \boldsymbol{p}^{\text{out}} = [(\boldsymbol{p}^{\text{in}} - \boldsymbol{p}^{\text{out}}) \cdot \boldsymbol{\nu}]\boldsymbol{\nu}$ on account of (4.40) and from the identity

$$\sum_{i}^{3} \nu_i h_{iab} = -|\boldsymbol{\nabla} f|^{-1} \sum_{i,j} \frac{\partial^2 f}{\partial x_i\, \partial x_j} h_{ia} h_{ib} \tag{4.46}$$

we derive a more convenient form of (4.44):

$$\sum_{i,j}^{3} S^{\text{out}}_{ij} h_{ia} h_{jb} = \left[\sum_{i,j}^{3} S^{\text{in}}_{ij} - (\boldsymbol{p}^{\text{in}} \cdot \boldsymbol{\nu} - \boldsymbol{p}^{\text{out}} \cdot \boldsymbol{\nu}) \sum_{i,j} R_{ij} \right] h_{ia} h_{jb}, \tag{4.47}$$

with $R_{ij} := |\boldsymbol{\nabla} f|^{-1}\, \partial^2 f / \partial x_i\, \partial x_j$ supplied by a model subroutine MODSGI. We then apply the solution of (4.45) and (4.47) (equation (104) of Hanyga, 1984b) and the reduction formula (4.6).

Plane-wave reflection/transmission coefficients a_B, $B = 1, \ldots, 6$, are determined from the equations

$$\left.\begin{aligned} &\sum_{B}^{6} \iota_B a_B \boldsymbol{r}_B = \boldsymbol{r}^{\text{in}}, \\ &\sum_{B}^{6} \iota_B \boldsymbol{B}(\boldsymbol{\nu}, \boldsymbol{p}_B) a_B \boldsymbol{r}_B = \boldsymbol{B}(\boldsymbol{\nu}, \boldsymbol{p}^{\text{in}}) \boldsymbol{r}^{\text{in}}, \end{aligned}\right\} \tag{4.48}$$

where $\boldsymbol{r}^{\text{in}}$ is the unit polarization vector of the incident wave, $\boldsymbol{r}_B$ are the unit polarization vectors of the outgoing waves $\boldsymbol{p}_B$, $\iota_B = -1$ for reflected waves and $\iota_B = 1$ for transmitted waves.

For an incident S wave $\boldsymbol{r}^{\text{in}}$ is calculated from (4.39); we note that the final value of the polarization angle ψ is taken into account in the expression for the outgoing amplitude; the initial value of ψ for an outgoing S wave is 0.

The 3×3 matrix $\boldsymbol{B}(\boldsymbol{v}, \boldsymbol{p})$ is a linear function of $\boldsymbol{p}$ and of $\boldsymbol{v}$ defined in terms of the elastic moduli (see Section 2.4 of Hanyga *et al.* (1984) for a general explanation and Appendices A and B for specific expressions).

If a_B is the plane-wave coefficient appropriate for the continuation of the ray then the amplitude A in the common block /AMPL/ and the phase α_1 in the common block /PHASE/ are updated according to the rule

$$A \rightarrow A\,|a_B|\,|J_B|^{-1/2}\,|J|^{1/2} \qquad \alpha_1 \rightarrow \alpha_1 + \arg a_B. \tag{4.49}$$

For an outgoing S wave the solution of (4.48) is given by a complex vector $a_1\boldsymbol{r}_1 + a_2\boldsymbol{r}_2$, with $a_1 = g_1 \mathrm{e}^{\mathrm{i}\gamma}$ and $a_2 = g_2 \mathrm{e}^{\mathrm{i}(\gamma+\delta)}$. The amplitude A and phase α_1 are then updated according to the rule $A \rightarrow A(g_1^2 + g_2^2)^{1/2}$, $\alpha_1 \rightarrow \alpha_1 + \gamma$, while $\tilde{g}_1 = g_1(g_1^2 + g_2^2)^{-1/2}$, $\tilde{g}_2 = g_2(g_1^2 + g_2^2)^{-1/2}$ and δ are stored in the common block /SPOL/.

The dynamical boundary conditions in the initial DRT are obtained by transformation to the variables of the regular DRT, application of the boundary conditions described above and the following transformation:

$$Q_{k2}^{\text{out}} = Q_{k2}^{\text{in}} F, \qquad \boldsymbol{\Pi}^{\text{out}} = \boldsymbol{C}^{\text{out}} \boldsymbol{Q}^{\text{out}}, \qquad \text{with } F = J^{\text{out}} \Big/ \frac{\partial H}{\partial p_I}. \tag{4.50}$$

The transformation (4.50) is consistent with (4.33) and (4.34). It is *not* consistent with equations (85) and (86) of Hanyga (1984c). For consistency with the latter equations, equations derived in Hanyga (1987) should replace (4.50).

4.14 Accuracy and consistency control in DRT

The accuracy of integration (in terms of the number of significant digits) as well as consistency of model subprograms computing the propagation speeds and their derivatives can be checked by computing an integral of motion $F(\boldsymbol{w})$ or the left-hand side of an *a priori* identity $F(\boldsymbol{w}, \partial \boldsymbol{w}/\partial t) = 0$. The following functions can be optionally calculated by DYNTRAC:

(1) $H(\boldsymbol{x}, \boldsymbol{p}; l, \sigma)$ (correct value 1), option IPL = 2 or 3;

(2) $\sum_j \partial H/\partial p_j S_{ij} + \partial H/\partial x_i$, $i = 1, 2, 3$ (correct value 0) in singular DRT, option IOP = 6;

(3) $\int \sum_j S_{ij}\, \mathrm{d}x_j/\mathrm{d}t - p_i$, $i = 1, 2, 3$ (expected to be a constant) in regular DRT, option IOP = 6;

(4) $\int \sum_j \partial^2 e_r/\partial x_i\, \partial x_j\, \mathrm{d}x_j/\mathrm{d}t - \partial c_r/\partial x_i$, $i = 1, 2, 3$ (expected to be a constant), option IOP = r, with e_1, e_2 = P, S wave speed if $|\rho| \leqslant 1$ and

e_r are the five elasticities N, L, A, C, F of transverse isotropy divided by density of $|\rho| > 1$.

(1) allows a check of the model subroutine MODELV1 providing values of propagation speeds or elasticities against MODELV1 supplying gradients of the same quantities, while (4) allows a check of their second-order derivatives provided by MODSGE against their gradients from MODELG1. On account of (2.1), a constant value of (3) is equivalent to the validity of the Hamilton–Jacobi equation $H(\boldsymbol{x}, \nabla S(\boldsymbol{x})) = \text{const}$ along the ray.

4.15 Regular and linear DRT in homogeneous blocks

Replacing integration by closed-form solutions of the DRT equations results in considerable speed-up of tracing. In particular closed-form solutions of the linear and regular DRT curves in a homogeneous block are given by fairly simple expressions:

$$\boldsymbol{Q}(t) = \boldsymbol{Q}(0) + t\boldsymbol{G}(p_0)\boldsymbol{\Pi}(0), \qquad \boldsymbol{\Pi}(t) = \boldsymbol{\Pi}(0), \qquad \boldsymbol{p}(t) = \boldsymbol{p}(0) = \boldsymbol{p}_0 \quad (4.51)$$

(for initial DRT, on account of (4.27) with $\boldsymbol{W} = \boldsymbol{K} = 0$);

$$\boldsymbol{C}(t) = \boldsymbol{C}(0)[\boldsymbol{E} + t\boldsymbol{G}(p_0)\boldsymbol{C}(0)]^{-1}, \qquad J(t) = J(0)\det[\boldsymbol{E} + t\boldsymbol{G}(p_0)\boldsymbol{C}(0)] \quad (4.52)$$

(for regular DRT, on account of (4.51), (4.33) and (4.34)).

Homogeneous isotropic blocks are assigned $\rho = 0$, while homogeneous transversely isotropic blocks with vertical/non-vertical isotropy axes are assigned $\rho = -2$ and -3. For these values of ρ the module CHINTR2 (Section 2.2) is called. The arc is defined by the modules DRTICUR or DRTRCUR.

For isotropic blocks with propagation speed $c(\boldsymbol{x}) = (C + \sum_i^3 a_i x_i)^{-1/2}$ closed-form formulac for $\boldsymbol{Q}(\zeta)$, $\boldsymbol{\Pi}(\zeta)$ and $\boldsymbol{C}(\zeta)$ can be derived by differentiating equations (13) of Červený (1987) with respect to the ray coordinates and applying equations (85) and (86) of Hanyga (1984c) and (B6) and (B7) of Hanyga (1986). The resulting formulae are quoted in Appendix E. We note here that the parameter ζ does not coincide with the travel time. The latter can be computed along the ray as a function of the former.

4.16 Dynamic tracing along edge-diffracted rays

In this case the input data are invariably obtained from two-point ray tracing. The module DYNTRAC is notified that the ray is edge-diffracted. The input data include the ordinal number j of the interface containing the diffracting edge; this number refers to the order of interfaces in the raycode. The edge is defined by the intersection of $f(\boldsymbol{x}; r(3j-1)) = 0$ and $g(\boldsymbol{x}) = 0$, where $r(k)$ denotes the raycode and $g(\boldsymbol{x}) = 0$ is a user-defined auxiliary function. DYNTRAC does not call any Fortran function to compute the values of $g(\boldsymbol{x})$. For the application of the edge boundary conditions two user-defined subroutines are called. The subroutine NORG(X, N) computes the unit normal $\boldsymbol{N} = |\nabla g(\boldsymbol{x})|^{-1} \nabla g(\boldsymbol{x})$, while the subroutine SECG(X, S) computes $S_{ij} = |\nabla g(\boldsymbol{x})|^{-1} \partial^2 g / \partial x_i \, \partial x_j$. These subroutines can be effectively equivalent to MODNOR and MODSGE if $g(\boldsymbol{x}) = f(\boldsymbol{x}; k_1)$ for some interface k_1.

In addition to the edge, a parameter θ defining the direction of slowness $\boldsymbol{p}_{\mathrm{d}}$ of the diffracted ray on the diffraction cone is specified:

$$\boldsymbol{p}_{\mathrm{d}} = (\boldsymbol{p}^{\mathrm{in}} \cdot \boldsymbol{t})\boldsymbol{t} + A(\cos\theta\, \boldsymbol{n}_1 + \sin\theta\, \boldsymbol{n}_2), \tag{4.53}$$

where $\boldsymbol{n}_1$ and $\boldsymbol{n}_2$ denote the unit normal vectors to $f(\boldsymbol{x}; r(j)) = 0$ and $g(\boldsymbol{x}) = 0$ respectively. Alternatively, the diffracted slowness vector $\boldsymbol{p}'_{\mathrm{d}}$ obtained from two-point ray tracing (by ITHDIF) is used. In the latter case DYNTRAC computes θ from (4.53) with $\boldsymbol{p}'_{\mathrm{d}}$ on the left-hand side and then calculates $\boldsymbol{p}_{\mathrm{d}}$ from (4.53), (4.57). This precaution avoids transfer of errors from two-point ray tracing to DRT (cf. Section 3.8).

The wave species σ_1 of the diffracted wave is specified in the reduced raycode.

Initial data for tracing the ray spreading J along the ray are calculated by the module INDAT1. Diffraction coefficients are not included in the package. They can be provided by the user by replacing the dummy subroutine DYNDAT by a subroutine that multiplies the amplitude by the diffraction coefficient, adds a phase shift to α_2 and calculates the polarization of a diffracted S wave. In particular, on the assumption that the diffracted pulse is calculated according to the rules of Hanyga (1988), the diffraction coefficients in DYNDAT should be set equal to the absolute value and argument of the reflection/transmission coefficient for the ray incident at the diffraction point.

The initial data for linear DRT are derived from the following formulae:

$$Q_{a2} = 0, \qquad Q_{a1} = \sum_j^3 R_{aj} \frac{\partial x_j}{\partial s}, \qquad \Pi_{a\alpha} = \sum_j^3 P_{ja} \frac{\partial p_j}{\partial u_\alpha}, \qquad a, \alpha = 1, 2, \tag{4.54}$$

with $u_1 = s$, $u_2 = \theta$ and

$$\boldsymbol{x}(t, s, \theta) = \boldsymbol{x}(s) + [t - S_{\mathrm{in}}(\boldsymbol{x}(s))] \frac{\partial H}{\partial \boldsymbol{p}} (\boldsymbol{x}(s), \boldsymbol{p}_d(s, \theta)), \tag{4.55}$$

$$\boldsymbol{p}_{\mathrm{d}}(s, \theta) = [\boldsymbol{p}^{\mathrm{in}}(s) \cdot \boldsymbol{t}(s)]\boldsymbol{t}(s) + A(s, \theta)(\cos\theta\, \boldsymbol{n}_1 + \sin\theta\, \boldsymbol{n}_2), \quad (4.56)$$

$$H(\boldsymbol{x}(s), \boldsymbol{p}_{\mathrm{d}}(s, \theta); l_1, \sigma_1) = 1, \quad (4.57)$$

$$\boldsymbol{p}(t, s, \theta) = \boldsymbol{p}_{\mathrm{d}}(s, \theta) - [t - S^{\mathrm{in}}(\boldsymbol{x}(s))] \frac{\partial H}{\partial \boldsymbol{x}}(\boldsymbol{x}(s), \boldsymbol{p}_{\mathrm{d}}(s, \theta)). \quad (4.58)$$

In these formulae H denotes the wave Hamiltonian corresponding to the wave species of the diffracted ray and the block entered by this ray, $\boldsymbol{t}(s)$ is the unit tangent to the edge at the point $\boldsymbol{x}(s)$ of the edge, S^{in} is the eikonal of the incident wave and $\boldsymbol{p}^{\mathrm{in}} = \nabla S^{\mathrm{in}}$ is the incident slowness. The transformation coefficient $R_{aj} = \partial y_a / \partial x_j$ is given by equation (B7) of Hanyga (1986). The expressions (4.54) are evaluated at $t = S^{\mathrm{in}}(\boldsymbol{x}(s))$, where $\boldsymbol{x}(s)$ is the point of incidence of the ray generating the edge-diffracted ray.

It is easy to see that symmetry of $\boldsymbol{C} = \boldsymbol{\Pi}\boldsymbol{Q}^{-1}$ is equivalent to the initial symmetry of $\boldsymbol{T} = \boldsymbol{Q}\boldsymbol{\Pi}^{-1}$. Since $T_{ab} = Q_{a1}(\boldsymbol{\Pi}^{-1})_{1b}$ and $\boldsymbol{\Pi}_{a2}$, $\boldsymbol{\Pi}^{-1}_{1a}$ are orthogonal in the plane, the symmetry condition assumes the form $\sum_a Q_{a1}\Pi_{a2} = 0$. On account of equation (B10) of Hanyga (1986), the left-hand-side is equal to

$$\sum_{i,j}^{3} P_{ij} \frac{\partial x_j}{\partial s} \frac{\partial p_i}{\partial \theta} = -\boldsymbol{p}_{\mathrm{d}} \cdot \boldsymbol{t} H^{-1} \frac{\partial}{\partial \theta} H(\boldsymbol{x}(s), \boldsymbol{p}_{\mathrm{d}}(s, \theta)) = 0,$$

by (4.57).

On account of errors inherent in the initial data, a diffracted ray does not exactly hit the edge. This often entails an additional intersection with the surface $g(\boldsymbol{x}) = 0$. If $g(\boldsymbol{x}) = 0$ is an interface then Snell's law is applied and the direction of the ray is significantly affected. This problem is circumvented in DYNTRAC by temporary deletion of $g(\boldsymbol{x}) = 0$ from the list of active interrupt surfaces.

Similar methods can be applied in DRT of head-wave and creeping rays. In the former case the horizontal leg of the head-wave ray does not originate at a caustic and linear DRT after critical reflection is not needed.

4.17 Exclusively linear DRT

Dynamic ray tracing can be based exclusively on linear DRT. Indeed, (4.27) and (4.28) are non-singular at the caustics. Whenever $|\det \boldsymbol{Q}|$ falls below a critical level $\varepsilon_3 > 0$ a phase shift is computed according to (4.33) above and equation (56) of Hanyga (1984a). The same phase shift is computed when $|\det \boldsymbol{Q}|$ crosses the same level in the opposite direction, and the difference between the two is added to the caustic phase shift $\frac{1}{4}\pi\alpha_2$. The intersections of caustics are monitored by defining the interrupt surface det $\boldsymbol{Q} = 0$.

Explicit expressions for the ray-centred coordinate transformations and interface boundary conditions in linear DRT are given in Hanÿga (1987).

In comparison with regular DRT, linear DRT involves integration of a larger system of differential equations (or computation of a larger number of coordinates specifying an arc in a higher-dimensional space if closed-form solutions are known). Computation of the right-hand sides, however, involves computation of the same matrices $\boldsymbol{G}$, $\boldsymbol{W}$ and $\boldsymbol{K}$. The matrix equations of linear and regular DRT involve the same number of 2×2 matrix multiplications, but regular DRT requires in addition computation of a scalar and the trace of the product of two matrices. The matrix solutions of linear DRT are more tame than wavefront curvature, but it must be borne in mind that linear DRT also involves integration of the non-linear ray equations (2.1), which cause some slowdown of integration in either case.

A relative advantage of regular DRT consists in the fact that it allows more direct control of the number of significant digits.

4.18 Closing remarks

In order to resolve the ambiguity inherent in the choice of outgoing wave species with a non-convex slowness surface $H = 1$ (Section 4.12), the module SNELLX eliminates the near-grazing outgoing waves. If two possible continuations still remain then the results of tracing are recorded in the common block /STOR/ and tracing is resumed along both ray continuations successively.

5 COMPUTATION OF BODY-WAVE PULSES

5.1 Introduction. Computation of a synthetic seismogram

In the first stage of construction of a synthetic seismogram at a recording point $\boldsymbol{x}_R$ the two-point boundary-value problem is solved for a selection of rays expected to yield a significant contribution. Dynamic ray tracing along each of these rays yields three numerical parameters for the computation of the pulses: travel time T_r, amplitude $b_r = c_r a_r$ and total phase shift $\beta_r = \alpha_r + \gamma_r$. The amplitude $a_r = A_r\,|J_r|^{-1/2}$ and the total phase shift α_r are obtained from DRT, while c_r and γ_r account for the source directivity and depend on the initial slowness $\boldsymbol{p}_r$ of the ray.

Within the set of all relevant rays there can be some groups of rays having the same raycode. Contributions of different raycodes are included in the synthetic seismogram by superposition. Contributions of rays having the same raycode RC have to be evaluated jointly if the offsets between their arrival times are small. Indeed, in this case the point $\boldsymbol{x}_R$ lies in the vicinity of a caustic of RC rays and singularity of their ART amplitudes a_r necessitates a uniformly asymptotic evaluation of the pulse. The pulse is then given by a vector-valued function $\boldsymbol{F}(t, \boldsymbol{x}_R; T_1, \ldots, T_n; a_1, \ldots, a_n; \alpha_1)$ involving the travel times and amplitudes of all the RC rays. The differences between their phase shifts α_r are implicitly accounted for by the function $\boldsymbol{F}$ (Hanyga and Seredyńska, 1988).

A similar situation arises if a group of reflected (or transmitted) body waves with the same raycode RC arrive nearly simultaneously with their edge-diffracted counterparts. In this case $\boldsymbol{x}_R$ lies in the Fresnel diffraction zone surrounding the shadow boundary of RC rays (Hanyga, 1988).

The notion of "small-time offset" can be expressed in terms of a numerical criterion related to the required accuracy of the pulse computation. Indeed, the *composite pulse* $\boldsymbol{F}(t, \boldsymbol{x}_R; T_1, \ldots, T_n; a_1, \ldots, a_n; \alpha_1)$ is a uniformly valid asymptotic solution of the elastodynamic equations. In particular, it is valid far from the caustics, and the straightforward ART solution is an asymptotic approximation of $\boldsymbol{F}$. The numerical criterion in question is therefore related to the error inherent in this approximation.

In order to shed more light on this relation, it is expedient to consider the field in the frequency domain. For the simple caustic and for the caustic cusp the field is given by

$$c_0 \operatorname{Ai}(z) + c_1 \operatorname{Ai}'(z) \tag{5.1}$$

and

$$d_0 \operatorname{Pe}(z_1, z_2) + d_1 \frac{\partial \operatorname{Pe}(z_1, z_2)}{\partial z_1} + d_2 \frac{\partial \operatorname{Pe}(z_1, z_2)}{\partial z_2} \tag{5.2}$$

respectively. The coefficients c_k and d_k are some linear functions of the amplitudes a_r, while z, z_1 and z_2 are algebraic functions of travel times. Ai and Pe denote the Airy function and the Pearcey function respectively. In particular, we quote the simple formula

$$z = -\tfrac{3}{4}\, |\omega(T_2 - T_1)|^{-2/3}. \tag{5.3}$$

For a given accuracy $\varepsilon > 0$ the condition of applicability of straightforward ART pulse expressions (1.9) assumes the following form (Abramowitz and Stegun, 1970):

$$|\mathrm{Ai}\,(z) - \pi^{-1/2} z^{-1/4} \sin\,[\omega(T_2 - T_1) + \tfrac{1}{4}\pi]| < \varepsilon, \tag{5.4}$$

$$|\mathrm{Ai}'\,(z) - \pi^{-1/2} z^{-1/4} \cos\,[\omega(T_2 - T_1) + \tfrac{1}{4}\pi]| < \varepsilon, \tag{5.5}$$

accompanied by similar conditions for the Pearcey functions. The inequalities (5.4) and (5.5) can alternatively be expressed in the form $|\omega(T_2 - T_1)| \geq d(\varepsilon)$ or, more generally,

$$|\omega(T_k - T_l)| \geq d(\varepsilon) \quad \text{for } k \neq l. \tag{5.6}$$

Given a lower bound ω_- on the frequency emitted by the source and recorded by the receiver, (1.8) is applicable provided that

$$|T_k - T_l| > d(\varepsilon)/\omega_- \quad \text{for } k \neq l. \tag{5.7}$$

If a collection of rays having the same raycode satisfies the opposite inequality then their contribution to the synthetic seismogram has to be expressed in terms of a composite pulse.

Grouping of rays of a given raycode according to (5.6) is achieved by incorporating a simple sorting algorithm in DYNTRAC.

5.2 Computation of composite signals in caustic regions

We now describe in brief outline the computation of a composite pulse for a group of nearly simultaneous arrivals in the vicinity of a caustic or caustic cusp (for details see Hanyga and Seredyńska, 1988).

For every caustic configuration the *maximally illuminated region* is defined by the condition that the number n of rays with a given raycode reaching a point is maximal. For a simple caustic the maximally illuminated region coincides with the oscillatory side of the caustic ($n = 2$), and for a caustic cusp it is the wedge-shaped region between the two simple caustics ($n = 3$, Fig. 4–5). The method described here applies to a receiver situated in a maximally illuminated region and it is assumed that data from DRT along all the n rays are available. For definiteness we postpone the S-wave case to Section 5.3.

Let $\tilde{s}(t)$ denote a signal (1.9) arriving at the caustic. $\tilde{s}(t)$ is one of the terms in the asymptotic decomposition of the composite signal at some distance from the caustic. For a finite-bandwidth signal $\tilde{s}(t)$ with a spectrum $\hat{s}(\omega)$, $\omega_- < \omega < \omega_+$, the composite signal can be computed according to the formula

$$\mathrm{Re} \int_{\omega_-}^{\omega_+} \mathrm{d}\omega\, \hat{s}(\omega) \hat{R}_\delta(\omega, z_1, z_2, \ldots, z_n, \boldsymbol{b}_1, \boldsymbol{b}_2, \ldots, \boldsymbol{b}_n) \mathrm{e}^{\mathrm{i}\alpha\, \mathrm{sgn}\, \omega} \mathrm{e}^{-\mathrm{i}\omega t}, \tag{5.8}$$

while for a signal of finite duration, $t_1 < t < t_2$,

$$\int_{t_-}^{t_+} \mathrm{d}t'\,\bar{s}(t')[R_\delta(t-t', z_1, z_2, \ldots, z_n, \boldsymbol{b}_1, \ldots, \boldsymbol{b}_n)\cos\alpha \\ - R_{H\delta}(t-t', z_1, z_2, \ldots, z_n, \boldsymbol{b}_1, \ldots, \boldsymbol{b}_n)\sin\alpha] \quad (5.9)$$

is recommended. In contrast with Section 1.1, R_δ here denotes the local response of a system of interacting caustics rather than the global wavefield. The functions $\hat{R}_\delta$, R_δ and $R_{\mathrm{H}\delta}$ are explicitly computable for simple caustics and caustic cusps and the parameters $z_1, \ldots, \boldsymbol{b}_n$ are determined from the DRT output data. The parameters $z_1, \ldots, z_n$ are given by a system of algebraic equations with some integer coefficients and with symmetric polynomials of travel times $T_1, \ldots, T_n$ on the right-hand side. The algebraic equations are solved by the iterative algorithm of Powell (1970). The iterations converge uniformly in the vicinity of caustics and at a distance from them.

The expressions (5.8) and (5.9) are linear functions of the vectors $\boldsymbol{b}_1, \ldots, \boldsymbol{b}_n$. The vectors $\boldsymbol{b}_1, \ldots, \boldsymbol{b}_n$ are given componentwise by a linear system of equations with the vectors $a_1\boldsymbol{r}_1, \ldots, a_n\boldsymbol{r}_n$ on the right-hand side, where a_i and $\boldsymbol{r}_i$ denote the amplitude and polarization vectors determined by DRT.

5.3 Computation of a composite S-wave pulse

For an isolated S-wave arrival the pulse is given by

$$a(\bar{s}_1(t)\boldsymbol{v}_1 + \bar{s}_2(t)\boldsymbol{v}_2) \quad (5.10)$$

where $\boldsymbol{v}_1$ and $\boldsymbol{v}_2$ are the two polarizations vectors returned by DYNTRAC (Section 4.11), the pulses $\bar{s}_1$ and $\bar{s}_2$ are given by (1.8) with the amplitude a and the phase shifts α and $\alpha + \delta$ defined by (1.7), (4.1) and (4.39).

For a composite S-wave pulse the elliptical polarization must be taken into account. This leads to two sets of vectors $\boldsymbol{b}_1, \ldots, \boldsymbol{b}_n$ obtained by solving the appropriate linear system of equations with the vector amplitudes $a_1\boldsymbol{v}_{11}, a_2\boldsymbol{v}_{12}, \ldots, a_n\boldsymbol{v}_{1n}$ and $a_1\boldsymbol{v}_{21}, \ldots, a_n\boldsymbol{v}_{2n}$ and the total phase shifts α and $\alpha + \delta$ respectively (the vectors $\boldsymbol{v}_1$ and $\boldsymbol{v}_2$ are defined in Section 4.11).

5.4 Seismic pulses at shadow boundaries

At a shadow boundary a reflected or transmitted ray coincides with its edge-diffracted counterpart.

In the vicinity of a shadow boundary the composite pulse is given by (5.8) or (5.9) with appropriately defined kernels $\hat{R}_\delta$, R_δ and $R_{H\delta}$. $\hat{R}_\delta$ is given in terms of a Fresnel integral and its derivative, while R_δ and $R_{H\delta}$ are given in terms of elementary functions. The signal $\bar{s}(t)$ here refers to the principal (direct, reflected or transmitted, according to the shadow under consideration) wave participating in setting up the composite pulse. As for caustic diffraction, it appears in the asymptotic decomposition of the composite pulse into its constituent principal and edge-diffracted components.

The coefficients $\boldsymbol{b}_1$, $\boldsymbol{b}_2$ and z_1, z_2 are now functions of amplitudes and travel times obtained by DRT along reflected (transmitted) and edge-diffracted rays. The theory also accounts for nearly simultaneous arrivals of several reflected and edge-diffracted rays (Hanyga, 1988).

5.5 Closing remarks

The main source of error in the computation of a composite pulse lies in the geometrical travel times. Composite pulses depend on travel-time offsets (cf. (5.3)). In the vicinity of a caustic or shadow boundary travel-time offsets are small numbers obtained by subtraction of travel times determined by two-point ray tracing. It is therefore necessary to determine the travel times to sufficiently high accuracy, depending on the distance of the receiver from the caustic.

ACKNOWLEDGEMENTS

The numerical codes for KRT and DRT were elaborated and tested during my stay at NTNF/NORSAR, Kjeller (Olso) in 1983–4. Many tests and improvements were introduced during my visit at the Seismological Observatory of the University of Bergen in 1985–6. Financial sponsorship was provided by NTNF (Royal Norwegian Council for Scientific and Industrial Research) in 1983–4 and by the Research Division of Norsk Hydro A/S in Bergen in 1985–6. Computer facilities of NORSAR, Seismological Observatory and Norsk Hydro in Bergen were exploited.

I am greatly indebted for the generous help and assistance of Professors E. S. Husebye, M. Sellevoll, E. Rygg, Dr Hans Helle and others. I acknowledge invaluable advice and criticisms of Dr Jan Pajchel (University of Bergen), Dr Hans Helle and his colleagues at the Research Division of Norsk Hydro, Professor D. Doornbos (University of Oslo) and Dr J. Havskov (University of Bergen).

Dr Jan Pajchel has run many tests of KRT and DRT on sophisticated models based on numerical data from geological and geophysical surveys (borehole and marine seismics).

For a critical review of the manuscript I am indebted to Drs J. Pajchel, M. Seredyńska, I. Pšenčík, D. Doornbos and to an anonymous referee.

APPENDICES

A. Isotropic blocks

Wave Hamiltonians for an isotropic block ($|\rho| \leqslant 1$) are defined by the formula

$$H(\boldsymbol{x}, \boldsymbol{p}; l, \sigma) = a_\sigma(\boldsymbol{x})[p_1^2 + p_2^2 + p_3^2]^{1/2} \tag{A1}$$

where a_1 and a_2 denote the P- and S-wave speeds respectively (Hanyga *et al.*, 1984).

Dynamic boundary conditions (4.48) involve the matrix function (Hanyga *et al.*, 1984)

$$\boldsymbol{B}(\boldsymbol{v}, \boldsymbol{p}) = \lambda \boldsymbol{v} \otimes \boldsymbol{p} + \mu \boldsymbol{p} \otimes \boldsymbol{v} + \mu(\boldsymbol{v} \cdot \boldsymbol{p})\boldsymbol{E}. \tag{A2}$$

Explicit expressions for the matrices $\boldsymbol{G}$, $\boldsymbol{K}$ and $\boldsymbol{W}$ can be found in Appendix E of Hanyga (1986).

B. Transversely isotropic blocks

The following types of anisotropic blocks are allowed:

(1) transversely isotropic blocks with the isotropy axis $\boldsymbol{v}(\boldsymbol{x})$ at $\boldsymbol{x}$ parallel to the tangent $\boldsymbol{\kappa}(\boldsymbol{x})$ to the x^3 coordinate line (*vertical alignment*, $\rho = 2$);

(2) transversely isotropic blocks with the isotropy axis $\boldsymbol{v}(\boldsymbol{x}) = \boldsymbol{R}(\boldsymbol{x})\boldsymbol{\kappa}(\boldsymbol{x})$; the rotation matrix $\boldsymbol{R}(\boldsymbol{x}) = \boldsymbol{R}_2(\varphi)\boldsymbol{R}_1(\theta)$ is defined by the formulae:

$$\boldsymbol{R}_1 = (\varphi) = \begin{bmatrix} \cos\varphi & -\sin\varphi & 0 \\ \sin\varphi & \cos\varphi & 0 \\ 0 & 0 & 1 \end{bmatrix}, \quad \boldsymbol{R}_2(\theta) = \begin{bmatrix} \cos\theta & 0 & \sin\theta \\ 0 & 1 & 0 \\ -\sin\theta & 0 & \cos\theta \end{bmatrix}; \tag{B1}$$

the block type $\rho = 3$ if θ and φ do not depend on $\boldsymbol{x}$, and $\rho = 4$ otherwise.

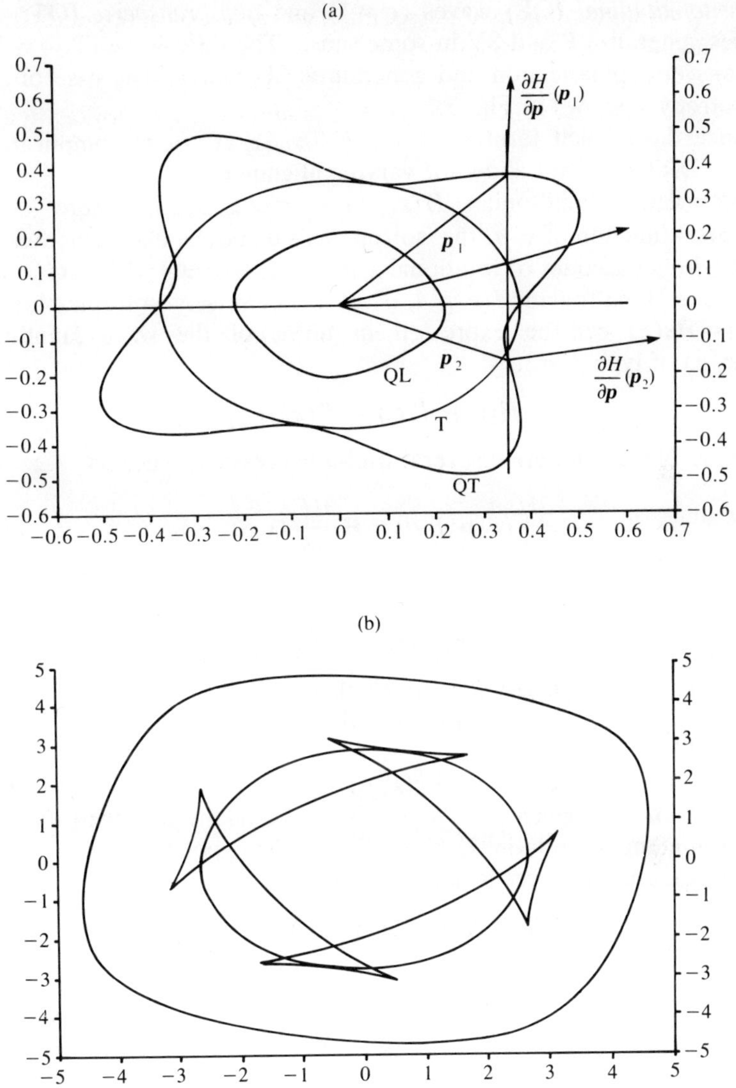

Fig. 3 Slowness surfaces (a) and wavefronts (b) in a transversely isotropic block. Case of strong anisotropy. The slowness surface of QT waves is non-convex and its wave surface presents four cuspidal edges lying in planes orthogonal to the isotropy axes. Two QT rays and one T ray enter the anisotropic block. The incident wave does not excite any QL wave. $N=7$, $L=8$, $A=20$, $C=22$, $F=15$.

Transverse isotropy allows for three distinct wave species. *Quasilongitudinal* (QL) waves ($\sigma = 1$) and *quasitransverse* (QT, $\sigma = 2$) waves generalize P and SV in some sense. The third wave (T, $\sigma = 3$) has a transverse polarization and generalizes SH waves. The case of strong anisotropy is shown in Fig. 3.

Since the explicit formulae for $\partial^2 H/\partial x_i\, \partial x_j$ are quite complicated for $\rho = 4$, DRT does not allow for varying alignment.

The wave Hamiltonian $H(\boldsymbol{x}, \boldsymbol{p}; l, \sigma) = G(\boldsymbol{x}, \boldsymbol{p}; \boldsymbol{v})$, where G is a universal function of $\boldsymbol{x}$, $\boldsymbol{p}$, the isotropy axis $\boldsymbol{v}$ and the elastic moduli N, L, A, C, F. On account of rotational symmetry, $G(\boldsymbol{x}, \mathbf{A}\boldsymbol{p}; \mathbf{A}\boldsymbol{v}) = G(\boldsymbol{x}, \boldsymbol{p}; \boldsymbol{v})$, the wave Hamiltonian $H(\boldsymbol{x}, \boldsymbol{p}; l, \sigma)$ of a transversely isotropic block with $\boldsymbol{v}(\boldsymbol{x}) = \mathbf{R}\boldsymbol{\kappa}(\boldsymbol{x})$ can be expressed in terms of the wave Hamiltonian $H^0(\boldsymbol{x}, \boldsymbol{p})$ of type $\rho = 2$:

$$H(\boldsymbol{x}, \boldsymbol{p}; l, \sigma) = H^0(\boldsymbol{x}, \mathbf{R}^{-1}\boldsymbol{p}). \tag{B2}$$

From (B2) we can derive several useful expressions, such as

$$\frac{\partial H}{\partial \boldsymbol{p}}(\boldsymbol{x}, \boldsymbol{p}; l, \sigma) = \mathbf{R}\frac{\partial H^0}{\partial \boldsymbol{p}}(\boldsymbol{x}, \mathbf{R}^{-1}\boldsymbol{p}), \tag{B3}$$

or, for the polarization vectors,

$$\boldsymbol{r}(\boldsymbol{x}, \boldsymbol{p}; l, \sigma) = \mathbf{R}\boldsymbol{r}^0(\boldsymbol{x}, \mathbf{R}^{-1}\boldsymbol{p}), \tag{B4}$$

reducing block type $\rho = 3, 4$ to block type $\rho = 2$. The transformations implied by (B2) and (B3) are carried out by the modules ROTAT and ROTAT11.

Explicit expressions for $H^0(\boldsymbol{x}, \boldsymbol{p})$, $\boldsymbol{r}^0(\boldsymbol{x}, \boldsymbol{p})$, $\mathbf{G}(\boldsymbol{x}, \boldsymbol{p})$ and $\mathbf{U}(\boldsymbol{x}, \boldsymbol{p})$, (4.7)–(4.9), are given in Appendix G of Hanyga (1986) (with U_{ab} corresponding to $w^a{}_b$ in Hanyga (1986). The polarization of inhomogeneous outgoing waves is defined by the formulae

$$\boldsymbol{r}^0{}_A = M[Rp_1 \quad Rp_2 \quad Sp_3]^{\mathrm{T}}, \quad A = 1, 2 \quad (\mathrm{QL}, \mathrm{QT}) \tag{B5}$$

$$\boldsymbol{r}^0{}_3 = M[-p_2 \quad p_1 \quad 0]^{\mathrm{T}} \quad (\mathrm{T}) \tag{B6}$$

where

$$R = H(\boldsymbol{x}, \boldsymbol{p})^2 - L'\boldsymbol{p}^2 - (C' - F' - 2L')p_3^2,$$

$$S = H(\boldsymbol{x}, \boldsymbol{p})^2 - L'\boldsymbol{p}^2 - (A' - F' - 2L')(p_1^2 + p_2^2),$$

$$L' = \frac{L}{d}, \qquad A' = \frac{A}{d}, \qquad C' = \frac{C}{d}, \qquad F' = \frac{F}{d},$$

d denotes the density and M is a normalization factor such that $\bar{\boldsymbol{r}}^0 \cdot \boldsymbol{r}^0 = 1$. In these expressions $H(\boldsymbol{x}, \boldsymbol{p})$ denotes the analytic continuation of $H(\boldsymbol{x}, \boldsymbol{p})$. For $p_1 = p_2 = 0$, $\boldsymbol{r}^0{}_2 = [0, 1, 0]^{\mathrm{T}}$ and $\boldsymbol{r}^0{}_3 = [1, 0, 0]^{\mathrm{T}}$.

The tracing programs assume that the elastic moduli N, L, A, C and F divided by density are passed through the common block /ELV/ in this order. In tracing through a block N, L should be specified for $\sigma = 3$ while L, A, C and F are required for $\sigma = 1, 2$. The corresponding gradients should be specified in the common block /ELG/. At interfaces all the five elastic moduli are required.

From (G4) of Hanyga (1986) we derive the matrix

$$\boldsymbol{B}(\boldsymbol{v}, \boldsymbol{p}) = \boldsymbol{B}_0 = \begin{bmatrix} Av_1p_1 + Nv_2p_2 + Lv_3p_3 & (A-2N)v_1p_2 + Nv_2p_1 & Fv_1p_3 + Lv_3p_1 \\ (A-2N)v_2p_1 + Nv_1p_2 & Nv_1p_1 + Av_2p_2 + Lv_3p_3 & Fv_2p_3 + Lv_3p_2 \\ Fv_3p_1 + Lv_1p_3 & Fv_3p_2 + Lv_2p_3 & L(v_1p_1 + v_2p_2) + Cv_3p_3 \end{bmatrix}$$

for block type $\rho = 2$. For $\rho \geqslant 3$, $\boldsymbol{B}(\boldsymbol{A}\boldsymbol{v}, \boldsymbol{A}\boldsymbol{p}) = \boldsymbol{A}\boldsymbol{B}_0\boldsymbol{A}^{-1}$, $\boldsymbol{A}^{\mathrm{T}} = \boldsymbol{A}^{-1}$.

We note that isotropic blocks correspond to

$$A = C = \lambda + 2\mu, \qquad L = N = \mu, \qquad F = \lambda. \tag{B7}$$

Dynamic ray tracing requires computation of (possibly complex-valued) slowness vectors of all the outgoing waves. Except for QL–QT waves, explicit solutions are used in this case. For QL–QT waves the quartic equation

$$(H^2 - m^2A' - n^2L')(H^2 - n^2C' - m^2L') - m^2n^2(F' + L')^2 = 0 \tag{B8}$$

is solved for μ, with $H = 1$, $m^2 = q_1^2 + q_2^2$, $n^2 = q_3^2$ and $\boldsymbol{q} = \boldsymbol{R}^{-1}(\boldsymbol{p}_{\mathrm{T}} + \mu\boldsymbol{v})$, where $\boldsymbol{v}$ is the unit normal to the interface and $\boldsymbol{p}_{\mathrm{T}}$ is the component of incident slowness tangent to the interface. It is easy to see that the expressions (G6) and (G7) for the wave Hamiltonian H of Hanyga (1986) are explicit solutions of (B8).

Equation (B8) is solved by the algorithm of Grant and Hitchins (1975), while the wave species of each root is identified by checking the value of $H(\boldsymbol{x}, \boldsymbol{p}; l, 1)$ and $H(\boldsymbol{x}, \boldsymbol{p}; l, 2)$.

In kinematic initial-value tracing the zeroin algorithm is applied to (4.40) at fixed wave-species (module SLWSOLV).

C. Isotropic constant-gradient blocks

The following closed-form formulae are applied in kinematic ray tracing through isotropic blocks in which the propagation speeds are given by the expression $c(\boldsymbol{x}) = a + b(z - z_1)$:

$$c = c(z) = [P \cosh(H_0 bs + \Delta)]^{-1}, \tag{C1}$$

$$\Delta = -\operatorname{sgn}(p_3^0) \ln |c_0^{-1} P^{-1} + (c_0^{-2} P^{-2} - 1)^{1/2}|, \tag{C2}$$

$$r = |(1 - c^2 P^2)^{1/2} \pm (1 - c_0^2 P^2)^{1/2}| / bP, \tag{C3}$$

$$H_0 = c_0[(p_1^0)^2 + (p_2^0)^2 + (p_3^0)^2]^{1/2}, \tag{C4}$$

where $c_0 = c(z_0)$, z_0 and $\boldsymbol{p}^0$ denote the initial depth and slowness (at $s = 0$), r is the horizontal ("epicentral") distance between the two ends of the ray arc and $P = [(p_1^0)^2 + (p_2^0)^2]^{1/2}/H_0$ is the ray parameter. The upper/lower sign applies in (C3) according to the sign of $H_0 bs + \Delta$.

The solution $z(s)$, $r(s)$ defined by (C1)–(C3) satisfies (3.2). The parameter s is related to the travel time by the formula $H_0 s = t$.

D. Derivation of (4.43)

Let $\tilde{\boldsymbol{x}}(t, u_1, u_2)$ denote the rayfield expressed in terms of the ray coordinates (t, u_1, u_2).

It is shown in Section 11 of Hanyga (1984c) that the area of a wavefront $t = \text{const}$ subtended by a ray tube is

$$\left(\left| \frac{\partial \tilde{\boldsymbol{x}}}{\partial u_1} \times \frac{\partial \tilde{\boldsymbol{x}}}{\partial u_2} \right| \right) du_1\, du_2 = |J|\, |\boldsymbol{p}|\, du_1\, du_2. \tag{D1}$$

Since the angle φ between the ray tangent $\partial H/\partial \boldsymbol{p}$ and the wavefront normal $\boldsymbol{p}$ is given by

$$\cos \varphi = \left(|\boldsymbol{p}| \left| \frac{\partial H}{\partial \boldsymbol{p}} \right| \right)^{-1} \tag{D2}$$

on account of the identity $\boldsymbol{p} \cdot \partial H/\partial \boldsymbol{p} = H = 1$ (Hanyga *et al.*, 1984), the orthogonal cross-section of the ray tube is

$$\frac{|J|}{|\partial H/\partial \boldsymbol{p}|} du_1\, du_2. \tag{D3}$$

By a simple geometric argument, (D3) implies (4.43). Equation (105) of Hanyga (1984c) is wrong.

E. Closed-form formulae for Π and Q in an analytically defined medium

The closed-form parametric solution of (2.1) for an isotropic medium with propagation speed $c(\boldsymbol{x}) = (C + \sum_i^3 a_i x_i)^{-1/2}$ (Červený, 1987)

$$x_i(\zeta) = x_i(\zeta_0) + p_i(\zeta_0)(\zeta - \zeta_0) + \tfrac{1}{4}a_i(\zeta - \zeta_0)^2, \tag{E1}$$

$$p_i(\zeta) = p_i(\zeta_0) + \tfrac{1}{2}a_i(\zeta - \zeta_0), \tag{E2}$$

$$t(\zeta) = t(\zeta_0) + c(\boldsymbol{x}(\zeta_0))^{-2}(\zeta - \zeta_0) + \tfrac{1}{2}\sum_i^3 a_i p_i(\zeta_0)(\zeta - \zeta_0)^2 + \tfrac{1}{12}\sum_i^3 a_i^2(\zeta - \zeta_0)^3 \tag{E3}$$

yields by Hanyga (1987) the closed-form solutions of the linear DRT equations

$$Q_{\pi(a),\alpha}(\zeta) = \sum_i R_{ai}(\zeta)\left[\sum_{b\neq I} P_{ib}(\zeta_0)Q_{\pi(b),\alpha}(\zeta_0) + (\zeta - \zeta_0)Z_{i\alpha}\right], \tag{E4}$$

$$\Pi_{\pi(a),\alpha}(\zeta) = \sum_i^3 P_{ia}(\zeta)Z_{i\alpha}, \tag{E5}$$

with $a \neq I$ and

$$Z_{i\alpha} = \sum_{a\neq I}\left[R_{ai}(\zeta_0)\Pi_{\pi(a),\alpha}(\zeta_0) + \frac{c^2}{2}\sum_j^3 a_j P_{ja}(\zeta_0)Q_{\pi(a),\alpha}(\zeta_0)p_i(\zeta_0)\right]. \tag{E6}$$

The cheapest method of tracing in such a medium is provided by linear DRT based on CHINTR2. Travel time is an additional dependent variable. In the case of linear interpolation of $c(\boldsymbol{x})^{-2}$ (Červený, 1987) the model can be defined in terms of tetragonal blocks with $c(\boldsymbol{x})$ continuous across those block boundaries that do not separate physically different blocks.

F. Algorithms INTFIND and CHINTR2 for determining intersections of curves and surfaces

The algorithms of INTFIND and CHINTR2 depend on the value assumed by an integer variable INT0 on entry. If the initial point $\boldsymbol{w}_0$ of the arc lies on an interrupt surface $f(\boldsymbol{w}; k) = 0$ such that $\text{SURIND}(k) \neq 0$ then INT0 should be assigned the value k. The sign is irrelevant for INTFIND, while for CHINTR2 sign (INT0) should be equal to the expected sign of $f(\boldsymbol{w}(t); k)$. If the initial point $\boldsymbol{w}_0$ does not lie on

$f(\boldsymbol{w}; k) = 0$ for any k such that SURIND $(k) \neq 0$ then INT0 should be assigned the value -100.

We now describe the algorithm of INTFIND.

If INT0 $\neq -100$ then INTFIND proceeds to Step 3. Otherwise INTFIND checks whether $f(\boldsymbol{w}_0; k) = 0$ for some k such that SURIND$(k) \neq 0$; if the result is negative then INTFIND proceeds to Step 3; otherwise it goes to Step 2.

In Step 2 INTFIND carries out one integration step of step length bounded by a small number ε (a value of order 10^{-6} is recommended; it must be smaller than the linear dimensions of blocks) and proceeds to Step 3.

In Step 3 INTFIND carries out integration by the Adams–Bashforth or Gear algorithm with variable step length and order. After estimating the step length for the next step, the signs of $f(\boldsymbol{w}; k)$ for such k that SURIND$(k) \neq 0$ are compared at both ends of the integration step. If a sign reversal is detected for $f(\boldsymbol{w}; k)$ then ARINT(k) is set equal to 1; otherwise ARINT$(k) = 0$. If at least one function $f(\boldsymbol{w}; k)$ such that SURIND$(k) \neq 0$ has changed sign then INTFIND goes to Step 5. If MARK $\neq 0$ then INTFIND goes to Step 4.

If the integer variable IPLOT $\neq 0$ and the curve parameter t has exceeded a value $n \times$ STEP at the current step then INTFIND proceeds to the output of $\boldsymbol{w}(t)$ (if IPLOT $= 1$ or 3) and control data (if IPLOT $= 2$ or 3) and integration is resumed.

If the value of t exceeds a maximum allowed value T then INTFIND exits with IFAIL $= -1$ and returns the value of $\boldsymbol{w}(T)$; INT0 is set equal to -100.

In Step 4 INTFIND exits with INT $=$ INT0 $= -100$ if MARK $\geqslant 0$, or else it calls TRANS(X) and goes to Step 3.

In Step 5 INTFIND calls the module INTRC1. INTRC1 combines Hermite cubic interpolation (Powell, 1981) of $\boldsymbol{w}(t)$ within the last integration step with the zeroin algorithm (Bus and Dekker, 1975) to find the exact values of t at the intersections detected in Step 3. On exit, INTFIND returns the reference number k of the first interrupt surface $f(\boldsymbol{w}; k) = 0$ intersected (in Common /A/) and the values of t and $\boldsymbol{w}(t)$ at the intersection point; INT0 is assigned the value k and sign (INT0) is the opposite of the sign $[f(\boldsymbol{w}(t); k)]$ before the intersection.

We note that the value of INT0 is correct for a next call of INTFIND or CHINTR2 if the arc is continued to the other side of $f(\boldsymbol{w}; k) = 0$. At reflections the sign of INT0 should be reversed by the calling program.

We now describe the algorithm of CHINTR2. Let $\boldsymbol{w}_0$ be the initial point of the arc $\boldsymbol{w}(t)$ defined by a generic subroutine RAY; for simplicity of notation we assume that $\boldsymbol{w}(0) = \boldsymbol{w}_0$.

At the start $T_{min} = 0$. If INTO $= -100$ on entry then CHINTR2 goes to Step 5; otherwise it goes to Step 1. In Step 1, T_{min} is increased by adding T_{max}/KL, where K and L denote two integer control parameters and T_{max} is the maximum length of the interval of the arc parameter specified in the argument of CHINTR2. If $T_{min} \geqslant T_{max}$ then CHINTR2 exits with IFAIL $= -1$. If $|f(\boldsymbol{w}(T_{min});\ |\text{INT0}|)|$ is smaller than a parameter η representing the upper limit of the error incurred in the computation of the functions $f(\cdot; k)$ then CHINTR2 returns to Step 2; otherwise it goes to Step 3.

In Step 3 the signs of INT0 and $f(\boldsymbol{w}(T_{min});\ |\text{INT0}|)$ are compared. If they are equal then CHINTR2 goes to Step 4; otherwise T_{min} is set equal to 0, the value of L is doubled and CHINTR2 goes to Step 1. Integer overflow of L results in a failure exit of CHINTR2 with IFAIL $= 6$. We note here that an inconsistency between the expected sign, sign (INT0), of $f(\boldsymbol{w}(t); |\text{INT0}|)$ and the sign of $f(\boldsymbol{w}(T_{min});\ |\text{INT0}|)$ can be due to intersection of the arc $\boldsymbol{w}(t)$, $0 < t < T_{min}$, with the surface $f(\boldsymbol{w}; |\text{INT0}|) = 0$.

In Step 4 the values of $f(\boldsymbol{w}(T_{min}); k)$ for all k such that SURIND$(k) \neq 0$ are stored. If for some $k \neq |\text{INT0}|$ such that SURIND$(k) \neq 0$ the signs of $f(\boldsymbol{w}(0); k)$ and $f(\boldsymbol{w}(T_{min}); k)$ are opposite then ARINT$(k) = 1$; otherwise ARINT$(k) = 0$. If for some k, ARINT$(k) = 1$ then CHINTR2 goes to Step 7; otherwise it goes to Step 5.

In Step 5 the integer j is set equal to 0 and CHINTR2 goes to Step 6. In Step 6, j is set equal to $j + 1$. If $j + 1 > K$ then CHINTR2 exits with IFAIL $= -1$; the input value of T is replaced by $T + T_{max}$ and the vector $\boldsymbol{w}(T_{max})$ is returned. Otherwise the signs of $f(\boldsymbol{w}(T_{min}); k)$ and $f(\boldsymbol{w}(T_{min} + jT_{max}/K)$ are compared for such k that SURIND$(k) \neq 0$. If they are opposite then ARINT$(k) = 1$; otherwise ARINT$(k) = 0$. If ARINT$(k) = 1$ for some k then CHINTR2 goes to Step 7; otherwise it goes to Step 6.

In Step 7 the zeroin algorithm is applied to all the functions $f(\boldsymbol{w}; k)$ such that ARINT$(k) = 1$, and the values of the arc parameter at the intersection points of $\boldsymbol{w}(t)$ with $f(\boldsymbol{w}; k) = 0$ are determined. CHINTR2 exits with IFAIL $= 0$. On exit, T is replaced by $T + t_0$, where $\boldsymbol{w}(t_0)$ is the first intersection point. The value of $\boldsymbol{w}(t_0)$ is also returned. The reference number k of the first interrupt surface $f(\boldsymbol{w}; k) = 0$ intersected is returned in the common block /A/ while INT0 is set equal to k, with the sign defined as in INTFIND.

G. Examples

The three examples presented below assume the simplest models giving rise to caustics. The receiver is located inside the caustic cusp of the source. Three rays with the same raycode join the source to the receiver, and one of them is tangent to a caustic, in agreement with the general theory (Hanyga, 1984b). The three rays are generated by a two-point tracer; the input ray is generated by initial-value tracing or by ITHGEN. For definiteness the initial secant matrix $\boldsymbol{G}_0 = 0$ in every two-point tracing.

Example 1: Caustic cusp generated by variation of the speed gradient (file MODEL5, Fig. 4)

The model consists of a single isotropic block. The propagation speed is defined by a hyperbola asymptotic to two constant gradients:

$$\mathrm{d}c/\mathrm{d}z = 0.5,\ 2.0$$

$$c(z) = \tfrac{1}{2}\{2.5z + 2.5 + [(1.5z - 7.5)^2 + 0.1]^{1/2}\}.$$

Model box: $-5 \leqslant x \leqslant 40,\ 0 \leqslant y \leqslant 5,\ -20 \leqslant z \leqslant 20.$

Source: (0, 2, 0); *receiver:* (23.3, 2, −8)

Initial slowness:

Ray 1: (0.132206470556, 0, 0.149893938273);

Ray 2: (0.143371460198, 0, 0.139252892209);

Ray 3: (0.049340640761, 0, 0.193680780557).

Ray 1 is tangent to a caustic at (11.5214335, 2, 4.7423695).

Amplitudes: 2.6769×10^{-3}, 2.3773×10^{-3}, 1.7898×10^{-3}.

Travel times: 6.95136 ± 10^{-8}, 6.94390 ± 10^{-8}, 6.59704 ± 10^{-9}.

Phases: $-\frac{1}{2}\pi, 0, 0$

Polarization (*P waves*):

(0.13238, 0, −0.99120), (0.14356, 0, −0.98964), (0.04940, 0, −0.99877).

The initial data were generated by running ITIL. The input rays were generated by running TRGRDC with take-off angles $(0.23\pi, 0)$, $(0.255\pi, 0)$ and $(0.08\pi, 0)$. The tolerance $\gamma = 10^{-14}$.

Number of iterations: 2 failed + 2 successful, 2, 3.

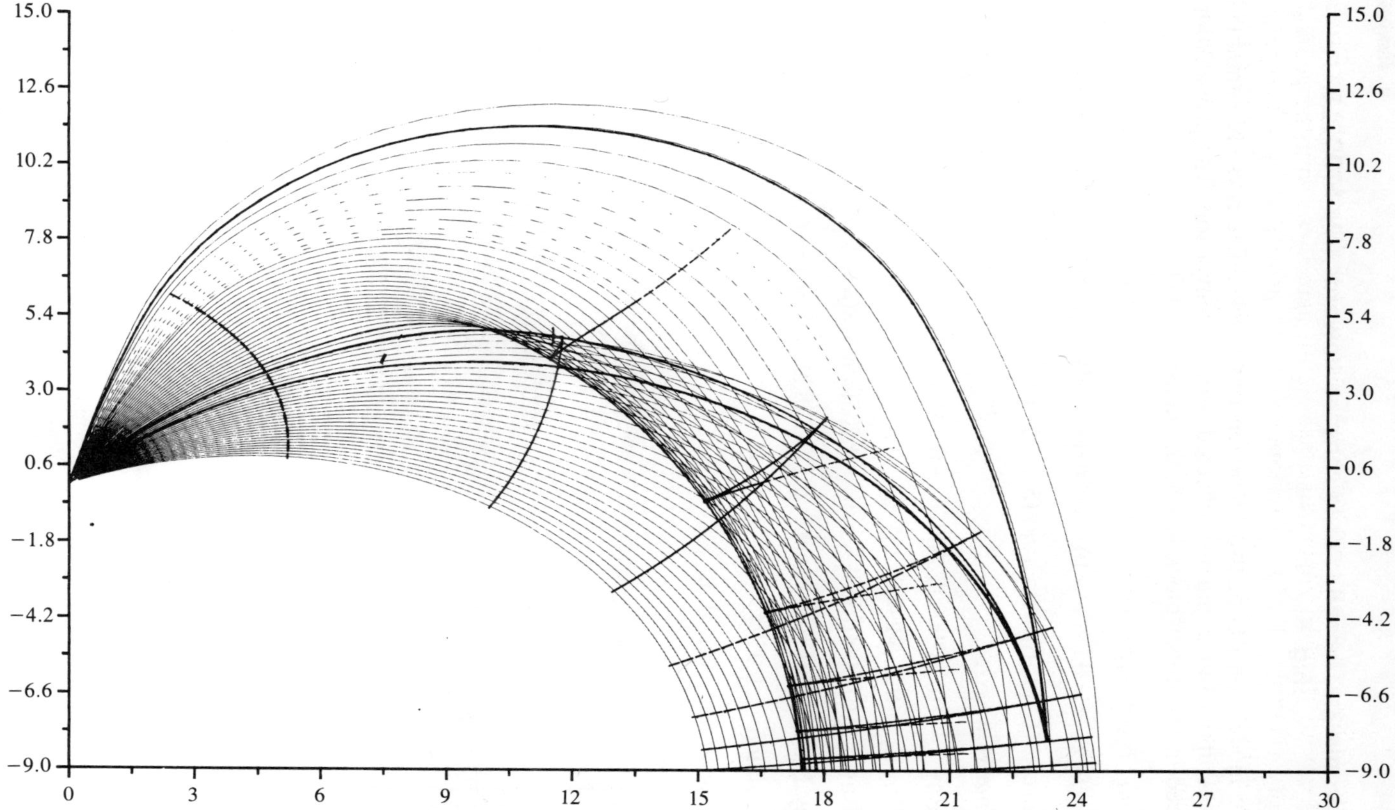

Fig. 4 Example 1. Rays 1, 2, 3 are marked by heavy solid lines against the rayfield marked by light solid lines. Wavefronts are marked by dashed lines. Note the infinite curvature of the wavefronts at the two caustics. The point of tangency of ray 1 to the caustic (as determined by DRT) is marked by a cross.

Example 2: The reflector is concave upwards (file MODEL8, Fig. 5)

The model consists of two homogeneous blocks. The rays are contained in the upper block and are reflected without conversion. The propagation speed in the upper block is 3.5; the density is 1.0.

Model box: $-10 \leqslant x \leqslant 20,\ 0 \leqslant y \leqslant 20,\ 0 \leqslant z \leqslant 20.$

Interface: $z = 0.01(x-10)^3 + 0.05(x-10)^2 + 5.$

Source: (6, 2, 16), *receiver*: (5.12, 2, 19.5)

Initial slowness:

Ray 1: $(0.167909902680, 0, -0.231168591386)$;

Ray 2: $(0.276854607265, 0, -0.070598721643)$;

Ray 3: $(0.006076681326, 0, -0.285649657807)$.

Ray 1 is tangent to a caustic at $(11.3788916, 2, 9.1369231)$.

Amplitudes: 1.9594×10^{-3}, 2.9052×10^{-3}, 1.6535×10^{-3}.

Phase shifts: $-\frac{1}{2}\pi$, 0, 0.

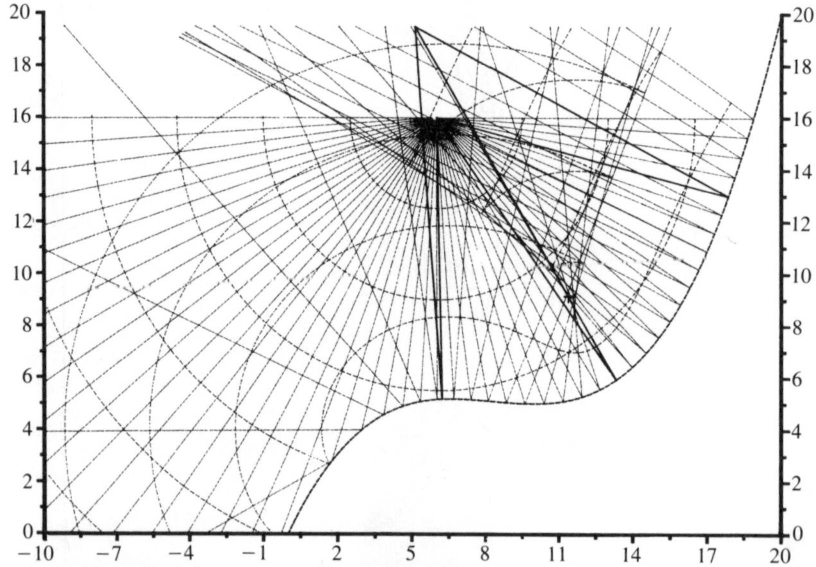

Fig. 5 Example 2. Rays 1, 2, 3 are marked by heavy solid lines against the rayfield marked by light solid lines. Wavefronts are marked by dashed lines. The point of tangency of ray 1 to the caustic is marked by a cross.

Travel times: 8.09316 ± 10^{-8}, 7.59275 ± 10^{-6}, $7.19878 \pm 3 \times 10^{-17}$.

Polarizations:

$(-0.51699, 0, 0.85599)$, $(-0.89014, 0, 0.45566)$,

$(-0.07724, 0, 0.99700)$.

The initial slowness vectors were obtained by running ITH with the tolerance $\gamma = 10^{-12}$. The input rays were generated by running TRHOM21 with the take-off angles $(\theta, \varphi) = (0.8\pi, 0)$, $(0.67\pi, 0)$, $(0.97\pi, 0)$.

Number of secant iterations: 1, 5, 3; *initial secant matrix:* 0.

Example 3: Caustic cusp of edge-diffracted rays (file MODEL4, Fig. 6)

The model consists of two homogeneous blocks. Block 2 is defined by the inequalities $f(\boldsymbol{x}; k) < 0$, $k = 1, 2$. The interfaces $f(\boldsymbol{x}; 2) = x = 0$ and

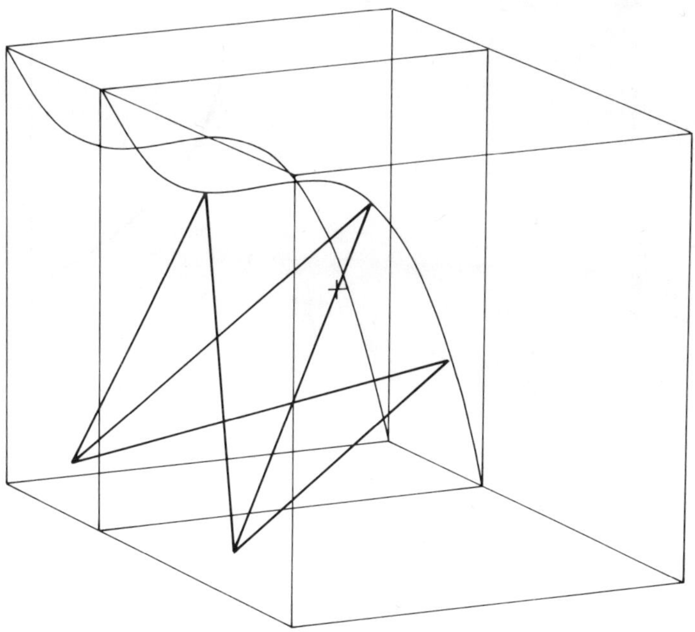

Fig. 6 Example 3. Rays 1, 2, 3 are marked by heavy solid lines against the ray field marked by light solid lines. Wavefronts are marked by dashed lines. The point of tangency of ray 1 to the caustic is marked by a cross. The caustic cusp is generated by a curved diffracting edge.

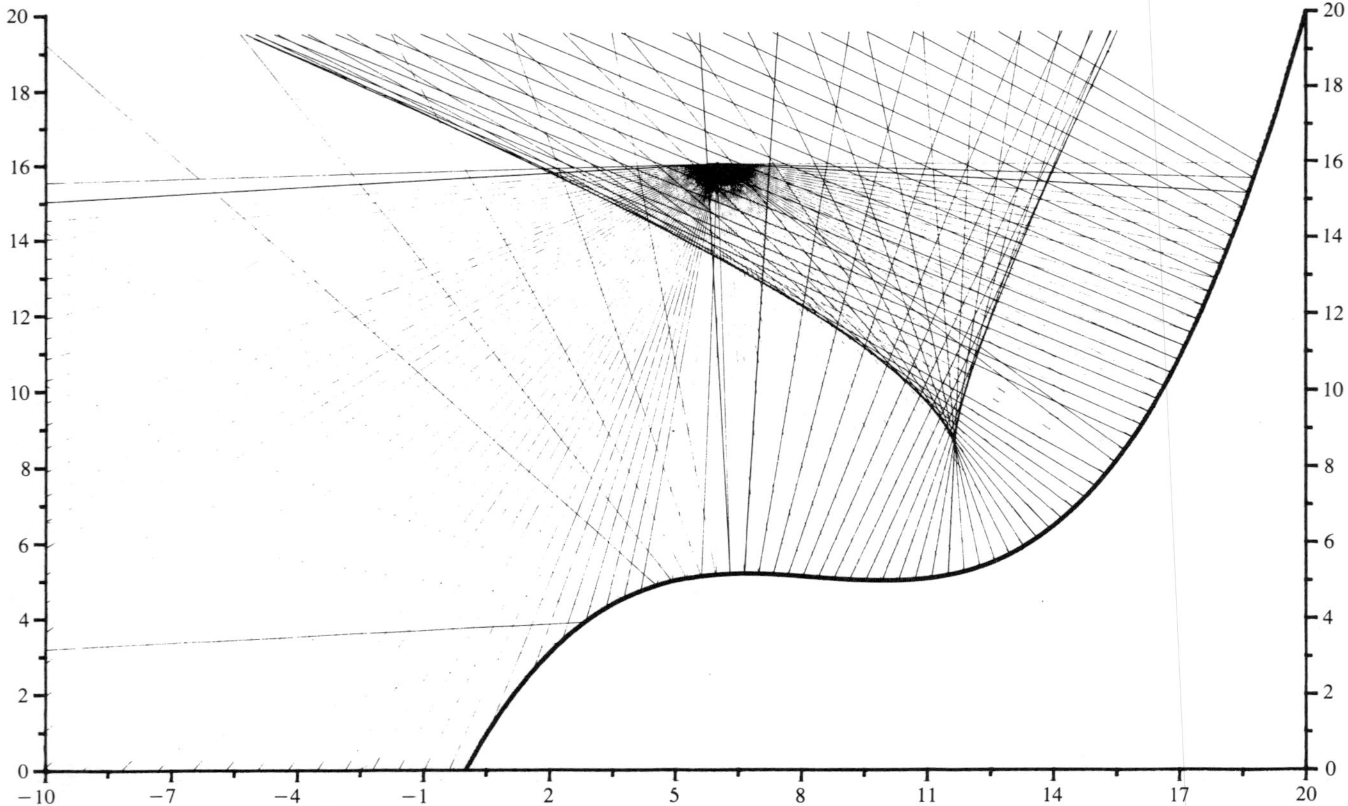

Fig. 7 The caustic cusp as determined by DRT is marked by light solid lines. It coincides with the envelopes of the rayfield. The model and source position are those of Fig. 5.

$f(\boldsymbol{x}; 1) = z - 0.01(y-10)^3 - 0.05(y-10)^2 - 5 = 0$ intersect at an edge $\mathscr{E}$. The rays are contained in block 1 (the speed is 3.5).

Source: $(-9, 3, 19)$; *receiver:* $(8, 3, 19)$.

Initial slowness and initial slowness of edge-diffracted leg (as determined by two-point tracing):

Ray 1: $(0.135591244775, 0.168425968687, -0.186762845538)$;
$(0.119473998569, -0.181890748171, 0.185133390984)$;

Ray 2: $(0.153543809397, 0.038667633446, -0.237827176287)$;
$(0.139362029343, -0.056903446602, 0.242843314925)$;

Ray 3: $(0.139811274652, 0.232880271070, -0.088612865249)$;
$(0.121785223959, -0.243434704911, 0.086836379077)$.

Ray 1 is tangent to a caustic at $(2.20928, 10.81596, 10.02687)$.

Amplitudes: $7.3231 \times 10^{-2}, 5.0386 \times 10^{-2}, 1.1159 \times 10^{-1}$.

Phases: $-\frac{1}{2}\pi, 0, 0$.

Travel times: $10.88458 \pm 10^{-18}, 9.47099 \pm 10^{-9}, 10.061730 \pm 10^{-18}$

Polarizations (P waves):

$(0.41816, -0.63662, 0.64797)$, $(0.48777, -0.19916, 0.84995)$, $(0.42625, -0.85202, 0.30393)$.

The initial data were obtained by running ITHDIF with $\gamma = 10^{-13}$. The input rays were generated by running the autonomous tracer ITHGEN with the endpoints at $(-9, 3, 19)$, $(5, 3, 19)$ to generate three rays reflected at $f(\boldsymbol{x}; 1) = 0$. In order to obtain three different solutions, the following three "shot vectors" $\boldsymbol{N}'$ (Section 3.7) were chosen: $(0, -1, 1)$, $(0, 1, 1)$, $(0, -2, 1)$. The tolerance $\gamma = 10^{-9}$.

Number of secant iterations:

ITHGEN: 2, 4, 5; ITHDIF: 5, 2, 2.

REFERENCES

Abramowitz, M. and Stegun, I. A. (1970). *Handbook of Mathematical Functions,* 10.4.59–60. Dover, New York.

Bertsekas, D. P. (1982). *Constrained Optimization and Lagrange Multiplier Methods.* Academic Press, New York.

Bus, J. C. P. and Dekker, T. J. (1975). Two efficient algorithms with guaranteed convergence for finding a zero of function. *ACM Trans. Math. Software* **1,** 330–345.

Červený, V. (1987). Ray tracing algorithms in 3-dimensional laterally varying layered structures. *Seismic Tomography* (ed. G. Nolet). Reidel, Dordrecht.

Červený, V. and Pšenčík, I. (1983). Gaussian beams and paraxial ray approximation in three-dimensional elastic inhomogeneous media, *J. Geophys.* **53,** 1–15.

Courant, R. and Hilbert, D. (1962). *Methods of Mathematical Physics,* Vol. 2. Interscience, New York.

Gel'fand, I. M. and Shilov, G. E. (1964). *Generalized Functions,* Vol. 1: *Properties and Operations.* Academic Press, New York.

Gjøystdal, H., Reinhardsen, J. E. and Åstebøl, K. (1984). A flexible user-oriented system for 3D dynamic ray tracing in complex geological structures. NORSAR Contribution 331.

Grant, J. A. and Hitchins, G. D. (1975). Two algorithms for the solution of polynomial equations to limiting machine precision. *Comp. J.* **18,** 258–264.

Hanyga, A. (1980). On the definition of outgoing waves and boundary value problems in anisotropic elasticity. *Bull. Acad. Polon. Sci., Ser. Sci. Terre* **28,** 267–279.

Hanyga, A. (1982). Dynamic ray tracing in anisotropic media. *Tectonophysics* **90,** 243–251.

Hanyga, A. (1983). Linear and non-linear dynamic ray tracing systems and their complex-valued solutions. *Acta Geophys. Polon.* **31,** 243–254.

Hanyga, A. (1984a). Numerical computation of elastic wavefields in anisotropic elastic media in the presence of caustics. *Hybrid Formulation of Wave Propagation and Scattering* (ed. L. B. Felsen), pp. 153–168. Marinus Nijhoff, The Hague.

Hanyga, A. (1984b). Dynamic ray tracing on Lagrangian manifolds. *Geophys. J. R. Astron. Soc.* **79,** 51–63.

Hanyga, A. (1984c). Transport equations for an anisotropic elastic medium in the presence of caustics. *Gerlands Beitr. Geophys.* (*Leipzig*) **93,** 261–286.

Hanyga, A. (1984d). Point source solution in anisotropic elastic medium. *Gerlands Beitr. Geophys.* (*Leipzig*) **93,** 463–479.

Hanyga, A. (1986). Gaussian beams in anisotropic media. *Geophys. J. R. Astron. Soc.* **85,** 473–504.

Hanyga, A. (1987). *Linear and Nonlinear DRT.* Report Seismological Observatory. University of Bergen Seismo-Series.

Hanyga, A. (1988) Boundary effects in asymptotic diffraction theory (in preparation).

Hanyga, A. and Seredyńska, M. (1988). Diffracted seismic pulses in the vicinity of simple caustics and caustic cusps (in preparation).

Hanyga, A., Lenartowicz, E. and Pajchel, J. (1984). *Seismic Waves in the Earth.* Elsevier, Amsterdam.

Julian, B. and Gubbins, D. (1975). Two-point seismic ray tracing. *Trans. AGU* **56,** 1027–1032.

Kamel, A. and Felsen, L. B. (1981). Hybrid ray-mode formulation of SH motion in a two-layer half-space. *Bull. Seismol. Soc. Am.* **71,** 1763–1781.

Keller, H. B. and Perozzi, D. (1983). Fast seismic ray tracing. *SIAM J. Appl. Maths* **43,** 981–992.

Ortega, J. M. and Rheinboldt, W. C. (1970). *Iterative Solution of Nonlinear Equations in Several Variables.* Academic Press, New York.

Pereyra, V., Lee, W. H. K. and Keller, H. B. (1980). Solving two-point seismic-ray tracing problems in a heterogeneous medium, Part 1. *Bull. Seismol. Soc. Am.* **70,** 79–99.

Polak, E. (1974). A globally converging secant method with applications to boundary value problems. *SIAM J. Numer. Anal.* **11,** 529–537.

Popov, M. M. and Pšenčík, I. (1976). Ray amplitudes in inhomogeneous media with curved interfaces. *Geophys. Sb.*, Vol. 24 (ed. A. Zátopek), pp. 118–129. Academia, Praha.

Powell, M. J. D. (1970). A hybrid method for non-linear equations. *Numerical Methods for Non-linear Algebraic Equations* (ed. P. Rabinowitz). Gordon and Breach, New York.

Powell, M. J. D. (1981). *Approximation Theory and Methods,* Section 5.5. Cambridge University Press.

Roberts, S. M. and Shipman, J. (1972). *Two-Point Boundary Value Problems: Shooting Methods.* Elsevier, New York.

III

REFLECTIVITY METHODS FOR STRATIFIED MEDIA

III.1

Systematic Approximations to the Seismic Wavefield

B. L. N. KENNETT
Research School of Earth Sciences, Australian National University, Canberra, Australia

1 INTRODUCTION

Much of the emphasis in recent work on the computation of theoretical seismograms has been on the construction of the complete response of a stratified medium to excitation by a source. There are many formally equivalent analytic developments for the excitation problem. These transcribe into a variety of computer algorithms for the construction of the response in frequency/wavenumber space, followed by integration over wavenumber and Fourier synthesis to give a set of seismograms at different spatial locations. Recent reviews with a comprehensive set of references are given by Kennett (1983) and Chapman and Orcutt (1985). Many of the available methods can only give the complete response, but by taking a physically based approach based on a representation in terms of the reflection and transmission properties of the medium, it is possible to extract approximations to portions of the wavefield (Kennett and Kerry, 1979).

In contrast, generalized ray methods use a rather limited approximation to the seismic wavefield, with attention concentrated on either a small part of the seismogram or on the behaviour of the major seismic phases. For models composed of stacks of uniform layers, the Cagniard technique can be used (see e.g. Helmberger and Harkrider, 1978), and good results are obtained if sufficient multiple reflections in the layer stack are included. The WKBJ technique introduced by Chapman (1978) provides an economical development applicable to a smoothly varying

SEISMOLOGICAL ALGORITHMS
ISBN 0-12-220770-X

velocity profile. This is an excellent way of assessing the main character of the wavefield, but for waves interacting with interfaces at near grazing incidence or in the presence of shadow zones the WKBJ results can have significant errors (see Chapman and Orcutt, 1985). These generalized ray methods depend on the assumption of perfect elasticity, and the effect of attenuation in the model has to be introduced by convolving each ray response with an averaged attenuation operator. If one wishes to include such features as crustal multiples then each type of possible ray path has to be specified and some of the computational convenience is lost.

For many purposes we should like to be able to develop systematic approximations for parts of the seismic wavefield in an attenuative medium. The earliest approximation of this type was that employed by Fuchs and Müller (1971)—the original "reflectivity method" in which the complete response for a portion of the medium was coupled with simple transmission phase delays down from the source and up to the receiver. This approach has been very widely used (see e.g. Chapman and Orcutt, 1985), and with care in the choice of the reflecting region gives good results where surface reflections are unimportant.

However, in many situations the interaction of the wavefield with the free surface plays a major role in shaping the character of the seismograms. As an example we can consider calculating short-period seismograms for the distance range 1700–3000 km. Although the body wavefield is here dominated by the interference of energy returned from different levels in the upper mantle, surface reflections and crustal multiples make a large contribution to the complexity of the seismograms. In order to disentangle the details of the propagation process, we should like to have control over interconversions between wavetypes and the types of multiples included in the calculation. We can achieve this by working in the frequency/slowness domain with the reflection and transmission response of portions of the stratification. The results are suitable for integration to the spatial domain with either a real-axis slowness integration ("reflectivity method"; derived from the work of Fuchs and Müller, 1971) or a numerical contour integration in slowness ("full-wave method"; see e.g. Cormier and Richards, 1977).

The algorithms that we shall develop are derived in part from the treatment of approximations to the response of a stratified medium in Chapter 9 of Kennett (1983), but they incorporate some useful results on interconversion between wavetypes from Kennett (1986). The notation will follow that of Kennett (1983) (which we shall denote by K83), and no explicit discussion of the integration step will be made (see e.g. K83, Chapter 7).

2 REFLECTION AND TRANSMISSION OF SEISMIC WAVES

We shall build up our approximations to the seismic wavefield using reflection and transmission terms, and therefore have to define our convention for such quantitites.

Following K83, we use an invariant embedding approach, i.e. when we consider the reflection from a zone $z_A \leqslant z \leqslant z_C$ we take this to be bordered by uniform half-spaces with continuity of properties at z_A and z_C. In these uniform media we can make an unambiguous decomposition of the seismic waves into upward and downward components via an eigenvector decomposition (K83, Chapters 3 and 5). When a general downgoing field is incident on the region (z_A, z_C) at z_A the upgoing reflected waves at z_A are described by a reflection matrix $\boldsymbol{R}_D^{AC}$ and the downgoing transmitted waves at z_C by a transmission matrix $\boldsymbol{T}_D^{AC}$. If the medium is isotropic or transversely isotropic with a vertical symmetry axis then the wave propagation processes separate into two parts:

(i) coupled P–SV waves with 2×2 reflection and transmission matrices, for example

$$\boldsymbol{R}_D^{AB} = \begin{bmatrix} R_{PP} & R_{PS} \\ R_{SP} & R_{SS} \end{bmatrix}_D^{AB}, \tag{1a}$$

where (for example) the reflection coefficient R_{PS} indicates the P wave generated by an incident SV wave;

(ii) SH waves with single reflection and transmission coefficients R_{HH} and T_{HH}.

Booth and Crampin (1983) and Fryer and Frazer (1984) have shown how the treatment can be extended to anisotropic media with 3×3 matrices incorporating possible coupling between all three wavetypes.

When we have an incident upgoing wave at z_C, the downgoing reflected waves at z_C are described by the reflection matrix $\boldsymbol{R}_U^{AC}$ and the upgoing transmitted waves at z_A by $\boldsymbol{T}_U^{AC}$.

There is one further important case for reflection, which is when we have a region $(0, z_E)$ with its upper boundary formed by the free surface $z = 0$. In this case we denote the reflection matrix due to incident upgoing waves at z_E by $\boldsymbol{R}_U^{fE}$; the upgoing waves also generate displacements at the free surface, and these may be represented in matrix form by $\boldsymbol{W}_U^{fE}$, which is the analogue of a transmission term in this case. Each row of this matrix will correspond to the values of a particular component excited by one type of incident wave. Thus for P–SV waves incident at the level z_E, the surface-displacement matrix takes the form

$$\boldsymbol{W}_{\mathrm{U}}^{\mathrm{fE}} = \begin{bmatrix} W_{\mathrm{UP}} & W_{\mathrm{US}} \\ W_{\mathrm{VP}} & W_{\mathrm{VS}} \end{bmatrix}_{\mathrm{U}}^{\mathrm{fE}}, \tag{1b}$$

where, for example, W_{UP} is the vertical component of surface displacement generated by an incident P wave at z_{E}, and W_{VP} is the corresponding horizontal component.

From the point of view of representing the response of the medium, the most important properties of the reflection and transmission terms arise when we split up a region. Consider breaking the zone $(z_{\mathrm{A}}, z_{\mathrm{C}})$ at z_{B}; then the original reflection and transmission matrices are related to those for the subdivisions by (K83, equations (6.3) and (6.4))

$$\left.\begin{aligned} \boldsymbol{T}_{\mathrm{D}}^{\mathrm{AC}} &= \boldsymbol{T}_{\mathrm{D}}^{\mathrm{BC}}[\boldsymbol{I} - \boldsymbol{R}_{\mathrm{U}}^{\mathrm{AB}}\boldsymbol{R}_{\mathrm{D}}^{\mathrm{BC}}]^{-1}\boldsymbol{T}_{\mathrm{D}}^{\mathrm{AB}}, \\ \boldsymbol{R}_{\mathrm{D}}^{\mathrm{AC}} &= \boldsymbol{R}_{\mathrm{D}}^{\mathrm{AB}} + \boldsymbol{T}_{\mathrm{U}}^{\mathrm{AB}}\boldsymbol{R}_{\mathrm{D}}^{\mathrm{BC}}[\boldsymbol{I} - \boldsymbol{R}_{\mathrm{U}}^{\mathrm{AB}}\boldsymbol{R}_{\mathrm{D}}^{\mathrm{BC}}]^{-1}\boldsymbol{T}_{\mathrm{D}}^{\mathrm{AB}}, \end{aligned}\right\} \tag{2}$$

and

$$\left.\begin{aligned} \boldsymbol{T}_{\mathrm{U}}^{\mathrm{AC}} &= \boldsymbol{T}_{\mathrm{U}}^{\mathrm{AB}}[\boldsymbol{I} - \boldsymbol{R}_{\mathrm{D}}^{\mathrm{BC}}\boldsymbol{R}_{\mathrm{U}}^{\mathrm{AB}}]^{-1}\boldsymbol{T}_{\mathrm{U}}^{\mathrm{BC}}, \\ \boldsymbol{R}_{\mathrm{U}}^{\mathrm{AC}} &= \boldsymbol{R}_{\mathrm{U}}^{\mathrm{BC}} + \boldsymbol{T}_{\mathrm{D}}^{\mathrm{BC}}\boldsymbol{R}_{\mathrm{U}}^{\mathrm{AB}}[\boldsymbol{I} - \boldsymbol{R}_{\mathrm{D}}^{\mathrm{BC}}\boldsymbol{R}_{\mathrm{U}}^{\mathrm{AB}}]^{-1}\boldsymbol{T}_{\mathrm{U}}^{\mathrm{AB}}, \end{aligned}\right\} \tag{3}$$

where $\boldsymbol{I}$ is the identity matrix. The physical content of these expressions can be recognized by reading from right to left, once it is appreciated that the inverse term has the character of a reverberation operator corresponding to successive internal multiples between AB and BC, since

$$[\boldsymbol{I} - \boldsymbol{R}_{\mathrm{U}}^{\mathrm{AB}}\boldsymbol{R}_{\mathrm{D}}^{\mathrm{BC}}]^{-1} = \boldsymbol{I} + \boldsymbol{R}_{\mathrm{U}}^{\mathrm{AB}}\boldsymbol{R}_{\mathrm{D}}^{\mathrm{BC}} + \boldsymbol{R}_{\mathrm{U}}^{\mathrm{AB}}\boldsymbol{R}_{\mathrm{D}}^{\mathrm{BC}}\boldsymbol{R}_{\mathrm{U}}^{\mathrm{AB}}\boldsymbol{R}_{\mathrm{D}}^{\mathrm{BC}}[\boldsymbol{I} - \boldsymbol{R}_{\mathrm{U}}^{\mathrm{AB}}\boldsymbol{R}_{\mathrm{D}}^{\mathrm{BC}}]^{-1}, \tag{4}$$

and continued expansion yields further reverberation terms.

A sequence of approximations to the reflection and transmission matrices can be made by truncating the reverberation operator at different points. The most convenient forms for general use are either to replace the operator by the identity matrix and so leave no reverberations within AC, or to invert the matrix numerically and thereby include all reverberative terms. The truncation of (4) after one interaction term, i.e. to retain $\boldsymbol{I} + \boldsymbol{R}_{\mathrm{U}}^{\mathrm{AB}}\boldsymbol{R}_{\mathrm{D}}^{\mathrm{BC}}$ enables a partial inclusion of internal reflection at slightly less computational cost than constructing $[\boldsymbol{I} - \boldsymbol{R}_{\mathrm{U}}^{\mathrm{AB}}\boldsymbol{R}_{\mathrm{D}}^{\mathrm{BC}}]^{-1}$ with all internal reflections included.

Similar rules apply if we subdivide the zone extending to the free surface by breaking it at z_{D} (K83, equations (6.9) and (6.10)):

$$\left.\begin{aligned} \boldsymbol{R}_{\mathrm{U}}^{\mathrm{fE}} &= \boldsymbol{R}_{\mathrm{U}}^{\mathrm{DE}} + \boldsymbol{T}_{\mathrm{D}}^{\mathrm{DE}}\boldsymbol{R}_{\mathrm{U}}^{\mathrm{fD}}[\boldsymbol{I} - \boldsymbol{R}_{\mathrm{D}}^{\mathrm{DE}}\boldsymbol{R}_{\mathrm{U}}^{\mathrm{fD}}]^{-1}\boldsymbol{T}_{\mathrm{U}}^{\mathrm{DE}}, \\ \boldsymbol{W}_{\mathrm{U}}^{\mathrm{fE}} &= \boldsymbol{W}_{\mathrm{U}}^{\mathrm{fD}}[\boldsymbol{I} - \boldsymbol{R}_{\mathrm{D}}^{\mathrm{DE}}\boldsymbol{R}_{\mathrm{U}}^{\mathrm{fD}}]^{-1}\boldsymbol{T}_{\mathrm{U}}^{\mathrm{DE}}. \end{aligned}\right\} \tag{5}$$

The displacement matrix $\boldsymbol{W}_{\mathrm{U}}^{\mathrm{fE}}$ behaves like a transmission term with displacement conversion at the free surface itself. Müller (1985) works with precisely such a decomposition, and so would set

$$\mathbf{W}_{\mathrm{U}}^{\mathrm{fE}} = \tilde{\mathbf{W}}\mathbf{T}_{\mathrm{U}}^{\mathrm{fE}},$$

where $\mathbf{T}_{\mathrm{U}}^{\mathrm{fE}}$ is the transmission matrix up to the free surface and $\tilde{\mathbf{W}}$ is the displacement conversion matrix at the free surface itself (K83, equation (5.69), which includes the amplification effects of reflection):

$$\tilde{\mathbf{W}} = \mathbf{M}_{\mathrm{U0}} + \mathbf{M}_{\mathrm{D0}}\tilde{\mathbf{R}},$$

where $\mathbf{M}_{\mathrm{U0}}$ and $\mathbf{M}_{\mathrm{D0}}$ are the matrices that generate the displacement components from up- and downgoing waves in a uniform medium with the properties at the surface, and $\tilde{\mathbf{R}}$ is the reflection matrix at the free surface itself.

We note that the matrix relations (5), in the presence of a free surface, have exactly the same structure as those when the zone is bounded by half-spaces on both sides. We may therefore set up a calculation scheme in which free-surface reflections can be included or excluded at will, by the way in which we specify the boundary conditions at the level $z = 0$. The results for a transparent upper boundary are to be found directly by setting the reflection matrix $\tilde{\mathbf{R}}$ equal to zero. However, we can retain the displacement amplification effects by retaining $\tilde{\mathbf{W}}$ rather than using just the matrix for upgoing waves $\mathbf{M}_{\mathrm{U0}}$.

When the medium consists of a stack of uniform layers, equations (2) and (5) form the basis of an effective recursive scheme for calculating the reflection and transmission matrices with control of internal multiples, which is described in the Appendix. Kennett and Illingworth (1981) have shown that, with care, a similar scheme can be applied for gradient zones.

For P–SV waves (or full anisotropy) equations (2), (3) and (5) imply coupling between wavetypes if any of the off-diagonal elements of the reflection and transmission matrices are non-zero. These conversion terms are usually small unless a region contains a material interface with a large jump in the seismic parameters. The free surface is also very efficient at introducing conversion between wavetypes. In general, however, any process requiring a twofold conversion of the type P→S→P will be relatively inefficient and not a major contributor to the wavefield (the exceptions are mostly when source and receiver lie in a liquid overlayer). Consider then a reverberation term for P–SV waves

$$[\mathbf{I} - \mathbf{R}^{\mathrm{A}}\mathbf{R}^{\mathrm{B}}] = \begin{bmatrix} 1 - R_{\mathrm{PP}}^{\mathrm{A}}R_{\mathrm{PP}}^{\mathrm{B}} - R_{\mathrm{PS}}^{\mathrm{A}}R_{\mathrm{SP}}^{\mathrm{B}} & -R_{\mathrm{PP}}^{\mathrm{A}}R_{\mathrm{PS}}^{\mathrm{B}} - R_{\mathrm{PS}}^{\mathrm{A}}R_{\mathrm{SS}}^{\mathrm{B}} \\ -R_{\mathrm{SP}}^{\mathrm{A}}R_{\mathrm{PP}}^{\mathrm{B}} - R_{\mathrm{SS}}^{\mathrm{A}}R_{\mathrm{SP}}^{\mathrm{B}} & 1 - R_{\mathrm{SS}}^{\mathrm{A}}R_{\mathrm{SS}}^{\mathrm{B}} - R_{\mathrm{SP}}^{\mathrm{A}}R_{\mathrm{PS}}^{\mathrm{B}} \end{bmatrix}$$

$$= \begin{bmatrix} X_{\mathrm{PP}} - Y_{\mathrm{PP}} & -Y_{\mathrm{PS}} \\ -Y_{\mathrm{SP}} & X_{\mathrm{SS}} - Y_{\mathrm{SS}} \end{bmatrix}, \tag{6}$$

where $X_{PP} = 1 - R^A_{PP}R^B_{PP}$ and $X_{SS} = 1 - R^A_{SS}R^B_{SS}$ are the terms whose inverse would be the reverberation operators without wavetype coupling. We can recognise P-to-S coupling with a single conversion in Y_{PS}, Y_{SP} and a double conversion in Y_{PP}, Y_{SS}, Kennett (1986) has shown that a partitioned inverse of (6) can be found in a form where only the inverses of the diagonal elements $(X_{PP} - Y_{PP})^{-1}$, $(X_{SS} - Y_{SS})^{-1}$ are required. These expressions are relatively complex, but a significant simplification can be produced by ignoring all double conversions, to give an approximate operator

$$[\boldsymbol{I} - \boldsymbol{R}^A\boldsymbol{R}^B]^{-1} \approx \begin{bmatrix} X_{PP}^{-1} & X_{PP}^{-1}Y_{PS}X_{SS}^{-1} \\ X_{SS}^{-1}Y_{SP}X_{PP}^{-1} & X_{SS}^{-1} \end{bmatrix}, \tag{7}$$

in which only the reverberations corresponding to single wavetypes are retained. Such an approximation is reasonable when the time delay between successive conversions is long. If, however, many phases can arrive close together in time the full inverse must be used. A similar expression to (7) can be derived for the 3×3 matrix for anisotropic coupling (Kennett, 1986).

We make use of the approximation (7) to the reverberation operator to establish simple approximations for the reflection and transmission coefficients with a limited allowance for conversion. Consider an overall transmission matrix in the form

$$\boldsymbol{T} = \boldsymbol{T}^B[\boldsymbol{I} - \boldsymbol{R}^A\boldsymbol{R}^B]^{-1}\boldsymbol{T}^C, \tag{8}$$

where, for example, $\boldsymbol{T}^C$ stands for a matrix of the type $\boldsymbol{T}^{AB}_D$. The transmission coefficients including a single conversion are

$$\left.\begin{aligned} T_{PP} &= T^B_{PP}[1 - R^A_{PP}R^B_{PP}]^{-1}T^C_{PP}, \\ T_{PS} &= T^B_{PS}[1 - R^A_{SS}R^B_{SS}]^{-1}T^C_{SS} + T^B_{PP}[1 - R^A_{PP}R^B_{PP}]^{-1}T^C_{PS} \\ &\quad + T^B_{PP}[1 - R^A_{PP}R^B_{PP}]^{-1}(R^A_{PS}R^B_{SS} + R^A_{PP}R^B_{PS})[1 - R^A_{SS}R^B_{SS}]^{-1}T^C_{SS}. \end{aligned}\right\} \tag{9}$$

For a reflection matrix

$$\boldsymbol{R} = \boldsymbol{R}^C + \boldsymbol{T}^A\boldsymbol{R}^B[\boldsymbol{I} - \boldsymbol{R}^A\boldsymbol{R}^B]^{-1}\boldsymbol{T}^C, \tag{10}$$

the corresponding coefficients, with just one conversion, are

$$\begin{aligned} R_{PP} &= R^C_{PP} + T^A_{PP}R^B_{PP}[1 - R^A_{PP}R^B_{PP}]^{-1}T^C_{PP}, \\ R_{PS} &= R^C_{PS} + T^A_{PP}R^B_{PS}[1 - R^A_{SS}R^B_{SS}]^{-1}T^C_{SS} \\ &\quad + T^A_{PS}R^B_{SS}[1 - R^A_{SS}R^B_{SS}]^{-1}T^C_{SS} \\ &\quad + T^A_{PP}R^B_{PP}[1 - R^A_{PP}R^B_{PP}]^{-1}T^C_{PS}. \end{aligned} \tag{11}$$

Thus, if double conversions can be neglected, the PP and SS coefficients behave as if there was no coupling and conversions feed into the SP and PS coefficients without further interaction with the wavefield.

It is possible to extend the treatment to include up to two conversions, but the expressions rapidly become very complicated, and it is probably most effective to use the full matrix forms (2), (3) and (5). However, since we are usually interested in the total P and S wavefields at a particular receiver, it is valuable to be able to rapidly isolate the principal propagation terms contributing to these fields, i.e. those with a single conversion.

3 RESPONSE OF A MEDIUM TO A SOURCE

We consider a stratified half-space, normally derived by earth flattening from a spherical model (Müller, 1971), with a free surface at $z = 0$. It is convenient to truncate the structure below some level z_L and to replace it with a uniform half-space (see Fig. 1). The source is taken at a depth z_S ($<z_L$) and in the slowness–frequency domain its radiation may be calculated by introducing a discontinuity in the displacement and vertical tractions across the plane $z = z_S$ (K83, Chapter 4). These discontinuities can be transcribed into the up- and downgoing radiation ($\boldsymbol{\Sigma}_U^S(p, \omega)$, $\boldsymbol{\Sigma}_D^S(p, \omega)$) that would be radiated into a uniform medium with the properties at the source level. These radiated waves impinge on the regions above and below the source level and the waves are successively reflected and transmitted by portions of the stratification. In terms of the reflection and transmission matrices introduced in the previous section, the surface-displacement terms can be represented as (K83, Chapter 7)

$$\mathbf{w}_0(p, \omega) = \mathbf{W}_U^{fS}[\mathbf{I} - \mathbf{R}_D^{SL}\mathbf{R}_U^{fS}]^{-1}(\boldsymbol{\Sigma}_U^S + \mathbf{R}_D^{SL}\boldsymbol{\Sigma}_D^S), \tag{12}$$

where $\mathbf{R}_D^{SL}$ is the reflection matrix for the region below the source and $\mathbf{R}_U^{fS}$ corresponds to reflections from the region above the source bounded above by the free surface.

The expression (12) is readily evaluated with two separate recursive calculations of the type described in the Appendix. First, starting at the base of the stratification, we build up to the source level z_S and so construct $\mathbf{R}_D^{SL}$. Then, starting at the free surface, we iterate downwards to the source level again, simultaneously evaluating $\mathbf{W}_U^{fS}$ and $\mathbf{R}_U^{fS}$. The reverberation operator then has to be inverted, and subsequently the terms are multiplied out to give the coefficients in the vector-spherical-harmonic expansion of the displacement. The singular values of the reverberation operator correspond to the location of modal poles for slownesses greater than β_L^{-1}, lying just above the real slowness axis in an attenuative medium.

If a full calculation is made for each of the terms in (12) complete seismograms can be generated, as in Kennett (1980). But this treatment allows flexibility by the use of approximate reflection and transmission matrices (for example by suppressing internal multiples in part of the model), which will generate a more limited expression for the displacement. If the free surface is replaced by a transparent boundary, the formal structure of (12) is retained but $\boldsymbol{R}_U^{fS}$ and $\boldsymbol{W}_U^{fS}$ are replaced by $\boldsymbol{R}_U^{0S}$ and $\tilde{\boldsymbol{W}}\boldsymbol{T}_U^{0S}$, where $\tilde{\boldsymbol{W}}$ can be either the matrix generating displacement from an upgoing wave in a uniform medium or could include the free-surface amplification factors, as discussed in the previous section.

We can rewrite (12) as

$$\boldsymbol{w}_0 = \boldsymbol{W}_U^{fS}\boldsymbol{\Sigma}_U^S + \boldsymbol{W}_U^{fS}[\boldsymbol{I} - \boldsymbol{R}_D^{SL}\boldsymbol{R}_U^{fS}]^{-1}\boldsymbol{R}_D^{SL}(\boldsymbol{\Sigma}_D^S + \boldsymbol{R}_U^{fS}\boldsymbol{\Sigma}_U^S), \tag{13}$$

in which the first term separates out the waves transmitted upwards from the source and the second includes all the energy that returns to the surface after reflection below the source level. The effective downward radiation for P waves at z_S is

$$(\boldsymbol{\Sigma}_D^S)_P + (\boldsymbol{R}_U^{fS})_{PP}(\boldsymbol{\Sigma}_U^S)_P + (\boldsymbol{R}_U^{fS})_{PS}(\boldsymbol{\Sigma}_U^S)_S, \tag{14}$$

in which we can recognise direct P and the pP, sP phases with the possibility of intermediate reflections from the structure above the source, since from (5) we can write

$$\boldsymbol{R}_U^{fS} = \boldsymbol{R}_U^{0S} + \boldsymbol{T}_D^{0S}\tilde{\boldsymbol{R}}[\boldsymbol{I} - \boldsymbol{R}_D^{0S}\tilde{\boldsymbol{R}}]^{-1}\boldsymbol{T}_U^{0S}, \tag{15}$$

where $\tilde{\boldsymbol{R}}$ is the reflection matrix for the free surface itself and $\boldsymbol{R}_U^{0S}$ refers to the structure from the source level to just below the surface. Both (12) and (13) include all possible propagation effects if full calculations are made for the reflection and transmission terms. We shall use them as the basis for developing alternative forms that are suitable for approximation, for example by truncating the appropriate reverberation operators.

We shall take a specific and important example to illustrate the construction of alternative representations of the seismic wavefield that can then be approximated in a variety of ways. We shall consider the case where the stratification is divided into an upper and a lower zone, but similar ideas can be used in other situations.

4 SPLIT STRATIFICATION

In many situations the nature of the seismic wavespeed distribution is such that it naturally divides into a sequence of depth zones with different character. On the scale of the whole Earth, such a division occurs into crust, upper mantle, lower mantle, and core. On smaller scales there is commonly a low-wavespeed zone at the surface formed, for example, by sedimentary layers. It is useful to be able to separate the portion of the wavefield dominated by propagation in the shallower regions from the waves that penetrate to greater depth.

Such a split can be achieved by introducing a separation level z_J and then working with the properties of the regions above and below z_J (Fig. 1). The split is most useful when the source lies above z_J, i.e. $z_S < z_J$. In this case the surface displacement takes the form (K83, Section 9.2)

$$\begin{aligned} \mathbf{w}_0 = \mathbf{W}^{fS}_U[\mathbf{I} - \mathbf{R}^{SJ}_D\mathbf{R}^{fS}_U]^{-1}(\boldsymbol{\Sigma}^S_U + \mathbf{R}^{SJ}_D\boldsymbol{\Sigma}^S_D) \\ + \mathbf{W}^{fJ}_U(\mathbf{I} - \mathbf{R}^{JL}_D\mathbf{R}^{fJ}_U]^{-1}\mathbf{R}^{JL}_D\mathbf{C}^J_D, \end{aligned} \tag{16}$$

where $\mathbf{C}^J_D$ is the net downward radiation at z_J:

$$\mathbf{C}^J_D = \mathbf{T}^{SJ}_D[\mathbf{I} - \mathbf{R}^{fS}_U\mathbf{R}^{SJ}_D]^{-1}(\boldsymbol{\Sigma}^S_D + \mathbf{R}^{fS}_U\boldsymbol{\Sigma}^S_U). \tag{17}$$

The first expression on the right-hand side of (16) corresponds directly to the response of the structure truncated at the level z_J (cf. (12)). The second expression represents those waves that are reflected back from beneath z_J, with the possibility of multiple deep reflections via the operator $[\mathbf{I} - \mathbf{R}^{JL}_D\mathbf{R}^{fJ}_U]^{-1}$ and shallow reverberations in $\mathbf{C}^J_D$ (17), $\mathbf{R}^{fJ}_U$ and the displacement transfer term $\mathbf{W}^{fJ}_U$, since from (5)

$$\left.\begin{aligned} \mathbf{R}^{fJ}_U &= \mathbf{R}^{SJ}_U + \mathbf{T}^{SJ}_D\mathbf{R}^{fS}_U[\mathbf{I} - \mathbf{R}^{SJ}_D\mathbf{R}^{fS}_U]^{-1}\mathbf{T}^{SJ}_U, \\ \mathbf{W}^{fJ}_U &= \mathbf{W}^{fS}_U[\mathbf{I} - \mathbf{R}^{SJ}_D\mathbf{R}^{fS}_U]^{-1}\mathbf{T}^{SJ}_U. \end{aligned}\right\} \tag{18}$$

The expression (17) for the downward radiation at z_J has the merit that it

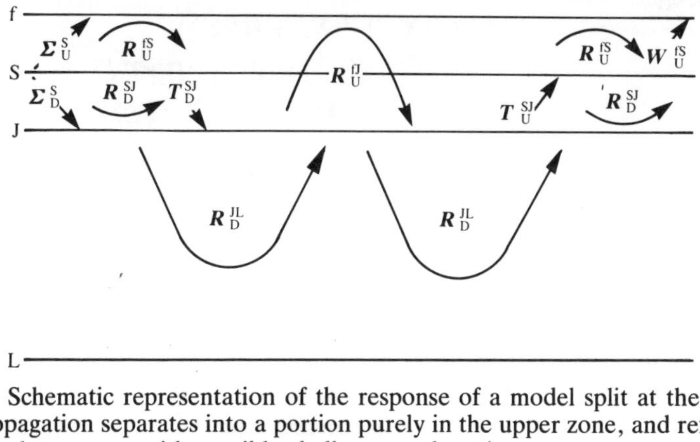

Fig. 1 Schematic representation of the response of a model split at the level J. The propagation separates into a portion purely in the upper zone, and reflections from the lower zone with possible shallow reverberations.

emphasizes the effect of surface reflections, but with a slight rearrangement it can be cast in a form where it has a common reverberation operator with the other terms for the upper zone:

$$\boldsymbol{C}_{\mathrm{D}}^{\mathrm{J}} = \boldsymbol{T}_{\mathrm{D}}^{\mathrm{SJ}}\{\boldsymbol{\Sigma}_{\mathrm{D}}^{\mathrm{S}} + \boldsymbol{R}_{\mathrm{U}}^{\mathrm{fS}}[\boldsymbol{I} - \boldsymbol{R}_{\mathrm{D}}^{\mathrm{SJ}}\boldsymbol{R}_{\mathrm{U}}^{\mathrm{fS}}]^{-1}(\boldsymbol{\Sigma}_{\mathrm{U}}^{\mathrm{S}} + \boldsymbol{R}_{\mathrm{D}}^{\mathrm{SJ}}\boldsymbol{\Sigma}_{\mathrm{D}}^{\mathrm{S}})\}. \tag{19}$$

The shallow reverberations may then be represented in terms of the operator

$$\begin{aligned}[\boldsymbol{X}^{\mathrm{fSJ}}]^{-1} &= [\boldsymbol{I} - \boldsymbol{R}_{\mathrm{D}}^{\mathrm{SJ}}\boldsymbol{R}_{\mathrm{U}}^{\mathrm{fS}}]^{-1},\\ &= \boldsymbol{I} + \boldsymbol{R}_{\mathrm{D}}^{\mathrm{SJ}}\boldsymbol{R}_{\mathrm{U}}^{\mathrm{fS}}[\boldsymbol{I} - \boldsymbol{R}_{\mathrm{D}}^{\mathrm{SJ}}\boldsymbol{R}_{\mathrm{U}}^{\mathrm{fS}}]^{-1}.\end{aligned} \tag{20}$$

Multiple reflections in the full half space are governed by

$$\begin{aligned}[\boldsymbol{Z}^{\mathrm{fJL}}]^{-1} &= [\boldsymbol{I} - \boldsymbol{R}_{\mathrm{D}}^{\mathrm{JL}}\boldsymbol{R}_{\mathrm{U}}^{\mathrm{fJ}}]^{-1},\\ &= [\boldsymbol{I} - \boldsymbol{R}_{\mathrm{D}}^{\mathrm{JL}}(\boldsymbol{R}_{\mathrm{U}}^{\mathrm{SJ}} + \boldsymbol{T}_{\mathrm{D}}^{\mathrm{SJ}}\boldsymbol{R}_{\mathrm{U}}^{\mathrm{fS}}\boldsymbol{T}_{\mathrm{U}}^{\mathrm{SJ}})\\ &\quad - \boldsymbol{R}_{\mathrm{D}}^{\mathrm{JL}}\boldsymbol{T}_{\mathrm{D}}^{\mathrm{SJ}}\boldsymbol{R}_{\mathrm{U}}^{\mathrm{fS}}\boldsymbol{R}_{\mathrm{D}}^{\mathrm{SJ}}\boldsymbol{R}_{\mathrm{U}}^{\mathrm{fS}}[\boldsymbol{X}^{\mathrm{fSJ}}]^{-1}\boldsymbol{T}_{\mathrm{U}}^{\mathrm{SJ}}]^{-1},\end{aligned} \tag{21}$$

where we have isolated those terms corresponding to a simple passage through the upper zone from the sequence of shallow reverberations. Such a multiple-reflection operator can be represented in a form that separates the dominant and minor parts of the propagation process, since

$$[\boldsymbol{Z} - \Delta\boldsymbol{Z}]^{-1} = [\boldsymbol{Z}]^{-1} + [\boldsymbol{Z}]^{-1}\Delta\boldsymbol{Z}[\boldsymbol{Z}]^{-1} + \dots \quad . \tag{22}$$

In the present context we should identify $\boldsymbol{Z}$ with the major multiple reflections from depth and $\Delta\boldsymbol{Z}$ with the term involving crustal reverberations.

Up to this point, we could, in principle, construct the entire response from (16), which we shall rewrite to explicitly display the reverberation operators:

$$\begin{aligned}w_0 = &\ \boldsymbol{W}_{\mathrm{U}}^{\mathrm{fS}}[\boldsymbol{X}^{\mathrm{fSJ}}]^{-1}(\boldsymbol{\Sigma}_{\mathrm{U}}^{\mathrm{S}} + \boldsymbol{R}_{\mathrm{D}}^{\mathrm{SJ}}\boldsymbol{\Sigma}_{\mathrm{D}}^{\mathrm{S}})\\ &+ \boldsymbol{W}_{\mathrm{U}}^{\mathrm{fS}}[\boldsymbol{X}^{\mathrm{fSJ}}]^{-1}\boldsymbol{T}_{\mathrm{U}}^{\mathrm{SJ}}[\boldsymbol{Z}^{\mathrm{fJL}}]^{-1}\boldsymbol{R}_{\mathrm{D}}^{\mathrm{JL}}\boldsymbol{T}_{\mathrm{D}}^{\mathrm{SJ}}\\ &\cdot\{\boldsymbol{\Sigma}_{\mathrm{D}}^{\mathrm{S}} + \boldsymbol{R}_{\mathrm{U}}^{\mathrm{fS}}[\boldsymbol{X}^{\mathrm{fSJ}}]^{-1}(\boldsymbol{\Sigma}_{\mathrm{U}}^{\mathrm{S}} + \boldsymbol{R}_{\mathrm{D}}^{\mathrm{SJ}}\boldsymbol{\Sigma}_{\mathrm{D}}^{\mathrm{S}})\}.\end{aligned} \tag{23}$$

We recall that $\boldsymbol{R}_{\mathrm{U}}^{\mathrm{fJ}}$, which appears in $\boldsymbol{Z}^{\mathrm{fJL}}$, can be written as

$$\boldsymbol{R}_{\mathrm{U}}^{\mathrm{fJ}} = \boldsymbol{R}_{\mathrm{U}}^{\mathrm{SJ}} + \boldsymbol{T}_{\mathrm{D}}^{\mathrm{SJ}}\boldsymbol{R}_{\mathrm{U}}^{\mathrm{fS}}[\boldsymbol{X}^{\mathrm{fSJ}}]^{-1}\boldsymbol{T}_{\mathrm{U}}^{\mathrm{SJ}}. \tag{24}$$

The physical processes described by (23) are represented schematically in Fig. 1. When we consider making approximations to (23), the actual physical processes to be included will depend in large part on what numerical implementation is used for the reverberation operators. We can also control which wavetypes are radiated by the source via $\boldsymbol{\Sigma}_{\mathrm{U}}^{\mathrm{S}}$ and $\boldsymbol{\Sigma}_{\mathrm{D}}^{\mathrm{S}}$ and which are included in the construction of the surface displacement through $\boldsymbol{W}_{\mathrm{U}}^{\mathrm{fS}}$.

We have cast (23) in a form where only a limited number of propagation elements are needed to synthesize the response. Source effects are described through $\boldsymbol{\Sigma}_{\mathrm{D}}^{\mathrm{S}}$ and $\boldsymbol{\Sigma}_{\mathrm{U}}^{\mathrm{S}} + \boldsymbol{R}_{\mathrm{D}}^{\mathrm{SJ}}\boldsymbol{\Sigma}_{\mathrm{D}}^{\mathrm{S}}$. The reflection matrix $\boldsymbol{R}_{\mathrm{U}}^{\mathrm{fS}}$ and the displacement transfer term $\boldsymbol{W}_{\mathrm{U}}^{\mathrm{fS}}$ are required from the source level to the surface. We also need the transmission and reflection matrices from the source level to z_{J}, and just the reflection matrix $\boldsymbol{R}_{\mathrm{D}}^{\mathrm{JL}}$ for the region below z_{J}. These quantities can be found with three iterative calculations over portions of the stratification. Although slightly more computation of reflection and transmission terms is required compared with (12), we can see that we now have much greater flexibility in the construction of approximations to the response.

5 THE CONSTRUCTION OF APPROXIMATIONS TO THE RESPONSE

In constructing specific approximations to (23), we have to be guided by the nature of the seismic wavefield and the particular velocity model we are investigating. Once we have decided to be selective, we can construct approximations in three ways: firstly, by defining the class of physical processes that are included in the response; secondly, by the choice of the dominant wavetype to be followed; and, thirdly, through the use of numerical integration over restricted intervals in slowness, to produce results appropriate to a limited distance range. We shall consider these procedures in turn.

5.1 Choice of propagation processes

The representation of the seismic wavefield in terms of reflection and transmission matrices gives a direct physical interpretation, and so allows considerable scope for selecting particular styles of wave propagation within one basic computational scheme. Such choices can avoid the generation of very long time series for theoretical seismograms, and so reduce computational expense. Alternatively, a selective response may be used to try to understand what features of the wave propagation contribute to particular parts of seismograms. As applied to (23), we have a hierarchy of possible approximations depending on how we treat the upper and lower zones of the model.

5.1.1 Inclusion of the shallow response

As we have noted above, the term $\boldsymbol{W}_{\mathrm{U}}^{\mathrm{fS}}[\boldsymbol{X}^{\mathrm{fSJ}}]^{-1}(\boldsymbol{\Sigma}_{\mathrm{U}}^{\mathrm{S}} + \boldsymbol{R}_{\mathrm{D}}^{\mathrm{SJ}}\boldsymbol{\Sigma}_{\mathrm{D}}^{\mathrm{S}})$ corresponds to propagation purely within the zone above z_{J}. This region is typically of relatively low wavespeed and the corresponding wavetrains are composed of multiple reflections with considerable mutual interference. Such guided waves, trapped between the free surface and the general trend of increasing wavespeed with depth, can build up to considerable amplitude and play an important role on the seismograms. For example, large "ground-roll" trains are seen for near-surface sources in seismic reflection work. Guided waves in the crust (Pg, Sg, Lg) are also very important on high-frequency seismograms at regional distances (about 400–1200 km). If the offset between source and receiver is of the same order as the total depth of the model under consideration then this shallow propagation term should be retained in (23). When, however, the receivers are well separated from the source, the relatively low group velocities of these phases mean that rather long time intervals have to be considered to encompass energy returned from depth as well as the guided waves. It is therefore most convenient to evaluate this shallow contribution separately and then add it to the final seismograms.

5.1.2 Shallow reverberations

The reverberation operator $[\boldsymbol{X}^{\mathrm{fSJ}}]^{-1} = [\boldsymbol{I} - \boldsymbol{R}_{\mathrm{D}}^{\mathrm{SJ}}\boldsymbol{R}_{\mathrm{U}}^{\mathrm{fS}}]^{-1}$ plays an important role in (23) as it appears in both the near-receiver and near-source contributions as well as being involved with the multiple deep reflections. We recall that we are able to adjust the boundary conditions at $z = 0$, which affect $\boldsymbol{W}_{\mathrm{U}}^{\mathrm{fS}}$ and $\boldsymbol{R}_{\mathrm{U}}^{\mathrm{fS}}$, and replace the free surface by a transparent boundary.

In order of increasing complexity, the approximations we can make in the upper zone are as follows.

(a) Transparent upper boundary: no free surface reflections are allowed, the surface amplification effects on $\boldsymbol{W}_{\mathrm{U}}^{\mathrm{fS}}$ can be included if desired.

(b) Free-surface reflections included but shallow reverberations suppressed: we replace $[\boldsymbol{X}^{\mathrm{fSJ}}]^{-1}$ by the identity matrix; this leaves surface reflections near the source for energy travelling to depth, so that

$$\boldsymbol{w}_0^{\mathrm{de}} = \boldsymbol{W}_{\mathrm{U}}^{\mathrm{fS}}\boldsymbol{T}_{\mathrm{U}}^{\mathrm{SJ}}[\boldsymbol{Z}^{\mathrm{fJL}}]^{-1}\boldsymbol{R}_{\mathrm{D}}^{\mathrm{JL}}\boldsymbol{T}_{\mathrm{D}}^{\mathrm{SJ}} \times \{\boldsymbol{\Sigma}_{\mathrm{D}}^{\mathrm{S}} + \boldsymbol{R}_{\mathrm{U}}^{\mathrm{fS}}(\boldsymbol{\Sigma}_{\mathrm{U}}^{\mathrm{S}} + \boldsymbol{R}_{\mathrm{D}}^{\mathrm{SJ}}\boldsymbol{\Sigma}_{\mathrm{D}}^{\mathrm{S}})\}. \tag{25}$$

The composite source term includes all primary reflections in the structure above z_{J}.

(c) Allowance for free-surface multiples: the reverberation operator $[\boldsymbol{X}^{\mathrm{fSJ}}]^{-1}$ is reinstated and multiple reflections in the upper zone are included. If the full numerical inverse $[\boldsymbol{X}^{\mathrm{fSJ}}]^{-1}$ is included then all shallow reverberations are retained. In perfectly elastic models these reverberations may decay somewhat slowly, with the result that it is difficult to avoid aliasing in time when the full response is included. In these circumstances a useful result may be obtained by truncating the representation of the reverberation operator and so setting

$$[\boldsymbol{X}^{\mathrm{fSJ}}]^{-1} \approx \boldsymbol{I} + \boldsymbol{R}_{\mathrm{D}}^{\mathrm{SJ}} \boldsymbol{R}_{\mathrm{U}}^{\mathrm{fS}}, \tag{26}$$

to include only one extra surface reflection at each appearance of $[\boldsymbol{X}^{\mathrm{fSJ}}]^{-1}$. The approximation (26) requires no extra computation, since the matrix product is needed to form the full inverse.

5.1.3 Multiple reflections from depth

These are controlled by the operator $[\boldsymbol{Z}^{\mathrm{fJL}}]^{-1} = [\boldsymbol{I} - \boldsymbol{R}_{\mathrm{D}}^{\mathrm{JL}} \boldsymbol{R}_{\mathrm{U}}^{\mathrm{fJ}}]^{-1}$ where the form of $\boldsymbol{R}_{\mathrm{U}}^{\mathrm{fJ}}$ has been given in (24). As in the case of the shallow reverberations, we can control the style of multiples both by changing the surface boundary condition and by varying the expansion level of the operator $[\boldsymbol{Z}^{\mathrm{fJL}}]^{-1}$.

When attention is to be concentrated on just the P or S phases, the operator $[\boldsymbol{Z}^{\mathrm{fJL}}]^{-1}$ is replaced by the identity, with a consequent approximate displacement field

$$\boldsymbol{w}_0^{\mathrm{de}} \approx \boldsymbol{W}_{\mathrm{U}}^{\mathrm{fS}} [\boldsymbol{X}^{\mathrm{fSJ}}]^{-1} \boldsymbol{T}_{\mathrm{U}}^{\mathrm{SJ}} \boldsymbol{R}_{\mathrm{D}}^{\mathrm{JL}} \boldsymbol{T}_{\mathrm{D}}^{\mathrm{SJ}} \{\boldsymbol{\Sigma}_{\mathrm{D}}^{\mathrm{S}} + \boldsymbol{R}_{\mathrm{U}}^{\mathrm{fS}} [\boldsymbol{X}^{\mathrm{fSJ}}]^{-1} (\boldsymbol{\Sigma}_{\mathrm{U}}^{\mathrm{S}} + \boldsymbol{R}_{\mathrm{D}}^{\mathrm{SJ}} \boldsymbol{\Sigma}_{\mathrm{D}}^{\mathrm{S}})\}. \tag{27}$$

This expression is appropriate for the generation of teleseismic records. Douglas *et al.* (1973), working at short periods, concentrate attention on the near-source and near-receiver reverberations, while for long-period synthesis Langston and Helmberger (1975) have ignored such effects. The original "reflectivity" method of Fuchs and Müller (1971) was based on (27) with only transmission through the upper zone.

When surface multiples are required, the long time delays associated with multiple reflections beneath z_{J} mean that it is difficult to include the full operator for calculations at large ranges. For most purposes it is sufficient to use the truncation of the operator after one surface interaction, so that the terms associated with deep reflections take the form

$$\boldsymbol{R}_{\mathrm{D}}^{\mathrm{JL}} + \boldsymbol{R}_{\mathrm{D}}^{\mathrm{JL}} (\boldsymbol{R}_{\mathrm{U}}^{\mathrm{SJ}} + \boldsymbol{T}_{\mathrm{D}}^{\mathrm{SJ}} \boldsymbol{R}_{\mathrm{U}}^{\mathrm{fS}} [\boldsymbol{X}^{\mathrm{fSJ}}]^{-1} \boldsymbol{T}_{\mathrm{U}}^{\mathrm{SJ}}) \boldsymbol{R}_{\mathrm{D}}^{\mathrm{JL}}. \tag{28}$$

The first term includes the P and S phases, and the second PP, PS, SP and SS as well as the possibility of shallow interactions depending on the treatment of $[\boldsymbol{X}^{\mathrm{fSJ}}]^{-1}$. If the separation level z_{J} is set rather shallow, it may be necessary to include further multiple terms to give a representation of the major part of the wavefield.

5.1.4 The nature of deep reflections

The reflection matrix $\boldsymbol{R}_{\mathrm{D}}^{\mathrm{JL}}$ from the lower zone plays a very important role in the construction of the response. For a model composed of a stack of uniform layers, the iterative scheme described in the Appendix can be used (with optional choices as to the level of internal multiples). In applications to the Earth's mantle, however, a more natural description is in terms of piecewise smooth velocity segments. Kennett and Illingworth (1981) have shown that the iterative scheme can be adapted to this case, and this is the procedure used for the examples in the next section.

Alternatively, somewhat simpler approximations may be used to represent the reflections if the detailed character of the deep reflections is not needed. Baag and Langston (1985) have used a WKBJ approximation for deep reflections from an attenuative mantle, coupled with uniform-layer calculations in the upper part of their model. The work of Douglas *et al.* (1973), designed to generate teleseismic records, assumes perfect reflection from depth and merely introduces a phase correction for the passage time in the deep structure.

5.2 Choice of wavetype

The expressions that we have generated for the surface displacement are cast in general matrix form. For SH waves they can be applied directly to the reflection and transmission coefficients. For the P–SV system, or in the more general case of anisotropy, the matrix expressions imply multiplication of 2×2 or 3×3 matrices. We have seen above in (9) and (11) that it is possible to generate specific approximations concentrating on one wavetype while allowing limited conversion to other forms. When we wish to retain maximum generality in an algorithm it is generally more convenient to use the full matrix development for the bulk of the propagation terms and impose a choice of wavetype at the source and receiver.

5.2.1 Source Radiation

The up- and downgoing radiation terms $\boldsymbol{\Sigma}_{\mathrm{U}}^{\mathrm{S}}(p, \omega)$ and $\boldsymbol{\Sigma}_{\mathrm{D}}^{\mathrm{S}}(p, \omega)$ can be split into a contribution for each wavetype. If we wish to have the most general behaviour then we retain all these terms. Frequently we know what are the major propagation paths of interest and so can try to make a specific choice of wavetype. However, this has to be done carefully. For teleseismic P waves the obvious contributions are the direct wave and the surface reflection pP, but for earthquake sources a very prominent phase is sP, which arises from waves radiated upwards from the source as S and then reflected with conversion to P at the free surface. Although less likely, conversion to P can also arise from downgoing waves.

A choice of wavetype is therefore of greatest value in the analysis of a seismic record, by allowing the association of specific features and phases with particular wavetypes radiated from the source.

5.2.2 Receiver response

The surface-displacement matrix $\boldsymbol{W}_U^{fS}$ takes the upgoing wavefield and generates the vertical- and horizontal-displacement components. The elements in one column of the matrix correspond to a particular wavetype. Thus, by choosing the columns in turn, it is possible to restrict attention to one wavetype at the receiver. This is often a very convenient way of ensuring a particular type of propagation path.

5.2.3 Deep reflections

If free-surface reflections are ignored as in (27) then the reflection matrix $\boldsymbol{R}_D^{JL}$ from the lower zone appears only once in the calculation. The dominant mode of propagation in the lower part of the model can therefore be imposed by selecting only one reflection coefficient (e.g. R_{pp}) from the matrix, which would still allow internal conversions (cf. Fuchs and Müller, 1971). Alternatively, if, for example, interest is focused on P waves and associated conversions then we can exclude the SS coefficient and so avoid the much longer time delays associated with purely S propagation.

5.3 Choice of integration interval

Once we have constructed a representation of the seismic wavefield in the frequency–slowness domain (e.g. (12), (23) and (27)), we have to integrate over slowness in order to generate the theoretical seismograms. This integration stage may also be used to limit the range of propagation phenomena represented in the final synthetic seismograms. If we take a slowness window from 0.06 to 0.12 s km^{-1} then this will correspond to P waves with turning points in the upper mantle, but S turning points lie in the lower mantle. In the interval 0.2–0.13 s km^{-1}, P waves will be confined to the crust, but S waves will be turned back from the mantle. When we use such information on the character of the wavefield in combination with the selective approximations we have just discussed, we are able to effectively isolate particular types of wave propagation.

Such a device was used in the "reflectivity" method of Fuchs and Müller (1971) to work with refracted P waves, but the approach can be applied to more general cases as well. Thus, in a study of regional S waves, Kennett (1985) used the full response (12) integrated over the slowness interval 0.2–0.4 s km^{-1} to concentrate attention on the Sn and Lg phases over the distance range 100–900 km. For the shield crustal model, all P waves were evanescent in this slowness interval, and so only a limited time interval around the S waves need be considered; this allowed computation to relatively high frequency (5 Hz) without excessive expense.

6 EXAMPLES OF APPROXIMATIONS

In the previous section we have discussed a wide range of approximations that can be employed on representations of the seismic wavefield, and have used the split stratification forms as a specific illustration. The way in which such approximations can be used is best seen by an example, and we shall consider short-period P-wave propagation in an upper-mantle model.

The calculations have been carried out for a flattened model of the upper mantle, derived from model SNA (Grand and Helmberger, 1984), which is shown in Fig. 2(a). The source was set at 15 km and the frequency–slowness response was constructed using the split-stratification scheme (23). The separation level was taken at 50 km, and so all crustal propagation is included in the terms for the upper zone. The response in the lower part of the medium was restricted to a single reflection, but all possible reflections were allowed in the upper zone. The moment-tensor source was taken from event 35 of Dziewonski *et al.* (1986), with tensor components normalized by the largest principal value (Fig. 2b). This source was chosen because it gives strong radiation into P waves returned from the mantle, but also has significant SV radiation, which give rise to converted phases and so increases the complexity of the seismograms.

In order to restrict attention to the P wavefield, we have included only P-wave terms in the receiver response and also have taken a slowness interval 0.078–0.12 s km^{-1} in the integration. This slowness range corresponds to P-wave turning points from the Moho down to 980 km; no S

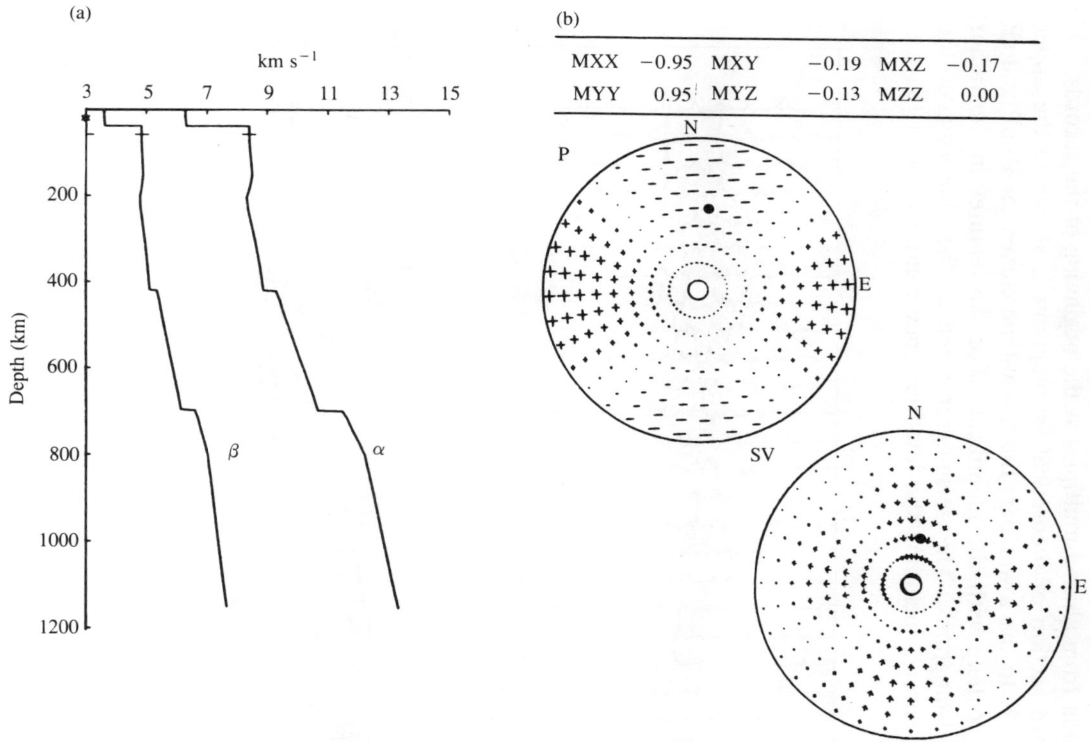

MXX	−0.95	MXY	−0.19	MXZ	−0.17
MYY	0.95	MYZ	−0.13	MZZ	0.00

Fig. 2 (a) The flattened velocity model used for the theoretical seismogram calculations. The S model is derived from SNA (Grand and Helmberger, 1984) and the P-wave velocities are scaled up by a factor of 1.74. The source depth is at 15 km, and the separation level of 50 km is indicated. (b) The moment tensor of the source and the associated radiation patterns for P and SV waves in equal-area projection. The receivers lie along the 5° azimuth, and the take-off angles are indicated by solid circles. The size of the symbols scales with the radiated amplitude.

waves turn within the model. The asymptotic forms of the vector-spherical-harmonic terms were used in the integration since we are well into the far field for even the lowest frequency considered. A time window 40.96 s long was used in the calculation and there is some slight contamination from aliased multiples at the beginning of the records.

The record section of theoretical seismograms is shown in the upper part of Fig. 3. In the lower part the travel-time curves for P, pP and sP are shown to help with the identification of the features in the rather complex seismograms. The seismograms reflect the character of the travel-time curves, including the low-frequency continuation of the first

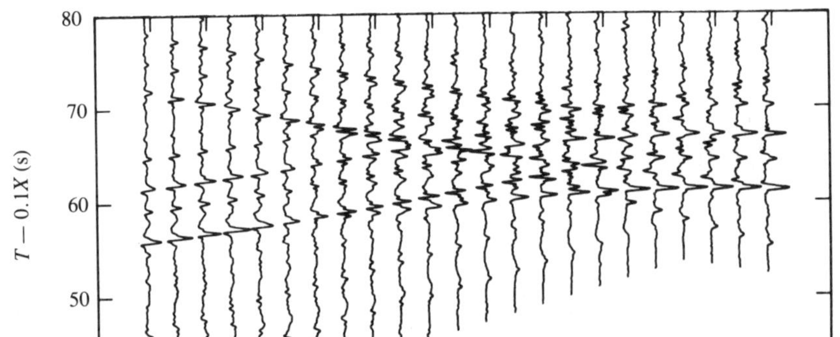

Fig. 3 Vertical-component theoretical seismograms for the structure and source shown in Fig. 2. The moment rate spectrum was flat from 0.1 to 2.0 Hz. The travel-time curves for this upper-mantle model are shown below for P and the two surface reflections pP and sP. The sP phase is particularly prominent in the theoretical seismograms.

P-wave branch into the shadow created by the low-velocity zone and the complex effects associated with the multiple arrivals from the upper-mantle transition zone. The very strong contrast in properties at the Moho means that there are conversions and multiples from this interface in addition to those from the free surface. The sP phases are more prominent than pP because for this slowness range the PS reflection at the surface dominates the PP reflection. At the smallest slownesses pP is more prominent and can be seen most clearly around 2900 km.

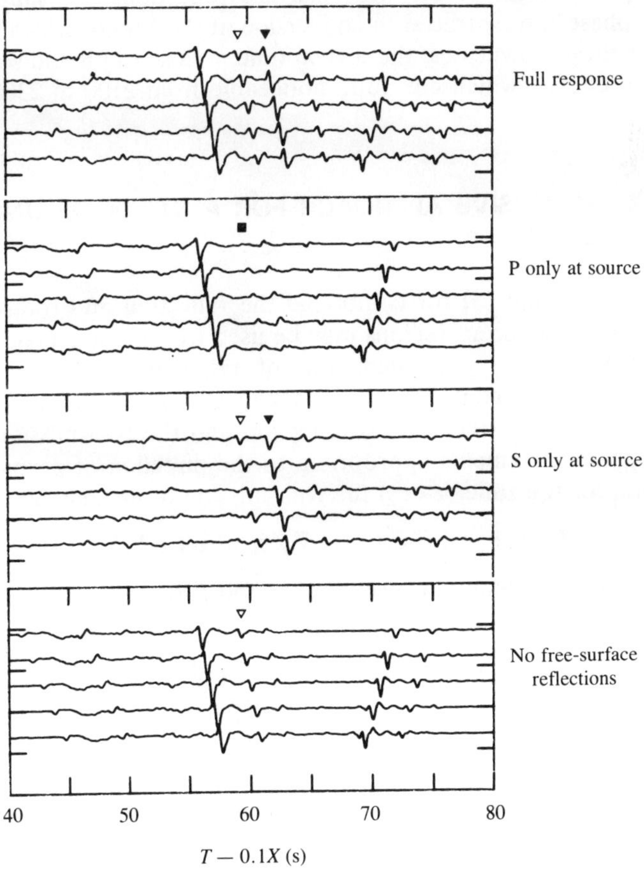

Fig. 4 Illustrations of theoretical seismograms for a sequence of approximations to the seismic wavefield for the range 1800–2000 km using the model shown in Fig. 2: ■, pP; ▼, sP; ▽, conversion from S to P at the Moho.

At long period all the detail we see in these seismograms is swamped by the duration of the instrument response. However, we should like to unravel the nature of the seismograms, and we can do this by selective approximations, as illustrated in Fig. 4 for the distance range 1800–2000 km. We show the full response transcribed from Fig. 3, and, in addition, various subsets of the response with control of the source radiation and the boundary condition at the free surface.

By removing the SV radiation at the source, we are able to reveal the weak pP phase marked by a solid square. With only SV radiation from the source, we see not only the expected surface conversion sP marked by the solid triangle, but also an earlier converted phase, indicated by the open triangle, which persists when surface reflections are suppressed. This extra phase can be traced to conversion at the Moho, which is quite efficient for these slownesses, the arrival time is close to pP and so we get interference effects, which are quite noticeable from 2100 to 2300 km in Fig. 3.

APPENDIX: RECURSIVE APPROACH FOR A STACK OF UNIFORM LAYERS

The relations (2) and (5) for combining the reflection and transmission properties of two adjacent regions may be used to construct a convenient recursive approach to the calculation of reflection and transmission matrices in a medium composed of a stack of uniform layers.

In a uniform region there is no reflection and the transmission terms reduce to phase increments appropriate to the direction of propagation; for example, for the zone (z_1, z_2) the transmission terms may be written

$$\mathbf{T}_{\mathrm{U}}(z_1+, z_2) = \mathbf{E}_{\mathrm{U}}^{12}, \qquad \mathbf{T}_{\mathrm{D}}(z_1+, z_2) = \mathbf{E}_{\mathrm{D}}^{12}. \tag{A1}$$

For an isotropic medium with velocities α and β these phase terms take the form

$$\mathbf{E}_{\mathrm{U}}^{12} = \mathbf{E}_{\mathrm{D}}^{12} = \begin{pmatrix} \exp(\mathrm{i}\omega q_\alpha h) & 0 \\ 0 & \exp(\mathrm{i}\omega q_\beta h) \end{pmatrix}, \tag{A2}$$

where $h = z_2 - z_1$ and

$$q_\alpha = (\alpha^{-2} - p^2)^{1/2}, \qquad q_\beta = (\beta^{-2} - p^2)^{1/2}, \qquad \mathrm{Im}\,\omega q_\alpha \geq 0, \qquad \mathrm{Im}\,\omega q_\beta \geq 0,$$

for slowness p.

Consider a uniform layer in $z_1 < z < z_2$ overlying an inhomogeneous region in $z_2 < z < z_3$. We suppose that the reflection and transmission matrices $\mathbf{R}_{\mathrm{D}}(z_2-) \equiv \mathbf{R}_{\mathrm{D}}(z_2-, z_3+)$ etc. are known (see Fig. 5). Then we

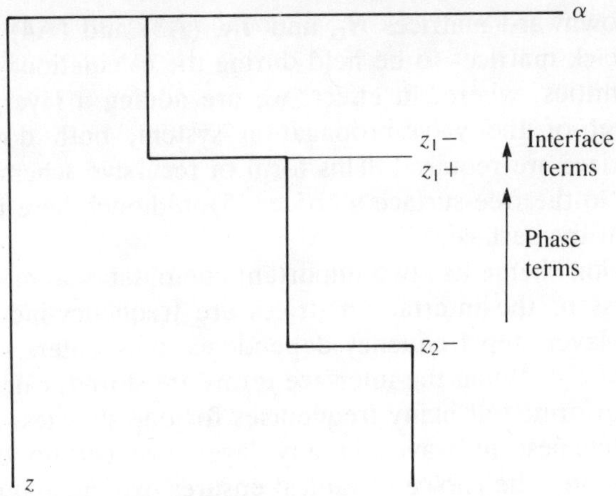

Fig. 5 Illustration of the recursive procedure for constructing reflection and transmission matrices.

may add in the propagation through the uniform layer, using (2) and (A1),

$$\left.\begin{aligned} \boldsymbol{R}_D(z_1+) &= \boldsymbol{E}_U^{12}\boldsymbol{R}_D(z_2-)\boldsymbol{E}_D^{12}, \\ \boldsymbol{T}_D(z_1+) &= \boldsymbol{T}_D(z_2-)\boldsymbol{E}_D^{12}, \\ \boldsymbol{T}_U(z_1+) &= \boldsymbol{E}_U^{12}\boldsymbol{T}_U(z_2-). \end{aligned}\right\} \qquad \text{(A3)}$$

A second application of (2) now allows us to include the effect of the interface $z = z_1$, to calculate the set of reflection matrices at z_1-. We shall write, for example, $\boldsymbol{r}_D(z_1) \equiv \boldsymbol{R}_D(z_1-, z_1+)$ for the interfacial matrices, and then

$$\left.\begin{aligned} \boldsymbol{R}_D(z_1-) &= \boldsymbol{r}_D(z_1) + \boldsymbol{t}_U(z_1)\boldsymbol{R}_D(z_1+)[\boldsymbol{I} - \boldsymbol{r}_U(z_1)\boldsymbol{R}_D(z_1+)]^{-1}\boldsymbol{t}_D(z_1), \\ \boldsymbol{T}_D(z_1-) &= \boldsymbol{T}_D(z_1+)[\boldsymbol{I} - \boldsymbol{r}_U(z_1)\boldsymbol{R}_D(z_1+)]^{-1}\boldsymbol{t}_D(z_1), \\ \boldsymbol{R}_U(z_1-) &= \boldsymbol{R}_U(z_2-) + \boldsymbol{T}_D(z_1+)\boldsymbol{r}_U(z_1)[\boldsymbol{I} - \boldsymbol{R}_D(z_1+)\boldsymbol{r}_U(z_1)]^{-1}\boldsymbol{T}_U(z_1+), \\ \boldsymbol{T}_U(z_1-) &= \boldsymbol{t}_U(z_1)[\boldsymbol{I} - \boldsymbol{R}_D(z_1+)\boldsymbol{r}_U(z_1)]^{-1}\boldsymbol{T}_U(z_1+). \end{aligned}\right\} \qquad \text{(A4)}$$

The two sets of equations (A3) and (A4) constitute our recursive scheme.

We start at the base of the layering, z_3 say, with the interfacial matrices at z_3, $\boldsymbol{r}_D(z_3)$ etc. These we now recognize as $\boldsymbol{R}_D(z_3-)$ etc.; we may then use (A3) to step the stack reflection and transmission matrices to the top of the lowest layer in the stack. Then, using (A4), we can bring our stack matrices to the upper side of this interface. The cycle (A3) followed by (A4) allows us to work up through the stack, a layer at a time, through an arbitrary number of layers.

For the downward matrices $\boldsymbol{R}_D$ and $\boldsymbol{T}_D$, (A3) and (A4) require only downward stack matrices to be held during the calculation. However for upward quantities, where, in effect, we are adding a layer at the most complex level of the wave-propagation system, both downward and upward matrices are required. This form of recursive scheme may easily be extended to the free-surface matrices (5), although here the recursion would start at the surface.

The recursion sheme has two important computational merits. First, at fixed slowness p, the interface matrices are frequency-independent, so that at each layer step frequency dependence only enters via the phase terms $\boldsymbol{E}_U^{12}$ and $\boldsymbol{E}_D^{12}$. When the interface terms are stored, calculations may be rapidly performed at many frequencies for one slowness p. Secondly, if we have evanescent waves in any layer, we get no exponentially growing solutions; the choice of radical ensures exponential decay.

The basic recursive scheme may be modified to include restricted approximations to propagation properties in a particular layer. Thus, for example, if the reverberation operator is replaced by the identity, no internal reflections are included in that layer. Truncation of the expansion of the reverberation operator after the inclusion of one reflection term can also be used to allow a partial treatment of internal multiples. However, it is difficult to enumerate what phases are included when this approximation is used for many layers.

REFERENCES

Baag, C.-E. and Langston, C. A. (1985). A WKBJ spectral method for the computation of SV synthetic seismograms in a cylindrically symmetric medium. *Geophys. J. R. Astron. Soc.* **80,** 387–417.

Booth, D. C. and Crampin, S. (1983). The anisotropic reflectivity technique: theory. *Geophys. J. R. Astron. Soc.* **72,** 755–766.

Chapman, C. H. (1978). A new method of computing synthetic seismograms. *Geophys. J. R. Astron. Soc.* **54,** 431–518.

Chapman, C. H. and Orcutt, J. A. (1985). The computation of body wave synthetic seismograms in laterally homogeneous media. *Rev. Geophys.* **23,** 105–163.

Cormier, V. and Richards, P. G. (1977). Fullwave theory applied to a discontinuous velocity increase: the inner core boundary. *J. Geophys.* **43,** 3–31.

Douglas, A., Hudson, J. A. and Blamey, C. (1973). A quantitative estimate of teleseismic radiation III—Computed P and Rayleigh wave seismograms. *Geophys. J. R. Astron. Soc.* **28,** 345–410.

Dziewonski, A. M., Franzen, J. E. and Woodhouse, J. H. (1986). Centroid-moment tensor solutions for July–September 1985. *Phys. Earth. Planet. Int.* **42,** 205–214.

Fryer, G. J. and Frazer, L. N. (1984). Seismic waves in stratified anisotropic media. *Geophys. J. R. Astron. Soc.* **78,** 691–710.

Fuchs, K. and Müller, G. (1971). Computation of synthetic seismograms with the reflectivity method and comparison with observations. *Geophys. J. R. Astron. Soc.* **23,** 417–433.

Grand, S. and Helmberger, D. V. (1984). Upper mantle shear structure of North America. *Geophys. J. R. Astron. Soc.* **76,** 399–438.

Helmberger, D. V. and Harkrider, D. (1978). Modelling earthquakes with generalised ray theory. *Modern Problems in Elastic Wave Propagation* (ed. J. Miklowitz and J. D. Achenbach), pp. 455–518, Wiley, New York.

Kennett, B. L. N. (1980). Seismic waves in a stratified half space II—Theoretical seismograms. *Geophys. J. R. Astron. Soc.* **61,** 1–10.

Kennett, B. L. N. (1983). *Seismic Wave Propagation in Stratified Media.* Cambridge University Press.

Kennett, B. L. N. (1985). On regional S. *Bull. Seismol. Soc. Am.* **75,** 1077–1086.

Kennett, B. L. N. (1986). Wavenumber and wavetype coupling in laterally heterogeneous media. *Geophys. J. R. Astron. Soc.* **87,** 313–331.

Kennett, B. L. N. and Kerry, N. J. (1979). Seismic waves in a stratified half space. *Geophys. J. R. Astron. Soc.* **57,** 557–583.

Kennett, B. L. N. and Illingworth, M. R. (1981). Seismic waves in a stratified half space III—Piecewise smooth models. *Geophys. J. R. Astron. Soc.* **66,** 633–675.

Langston, C. A. and Helmberger, D. V. (1975). A procedure for modelling shallow dislocation sources. *Geophys. J. R. Astron. Soc.* **42,** 117–130.

Müller, G. (1971). Approximate treatment of elastic body waves in media with spherical symmetry. *Geophys. J. R. Astron. Soc.* **23,** 435–449.

Müller, G. (1985). The reflectivity method: a tutorial. *J. Geophys.* **58,** 153–174.

III.2

Hybrid Methods in Reflectivity Calculations

URI ASCHER
Department of Computer Science, The University of British Columbia, Vancouver, Canada

I INTRODUCTION

One basic problem in computational seismology is to calculate displacements at the Earth's surface due to a given moment-tensor source (modelling an earthquake or a man-made explosion) for a given Earth model. The governing system of PDEs (in time and three space variables) is rather complex and often some restrictions are placed on the model. One such restriction, which has been considered by many, is to allow seismic velocities, density and attenuative factors to be functions of depth only (see e.g. Aki and Richards, 1980; Kennett, 1983). Then the displacement and the traction across a horizontal plane (i.e. the z component of the stress tensor) can be expanded in a cylindrical coordinate system, as in Takeuchi and Saito (1972). Owing to the assumed horizontal material homogeneity, this results in the seismogram being expressed as a double integral (a Fourier–Bessel transform) over frequency and wavenumber, with the azimuthal dependence essentially disappearing. The efficient evaluation of this double integral is discussed elsewhere in this book. Each integrand evaluation in the frequency–wavenumber plane corresponds to two linear systems of coupled boundary-value ordinary differential equations (two-point problems) describing SH and P–SV waves as a function of depth z.

In this chapter we concentrate on the numerical solution of these two-point problems. In a typical calculation there may sometimes be 20 000 such problems to solve. Care must therefore be taken to solve

SEISMOLOGICAL ALGORITHMS
ISBN 0-12-220770-X

them very efficiently. A saving grace is that their solution is independent and can be done entirely *in parallel.*

In designing the method and routines discussed here, we have emphasized generality, being able to calculate the complete response (theoretical seismograms) for media that may have strong gradients and multiple turning points. The material presented here is essentially contained in Spudich and Ascher (1983) or in Ascher and Spudich (1986).

Most of the traditional work for solving these two-point problems has followed two approaches. In one approach, initial-value numerical techniques are used. But these techniques may become unstable for high frequencies. In the other approach, transformations of the boundary-value problem are employed. This approach has been quite successful for certain models. For example, at high frequencies and in regions having small gradients in material properties, asymptotic expansions have been successfully used to describe P- and S-wave propagation (Richards, 1971; Chapman, 1976; Richards and Frasier, 1976). However, in some cases the accuracy of the approximation deteriorates (or, alternatively, good approximations are harder to obtain) when the medium is not sufficiently close to being a stack of uniform layers (i.e. when there are regions of large velocity gradients). Also, the uniform quality of the approximation is hard to maintain and quantitative error control has rarely been attempted.

The approach described here relies on a basic numerical method that in principle does not need any special assumptions (like high frequencies or gentle gradients) to work. Then we may have a menu of methods, including the basic one and also some special methods that may be more efficient in certain circumstances. A hybrid scheme is subsequently constructed where appropriate special methods are used in regions where this is deemed possible and desirable on efficiency grounds, while it is always possible to fall back to the basic method in cases where none of the special methods is worthwhile.

Let us specify the two-point problems under consideration. Consider a half-space in which P-wave velocity α, S-wave velocity β and density ρ are piecewise-continuous functions of depth z into the half-space only. For $0 \leq z < z_>$, α, β and ρ are arbitrary, but for $z \geq z_>$ they are constant, $\alpha(z) = \alpha_>$, $\beta(z) = \beta_>$, $\rho(z) = \rho_>$, as depicted in Fig. 1. We allow α and β to be complex functions of z, thereby modelling the effects of anelastic attenuation (see Schwab and Knopoff, 1972; Kennett, 1983). We also allow ω to be complex, following Phinney (1965).

Applying a Fourier–Bessel transform, let ω be a given frequency and k a horizontal wavenumber and define the slowness

$$p = k/\omega. \tag{1.1}$$

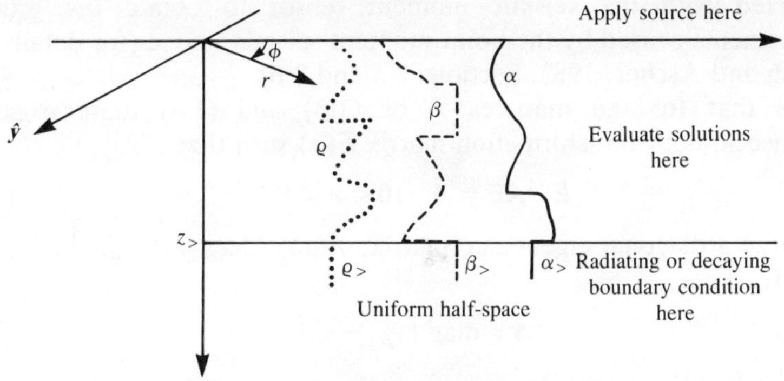

Fig. 1 General problem geometry.

Then we get differential equations of the form

$$\partial_z \boldsymbol{b} = \omega \boldsymbol{A}(z)\boldsymbol{b} \quad (0 < z < z_>), \tag{1.2}$$

where $\boldsymbol{A}$ depends on p but is independent of ω. For SH waves we have a problem of order $n = 2$,

$$\boldsymbol{b} \equiv \boldsymbol{b}_{\mathrm{H}}, \qquad \boldsymbol{A} \equiv \boldsymbol{A}_{\mathrm{H}} = \begin{bmatrix} 0 & \mu^{-1} \\ \mu p^2 - \rho & 0 \end{bmatrix}, \quad \mu = \rho\beta^2, \tag{1.3}$$

whereas for P–SV waves we get $n = 4$ and

$$\boldsymbol{b} \equiv \boldsymbol{b}_p, \qquad \boldsymbol{A} \equiv \boldsymbol{A}_{\mathrm{p}} = \begin{bmatrix} 0 & (\rho\alpha^2)^{-1} & q_1 & 0 \\ -\rho & 0 & 0 & p \\ -p & 0 & 0 & \mu^{-1} \\ 0 & -q_1 & q_2 & 0 \end{bmatrix}, \tag{1.4}$$

$$q_1 = p\left(\frac{1 - 2\beta^2}{\alpha^2}\right), \qquad q_2 = 4\mu p^2\left(\frac{1 - \beta^2}{\alpha^2}\right) - \rho.$$

The unknown functions are $\boldsymbol{b}_{\mathrm{H}} = [a_5 \;\; \omega^{-1}a_6]^{\mathrm{T}}$ and $\boldsymbol{b}_{\mathrm{p}} = [a_1 \;\; \omega^{-1}a_2 \;\; a_3 \;\; \omega^{-1}a_4]^{\mathrm{T}}$, where $a_1, \ldots, a_6$ are the coefficients of the expansions of displacement and traction mentioned in the first paragraph.

The system (1.2) is supplemented by $\frac{1}{2}n$ boundary conditions at each of the two ends, $z = 0$ and $z = z_>$, as described in Spudich and Ascher (1983). Basically, at $z_>$ we impose radiation conditions, requiring only downgoing waves for $z \geqslant z_>$. At $z = 0$ we apply three orthogonal forcing term conditions following Olson *et al.* (1984), which allow us to construct the necessary Green functions for the medium, using their symmetry to exchange the locations of the observer and the source. This trades an unpleasant (explosion) source term in the middle of our domain for a harmless inhomogeneity in the boundary values. From these Green functions one can obtain the strain tensor in the medium, which is then contracted with the seismic moment tensor to obtain the ground displacements caused by the point-moment seismic source (for details see Spudich and Ascher, 1983, Sections 2.5 and 2.6).

Note that for the matrices $\boldsymbol{A}$ of (1.3) and (1.4) there exists a piecewise-smooth transformation matrix $\boldsymbol{E}(z)$ such that

$$\boldsymbol{E}^{-1}\boldsymbol{A}\boldsymbol{E} = \boldsymbol{\Lambda} \quad (0 < z < z_>), \tag{1.5}$$

with $\boldsymbol{\Lambda}(z)$ a diagonal eigenvalue matrix, $\boldsymbol{\Lambda}(z) = \text{diag}\,\{\lambda_1(z), \ldots, \lambda_n(z)\}$. For SH,

$$\boldsymbol{\Lambda} = \text{diag}\,\{\nu_\beta, -\nu_\beta\}, \tag{1.6}$$

while for P–SV

$$\boldsymbol{\Lambda} = \text{diag}\,\{\nu_\alpha, \nu_\beta, -\nu_\alpha, -\nu_\beta\}, \tag{1.7}$$

with

$$\nu_\alpha = (p^2 - \alpha^{-2})^{1/2}, \qquad \nu_\beta = (p^2 - \beta^{-2})^{1/2}, \tag{1.8}$$

and we choose

$$\text{Re}\,(\omega\nu) \geq 0; \qquad \text{Im}\,(\omega\nu) \leq 0 \quad \text{when} \quad \text{Re}\,(\omega\nu) = 0. \tag{1.9}$$

Note that $\nu_\alpha(z) \neq \nu_\beta(z)$ for all z, because $\text{Re}\,(\alpha(z)) > \text{Re}\,(\beta(z))$ by physical considerations.

2 METHODS FOR SOLVING TWO-POINT PROBLEMS

In this section we briefly describe the methods that are on our menu to solve the two-point problems for (1.2)–(1.4). But first we give a general framework for the numerical method used.

2.1 A general framework for the numerical method

For given values of ω and p, the interval in z is subdivided by a mesh

$$0 = z_1 < z_2 < \ldots < z_N < z_{N+1} = z_>, \tag{2.1}$$

which is constructed to satisfy the following properties.

(a) All points of material discontinuity (i.e. interfaces between layers, where $\alpha(z)$, $\beta(z)$, $\rho(z)$ or their derivatives may vary discontinuously) are included in the mesh. Thus we may assume, for example that $\boldsymbol{A}(z)$ is twice continuously differentiable on each subinterval, viz

$$\boldsymbol{A}(z) \in C^2[z_i, z_{i+1}) \quad (1 \leqslant i \leqslant N). \tag{2.2}$$

(b) On each subinterval (or element) (z_i, z_{i+1}), precisely one of the cases of solution behaviour, to be discussed below, holds.

(c) Each element size

$$h_i := z_{i+1} - z_i \quad (1 \leqslant i \leqslant N) \tag{2.3}$$

is sufficiently small so that the numerical method used on it produces the solution to a desired accuracy.

An algorithm for constructing such a mesh is described in Section 3. Here, assuming that we have one, we seek a numerical solution $\{\boldsymbol{b}_j\}_{j=1}^{N+1}$ such that for a given tolerance δ,

$$|\boldsymbol{b}_j - \boldsymbol{b}(z_j)| \leqslant \delta \quad (1 \leqslant j \leqslant N+1). \tag{2.4}$$

This numerical solution is obtained by requiring that $\boldsymbol{b}_1$ and $\boldsymbol{b}_{N+1}$ satisfy the boundary conditions for $\boldsymbol{b}(0)$ and $\boldsymbol{b}(z_>)$ respectively, and that on each mesh element a relation of the form

$$\boldsymbol{R}_j \boldsymbol{b}_{j+1} = \boldsymbol{S}_j \boldsymbol{b}_j \quad (1 \leqslant j \leqslant N) \tag{2.5}$$

hold. Here $\boldsymbol{R}_j$ and $\boldsymbol{S}_j$ are $n \times n$ matrices and $\boldsymbol{R}_j$ is non-singular. Letting

$$\boldsymbol{\Gamma}_j = \boldsymbol{R}_j^{-1} \boldsymbol{S}_j, \tag{2.6}$$

we have

$$\boldsymbol{b}_{j+1} = \boldsymbol{\Gamma}_j \boldsymbol{b}_j. \tag{2.7}$$

Thus we expect $\boldsymbol{\Gamma}_j$ to approximate the propagator matrix $\boldsymbol{B}(z_{j+1}; z_j)$, where

$$\boldsymbol{b}(z) = \boldsymbol{B}(z; \zeta)\boldsymbol{b}(\zeta), \tag{2.8}$$

unless the solution on the element (z_j, z_{j+1}) contains highly evanescent waves. Conversely, any method that generates approximations to $\boldsymbol{B}(z_{j+1}; z_j)$ for some subdivision of the form (2.1) can obviously be written in the form (2.5).

The system of $n(N+1)$ linear equations (2.5) plus the boundary equations has a familiar form and is solved, for example, as described in Spudich and Ascher (1983, Section 2.4).

2.2 Collocation

Our basic method is collocation at Gaussian points, as described in detail in Spudich and Ascher (1983) and Ascher and Weiss (1984) and references therein. Given a mesh like (2.1) and l points $0 \leqslant \xi_1 < \xi_2 < \ldots < \xi_l \leqslant 1$ (e.g. $l = 3$), the basic idea is to construct an approximate solution $\boldsymbol{b}^h(z)$ that is continuous throughout $[0, z_>]$, reduces to a polynomial of degree at most l on each element $[z_j, z_{j+1}]$, satisfies the boundary conditions, and satisfies the differential equations (1.2) at the lN collocation points $z_{jr} = z_j + h_j\xi_r$, viz

$$\partial_z \boldsymbol{b}(z_{jr}) = \omega \boldsymbol{A}(z_{jr})\boldsymbol{b}^h(z_{jr}) \quad (1 \leqslant j \leqslant N, 1 \leqslant r \leqslant k). \tag{2.9}$$

We choose the points ξ_r to be Gaussian points (zeros of a Legendre polynomial), so this collocation method corresponds to a composite Gaussian quadrature formula.

It can be shown that the method can be reduced to a symmetric high-order compact finite-difference method of the form (2.5) (see e.g. Spudich and Ascher, 1983, Section 2.4). It has the following properties.

(i) It is accurate for low and intermediate frequencies, because

$$\max_{1 \leqslant j \leqslant N} |\boldsymbol{b}(z_j) - \boldsymbol{b}^h(z_j)| = O(h^{2l}). \tag{2.10}$$

(For high-frequency body waves (2.10) still holds, but the constant of proportionality has ω^{2l} in it and becomes large.)

(ii) It performs well if used properly for highly evanescent waves. This holds even though no explicit separation of up- and downgoing waves is done. (The separation is done implicitly and weakly, but sufficiently so as to have a stable method.)

As it turns out, the situations where the collocation method works well are precisely those where the special methods mentioned earlier (and in Sections 2.3 and 2.4 below) are inapplicable or inefficient. On the other hand, the collocation method does not perform efficiently when the solution is highly oscillatory, i.e. when the eigenvalues of $\boldsymbol{A}(z)$ ($\nu_\alpha(z)$ or $\nu_\beta(z)$) are (almost) purely imaginary and ω is large. (This is just like Gaussian quadrature for highly oscillatory integrals.)

The simplest instance of such a collocation scheme is with one collocation point taken at the midpoint of each element. This is known as the midpoint scheme (or the box scheme) and we have for (1.2)

$$\left.\begin{aligned} \boldsymbol{R}_j &= \boldsymbol{I} - \tfrac{1}{2}\omega h_j \boldsymbol{A}(z_{j+1/2}), \quad (2.11a)\\ \boldsymbol{S}_j &= \boldsymbol{I} + \tfrac{1}{2}\omega h_j \boldsymbol{A}(z_{j+1/2}), \quad (2.11b) \end{aligned}\right\} \; z_{j+1/2} = \tfrac{1}{2}(z_j + z_{j+1}) \quad (1 \leqslant j \leqslant N).$$

Compared with a higher-order collocation scheme, the construction of $\boldsymbol{\Gamma}_j$ (of (2.6)) here is much simpler (and cheaper); however, the precision is lower, so smaller step sizes h_j are needed to achieve a given accuracy.

Consider next the selection of a mesh (2.1) in order to approximately satisfy the uniform accuracy requirements (2.4) using collocation. The principle behind a successful mesh selection (which attempts to make the number of mesh elements N as small as possible) is that of error equidistribution, i.e. choosing h_j such that the error contribution from the jth element is approximately constant for each j, $1 \leq j \leq N$. Such a selection process requires some knowledge of the solution profile and so in general can only be performed adaptively. (See Ascher *et al.* (1979) for an instance of such an adaptive algorithm.) However, for the P–SV and SH problems we can obtain some a priori knowledge of the solution, which allows us to select an appropriate mesh in advance and hence to compute the solution in a cheaper way, as described below.

The properties of the matrix $\boldsymbol{A}(z)$ of (1.2) plus the boundary conditions at $z_>$ allow us to consider just downgoing waves. Thus we consider one differential equation of the form

$$\partial_z y = -\omega \nu_\beta(z) y, \qquad y(0) = 1; \qquad \mathrm{Re}\,(\omega \nu_\beta) \geq 0, \tag{2.12}$$

and we further assume that $\nu_\beta(z)$ is constant on each element $[z_j, z_{j+1})$ of the mesh to be constructed. These simplifying assumptions are particularly appropriate when $\mathrm{Re}\,(\omega\nu_\beta)$ is large, which is also the case when the collocation method should be used and a good mesh selection is important.

The solution of (2.12) is

$$y(z) = \exp\left\{-\sum_{i=2}^{j} \omega[\nu_\beta(z_{i-1}) - \nu_\beta(z_i)]z_i - \omega\nu_\beta(z_j)z\right\} \quad (z_j \leq z \leq z_{j+1}). \tag{2.13}$$

So we can estimate

$$\left.\begin{aligned} |y(z_j)| &\approx \exp\left\{-\sum_{i=1}^{j-1} h_i \,\mathrm{Re}\,[\omega\nu_\beta(z_i)]\right\}, \\ |y^{(2l)}(z_j)| &\equiv |\partial_z^{2l} y(z_j)| = |\omega\nu_\beta(z_j)|^{2l}\,|y(z_j)|. \end{aligned}\right\} \tag{2.14}$$

Now the error contribution from the jth element is (as in Gaussian quadrature) proportional to

$$h_j^{2l+1}\,|y^{(2l)}(z_j)|,$$

so the principle of error equidistribution per unit step requires choosing h_j so that

$$h_j^{2l}\,|\nu_\beta(z_j)|^{2l} \exp\left\{-\sum_{i=1}^{j-1} h_i \,\mathrm{Re}\,[\omega\nu_\beta(z_i)]\right\} = h_1^{2l}\,|\nu_\beta(0)|^{2l}. \tag{2.15}$$

In practice we choose for the SH equations

$$h_j := h_1 \min \left\{1, \left|\frac{\nu_\beta(0)}{\nu_\beta(z_j)}\right|\right\} \exp\left\{\frac{1}{2l}\sum_{i=1}^{j-1} h_i \operatorname{Re}\left[\omega \nu_\beta(z_i)\right]\right\}. \tag{2.16}$$

The value of h_1 is set by the required tolerance δ to

$$h_1 := \text{cons}_l \, |\omega \nu_\beta(0)|^{-1} \, \delta^{1/2l}, \tag{2.17}$$

where cons_l is a known constant, representing the Gaussian quadrature error constant with l points. For the P–SV equations we calculate quantities as in (2.16) and (2.17) also with ν_α replacing ν_β and choose for h_j the minimum of the two obtained values. In addition, if min [Re $\nu_\alpha(z_j)$, Re $\nu_\beta(z_j)$] < 0.1 then, having less trust in this mesh selection process, we set $h_j := \frac{1}{2}h_j$.

From (2.17) it can be seen that h_1 may be taken larger for larger l and that this advantage of higher-order collocation methods is particularly important when δ is very small. On the other hand, lower-order (smaller-l) methods are cheaper to construct per element. When selecting a mesh for given δ, ω and p, the most appropriate collocation scheme at a point z_j is the one which can cover a fixed size interval $(z_j, \bar{z})$ at the least expense. In practice we consider only two alternatives: a higher-order scheme with l collocation points (usually $l = 3$ or 4) or the midpoint scheme ($l = 1$). Calculating h_j by (2.16) and (2.17), both for l collocation points and for the midpoint scheme, we would prefer the latter if

$$h_j(\text{collocation}) < h_j(\text{midpoint})\epsilon_M$$

where ϵ_M is the relative efficiency of the midpoint scheme, as is determined in Ascher and Spudich (1986).

2.3 A numerical method for rapidly oscillating solutions (body waves)

Consider the problem (1.2) for $z_j \leq z \leq z_{j+1}$, with j fixed, $1 \leq j \leq N$, assuming that ω is (positive and) large, and that $\nu_\alpha(z)$ and $\nu_\beta(z)$ are almost purely imaginary. Thus there may be many wavelengths between z_j and z_{j+1} and we want a method that would connect the information at z_j to that at z_{j+1}, without closely following all the oscillations in between. General finite-difference or collocation methods are inadequate for this purpose when applied directly to (1.2).

On the other hand, the situation here is easier than in the highly evanescent case in that approximating the propagator matrix $\boldsymbol{B}(z_{j+1}; z_j)$ numerically makes sense: the growth factor of the fundamental solution components is at most

$$\exp\left\{h_j \max_{z_j \leq z \leq z_{j+1}} [\mathrm{Re}\,(\omega \nu_\beta)]\right\},$$

which is bounded in this case by a constant of moderate size. In other words, the conditioning of the initial-value problem that the propagator matrix solves is not significantly worse than that of the given boundary-value problem.

Thus we seek to construct a matrix $\boldsymbol{\Gamma}_j$ to approximate $\boldsymbol{B}(z_{j+1}; z_j)$ using the method of Richards (1971) and Chapman (1974). This yields an $O(\omega^{-2})$ approximation with

$$\boldsymbol{R}_j = [\boldsymbol{I} - \omega^{-1}\boldsymbol{P}_1(z_{j+1}^-)]\boldsymbol{E}^{-1}(z_{j+1}^-) \tag{2.18a}$$

$$\boldsymbol{S}_j = \exp\,[\omega \boldsymbol{Q}_0(z_{j+1}^-) + \boldsymbol{Q}_1(z_{j+1}^-)][\boldsymbol{I} - \omega^{-1}\boldsymbol{P}_1(z_j^+)]\boldsymbol{E}^{-1}(z_j^+), \tag{2.18b}$$

where

$$\boldsymbol{Q}_0(z) = \int_{z_j}^{z} \boldsymbol{\Lambda}(\zeta)\,\mathrm{d}\zeta, \tag{2.19}$$

$$\tilde{\boldsymbol{E}}(z) = -\boldsymbol{E}^{-1}(z)\,\partial_z \boldsymbol{E}(z), \tag{2.20}$$

$$\partial_z \boldsymbol{Q}_1 = \mathrm{diag}\,\{\tilde{E}_{11}, \ldots, \tilde{E}_{nn}\}, \qquad Q_1(z_j) = 0, \tag{2.21}$$

$$(\boldsymbol{P}_1)_{ik} = (\lambda_k - \lambda_i)^{-1}\tilde{E}_{ik} \quad (i \neq k), \tag{2.22}$$

$$(\boldsymbol{P}_1)_{ii} = \int_{z_j}^{z} \sum_{k \neq i} \tilde{E}_{ik}(\boldsymbol{P}_1)_{ki} = \int_{z_j}^{z} \sum_{k \neq i} (\lambda_k - \lambda_i)(\boldsymbol{P}_1)_{ik}(\boldsymbol{P}_1)_{ki}. \tag{2.23}$$

The feasibility of using the approximation (2.18) is determined as follows. Given an accuracy tolerance δ, we require that

$$\mathrm{Re}\,(\omega\nu_\beta)\,h_j \leq 4, \qquad |\omega\nu_\beta| \geq |\partial_z\beta|\,\delta^{-1/2}, \qquad |\omega\rho^{-1}| \geq |\partial_z\rho|\,\delta^{-1/2}, \tag{2.24a}$$

$$|\omega\nu_\alpha| \geq |\partial_z\alpha|\,\delta^{-1/2}, \tag{2.24b}$$

$$\mathrm{Re}\,(\omega\nu_\alpha)\,h_j \leq 6, \qquad \mathrm{Re}\,(\nu_\alpha) \leq \mathrm{Im}\,(\nu_\alpha) \tag{2.24c}$$

hold both at z_j and at z_{j+1}. If these conditions are satisfied, and there is no turning point between z_j and z_{j+1} (which is easy to check: we simply verify that $\mathrm{Re}\,(\beta - p^{-1})$ and $\mathrm{Re}\,(\alpha - p^{-1})$ do not change sign), then we may use (2.18). Whether or not we actually use it depends on the element size h_j that it would require compared with the element size that a collocation method would need at the same depth z_j (see Section 3). These conditions effectively prevent an unacceptable round-off error accumulation.

2.4 Turning points and leaky modes

Here we consider the possibility that $\nu_\alpha(z) \approx 0$ or $\nu_\beta(z) \approx 0$, $z_j \leq z < z_{j+1}$. For simplicity of presentation, let us assume that the velocities α and β are real. Then, comparing them with the phase velocity p^{-1} (recall (1.8)) we distinguish three cases.

(i) $p^{-1} \approx \beta(z)$ (S-wave turning point)

Then $\nu_\beta(z) \approx 0$ and $\nu_\alpha^2(z) > 0$. This case, the only one occurring for the SH problem, is handled quite well by the collocation scheme, because $|\omega\nu_\beta|$ is not large, while ν_α is real.

(ii) $\beta(z) < p^{-1} < \alpha(z)$ ("leaky mode")

This situation occurs between the P- and S-wave turning points. Then $\nu_\beta^2(z) < 0$ and $\nu_\alpha^2(z) > 0$. For large ω we must decouple the S- and P-wave components, because the fundamental-matrix method of Section 2.3 is suitable only for the oscillatory S waves, while the collocation method is suitable only for the evanescent P waves. Without separating these modes, neither method can be effectively applied. We therefore consider a transformed ordinary differential equation (1.2), where $\boldsymbol{A}(z)$ is brought into diagonal form by a similarity transformation. This approximately yields the diagonal differential system

$$\partial_z \hat{\boldsymbol{U}} = [\omega \boldsymbol{\Lambda}(z) + \partial_z \boldsymbol{Q}_1(z)]\hat{\boldsymbol{U}} \tag{2.25}$$

(see Ascher and Spudich, 1986), so each row of $\hat{\boldsymbol{U}}$ can be dealt with separately.

For the P–SV problem we see from (1.7) that the second and fourth rows of $\hat{\boldsymbol{U}}$ can be dealt with as in Section 2.3, while the first and third rows of $\hat{\boldsymbol{U}}$ can be approximated well by a collocation scheme. Applying the appropriate discretization to each row of (2.25), we then transform back to the original variables to obtain (2.5). The feasibility of using this approximation is determined by requiring that conditions (2.24a,b) (but not (2.24c)) hold at both z_j and z_{j+1} and that there be no turning point there. The element size h_j is the minimum of the step size allowed by the collocation scheme for the P wave and the step size allowed by the body-wave approximation for the S wave, for a given accuracy tolerance δ.

(iii) $p^{-1} \approx \alpha(z)$ (P-wave turning point)

Then $\nu_\beta^2(z) < 0$ and $\nu_\alpha(z) \approx 0$. Here again we wish to decouple P and S waves at high frequencies and use collocation for the P wave and the approximate fundamental matrix for the S wave. However, unlike case (ii) above, we cannot use the eigenvector matrix $\boldsymbol{E}(z)$ of $\boldsymbol{A}(z)$ for a transformation of (1.2) into a diagonal form like (2.25), because $\|\boldsymbol{E}(z)\| \, \|\boldsymbol{E}^{-1}(z)\|$ is not bounded. Instead, the decoupling can be done in the form

$$\boldsymbol{D}^{-1}\boldsymbol{A}\boldsymbol{D} = \begin{bmatrix} 0 & p & 0 & 0 \\ v_\alpha^2/p & 0 & 0 & 0 \\ 0 & 0 & v_\beta & 0 \\ 0 & 0 & 0 & -v_\beta \end{bmatrix} \equiv \hat{\boldsymbol{A}} \equiv \begin{bmatrix} \boldsymbol{H}_\alpha & 0 \\ 0 & \boldsymbol{\Lambda}_s \end{bmatrix}, \qquad (2.26)$$

where $\boldsymbol{D}^{-1}$ has terms like v_β^{-1} but not v_α^{-1} and so stays well-defined and smooth in the passage through an α turning point.

This yields a transformed approximated system

$$\partial_z \boldsymbol{U} = [\omega \hat{\boldsymbol{A}} + \tilde{\boldsymbol{A}} + O(\omega^{-1})]\boldsymbol{U}, \qquad (2.27)$$

where $\tilde{\boldsymbol{A}}$ has zero 2×2 off-diagonal blocks and a diagonal lower-right block. Then we neglect the $O(\omega^{-1})$ term and (2.27) decouple. This allows us again to apply the appropriate scheme to each row of the decoupled system. The remaining construction is similar to case (ii) for leaky modes discussed earlier; for details see again Ascher and Spudich (1986). The feasibility of using this approximation is determined by requiring that conditions (2.24a) (but not (2.24b)) hold at both z_j and z_{j+1}.

Note that in (2.26) we are using an unusual transformation. Thus it is not growing and decaying solution components that are decoupled here: the P-wave components are handled together implicitly by the collocation scheme. Rather, the P wave is decoupled from the highly oscillatory S wave.

3 A HYBRID METHOD

For given values of frequency ω, slowness p and error tolerance δ, the hybrid method consists of a constructed mesh (2.1) and, on each element of the mesh, of a specified numerical scheme. This hybrid scheme is to be used to obtain the solution $\{\boldsymbol{b}_j\}_{j=1}^{N+1}$ to within the specified accuracy. In the preceding section we have presented for each mesh element five possible schemes:

(1) collocation with l points ($l > 1$, an input parameter; default value $l = 3$);
(2) the midpoint scheme (collocation with $l = 1$);

(3) the body-wave method of Section 2.3;
(4) the leaky-mode method (case (ii) of Section 2.4);
(5) the P-wave turning-point method (case (iii) of Section 2.4).

Let

$$0 = d_1 < d_2 < \ldots < d_M = z_> \tag{3.1}$$

be a set of depths that must be included in the mesh (recall requirement (a) on (2.1) in Section 2.1. Also d_i may be a point where $\boldsymbol{b}(d_i)$ values are desired). The algorithm for mesh and method selection produces the mesh size N, an array $\{z_1, z_2, \ldots, z_{N+1}\}$ of mesh points (2.1) and an array $\{\text{method}_1, \text{method}_2, \ldots, \text{method}_N\}$, where method_j indicates the scheme to be used on (z_j, z_{j+1}), $1 \leq \text{method}_j \leq 5$, $1 \leq j \leq N$.

The rationale behind the algorithm is quite simple: At each step, the more efficient of schemes (1) and (2) is used, unless one of the schemes (3)–(5) is both applicable and more efficient. Note that schemes (3), (4) and (5) are checked for feasibility in this order and at most one of them may be found feasible at any step.

Algorithm: Mesh and method selection

$N := 1$; $z_1 := 0$
FOR $i := 1$ to $M - 1$ DO {construct mesh between each two given depths}
  $hh := d_{i+1} - d_i$
  WHILE $hh > 0$ DO
    determine $h_c = h_j$ from (2.16) as described in Section 2.2
    determine h_M similarly for the midpoint scheme
    IF $h_M \geq hh$ THEN {use the midpoint scheme}
      $\text{method}_N := 2$; $h := hh$
    ELSE
      $h := \max\{h_c, \epsilon_M h_M\}$
      {check feasibility of using schemes (3), (4), (5)}
      IF $h < hh$ AND for some j, $3 \leq j \leq 5$, scheme (j) is feasible with step size $h^* > h$ {as described in Sections 2.3, 2.4; at most one j can qualify} THEN
        $\text{method}_N := j$; $h := h^*$ {use scheme (j)}
      ELSE {use collocation}
        IF $h = h_c$ THEN
          $\text{method}_N := 1$; $h := \min\{h, hh\}$

```
            ELSE
               method_N := 2; h := h_M
            END IF
         END IF
      END IF
      z_{N+1} := z_N + h; hh := hh − h; N := N + 1
   END WHILE
END FOR
```

4 AN EXAMPLE

In order to demonstrate how our hybrid method works, we calculated in Ascher and Spudich (1986) synthetic seismograms for a test velocity structure whose parameters are given in Table 1. The structure is perfectly elastic (α and β are purely real), and β has been derived from α assuming a Poisson solid. All material properties vary linearly with depth except at the points given in Table 1.

A set of Fortran routines has been written that implements the hybrid method described above. This is a stand-alone package that requires a driving program where the input parameters for the package are defined and where the output from it is handled upon return, and a subroutine describing the velocity structure used. Details like those in Table 1 are specified in this latter routine, which is expected to return for each given value of z the values of $\rho(z)$, $\beta(z)$ and $\alpha(z)$.

The test model has been chosen to have two regions having rather different velocity gradients ($\partial_z\alpha = 2\,\mathrm{s}^{-1}$ versus $\partial_z\alpha = 0.33\,\mathrm{s}^{-1}$) in order to demonstrate how the computational algorithm of Section 3 chooses mesh points and solution schemes on each mesh subinterval depending upon ω, k and the velocity gradient. Figure 2 shows mesh points and solution methods chosen for various values of wavenumber k, holding frequency fixed at 9.0 Hz. The cases displayed correspond, from top to bottom, to a highly evanescent wave, an S-wave turning point (case (i) of Section 2.4), a leaky mode (case (ii) of Section 2.4), a P-wave-turning point (case (iii)

Table 1 Test-velocity model.

Depth (km)	α (km s^{-1})	β (km s^{-1})	ρ (g, cm^{-3})
0	4.00	2.31	2.37
0.50	5.00	2.89	2.54
2.00	5.50	3.18	2.63
>2.00	6.00	3.46	2.71

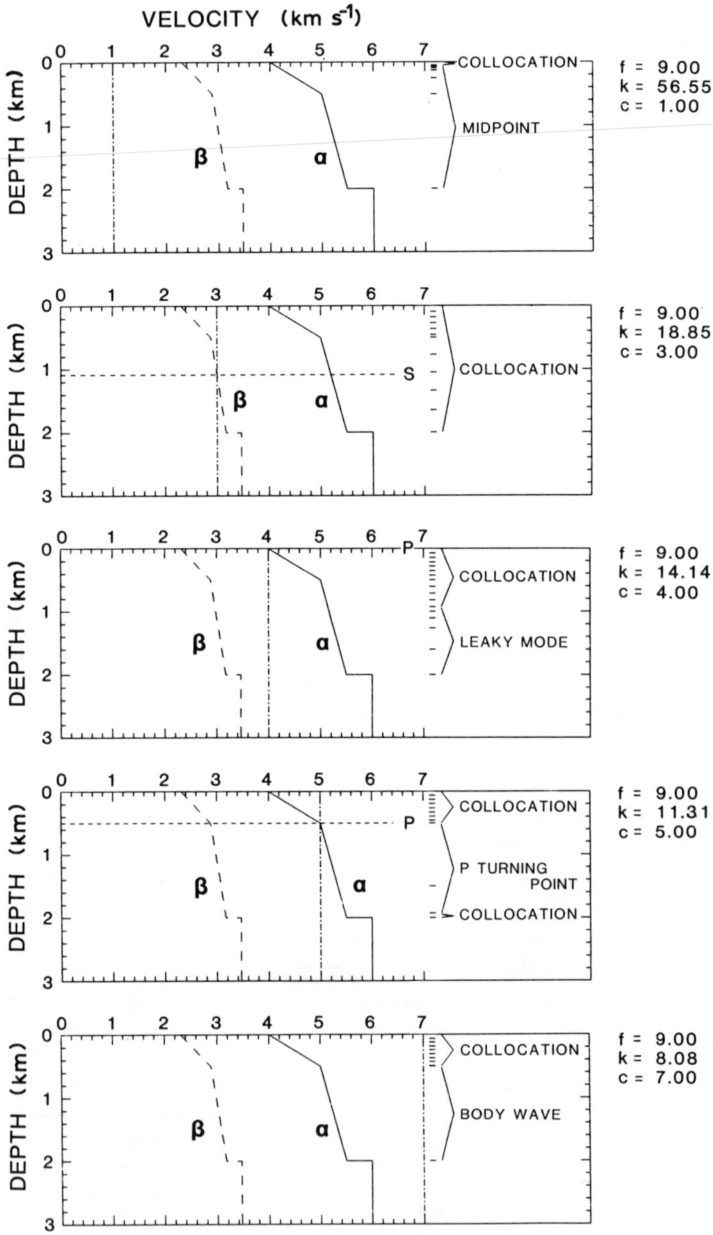

Fig. 2 Mesh points and solution schemes chosen by the hybrid method.

of Section 2.4) and a situation above both turning points (Section 2.3). Note the automatic choice of schemes for each mesh subinterval, resulting in a hybrid method tailored for each case. The collocation scheme successfully resolves the S-wave turning point with a reasonably sparse mesh. In the highly evanescent case, collocation points are densely placed near the surface, where the solution varies fast. Then, for larger depths, the mesh becomes sparse according to (2.16), and, in fact, the limit on the step size is eventually dictated by the next point of medium discontinuity, invoking the choice of the cheaper midpoint scheme.

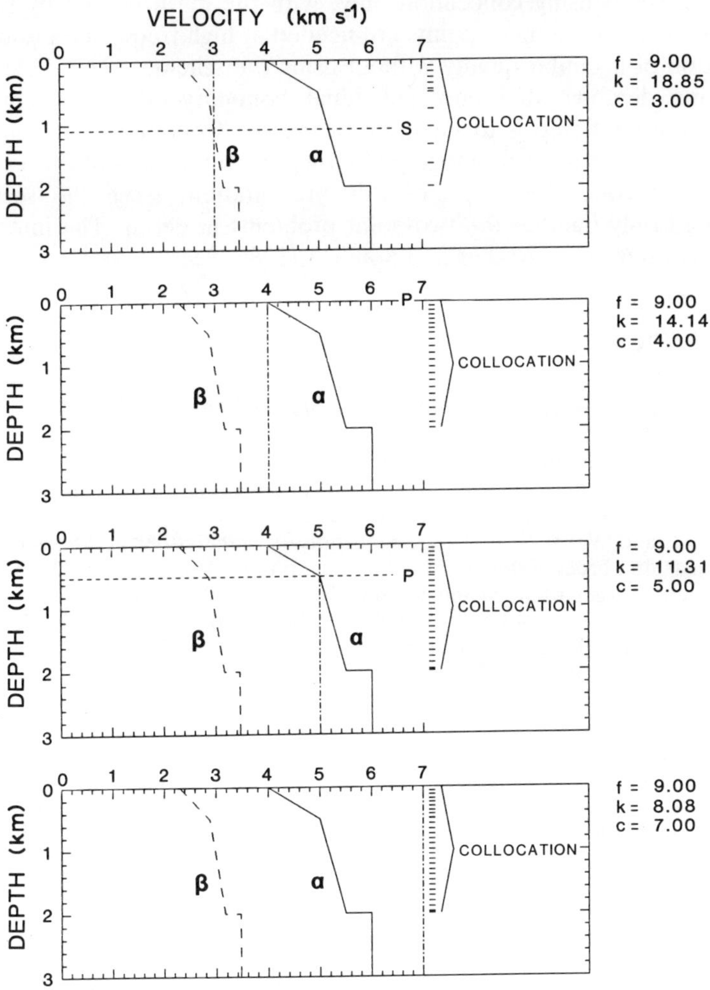

Fig. 3 Mesh points for the P–SV problem when using collocation only.

The velocity gradients for $0<z<0.5$ are so large in this model that collocation had to be used in all cases for that part of the domain in Fig. 2. When $\nu_\beta^2<\nu_\alpha^2<0$ it may happen that (2.24a) holds but (2.24b) does not; hence earlier use of the body-wave approximation for the SH problem than for the P–SV problem is feasible. The use of schemes (3)–(5) for the P–SV problem in $0<z<0.5$ is possible only at a higher frequency.

To show that the effort of constructing the hybrid method with the five schemes is worthwhile, we display in Fig. 3 corresponding results to those of Fig. 2 when using collocation only, with the mesh chosen by (2.16). Clearly, many more mesh points are needed at high frequencies to obtain a solution of a similar quality without using the schemes (3)–(5).

Having demonstrated how individual boundary-value problems are solved, the next step is to integrate in the (ω, k) domain and construct the resulting synthetic seismograms. However, full details of this calculation are beyond the scope of this presentation, since the software presented only handles the two-point problems in depth. The interested reader is referred to Ascher and Spudich (1986, Section 4).

REFERENCES

Aki, K. and Richards, P. G. (1980). *Quantitative Seismology,* Vol. 1. W. H. Freeman, San Francisco.

Ascher, U. and Spudich, P. (1986). A hybrid collocation method for calculating complete theoretical seismograms in vertically varying media. *Geophys. J. R. Astron. Soc.* **86,** 19–40.

Ascher, U. and Weiss, R. (1984). Collocation for singular perturbation problems II: linear first order systems without turning points. *Math. Comp.* **43,** 157–187.

Ascher, U., Christiansen, J. and Russell, R. D. (1979). A collocation solver for mixed order systems of boundary value problems. *Math. Comp.* **33,** 659–679.

Chapman, C. H. (1974). The turning point of elastodynamic waves. *Geophys. J. R. Astron. Soc.* **39,** 613–621.

Chapman, C. H. (1976). Exact and approximate generalized ray theory in vertically inhomogeneous media. *Geophys. J. R. Astron. Soc.* **46,** 201–234.

Kennett, B. L. N. (1983). *Seismic Wave Propagation in Stratified Media.* Cambridge University Press.

Olson, A. H., Orcutt, J. A. and Frazier, G. A. (1984). The discrete wavenumber/finite element method for synthetic seismograms. *Geophys. J. R. Astron. Soc.* **77,** 421–460.

Phinney, R. A. (1965). Theoretical calculations of the spectra of first arrivals in layered elastic mediums. *J. Geophys. Res.* **70,** 5107–5123.

Richards, P. G. (1971). Elastic wave solutions in stratified media. *Geophysics* **36,** 798–809.

Richards, P. G. and Frasier, C. W. (1976). Scattering of elastic waves from depth-dependent inhomogeneities. *Geophysics* **41,** 441–458.

Schwab, F. and Knopoff, L. (1972). Fast surface wave and free mode computations. *Methods of Computational Physics,* Vol. 11 (ed. B. A. Bolt), pp. 87–180. Academic Press, New York.

Spudich, P. and Ascher, U. (1983). Calculation of complete theoretical seismograms in vertically varying media using collocation methods. *Geophys. J. R. Astron. Soc.* **75,** 101–124.

Takeuchi, H. and Saito, M. (1972). Seismic surface waves. *Methods of Computational Physics,* Vol. 11 (ed. B. A. Bolt), pp. 217–295. Academic Press, New York.

III.3

Quadrature of Wavenumber Integrals

L. NEIL FRAZER

Hawaii Institute of Geophysics, University of Hawaii at Manoa, Honolulu, Hawaii, USA

1 INTRODUCTION

A Filon-type trapezoidal rule can be used to speed the evaluation of most of the oscillatory integrals that arise in computational seismology. Here it is shown how to derive the rule and apply it to the integrals that arise in the reflectivity method, the phase-integral method and the Kirchhoff–Helmholtz method. Some recently developed quadrature methods that are more efficient than trapezoidal Filon are also reviewed.

In the computation of synthetic seismograms the motion u is often obtained in the form of an integral

$$u(x, t) = \int \mathrm{d}\omega \int \mathrm{d}k\, \mathrm{e}^{\mathrm{i}(kx-\omega t)}\, \tilde{u}(k, \omega), \tag{1}$$

in which x is a space coordinate, t is time, ω is temporal frequency and k is wavenumber. Methods in which the k integral is evaluated first and the ω integral is evaluated second are sometimes called *spectral methods,* whereas methods in which the ω integral is evaluated first and the k integral is evaluated second are sometimes called *slowness methods* (Chapman, 1978). In the Cagniard–de Hoop method (de Hoop, 1960) the ω integral and the k integral are made to cancel by use of a mapping $k = k(t)$. Spectral generalized ray methods are slower to compute than slowness generalized ray methods, and the Cagniard–de Hoop method, but all spectral methods have the great advantage of allowing frequency-dependent seismic velocities (Frazer, 1983). It has been demonstrated many times that the frequency dependence of seismic velocities, which is

SEISMOLOGICAL ALGORITHMS
ISBN 0-12-220770-X

a consequence of causality, can profoundly affect the waveforms and arrival times of body waves (Jeffreys, 1965, 1967; Strick, 1970). Consideration of this frequency dependence is especially important when comparing velocity models obtained from seismic data in different frequency bands (Liu *et al.*, 1976). Here we treat only those wavenumber integrals that arise in spectral methods. The quadrature of wavenumber integrals that arise in slowness methods is treated in Wiggins (1974), Chapman (1978), Dey-Sarkar and Chapman (1978), Frazer and Phinney (1980), Fryer (1981), Sinton and Frazer (1982), Chapman and Drummond (1982) and Burdick and Salvado (1986). Quadrature techniques suitable for slowness integrals are also of use in time-domain versions of the Kirchhoff–Helmholtz method (Hilterman, 1970, 1976; Berryhill, 1979; Kinney *et al.*, 1983; Spudich and Frazer, 1984) and in the time-domain multifold path-integral method (Frazer, 1986, 1987; Sen and Frazer, 1987). The spectral WKBJ method (Richards, 1973; Choy, 1977; Cormier and Richards, 1977; Frazer, 1977) is here referred to as the phase-integral method.

2 TRAPEZOIDAL FILON

Consider first the integral

$$I_{\mathrm{W}} = \int_{\Gamma} \mathrm{d}p\, f(p) \mathrm{e}^{sg(p)}. \tag{2}$$

This integral arises in the phase-integral method, where

$$g(p) = T(p) + p\Delta - X(p). \tag{3}$$

As shown in Fig. 1(a), $T(p)$ and $X(p)$ are the travel-time and distance functions respectively of the generalized ray with parameter p, and Δ is the horizontal distance between source and receiver. The integral (2) also arises in the spectral Kirchhoff–Helmholtz method, where $g(p)$ is the travel time from the source down to an interface and back up to the receiver, and p is the position of the reflecting point on the interface. In both the phase-integral method and the spectral Kirchhoff–Helmholtz method, s is given by

$$s = \mathrm{i}\omega, \tag{4}$$

where ω is the radian temporal frequency.

It often happens that ω is large, so that, as shown in Fig. 2, the variation of $\mathrm{e}^{sg(p)}$ with p is much more rapid than the variation of either

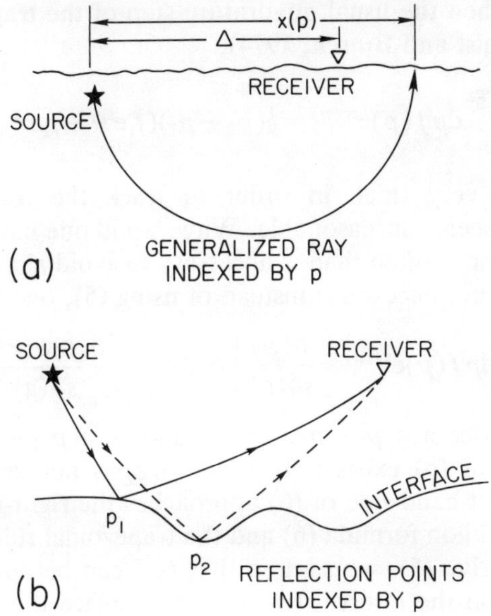

Fig. 1 (a) In the phase-integral method the *p* of integral (2) is a slowness; whereas (b) in the spectral Kirchhoff-Helmholtz method *p* is the position of the reflection point on the interface.

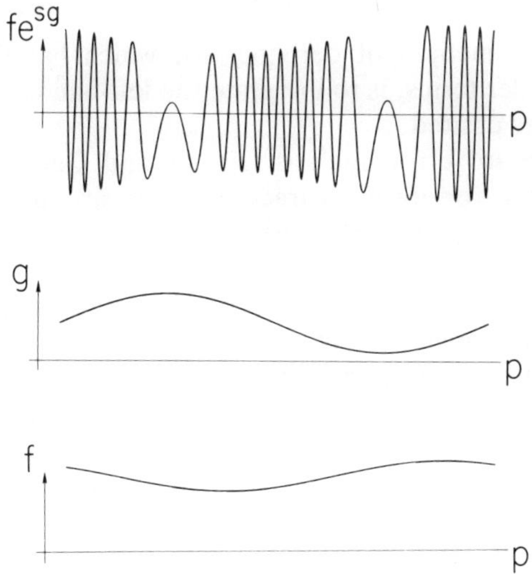

Fig. 2 When Im (s) is large fe^{sg} may vary much more rapidly than g and f.

$f(p)$ or $g(p)$. Then the usual quadrature step of the trapezoidal method (see e.g. Dahlquist and Bjorck, 1974),

$$\int_{p_1}^{p_2} \mathrm{d}p\, f(p)\mathrm{e}^{sg(p)} \approx \tfrac{1}{2}(p_2 - p_1)(f_1\mathrm{e}^{sg_1} + f_2\mathrm{e}^{sg_2}), \tag{5}$$

must be taken very small in order to track the oscillations of e^{sg}. Intuitively, this seems unreasonable. Why should one have to numerically sample f and g more often than is necessary to avoid aliasing them? Well, of course this is not necessary: instead of using (5), one can use

$$\int_{p_1}^{p_2} \mathrm{d}p\, f(p)\mathrm{e}^{sg(p)} \approx \frac{\delta(p)}{s\delta(g)}\left[\delta(f\mathrm{e}^{sg}) - \frac{\delta(f)\delta(\mathrm{e}^{sg})}{s\delta(g)}\right], \tag{6}$$

in which $\delta(p)$ denotes $p_2 - p_1$, $\delta(g)$ denotes $g(p_2) - g(p_1)$, etc. The right-hand side of (6) exists only when $\delta(g)$ is non-zero; however, as $\delta(g)\rightarrow 0$ the right-hand side of (6) approaches the right-hand side of (5). The trapezoidal Filon formula (6) and the trapezoidal rule (5) are derived similarly. To derive (5), we assume that $f\mathrm{e}^{sg}$ can be approximated by a linear function on the interval (p_1, p_2) and replace the integrand in the left-hand side of (5) by $f_1\mathrm{e}^{sg_1} + (p - p_1)\delta(f\mathrm{e}^{sg})/\delta(p)$; to derive (6), we assume that f and g can be approximated by linear functions on the interval (p_1, p_2) and replace the integrand in the left-hand side of (6) by

$$\left[f_1 + (p - p_1)\frac{\delta(f)}{\delta(p)}\right] \exp\left\{s\left[g_1 + (p - p_1)\frac{\delta(g)}{\delta(p)}\right]\right\}.$$

An alternative method of deriving (6), which has the advantage of giving higher-order terms, is to integrate the left-hand side of (6) by parts as many times as desired.

A straightforward error analysis (Frazer and Gettrust, 1984) can be used to show that, for given accuracy, the step size in (5) is proportional to $|s|^{-1}$ whereas the step size in (6) is proportional to $|s|^{-1/2}$. Numerical experience in many different computer codes seems to bear this out. When using (6), it should be remembered that the step size $\delta(p)$ must always be small enough so that neither $f(p)$ nor $g(p)$ is aliased. If the variation of f with p is as great as the variation of e^{sg} with p then there is no point in using (6) instead of (5). On the other hand, (6) is always at least as accurate as (5) and takes only about ten per cent longer to evaluate. Both (5) and (6) give a code that vectorizes well on computers such as the Cray.

The trapezoidal Filon formula (6) was derived and applied to the phase-integral method in Frazer (1977) and to the spectral Kirchhoff–Helmholtz method in Sen and Frazer (1985) and Frazer and Sinton

(1984). In Frazer and Gettrust (1984) the trapezoidal Filon formula was referred to as the "generalized Filon method" after Filon (1928). It was hoped that this rather grandiose name would induce seismologists to begin using (6). In the event, the response of most seismologists was similar to the response of the applied mathematician who reviewed Frazer and Gettrust (1984) when it was submitted for publication. "I didn't believe a word of it", he confessed, "until I coded it up and tried it myself." So many variations and extensions of Filon's method have by now been developed that "trapezoidal Filon" seems a better name for (6). A version of the trapezoidal Filon rule suitable for triple integrals is given in Sen and Frazer (1987).

Consider next the integral

$$u(\omega, x) = \int_0^{\infty - \mathrm{i}0} k \, \mathrm{d}k \, \tilde{u}(\omega, k) J_n(kx), \tag{7}$$

in which J_n is the Bessel function of order n, and k is the wavenumber. This integral arises in the reflectivity method (Fuchs and Müller, 1971; Kennett, 1983). Its quadrature can be simply accomplished in the following two ways.

The first method is suitable for use when x is large and also when x is small or zero. Here advantage is taken of the relation (Olver, 1972, formula (9.2.19))

$$J_n = M_n \cos \theta_n \tag{8}$$

and the asymptotic expansion (Olver, 1972, formula (9.2.29))

$$\theta_n = kx - \alpha_n + O(|kx|^{-1}), \tag{9}$$

in which

$$\alpha_n = \tfrac{1}{4}(2n + 1)\pi. \tag{10}$$

Now (7) is written in the form

$$u(\omega, x) = \int_0^{\infty - \mathrm{i}0} \mathrm{d}k \, f_1(k) \mathrm{e}^{\mathrm{i}xk} + \int_0^{\infty - \mathrm{i}0} \mathrm{d}k \, f_2(k) \mathrm{e}^{-\mathrm{i}xk}, \tag{11}$$

where the functions f_1 and f_2 are given by

$$f_1(k) = \tfrac{1}{2} k \tilde{u}(\omega, k) M_n \mathrm{e}^{\mathrm{i}(\theta_n - kx)}, \tag{12}$$

$$f_2(k) = \tfrac{1}{2} k \tilde{u}(\omega, k) M_n \mathrm{e}^{-\mathrm{i}(\theta_n - kx)}. \tag{13}$$

Equation (11) is exact, and yet the functions f_1 and f_2 are relatively non-oscillatory because of (9). Each integral on the right-hand side of (11) is seen to be of the form (2), with $g(k) = k$ and $s = \pm \mathrm{i}x$, and so it can be evaluated using the quadrature formula (6).

Equations (11)–(13) illustrate an important general technique that is often useful in computing integrals of oscillatory functions. An inexact asymptotic expansion is used to remove most of the phase variation from part of the integrand, but the accuracy of the end result does not depend on the accuracy of the asymptotic expansion. The same technique can often be used (Frazer and Gettrust, 1984; Sen *et al.*, 1987) to reduce the variability of $\tilde{u}(\omega, k)$.

Because of the time dependence $e^{-i\omega t}$, the second integral on the right-hand side of (11) is never significant when x is more than a few wavelengths, and so the second integral is often neglected. The functions M_n and θ_n are available as polynomial approximations (Allen, 1954; Olver, 1972, formulae (9.4.3) and (9.4.6)) for values of kx greater than three. For values of kx less than three, exact values of M_n and θ_n *could* be computed from their definitions (Olver, 1972, formula (9.2.17))

$$\theta_n \equiv \tan^{-1}(Y_n/J_n), \tag{14}$$

$$M_n \equiv (J_n^2 + Y_n^2)^{1/2}, \tag{15}$$

but this would be inconvenient because M_n is singular when kx is zero, and because (14) and (15) require that both J_n and Y_n be computed using separate polynomials. A better way to get values of θ_n and M_n for kx less than three is to *redefine* them so that they are more convenient to compute but still satisfy (8). The following is one scheme for doing this:

$$\theta_0(kx<3) \equiv \frac{\pi}{2}\frac{kx(kx-3)}{j_{0,1}(j_{0,1}-3)} + \theta_0(3)\frac{kx(kx-j_{0,1})}{3(3-j_{0,1})}, \tag{16a}$$

$$\theta_n(kx<3) \equiv -\tfrac{1}{2}\pi + \epsilon + \tfrac{1}{3}kx(\theta_n(3) + \tfrac{1}{2}\pi - \epsilon) \quad (n \geqslant 1), \tag{16b}$$

$$M_0(kx<3) \equiv \begin{cases} \dfrac{\text{POLY}\,[J_0(kx)]}{\cos\theta_0(kx)} & (|kx - j_{0,1}| \geqslant \delta), \\[2ex] M_0(a_1) + \dfrac{kx - a_1}{a_2 - a_1} M_0(a_2) & (|kx - j_{0,1}| < \delta), \end{cases} \tag{17a}$$

$$M_n(kx<3) \equiv \frac{\text{POLY}\,[J_n(kx)]}{\cos\theta_n(kx)} \quad (n \geqslant 1). \tag{17b}$$

In these relations $j_{0,1} = 2.40482$ is the first zero of J_0; $\epsilon = 0.5$ is for the purpose of moving $\theta_n(0)$ away from $-\frac{1}{2}\pi$ when $n \geqslant 1$ so that M_n will not be singular there; $\delta = 0.01$; $a_2 = j_{0,1} + \delta$; $a_1 = j_{0,1} - \delta$; and POLY [. . .] means that the quantity [. . .] is to be computed by means of a polynomial expansion valid for $0 \leqslant kx \leqslant 3$. The definition of M_0 is more complicated than the definition of M_n because the first zero of J_0 is less than three whereas all the zeroes of J_n, $n \geqslant 1$, are greater than three.

The reason for the "definitions" (16) and (17) is that they enable one to use the forms (11)–(13), and the trapezoidal Filon rule (6), right from $k = 0$ without sacrifice of accuracy—recall that (8) is preserved by (16) and (17). The alternative is to evaluate (7) by integrating from $k = 0$ to $3/x$ by use of the trapezoidal rule, and then from $k = 3/x$ using (11)–(13) with the lower limits on the integrals in (11) changed from 0 to $3/x$. The alternative is unsatisfying to code because (a) two different quadrature schemes must be coded and (b) the limits of integration depend on source–receiver distance x. If (16) and (17) are used then only one quadrature must be coded; the seeming difficulties caused by small values of kx are all disposed of in the subroutine that returns $\theta_n(kx)$ and $M_n(kx)$.

The procedure just given for the reflectivity integral (7) is somewhat complicated because of a need to accommodate both large and small, or zero, values of source–receiver offset x. In the modeling of refraction data x is almost always greater than four or five wavelengths, and then the following simpler procedure can be used. In (12) replace θ_n by $kx - \alpha_n$ and M_n by $(2/\pi kx)^{1/2}$. Now neglect the second integral on the right-hand side of (11) and change the variable of integration in the first integral on the right-hand side of (11) to $p = k/\omega$. The result is

$$u(\omega, x) \equiv \int_0^{\infty - \mathrm{i}0} \mathrm{d}p\, f(p) \mathrm{e}^{\mathrm{i}\omega x p}, \tag{18}$$

in which

$$f(p) = \left(\frac{\omega p}{2\pi}\right)^{1/2} \tilde{u}(\omega, \omega p) \mathrm{e}^{-\mathrm{i}(2n+1)\pi/4}. \tag{19}$$

The integral (18) can now be evaluated using the trapezoidal Filon quadrature formula (6) with $g(p) = p$ and $s = \mathrm{i}\omega x$.

Of course the upper limits of the integrals in (11) and (18) are not really infinite. If the source and receiver differ in depth by more than a few wavelengths then the modulus of $\tilde{u}(\omega, \omega p)$ will rapidly decay to zero as p increases past β_s^{-1}, where β_s is the shear velocity at the source. Also it is often desirable to exclude slow events from the synthetic seismograms by smoothly tapering $f(p)$ to zero past $p_{\max} = 1/c_{\min}$, where $c_{\min}$ is the lowest phase velocity of interest. Frazer and Gettrust (1984) give numerous examples of the improvements that result when (18) is evaluated with the trapezoidal Filon rule (6) instead of the trapezoidal rule (5).

The contour of integration in the reflectivity integral (7) should be kept on the Re (k) axis if possible because $J_n(kx)$ grows exponentially if k is moved away from there. Unfortunately $\tilde{u}(\omega, k)$ often has singularities just above the Re (k) axis. It has often been suggested that to get away from these singularities the contour of integration should be taken slightly below the Re (k) axis. A better procedure (Phinney, 1965; Bouchon and Aki, 1977; Spudich and Ascher, 1983) is to keep the contour of integration on the Re (k) axis but to use a complex value of ω with fixed imaginary part a. Then the singularities of $\tilde{u}(\omega, k)$ are moved farther up into the first quadrant away from the Re (k) axis. The resulting synthetic seismograms must be multiplied by e^{at}, where t is time, to restore their correct amplitudes. Use of a complex ω also suppresses wraparound caused by too small a step size $d\omega$ in the inverse Fourier transform, the outside integral in (1). The use of a complex frequency is discussed at some length in Mallick and Frazer (1987). One should choose a, the fixed imaginary part of ω, according to the formula

$$a = \frac{1}{T} \ln 50,$$

in which T is the length of the time series being synthesized. Then the amplitude of wrap-around arrivals will be reduced by at least a factor of 50. If numerical problems result, use 40 instead of 50, or increase T.

If a complex frequency is used then, in order to keep k real, the contour of integration in (18) should be rotated down into the fourth quadrant of the complex p plane. The upper limit of the integral (18) is changed from $\infty - i0$ to $\exp[-i \arg(\omega)]\infty$, where $\arg(\omega)$ is given by

$$\arg(\omega) = \tan^{-1} \frac{\text{Im}(\omega)}{\text{Re}(\omega)}.$$

Note that the quadrature formulae (5) and (6) remain valid when all the quantities in them are complex. In a reflectivity computer code if the step size dp in (18) was formerly ϵ then with a complex ω the new step size dp is just $\epsilon \exp[-i \arg(\omega)]$.

The trapezoidal Filon formula (6) is valuable for reflectivity syntheses if the source–receiver offset x is large because it is easily coded and it vectorizes well. However, (6) is not helpful when x is small because then the variation of the reflectivity integrand with k is mostly due to $\tilde{u}(\omega, k)$. In applications where x is always going to be small then either (5) should be coded instead of (6), or else Clenshaw–Curtis quadrature (see below) should be used.

3 ALTERNATIVE METHODS

The two methods discussed in this section are newer variations of Filon's method that are more difficult to code and more difficult to vectorize than the simple formulae discussed above. However, it is likely that on serial processors they are more efficient than (6). For the evaluation of reflectivity integrals they are almost certainly more efficient than (6) when the source–receiver distance is less than a few wavelengths.

The first alternative scheme is the adaptive version of Clenshaw–Curtis quadrature (Clenshaw and Curtis, 1960) proposed by Xu and Mal (1985). This scheme is most applicable to the reflectivity integral (11). In Clenshaw–Curtis quadrature the function $k\tilde{u}(\omega, k)$ in the reflectivity integral is approximated not by a polynomial, as in the original Filon method, but instead by a series of Chebyshev polynomials. The Xu–Mal scheme is adaptive, in the sense that the number of points at which $k\tilde{u}$ must be evaluated is successively doubled in such a way that no previous evaluations of $k\tilde{u}$ are wasted. Additional material on Clenshaw–Curtis quadrature is given in Piessens and Branders (1983), Oliver (1971) and O'Hara and Smith (1969).

The second alternative scheme is that of Levin (1982), discussed briefly in Davis and Rabinowitz (1984). This scheme is directly applicable to integrals of the form (2). Suppose that integral (2) has lower limit p_1 and upper limit p_2. If $f(p)$ in (2) can be written as

$$f(p) = sg'(p)q(p) + q'(p), \tag{20}$$

where $q(p)$ is a function to be determined, then (2) can be evaluated directly as

$$I_{\mathrm{W}} = q(p_2)\mathrm{e}^{sg(p_2)} - q(p_1)\mathrm{e}^{sg(p_1)}. \tag{21}$$

One way to get the values of $q(p_1)$ and $q(p_2)$ is to assume that q is given by the series of basis functions

$$q(p) = \sum_{m=1}^{n} \alpha_m \varphi_m(p). \tag{22}$$

The basis functions φ_m could be the monomials p^{m-1}, for example. Now evaluation of $f(p)$, $g(p)$ and the $\varphi_m(p)$ at the points $p_1, p_2, \ldots, p_n$ turns (20) into a system of n linear equations in the n unknown coefficients $\alpha_1, \alpha_2, \ldots, \alpha_n$. When this system is solved $q(p_1)$ and $q(p_2)$ are computed from (22) and inserted into (21). To avoid the difficulties of large linear systems, one should keep n low by subdividing the interval of integration and applying the method separately on each subinterval.

Levin's (1982) method is the most general extension of Filon's method that I have seen. Formula (6) appears to be the version of Levin's method that results when $q(p)$ and $g(p)$ are assumed to be linear functions. Recall that the trapezoidal Filon formula (6) was derived directly by approximating f and g by linear functions. Attempts have been made to derive higher-order Filon formulae by approximating f and

g by higher-order polynomials. Unfortunately the resulting quadrature formulae contain special functions whose evaluation makes their efficiency difficult to judge. Levin's method avoid these difficulties. It also has the advantage of being straightforward to use on double integrals.

ACKNOWLEDGEMENTS

M. K. Sen, S. Mallick, J. B. Sinton, F. A. Dahlen and R. A. Phinney are thanked for helpful comments present and past. Alex Tolstoy and Ed Berg are thanked for pointing out how integration by parts can be used to give simple derivations of Filon quadrature formulae. Much of the work summarized here was supported by the seismology division of the National Science Foundation and by the Office of Naval Research. Hawaii Institute of Geophysics Contribution 1992.

REFERENCES

Allen, E. E. (1954). Analytical approximations. *MTAC* **8,** 240–241.

Berryhill, J. R. (1979). Wave equation datuming. *Geophysics* **44,** 1329–1344.

Bouchon, M. and Aki, K. (1977). Discrete wavenumber representation of seismic source wavefields. *Bull. Seismol. Soc. Am.* **67,** 257–277.

Burdick, L. J. and Salvado, C. A. (1986). Modeling body wave amplitude fluctuations using the three-dimensional slowness method. *J. Geophys. Res.* **91,** 12482–12496.

Chapman, C. H. (1978). A new method for computing synthetic seismograms. *Geophys. J. R. Astron. Soc.* **54,** 481–518.

Chapman, C. H. and Drummond, R. (1982). Body wave seismograms in inhomogeneous media using Maslov asymptotic theory. *Bull. Seismol. Soc. Am.* **72,** 5277–5317.

Choy, G. L. (1977). Theoretical seismograms of core phases calculated by frequency-dependent full wave theory and their implications. *Geophys. J. R. Astron. Soc.* **51,** 275–311.

Clenshaw, C. W. and Curtis, A. R. (1960). A method for numerical integration on an automatic computer. *Numer. Maths* **2,** 197–205.

Cormier, V. F. and Richards, P. G. (1977). Full wave theory applied to a discontinuous velocity increase: the inner-core boundary. *J. Geophys.* **43,** 3–22.

Dahlquist, G. and Bjorck, A. (1974). *Numerical Methods.* Prentice-Hall, Englewood Cliffs, N.J.

Davis, J. and Rabinowitz, P. (1984). *Methods of Numerical Integration,* 2nd edn. Academic Press, New York.

de Hoop, A. T. (1960). A modification of Cagniard's method for solving seismic pulse problems. *Appl. Sci. Res.* **B8,** 349–356.

Dey-Sarkar, S. K. and Chapman, C. H. (1978). A simple method for the computation of body wave synthetic seismograms. *Bull. Seismol. Soc. Am.* **68,** 1577–1593.

Filon, L. N. G. (1928). On a quadrature method for trigonometric integrals. *Proc. R. Soc. Edin.* **49,** 38–47.

Frazer, L. N. (1977). Synthesis of shear-coupled PL. Ph.D. thesis, Princeton University.

Frazer, L. N. (1983). Two problems in WKBJ theory: the interpolation of sampled velocity profiles and the use of frequency-dependent, complex velocities. *Geophys. J. R. Astron. Soc.* **72,** 193–211.

Frazer, L. N. (1986). Applications of multifold Kirchhoff–Helmholtz path integrals to sound propagation in the ocean. Part I: Theory. *J. Acoust. Soc. Am.* **79,** 1748–1759.

Frazer, L. N. (1987). Synthetic seismograms using multi-fold path integrals—I. Theory. *Geophys. J. R. Astron. Soc.* **77,** 621–646.

Frazer, L. N. and Gettrust, J. F. (1984). On a generalization of Filon's method and the computation of the oscillatory integrals of seismology. *Geophys. J. R. Astron. Soc.* **76,** 461–481.

Frazer, L. N. and Phinney, R. A. (1980). A theory for the computation of finite frequency synthetic seismograms in an inhomogeneous medium. *Geophys. J. R. Astron. Soc.* **63,** 691–717.

Frazer, L. N. and Sinton, J. B. (1984). A Kirchhoff method for the computation of finite-frequency body wave synthetic seismograms in laterally inhomogeneous media. *Geophys. J. R. Astron. Soc.* **78,** 413–429.

Fryer, G. J. (1981). A slowness approach to the reflectivity method of seismogram synthesis. *Geophys. J. R. Astron. Soc.* **63,** 747–758.

Fuchs, K. and Müller, G. (1971). Computation of synthetic seismograms by the reflectivity method and comparison with observations. *Geophys. J. R. Astron. Soc.* **23,** 417–433.

Hilterman, F. J. (1970). Three-dimensional seismic modeling. *Geophysics* **35,** 1020–1037.

Hilterman, F. J. (1975). Amplitudes of seismic waves—a quick look. *Geophysics* **40,** 745–762.

Jeffreys, H. (1965). Damping of S-waves. *Nature* **208,** 675.

Jeffreys, H. (1967). Radius of the Earth's core. *Nature* **215,** 1365–1366.

Kennett, B. L. N. (1983). *Seismic Wave Propagation in Stratified Media.* Cambridge University Press.

Kinney, W. A., Clay, C. S. and Sandness, G. A. (1983). Scattering from a corrugated surface: comparison between experiment, Helmholtz–Kirchhoff theory, and the facet ensemble method. *J. Acoust. Soc. Am.* **73,** 183–194.

Levin, D. (1982). Procedures for computing one- and two-dimensional integrals of functions with rapid irregular oscillations. *Math. Comp.* **38,** 531–538.

Liu, H.-P., Anderson, D. L. and Kanamori, H. (1976). Velocity dispersion due to anelasticity: implications for seismology and mantle composition. *Geophys. J. R. Astron. Soc.* **47,** 41–58.

Mallick, S. and Frazer, L. N. (1987). Practical aspects of reflectivity modeling. *Geophysics* **52,** 1355–1364.

O'Hara, H. and Smith, F. J. (1969). The evaluation of definite integrals by interval subdivision. *Comput. J.* **12,** 179–182.

Oliver, J. (1971). A practical strategy for the Clenshaw–Curtis quadrature method. *J. Inst. Math. Applics* **8,** 53–56.

Olver, F. W. J. (1972). Bessel functions of integer order. *Handbook of Mathematical Functions* (ed. M. Abramowitz and I. A. Stegun). National Bureau of Standards.

Phinney, R. A. (1965). Theoretical calculations of the spectra of first arrivals in layered elastic mediums. *J. Geophys. Res.* **70,** 5107–5123.

Piessens, R. and Branders, M. (1983). Modified Clenshaw–Curtis method for the computation of Bessel function integrals. *BIT* **23,** 370–381.

Richards, P. G. (1973). Calculation of body waves for caustics and tunnelling in core phases. *Geophys. J. R. Astron. Soc.* **35,** 243–264.

Sen, M. K. and Frazer, L. N. (1985). Kirchhoff–Helmholtz reflection seismograms in a laterally inhomogeneous multi-layered elastic medium—II: Computations. *Geophys. J. R. Astron. Soc.* **82,** 415–437.

Sen, M. K. and Frazer, L. N. (1987). Synthetic seismograms using multifold path integrals—II. Computations. *Geophys. J. R. Astron. Soc.* **77,** 647–671.

Sen, M. K., Frazer, L. N. and Chapman, N. R. (1988). Analysis of multipath sound propagation in the ocean near 50°N, 128°W. *J. Acoust. Soc. Am.* (in press).

Sinton, J. B. and Frazer, L. N. (1982). A method for the computation of finite frequency body wave synthetic seismograms in laterally varying media. *Geophys. J. R. Astron. Soc.* **71,** 37–55.

Spudich, P. and Ascher, U. (1983). Calculation of complete theoretical seismograms in vertically varying media using collocation methods. *Geophys. J. R. Astron. Soc.* **75,** 101–124.

Spudich, P. and Frazer, L. N. (1984). Use of ray theory to calculate high-frequency radiation from earthquake sources having spatially variable rupture velocity and stress drop. *Bull. Seismol. Soc. Am.* **74,** 2061–2082.

Strick, E. (1970). A predicted pedestal effect for pulse propagation in constant-Q solids. *Geophysics* **35,** 387–402.

Wiggins, R. A. (1974). Body wave amplitude calculations—II. *Geophys. J. R. Astron. Soc.* **46,** 1–10.

Xu, P.-C. and Mal, A. K. (1985). An adaptive integration scheme for irregularly oscillatory functions. *Wave Motion* **7,** 235–243.

IV

MODE METHODS FOR STRATIFIED MEDIA

IV.1

DISPER80: A Subroutine Package for the Calculation of Seismic Normal-Mode Solutions

MASANORI SAITO
Department of Applied Physics, Tokyo Institute of Technology, Japan

1 INTRODUCTION

DISPER80 is a Fortran subroutine package for the calculation of normal modes of non-rotating perfectly elastic transversely isotropic and laterally homogeneous Earth models. It is essentially based on the algorithms described in Takeuchi and Saito (1972). Since then, several algorithmic improvements have been implemented in the package. This chapter describes the algorithms in outline, with special emphasis on the compound-matrix method.

2 TRANSVERSELY ISOTROPIC EARTH MODEL

The simplest form of anisotropy is one for which the stress–strain relations are symmetric about a specific axis, and is called "transverse isotropy". Takeuchi and Saito (1972) have shown that modal solutions for transversely isotropic media can be obtained by the method for isotropic media with minor modifications. In DISPER80 it is assumed that the Earth is transversely isotropic, with the symmetry axis taken as vertical. It is also assumed that the Earth is perfectly elastic, non-rotating and laterally homogeneous.

Elastic properties of a transversely isotropic material are specified by five elastic constants, *A, C, L, N* and *F,* according to Love's notation

SEISMOLOGICAL ALGORITHMS
ISBN 0-12-220770-X

(see Takeuchi and Saito, 1972). In the formulation below we shall use the five elastic constants. In the package, however, two elastic wave velocities and three anisotropy factors are used as independent parameters of elasticity:

$$\left.\begin{aligned} \alpha = \left(\frac{A}{\rho}\right)^{1/2}, \quad \beta = \left(\frac{L}{\rho}\right)^{1/2}, \\ \xi = \frac{N}{L}, \quad \varphi = \frac{C}{A}, \quad \eta = \frac{F}{A-2L} \end{aligned}\right\} \tag{1}$$

where α is the velocity of horizontally polarized P waves, β is the velocity of vertically polarized S waves and ρ is the density. ξ, φ and η are "anisotropy factors", all of which are unity when the material is isotropic.

DISPER80 has a flag (IA) that indicates whether the Earth model is isotropic or transversely isotropic. When the flag is off (IA = 0) the package assumes that the model is isotropic and neglects the anisotropy factors, even when they are given. In other words, one need not specify the anisotropy factors if the model is isotropic (of course dummy arguments are necessary). Another flag (IQ) indicates whether the Q model is given. The dissipation factor of normal modes will be computed when IQ = 1.

3 METHODS OF INTEGRATION AND SPECIFICATION OF THE MODEL

DISPER80 integrates the equations of motion in the form

$$\frac{\mathrm{d}\boldsymbol{y}}{\mathrm{d}r} = \boldsymbol{A}\boldsymbol{y}, \tag{2}$$

where $\boldsymbol{y}$ is a column vector and $\boldsymbol{A}$ is a square coefficient matrix. The Love-wave equations of motion and the Rayleigh-wave equations of motion for flat models may be integrated either by the matrix method or the Runge–Kutta–Gill method. The matrix-method formulae can be found in Takeuchi and Saito (1972). Equations for spherical models are integrated by the conventional Runge–Kutta–Gill method.

Since the conventional Runge–Kutta–Gill method is of fourth order, the relative truncation error per step is given approximately by

$$e = (vh)^5/5!, \tag{3}$$

where v denotes the largest eigenvalue of $\boldsymbol{A}$ and h is the step size in r. The cumulative relative error will be approximated by e times the number of integration steps:

$$E = \frac{(vh)^5}{5!}\frac{H}{h},$$

where H is the integration interval. To get a rough idea of the error, we define the nominal vertical wavelength λ by $2\pi/v$, and we have

$$E = \frac{(2\pi)^5}{120}\left(\frac{h}{\lambda}\right)^4 \frac{H}{\lambda}. \tag{4}$$

Suppose that we wish to integrate the equation for twice the vertical wavelength, $H = 2\lambda$, to six significant figures, $E = 10^{-6}$. We find from the above estimate that the step size h must be less than one hundredth of the vertical wavelength.

Despite its apparent inefficiency, the Runge–Kutta–Gill scheme has been adopted as the main integrator in DISPER80, since it is self-starting, and it requires simpler grid work than the more sophisticated integrators, such as Fehlberg's method; the Runge–Kutta–Gill scheme requires only one auxiliary grid point between the grid points where function values have to be evaluated. Thus the model parameters should be specified, starting from the surface, at depth points equally spaced in pairs. An example of a model specification is given below:

Depth	h_i	α_i	
0	7.5	5.5	...
7.5	7.5	6.0	...
15.0	7.5	6.5	...
22.5	7.5	7.0	...
30.0	0.0	7.5	...
30.0	10.0	8.2	...
40.0	10.0	...	...
50.0			

Note that h_i denotes the grid interval in the model, not the step size in the Runge–Kutta–Gill scheme. Note also that h_i measures the interval from the current grid point to the one below. Thus h_i is zero at a grid point that specifies the upper side of a discontinuity (the fifth point in the example). In this example, eigenfunctions will be computed at depth points 0, 15, 30, 50 km etc., and these values will be used to compute energy integrals and partial derivatives. Simpsom's formula is used to compute the energy integrals, and this formula requires that integrands be available at three successively equally spaced grid points. This is the reason why the first 30 km is divided into four 7.5 km thick layers in the above example.

The matrix method can be used to integrate Love-wave equations in a transversely isotropic layered model, and Rayleigh-wave equations in an *isotropic* layered model. "Layered" means that the model is represented by a stack of homogeneous layers. Thus the model parameters are specified for each homogeneous layer—not at each grid point as in the case of the Runge–Kutta–Gill method.

3.1 Units of parameters

In DISPER80 it is implicitly assumed that the unit of length is km, that of velocity is km s^{-1} and the unit of density is g cm^{-3}, in accordance with the current convention. Therefore the unit of time is s, that of angular frequency is rad s^{-1} and the unit of stress is 10^{10} c.g.s. However, a user may choose different units for length, velocity and density; the units of time, stress and other quantities are derived from them.

One must also specify the units of gravity and of the universal constant of gravitation G. The standard unit for gravity is km s^{-2} and the corresponding value of G is $6.673 \times 10^{-8}\ cm^3\ g^{-1}\ s^{-2}$.

4 TORSIONAL OSCILLATIONS

4.1 Definition of motion–stress vector

We define the radial functions, or, in the terminology of Aki and Richards (1980), the motion–stress vector $\boldsymbol{y}(r) = [y_1(r) \quad y_2(r)]^{\mathrm{T}}$ for torsional modes by

$$\left.\begin{aligned} \boldsymbol{u} &= y_1(r)\boldsymbol{\nabla} \times [\hat{\boldsymbol{r}}Y_l(\theta, \phi)], \\ \boldsymbol{\sigma}_r &= y_2(r)\boldsymbol{\nabla} \times [\hat{\boldsymbol{r}}Y_l(\theta, \phi)], \end{aligned}\right\} \tag{5}$$

where (r, θ, ϕ) are spherical polar coordinates, $\boldsymbol{u}$ is the displacement vector, $\boldsymbol{\sigma}_r$ is the stress vector on a spherical surface $r = \text{const}$, $Y_l(\theta, \phi)$ is a surface spherical harmonic of degree l, and $\hat{\boldsymbol{r}}$ is the unit vector in the radial direction. Since the eigenvalue of a laterally homogeneous non-rotating Earth model is degenerate with respect to the azimuthal wavenumber, it is not necessary to specify the azimuthal order number of $Y_l(\theta, \phi)$. The time dependence $e^{-i\omega t}$ is assumed and will not be specified. Note that there is a factor of r difference between the present definition and the conventional one. Thus $W(r)$ and $T(r)$ of Aki and Richards

(1980, p. 431) are related to the present $\boldsymbol{y}$ by

$$W(r) = \frac{y_1(r)}{r}, \qquad T(r) = \frac{y_2(r)}{r}.$$

We can write the equations of motion in terms of the motion–stress vector in the familiar form

$$\frac{\mathrm{d}}{\mathrm{d}r}\boldsymbol{y}(r) = \boldsymbol{A}(r)\boldsymbol{y}(r), \tag{6}$$

where

$$\boldsymbol{A}(r) = \begin{bmatrix} 2/r & 1/L \\ -\rho\omega^2 + \dfrac{(l-1)(l+2)N}{r} & -2/r \end{bmatrix}. \tag{7}$$

The factor r has been introduced to effect the antisymmetry of the coefficients, $a_{22} = -a_{11}$.

In DISPER80 the equation is integrated upwards from a specified depth by using the conventional Runge–Kutta–Gill scheme up to the free surface to see if the boundary conditions are satisfied (shooting method). The initial values of $\boldsymbol{y}$ are given by

$$y_1(b) = 1, \qquad y_2(b) = 0,$$

where b is the radius of the outer core. For higher frequencies we may wish to start the integration from somewhere inside the mantle. We then compute the initial values of $\boldsymbol{y}$ by assuming that the layer below the starting depths is isotropic and homogeneous. See Takeuchi and Saito for the formulae. A similar approach will be followed in the calculation of spheroidal, Rayleigh and Love modes to be presented in later sections.

The boundary condition at the free surface $r = a$, the vanishing of shear stress, requires

$$y_2(a) = 0 \tag{8}$$

Thus $y_2(a)$ may be used as the characteristic function of the torsional oscillation. We are to find zeros of $y_2(a)$ for a given angular order l. However, since the magnitude of $y_2(a)$ is strongly dependent on the depth from where we start the integration, we adopt

$$\Delta_{\mathrm{T}}(\omega, l) = \frac{y_2(a)}{y_1(a)} \tag{9}$$

as the characteristic function of the torsional mode. This introduces an artificial singularity at $y_1(a) = 0$, but, since zeros of $y_2(a)$ are well separated from those of $y_1(a)$, the denominator will not cause difficulties in the root-searching procedure. Another advantage of the above definition is that, since the characteristic function depends only on the ratio of ys, we can scale the ys in the course of integration to avoid them blowing up, without affecting the value of the characteristic function.

4.2 Energy integral relations

Once an eigenvalue is found, the corresponding average Lagrangian is computed by

$$\langle L \rangle = \omega^2 I_1 - I_2 = -[y_1 y_2]_a, \tag{10}$$

where

$$\begin{aligned} I_1 &= \int_0^a \rho y_1^2 \, \mathrm{d}r, \\ I_2 &= \int_0^a \left[\frac{1}{L} y_2^2 + \frac{(l-1)(l+2)N}{r} y_1^2 \right] \mathrm{d}r. \end{aligned} \tag{11}$$

The right-hand side of (10), a boundary value, is of course zero when ω is an eigenvalue.

In the earlier version of the package the group velocity U was computed via the Lagrangian energy integrals:

$$U = \frac{a^2 I_3}{c I_1}, \tag{12}$$

where

$$c = \frac{\omega a}{l + \frac{1}{2}} \tag{13}$$

is the phase velocity and

$$I_3 = \int_0^a \frac{N}{r} y_1^2 \, \mathrm{d}r. \tag{14}$$

In DISPER80, on the other hand, the group velocity is also computed via partial derivatives of ys, as will be described below.

4.3 Derivatives of ys with respect to l and ω

The group velocity of a surface wave corresponding to a free oscillation of degree l is defined by

$$U = -\frac{\partial y_2(a)}{\partial l} \bigg/ \frac{\partial y_2(a)}{\partial(\omega a)}, \tag{15}$$

where the derivatives are to be taken along a dispersion curve, $y_2(a)=0$. The formula shows that, instead of evaluating I_1 and I_3, we can compute the group velocity by evaluating partial derivatives of $y_2(a)$ with respect to l and ω.

First we differentiate the fundamental equation (6) with respect to l and ω, to get

$$\left.\begin{aligned}\frac{\mathrm{d}}{\mathrm{d}r}\left(l\frac{\partial \mathbf{y}(r)}{\partial l}\right) &= \boldsymbol{A}(r)\left(l\frac{\partial \mathbf{y}(r)}{\partial l}\right)+l\frac{\partial \boldsymbol{A}(r)}{\partial l}\mathbf{y}(r),\\ \frac{\mathrm{d}}{\mathrm{d}r}\left(\omega\frac{\partial \mathbf{y}(r)}{\partial \omega}\right) &= \boldsymbol{A}(r)\left(\omega\frac{\partial \mathbf{y}(r)}{\partial \omega}\right)+\omega\frac{\partial \boldsymbol{A}(r)}{\partial \omega}\mathbf{y}(r).\end{aligned}\right\} \tag{16}$$

Equations (6) and (16) constitute a set of first order differential equations with respect to $\mathbf{y}(r)$, $l\,\partial\mathbf{y}(r)/\partial l$ and $\omega\,\partial\mathbf{y}(r)/\partial\omega$, a sixth-order system in all. The partial derivatives of $\boldsymbol{A}$ can easily be computed using (7). Here we have chosen $l\,\partial\mathbf{y}/\partial l$ and $\omega\,\partial\mathbf{y}/\partial\omega$ for dependent variables instead of the partial derivative alone so that the newly introduced dependent variables will have the same dimension as the corresponding original ones.

The surface values of the partial derivatives, which are necessary for computing the group velocity (15), can be obtained by integrating the three sets of equations simultaneously. In order to integrate (16), we need initial values of the partial derivatives. When starting from the core–mantle boundary, they are trivial, because $\mathbf{y}(b)$ does not depend on l or ω. When starting the integration in the mantle, we evaluate the partial derivatives at the starting point by a forward finite-difference approximation

$$\frac{\mathrm{d}f(x)}{\mathrm{d}x}=\frac{f(x+\mathrm{d}x)-f(x)}{\mathrm{d}x},$$

where $f(x)$ is the analytical solution for an isotropic and homogeneous sphere, and x stands for a parameter l or ω. In DISPER80 $\mathrm{d}x$ is tentatively taken as one tenth of x, $\mathrm{d}x=\frac{1}{10}x$.

The energy integral I_1 can also be computed from the partial derivatives by differentiating (10) at constant l, we have

$$2\omega I_1=-\left[y_1\frac{\partial y_2}{\partial\omega}-y_2\frac{\partial y_1}{\partial\omega}\right]_a.$$

Along $y_2(a)=0$, we have

$$2\omega^2 I_1=-\left[y_1\omega\frac{\partial y_2}{\partial\omega}\right]_a. \tag{17}$$

To get I_1 consistent with the normalization $y_1(a)=1$, one must divide the right-hand side of (17) by $[y_1(a)]^2$.

4.4 Description of torsional programs

The core of the torsional part of DISPER80 consists of two subroutines: TORRKG and TOREFR. TORRKG integrates the basic equations of motion (6) and (16), while TOREFR computes eigenfunctions, energy integrals and partial derivatives of an eigenvalue with respect to the elastic parameters. To find an eigenvalue, one need only call TORDSP, which in turn calls TORRKG and does the necessary interpolations. The output of TORDSP consists of an eigenvalue, the corresponding group velocity according to (15), and the kinetic part of the average Lagrangian, $2\omega^2 I_1$ according to (17). To speed up the computation, (16) is integrated only once after an eigenvalue has been located within a prespecified error bound.

The output of TOREFR consists of eigenfunctions $y_1(r)$ and $y_2(r)$, the energy integral $2\omega^2 I_1$, the group velocity computed by energy integrals, (12), and partial derivatives of eigenfunctions. The eigenfunctions are normalized to

$$y_1(a) = 1 \tag{18}$$

and the energy integrals are computed following this normalization.

TOREFR has a flag (IQ) that indicates whether the user has supplied the program with a Q structure. When IQ is on ($\text{IQ} = 1$) the program computes the temporal dissipation factor of the mode.

For diagnostic purposes

$$e = 1 - \frac{I_2}{\omega^2 I_1} \tag{19}$$

is also computed. If an eigenvalue and the eigenfunctions are computed correctly, $e = 0$ (see (10)). However, e is not a good indicator of accuracy. Since the Lagrangian is stationary with respect to variation of the ys, e is apt to be zero even if the eigenvalue and eigenfunctions are not accurately calculated. Another indicator is the difference between the group velocity and kinetic energy computed by TORRKG and those computed by TOREFR.

5 SPHEROIDAL OSCILLATIONS OF A NON-GRAVITATING EARTH

5.1 Definition of motion–stress vector

Solid layer

The components of the motion–stress vector for the spheroidal oscillations in a solid layer of a non-gravitating Earth are defined by

$$\left.\begin{aligned} \boldsymbol{u} &= \frac{1}{r} y_1(r)[\hat{\boldsymbol{r}} Y_l(\theta, \phi)] + y_3(r) \boldsymbol{\nabla}\left[\frac{1}{[l(l+1)]^{1/2}} Y_l(\theta, \phi)\right], \\ \boldsymbol{\sigma}_r &= \frac{1}{r} y_2(r)[\hat{\boldsymbol{r}} Y_l(\theta, \phi)] + y_4(r) \boldsymbol{\nabla}\left[\frac{1}{[l(l+1)]^{1/2}} Y_l(\theta, \phi)\right]. \end{aligned}\right\} \tag{20}$$

As in (5), the present ys differ from the conventional ys by a factor of r. A factor $[l(l+1)]^{1/2}$ has also been introduced in accordance with recent conventions (see e.g. Aki and Richards, 1980).

The equations of motion in terms of the four-vector $\boldsymbol{y}$ are again given by (6), with the coefficient matrix $\boldsymbol{A} = [a_{ij}]$:

$$\left.\begin{aligned} & a_{11} = \frac{1}{r}\left(1 - \frac{2F}{C}\right), \quad a_{12} = \frac{1}{C}, \quad a_{13} = \frac{[l(l+1)]^{1/2}}{r}\frac{F}{C}, \\ & a_{21} = -\rho\omega^2 + \frac{4}{r^2}\left(A - \frac{F^2}{C} - N\right), \quad a_{22} = -a_{11}, \\ & a_{23} = -\frac{2[l(l+1)]^{1/2}}{r^2}\left(A - \frac{F^2}{C} - N\right), \quad a_{24} = \frac{[l(l+1)]^{1/2}}{r}, \\ & a_{31} = -a_{24}, \quad a_{33} = \frac{2}{r}, \quad a_{34} = \frac{1}{L}, \\ & a_{41} = a_{23}, \quad a_{42} = -a_{13}, \\ & a_{43} = -\rho\omega^2 + \frac{[l(l+1)]^{1/2}}{r^2}\left(A - \frac{F^2}{C}\right) - \frac{2N}{r^2}, \quad a_{44} = -a_{33}; \end{aligned}\right\} \tag{21}$$

all other elements are zero. Note that the coefficient matrix now has a number of symmetries.

Initially we shall assume that the Earth has no liquid core. The fourth-order equation (6) with coefficient matrix (21) has two independent solutions $\boldsymbol{y}_1(r)$ and $\boldsymbol{y}_2(r)$ that are regular at the centre of the Earth, $r = 0$. Thus any solution regular at the centre is a linear combination of these two solutions. Since the components of stress on the outermost surface of the Earth must vanish, the dispersion equation for the spheroidal oscillation is given by

$$\begin{aligned} y_2(a) &= Ay_{21}(a) + By_{22}(a) = 0, \\ y_4(a) &= Ay_{41}(a) + By_{42}(a) = 0, \end{aligned}$$

or

$$\begin{vmatrix} y_{21}(a) & y_{22}(a) \\ y_{41}(a) & y_{42}(a) \end{vmatrix} = 0, \tag{22}$$

where $y_{ij}(r)$ denotes the ith component of the jth solution; that is, $y_i(r) = [y_{i1}(r) \quad y_{i2}(r)]^{\mathrm{T}}$, and A and B are constants of integration.

Liquid layer

In a liquid layer ($L = 0$), the third equation of (6) becomes an identity, and from the fourth equation we get

$$y_3(r) = -\frac{[l(l+1)]^{1/2}}{\rho\omega^2 r} y_2(r), \tag{23}$$

where we have assumed isotropy ($A = C = F$). Using this relation, we can eliminate y_3 from the first equation of (6), and we get a second-order differential equation with the coefficient matrix

$$\mathbf{A}(r) = \begin{bmatrix} -\dfrac{1}{r} & \dfrac{1}{C} - \dfrac{l(l+1)}{\rho\omega^2 r^2} \\ -\rho\omega^2 & 1/r \end{bmatrix}. \tag{24}$$

Since the equation is of second order, there are two linearly independent solutions. But the condition of regularity at the centre of the Earth precludes one solution that is singular there, and we retain only one regular solution.

5.2 Compound-matrix method

The dispersion relation for the spheroidal oscillations of the Earth is given by a second-order determinant equation (22) on the solid free surface. It is well known that solving for the elementary solutions y_{ij} and forming the determinant in (22) is a numerically unstable method. Hence direct integration of the dispersion function has been proposed in various contexts (Gilbert and Backus, 1966; Dunkin, 1965; Abo-Zena, 1979). The methods are essentially equivalent to the method advocated by Gilbert and Backus (1966) and developed in the earlier version of the package (Takeuchi and Saito, 1972).

We define second-order minors obtained from two independent solutions of (6) with the coefficient matrix (21) by

$$Y_{ij}(r) = \begin{vmatrix} y_{i1}(r) & y_{i2}(r) \\ y_{j1}(r) & y_{j2}(r) \end{vmatrix}, \tag{25}$$

where $(ij) = (12)$, (13), (14), (23), (24) and (34). Clearly, $Y_{24}(a)$ is the characteristic function for the spheroidal oscillations (see (22)). By differentiating the above equation and using (21), we find that the $Y_{ij}(r)$

also satisfy a set of linear equations similar to (6):

$$\frac{d}{dr} Y_{ij}(r) = \sum_k [a_{ik}(r) Y_{kj}(r) + a_{jk}(r) Y_{ik}(r)]. \tag{26}$$

It has been shown that the first derivatives of the six Ys are not all independent (see Takeuchi and Saito, 1972). Since the ys are regular at $r = 0$, we have

$$Y_{12}(r) + Y_{34}(r) = 0, \tag{27}$$

which reduces (26) to a fifth-order system.

The fifth-order equation (26) may be integrated through solid layers. The initial value of the compounds $Y_{ij}(r)$ are readily computed from the corresponding uniform-sphere approximation for the elementary solutions $y_{ij}(r)$ (Takeuchi and Saito, 1972). There remains the problem of continuing the solution across a solid–liquid interface, since the compound-matrix scheme breaks down in a liquid layer; this will be discussed later.

There are several interesting relations between the elements of the compound matrix $\boldsymbol{Y}$ and the solution $\boldsymbol{y}$ of the original equation. Since any $\boldsymbol{y}$ may be written in terms of a linear combination of elementary solutions $\boldsymbol{y}_i$, we have

$$\left.\begin{aligned} y_i &= Ay_{i1} + By_{i2}, \\ y_j &= Ay_{j1} + By_{j2}, \end{aligned}\right\} \tag{28}$$

where A and B are constants of integration independent of r. Solving A and B from the above equations, and substituting back into similar equations, we obtain

$$\begin{bmatrix} y_k \\ y_l \end{bmatrix} = \frac{1}{Y_{ij}} \begin{bmatrix} Y_{kj} & Y_{ik} \\ Y_{lj} & Y_{il} \end{bmatrix} \begin{bmatrix} y_i \\ y_j \end{bmatrix}. \tag{29}$$

We therefore have the identity

$$\begin{bmatrix} Y_{il} & Y_{ki} \\ Y_{jl} & Y_{kj} \end{bmatrix} \begin{bmatrix} Y_{kj} & Y_{ik} \\ Y_{lj} & Y_{il} \end{bmatrix} = Y_{ij} Y_{kl} \boldsymbol{I}, \tag{30}$$

where $\boldsymbol{I}$ is the 2×2 identity matrix.

The relation (29) may be used in a variety of situations. Suppose that we impose the constraint $y_4(r) = 0$. This determines the ratio A/B, thus determining the relative magnitudes of two elementary solutions. From (29), we have

$$\left.\begin{aligned} y_1(r) &= \frac{Y_{14}(r)}{Y_{24}(r)} y_2(r), \\ y_3(r) &= \frac{Y_{34}(r)}{Y_{24}(r)} y_2(r) = \frac{Y_{34}(r)}{Y_{14}(r)} y_1(r). \end{aligned}\right\} \tag{31}$$

At a solid free surface $y_2(a) = 0 = y_4(a)$, and since the second condition has already been satisfied in (31), a dispersion relation for the spheroidal mode is

$$\Delta_{NG}(\omega, l) = \frac{Y_{24}(a)}{Y_{14}(a)}, \tag{32}$$

which is independent of the magnitude of the Ys. The second equation is used to define the initial value of the eigenfunction at the free surface.

5.3 Continuation of solutions across a solid–liquid interface

Since the compound $\boldsymbol{Y}$ is not defined in a liquid layer, we must switch from the $\boldsymbol{Y}$ to the $\boldsymbol{y}$ solution across a solid-to-liquid interface, and vice versa. First, consider a solid-to-liquid interface, where we switch from the $\boldsymbol{Y}$ solution to the $\boldsymbol{y}$ solution. Since the shear component of stress must vanish at the interface, and y_1 and y_2 are continuous at the interface, we have from (31)

$$y_1^{\mathrm{L}}(r) = \frac{Y_{14}^{\mathrm{S}}(r)}{Y_{24}^{\mathrm{S}}(r)} y_2^{\mathrm{L}}(r) \tag{33}$$

where the superscript S denotes the solution on the solid side of the interface and the superscript L indicates the solution on the liquid side of the interface. Since a common factor is immaterial for the evaluation of the dispersion function (see (32)), we may take

$$y_1^{\mathrm{L}}(r) = Y_{14}^{\mathrm{S}}(r), \qquad y_2^{\mathrm{L}}(r) = Y_{24}^{\mathrm{S}}(r). \tag{34}$$

On the other hand, when we are integrating from a liquid layer to a solid layer, we only have y_1 and y_2 on the liquid side of the interface. But, since the shear component of displacement may be non-zero at the interface, we have two independent solutions at the solid side:

$$y_{11}^{\mathrm{S}} = y_1^{\mathrm{L}}, \qquad y_{21}^{\mathrm{S}} = y_2^{\mathrm{L}}, \qquad y_{31}^{\mathrm{S}} = 0, \qquad y_{41}^{\mathrm{S}} = 0;$$

$$y_{12}^{\mathrm{S}} = 0, \qquad y_{22}^{\mathrm{S}} = 0, \qquad y_{32}^{\mathrm{S}} = 1, \qquad y_{42}^{\mathrm{S}} = 0.$$

From this set of solutions, we have components of minors in the solid layer:

$$Y_{13}^{\mathrm{S}} = y_1^{\mathrm{L}}, \qquad Y_{23}^{\mathrm{S}} = y_2^{\mathrm{L}}; \qquad \text{otherwise } Y_{ij}^{\mathrm{S}} = 0. \tag{35}$$

5.4 Computation of eigenfunctions

Roots of the characteristic equation can be located by the algorithm described above. But it remains to find the corresponding eigenfunctions. Ng and Reid (1979a, b) proposed a method that essentially depends on (29). We can eliminate two of the ys from (21) using (29). Thus the equation becomes of second order, with known coefficients in terms of the Ys, and we can integrate the equations upwards or downwards. However, the stability of the resulting differential equation depends on the choice of the dependent variables, and sometimes it is necessary to switch from one set of dependent variables to another to avoid singularities of coefficients. Hence we have chosen the method proposed by Harvey (1977).

We first integrate the compound matrices Y upwards. This is necessary because we have to integrate them upwards to find an eigenvalue. Once an eigenvalue is found, we have available the corresponding Ys at every other depth point. To find the corresponding eigenfunction, we integrate the original equations (6) and (21) from the free surface downwards, starting from the initial value given by (31). Now, the trick is that we do not integrate the equations all the way, but after every integration step we replace two of the four variables y_i with those computed by using (29):

$$\begin{bmatrix} y_2 \\ y_4 \end{bmatrix} = \frac{1}{Y_{13}} \begin{bmatrix} Y_{23} & Y_{12} \\ Y_{12} & Y_{14} \end{bmatrix} \begin{bmatrix} y_1 \\ y_3 \end{bmatrix}. \tag{36}$$

The point here is that, since the original equation is not stable when integrated downwards, we make corrections at every step using Y_{ij} obtained by upward integration. This scheme is found to be very effective and stable, and our experiences show that even with single-precision arithmetic we can calculate eigenfunctions very accurately.

The continuation scheme across a solid-to-liquid interface is obvious since only y_1 and y_2 are needed to resume the integration through a liquid layer. Continuation across a liquid-to-solid boundary is also simple. Remember that (31) was derived under the constraint $y_4 = 0$, which must be met at the interface. Thus

$$y_3^{\mathrm{S}} = -\frac{Y_{12}^{\mathrm{S}}}{Y_{14}^{\mathrm{S}}} y_1^{\mathrm{S}}, \tag{37}$$

with $y_1^{\mathrm{S}} = y_1^{\mathrm{L}}$, $y_2^{\mathrm{S}} = y_2^{\mathrm{L}}$ and $y_4^{\mathrm{S}} = 0$, completely determines the eigenfunction at the top of the solid layer.

5.5 Energy integral relations

Once an eigenfunction has been found, the average Lagrangian and the group velocity are computed from

$$\left.\begin{aligned}
\langle L\rangle &= \omega^2 I_1 - I_2 = -[y_1 y_2 + y_2 y_4]_a,\\
I_1 &= \int_0^a \rho(y_1^2 + y_3^2)\,\mathrm{d}r,\\
I_2 &= \int_0^a \left\{\frac{1}{C}y_2^2 + \frac{1}{L}y_4^2 + \frac{1}{r^2}\left(A - \frac{F^2}{C} - N\right)[2y_1 - [l(l+1)]^{1/2}y_3]^2\right.\\
&\quad \left. + \frac{(l-1)(l+2)N}{r^2}y_3^2\right\}\mathrm{d}r\\
U &= a^2 I_3/cI_1,\\
I_3 &= \int_0^a \left\{\frac{1}{r^2}\left(A - \frac{F^2}{C}\right)y_3^2 + \frac{1}{r[l(l+1)]^{1/2}}\left[y_1 y_4 - \frac{F}{C}y_2 y_3\right.\right.\\
&\quad \left.\left. - \frac{2}{r}\left(A - \frac{F^2}{C} - N\right)y_1 y_3\right]\right\}\mathrm{d}r.
\end{aligned}\right\} \tag{38}$$

Also, the first-order rotational splitting factor is computed from

$$\beta = \frac{I_5}{I_1}, \qquad I_5 = \int_0^a \rho y_3\{2[l(l+1)]^{1/2}y_1 + y_3\}\,\mathrm{d}r. \tag{39}$$

Note that the first-order perturbation $\delta\omega$ in the eigenfunction is given by

$$\frac{\delta\omega}{\omega} = m\frac{\beta}{l(l+1)}\frac{\Omega}{\omega}, \tag{40}$$

where Ω is the angular velocity of the Earth's diurnal rotation (Dahlen, 1976).

5.6 Partial derivatives of the compound matrix

As in the case of torsional oscillations, we can compute the group velocity and the average kinetic energy without recourse to the definite integrals in (38). We recall that our fundamental equation (28) is written in matrix form as

$$\frac{\mathrm{d}}{\mathrm{d}r}\boldsymbol{Y}(r) = \boldsymbol{B}(r)\boldsymbol{Y}(r) \tag{41}$$

where $\boldsymbol{Y}$ is a five-vector made up of the minors Y_{ij}. By differentiating the equation with respect to l and ω, we obtain equations similar to (16). Therefore, by integrating these equations, we get partial derivatives of $\boldsymbol{Y}$

with respect to l and ω. From these, we can compute

$$U = -\frac{\partial Y_{24}(a)}{\partial l} \bigg/ \frac{\partial Y_{24}(a)}{\partial(\omega a)}. \tag{42}$$

To derive a formula for the kinetic energy I_1, we differentiate (38) with respect to ω along $y_4(a) = 0$, and we obtain

$$2\omega I_1 = -\left[y_1 \frac{\partial y_2}{\partial \omega} - y_2 \frac{\partial y_1}{\partial \omega} \right]_a.$$

Along $y_4(a) = 0$ we have, from (29), $y_2(a) = [Y_{24}(a)/Y_{14}(a)]y_1(a)$. Therefore

$$2\omega^2 I_1 = -\left[y_1^2 \frac{\omega}{Y_{14}} \frac{\partial Y_{24}}{\partial \omega} \right]_a. \tag{43}$$

The factor $y_1^2(a)$ is immaterial because we usually normalize the eigenfunction $y_1(a) = 1$.

5.7 Description of non-gravitating spheroidal programs

The structure of the program package for the non-gravitating spheroidal modes is identical with that for the torsional mode, except for trivial differences (for instance, P-wave velocity must be specified). NOGDSP controls NOGRKG, and its output consists of an eigenvalue, group velocity and kinetic energy. To compute partial derivatives of the eigenvalue, we call NOGEFR, the output of which consists of the group velocity, kinetic energy, splitting factor as defined by (39) and temporal dissipation factor when specified.

6 SPHEROIDAL OSCILLATIONS OF A SELF-GRAVITATING EARTH

6.1 Definition of motion–stress vector

Solid layer

The definitions of $y_1, \ldots, y_4$ are identical with those for the non-gravitating spheroidal modes (see (20)). We further define y_5 and y_6 by

$$\left.\begin{aligned} \psi &= \frac{1}{r} y_5(r) Y_l(\theta, \phi), \\ y_6 &= (4\pi G)^{-1}\left(\frac{\mathrm{d}}{\mathrm{d}r} y_5 - 4\pi G \rho y_1 + \frac{l}{r} y_5\right), \end{aligned}\right\} \tag{44}$$

where ψ is the perturbation of the gravitational potential and G is the gravitational constant. y_6 is defined such that it vanishes at the free surface, $y_6(a)=0$. The factor of $(4\pi G)^{-1}$ in the definition of y_6 is introduced in order to improve the symmetry in the coefficients. The elements of the coefficient matrix $\boldsymbol{A}$ are as follows (all other elements are zero):

$$\left.\begin{aligned}
&a_{11}=\frac{1}{r}\left(1-\frac{2F}{C}\right), \qquad a_{12}=\frac{1}{C}, \qquad a_{13}=\frac{[l(l+1)]^{1/2}}{r}\frac{F}{C}\\
&a_{21}=-\rho\omega^2+\frac{4}{r^2}\left(A-\frac{F^2}{C}-N-\rho g r\right), \qquad a_{22}=-a_{11}\\
&a_{23}=-\frac{[l(l+1)]^{1/2}}{r^2}\left[2\left(A-\frac{F^2}{C}-N\right)-\rho g r\right], \qquad a_{24}=\frac{[l(l+1)]^{1/2}}{r},\\
&a_{25}=\frac{(l+1)\rho}{r}, \qquad a_{26}=-4\pi G\rho,\\
&a_{31}=-a_{24}, \qquad a_{33}=\frac{2}{r}, \qquad a_{34}=\frac{1}{L},\\
&a_{41}=a_{23}, \qquad a_{42}=-a_{13},\\
&a_{43}=-\rho\omega^2+\frac{1}{r^2}\left[l(l+1)\left(A-\frac{F^2}{C}\right)-2N\right], \qquad a_{44}=-a_{33}\\
&a_{45}=-\frac{[l(l+1)]^{1/2}\rho}{r},\\
&a_{51}=-a_{26}, \qquad a_{55}=-\frac{l}{r}, \qquad a_{56}=4\pi G,\\
&a_{61}=a_{25}, \qquad a_{63}=a_{45}, \qquad a_{66}=-a_{55}.
\end{aligned}\right\} \tag{45}$$

Here $g(r)$ denotes gravity. The system is of sixth order and it has three linearly independent solutions that are regular at $r=0$. As stated before, y_6 must vanish at the free surface. Thus the free-surface boundary conditions are

$$\begin{aligned}
y_2(a)&=Ay_{21}(a)+By_{22}(a)+Cy_{23}(a)=0,\\
y_4(a)&=Ay_{41}(a)+By_{42}(a)+Cy_{43}(a)=0,\\
y_6(a)&=Ay_{61}(a)+By_{62}(a)+Cy_{63}(a)=0,
\end{aligned}$$

or

$$\begin{vmatrix} y_{21}(a) & y_{22}(a) & y_{23}(a)\\ y_{41}(a) & y_{42}(a) & y_{43}(a)\\ y_{61}(a) & y_{62}(a) & y_{63}(a)\end{vmatrix}=0. \tag{46}$$

Liquid layer

In an isotropic liquid layer ($L = N = 0,\ A = C = F$) the third equation becomes an identity and the fourth equation of (6) with coefficient matrix (45) can be solved for y_3:

$$y_3 = \frac{[l(l+1)]^{1/2}}{\omega^2 r}\left(g y_1 - \frac{1}{\rho} y_2 - y_5\right). \tag{47}$$

When this relation is substituted into the remaining equation, we get a fourth-order equation whose coefficients are

$$\left.\begin{aligned}
&a_{11} = -\frac{1}{r} + \frac{l(l+1)g}{\omega^2 r^2}, \qquad a_{12} = \frac{1}{C} - \frac{l(l+1)}{\rho\omega^2 r^2}, \qquad a_{15} = -\frac{l(l+1)}{\omega^2 r^2},\\
&a_{21} = -\rho\omega^2 - \frac{4\rho g}{r} + \frac{l(l+1)\rho g^2}{\omega^2 r^2}, \qquad a_{22} = -a_{11},\\
&a_{25} = \frac{(l+1)\rho}{r} - \frac{l(l+1)\rho g}{\omega^2 r^2}, \qquad a_{26} = -4\pi G\rho,\\
&a_{51} = -a_{26}, \qquad a_{55} = -\frac{l}{r}, \qquad a_{56} = 4\pi G,\\
&a_{61} = a_{25}, \qquad a_{62} = -a_{15}, \qquad a_{65} = \frac{l(l+1)\rho}{\omega^2 r^2}, \qquad a_{66} = -a_{55}.
\end{aligned}\right\} \tag{48}$$

6.2 Compound matrix

Solid layer

Let us define third-order minors made of three six-vectors $\mathbf{y}_i = [y_{i1} \quad y_{i2} \quad y_{i3}]^{\mathrm{T}}$ by

$$Y_{ijk} = \begin{vmatrix} y_{i1} & y_{i2} & y_{i3} \\ y_{j1} & y_{j2} & y_{j3} \\ y_{k1} & y_{k2} & y_{k3} \end{vmatrix}. \tag{49}$$

In general, there are ${}_6C_3 = 20$ linearly independent minors, but only 14 minors are independent for spheroidal modes. It has been shown that any two linearly independent regular solutions satisfy the following identity (Takeuchi and Saito, 1972):

$$(y_{11}y_{21} - y_{12}y_{21}) + (y_{31}y_{42} - y_{32}y_{41}) + (y_{51}y_{62} - y_{52}y_{61}) = 0 \tag{50}$$

Note that each term in parentheses is a second-order minor in Y_{i12}, Y_{i34},

Y_{i56}. Therefore we get

$$Y_{i12} + Y_{i34} + Y_{i56} = 0, \tag{51}$$

or, explicitly,

$$\left.\begin{aligned} Y_{134} + Y_{156} &= 0, \\ Y_{234} + Y_{256} &= 0, \\ Y_{123} + Y_{356} &= 0, \\ Y_{124} + Y_{456} &= 0, \\ Y_{125} + Y_{345} &= 0, \\ Y_{126} + Y_{346} &= 0. \end{aligned}\right\} \tag{52}$$

By a direct calculation, it can be shown that the minors satisfy the linear differential equation

$$\frac{\mathrm{d}}{\mathrm{d}r} Y_{ijk} = \sum_{l} (a_{il} Y_{ljk} + a_{jl} Y_{ilk} + a_{kl} Y_{ijl}). \tag{53}$$

For the users' convenience, 14 independent equations are listed below:

$$\frac{\mathrm{d}}{\mathrm{d}r} Y_{123} = a_{33} Y_{123} + a_{34} Y_{124} - a_{24} Y_{134} - a_{25} Y_{135} - a_{26} Y_{136},$$

$$\frac{\mathrm{d}}{\mathrm{d}r} Y_{124} = a_{43} Y_{123} - a_{33} Y_{124} + a_{45} Y_{125} + a_{23} Y_{134} - a_{25} Y_{145} - a_{26} Y_{146} - a_{13} Y_{234},$$

$$\frac{\mathrm{d}}{\mathrm{d}r} Y_{125} = a_{55} Y_{125} + a_{56} Y_{126} + a_{26} Y_{134} + a_{23} Y_{135} + a_{24} Y_{145} - a_{13} Y_{235},$$

$$\frac{\mathrm{d}}{\mathrm{d}r} Y_{126} = a_{45} Y_{123} - a_{55} Y_{126} - a_{25} Y_{134} + a_{23} Y_{136} + a_{24} Y_{146} - a_{13} Y_{236},$$

$$\frac{\mathrm{d}}{\mathrm{d}r} Y_{134} = a_{13} Y_{123} + a_{11} Y_{134} + a_{45} Y_{135} + a_{12} Y_{234},$$

$$\frac{\mathrm{d}}{\mathrm{d}r} Y_{135} = (a_{11} + a_{33} + a_{55}) Y_{135} + a_{56} Y_{136} + a_{34} Y_{145} + a_{12} Y_{235},$$

$$\frac{\mathrm{d}}{\mathrm{d}r} Y_{136} = (a_{11} + a_{33} - a_{55}) Y_{136} + a_{34} Y_{146} + a_{12} Y_{236},$$

$$\frac{d}{dr}Y_{145} = -2a_{13}Y_{125} + a_{43}Y_{135} + (a_{11} - a_{33} + a_{55})Y_{145}$$

$$+ a_{56}Y_{146} + a_{12}Y_{245},$$

$$\frac{d}{dr}Y_{146} = -2a_{13}Y_{126} - 2a_{45}Y_{134} + a_{43}Y_{136}$$

$$+ (a_{11} - a_{33} - a_{55})Y_{146} + a_{12}Y_{246},$$

$$\frac{d}{dr}Y_{234} = a_{23}Y_{123} + a_{24}Y_{124} - a_{25}Y_{125} - a_{26}Y_{126}$$

$$+ a_{21}Y_{134} - a_{11}Y_{234} + a_{45}Y_{235},$$

$$\frac{d}{dr}Y_{235} = -2a_{26}Y_{123} + 2a_{24}Y_{125} + a_{21}Y_{125} + a_{21}Y_{135}$$

$$+ (-a_{11} + a_{33} + a_{55})Y_{235} + a_{56}Y_{236} + a_{34}Y_{245},$$

$$\frac{d}{dr}Y_{236} = 2a_{25}Y_{123} + 2a_{24}Y_{126} + a_{21}Y_{136}$$

$$+ (-a_{11} + a_{33} - a_{55})Y_{236} + a_{34}Y_{246},$$

$$\frac{d}{dr}Y_{245} = -2a_{26}Y_{124} - 2a_{23}Y_{125} + a_{21}Y_{145} + a_{43}Y_{235}$$

$$+ (-a_{11} - a_{33} + a_{55})Y_{245} + a_{56}Y_{246},$$

$$\frac{d}{dr}Y_{246} = 2a_{25}Y_{124} - 2a_{23}Y_{126} + a_{21}Y_{146} - 2a_{45}Y_{234}$$

$$- (a_{11} + a_{33} + a_{55})Y_{246}.$$

Here six compound-matrix elements have been eliminated using (52).

The relations between the compound elements and the motion–stress vector elements will be found as in the case of the non-gravitating spheroidal modes. Since an arbitrary regular solution is a linear combination of three elementary solutions, we have

$$\begin{bmatrix} y_i \\ y_j \\ y_k \end{bmatrix} = \begin{bmatrix} y_{i1} & y_{i2} & y_{i3} \\ y_{j1} & y_{j2} & y_{j3} \\ y_{k1} & y_{k2} & y_{k3} \end{bmatrix} \begin{bmatrix} A \\ B \\ C \end{bmatrix}. \tag{54}$$

Solving for the constants of integration $[A \quad B \quad C]$ and substituting the constants into an equation for y_l, we obtain

$$y_l = [Y_{ljk} \quad Y_{ilk} \quad Y_{ijl}][y_i \quad y_j \quad y_k]^{\mathrm{T}}/Y_{ijk}. \tag{55}$$

In particular, we shall make use of the following identities:

$$\begin{bmatrix} y_1 \\ y_3 \\ y_5 \end{bmatrix} = \frac{1}{Y_{246}} \begin{bmatrix} Y_{146} & -Y_{126} & Y_{124} \\ Y_{346} & Y_{236} & -Y_{234} \\ -Y_{456} & Y_{256} & Y_{245} \end{bmatrix} \begin{bmatrix} y_2 \\ y_4 \\ y_6 \end{bmatrix}, \tag{56}$$

$$\begin{bmatrix} y_2 \\ y_4 \\ y_6 \end{bmatrix} = \frac{1}{Y_{135}} \begin{bmatrix} Y_{235} & Y_{125} & -Y_{123} \\ -Y_{345} & Y_{145} & Y_{134} \\ Y_{356} & -Y_{156} & Y_{136} \end{bmatrix} \begin{bmatrix} y_1 \\ y_3 \\ y_5 \end{bmatrix}. \tag{57}$$

When (57) is substituted into the right-hand side of (56), we obtain an identity of the form

$$\begin{bmatrix} Y_{ljk} & Y_{ilk} & Y_{ijl} \\ Y_{mjk} & Y_{imk} & Y_{ijm} \\ Y_{njk} & Y_{ink} & Y_{ijn} \end{bmatrix} \begin{bmatrix} Y_{imn} & Y_{lin} & Y_{lmi} \\ Y_{jmn} & Y_{ljn} & Y_{lmj} \\ Y_{kmn} & Y_{lkn} & Y_{lmk} \end{bmatrix} = Y_{ijk} Y_{lmn} \boldsymbol{I}, \tag{58}$$

where $\boldsymbol{I}$ is the 3×3 identity matrix.

It is evident from (46) that $Y_{246}(a)$ may be used as the characteristic function of the gravitating spheroidal mode, but we prefer to define the characteristic function by

$$\Delta_{\mathrm{SPH}}(\omega, l) = Y_{246}(a)/Y_{146}(a), \tag{59}$$

which is $y_2(a)/y_1(a)$ along $y_4(a) = 0 = y_6(a)$ and is independent of scaling.

Liquid layer

The compound matrix in a liquid layer is defined again by (25) with $(ij) = (12), (15), (16), (25), (26), (56)$. A similar argument yields

$$Y_{12} + Y_{56} = 0, \tag{60}$$

and the governing equations are given by

$$\left.\begin{aligned} \frac{\mathrm{d}}{\mathrm{d}r} Y_{12} &= a_{25} Y_{15} + a_{26} Y_{16} - a_{15} Y_{25}, \\ \frac{\mathrm{d}}{\mathrm{d}r} Y_{15} &= (a_{11} + a_{55}) Y_{15} + a_{56} Y_{16} + a_{12} Y_{25}, \\ \frac{\mathrm{d}}{\mathrm{d}r} Y_{16} &= -2a_{15} Y_{12} + a_{65} Y_{15} + (a_{11} - a_{55}) Y_{16} + a_{12} Y_{26}, \\ \frac{\mathrm{d}}{\mathrm{d}r} Y_{25} &= 2a_{26} Y_{12} + a_{21} Y_{15} - (a_{11} - a_{55}) Y_{25} + a_{56} Y_{26}, \\ \frac{\mathrm{d}}{\mathrm{d}r} Y_{26} &= -2a_{25} Y_{12} + a_{21} Y_{16} + a_{65} Y_{25} - (a_{11} + a_{55}) Y_{26}. \end{aligned}\right\} \tag{61}$$

Equation (29) still holds, and the normalized characteristic function, which should vanish at a liquid free surface, is defined by

$$\Delta_{\mathrm{SPH}}(\omega, l) = Y_{16}(a)/Y_{26}(a). \tag{62}$$

6.3 Continuation across a solid–liquid boundary

Switching from one set of dependent variables to another is inevitable across a solid–liquid boundary. First, consider a boundary where a liquid layer is overlain by a solid layer. Integrating upwards, we would have only two elementary solutions on the liquid side of the boundary, but since horizontal components of displacement may have a jump discontinuity there, we have

$$Y^{\mathrm{S}}_{ijk} = \begin{vmatrix} y_{11} & y_{12} & 0 \\ y_{21} & y_{22} & 0 \\ y_{31} & y_{32} & 1 \\ 0 & 0 & 0 \\ y_{51} & y_{52} & 0 \\ y_{61} & y_{62} & 0 \end{vmatrix}^{\mathrm{L}} \tag{63}$$

Again, superscript S denotes the quantities at the solid side of the interface and L denotes those at the liquid side of the interface, and the equation indicates that Y_{ijk} is a third-order minor of the right-hand-side matrix. Expanding the right-hand side, we obtain

$$\left.\begin{matrix} Y^{\mathrm{S}}_{123} = Y^{\mathrm{L}}_{12}, & Y^{\mathrm{S}}_{135} = -Y^{\mathrm{L}}_{15}, & Y^{\mathrm{S}}_{136} = -Y^{\mathrm{L}}_{16}; \\ Y^{\mathrm{S}}_{235} = -Y^{\mathrm{L}}_{25}, & Y^{\mathrm{S}}_{236} = -Y^{\mathrm{L}}_{26}, & Y^{\mathrm{S}}_{356} = Y^{\mathrm{L}}_{56}; \end{matrix}\right\} \tag{64}$$

all other $Y^{\mathrm{S}}_{ijk} = 0$. If we have Y^{L}_{ij}, the compound matrix on the liquid side of the boundary, we can find the one on the solid side of the boundary using this equation, and we can resume integration further through the solild layer.

On the other hand, when we are integrating through a solid layer and find a solid-to-liquid boundary, we must use a different continuation scheme. First we form two independent solutions from three solutions on the solid side of the boundary in such a way that the shear stress of the solution vanishes:

$$\hat{y}^{\mathrm{S}}_{ij} = y^{\mathrm{S}}_{ij} - \frac{y^{\mathrm{S}}_{4j}}{y^{\mathrm{S}}_{43}} y^{\mathrm{S}}_{i3} \quad (j = 1, 2)$$

Evidently, two sets of solutions $\hat{y}_{ij}^{S}$ $(i = 1, 2, 5, 6)$ as defined above may be used as the initial values on the liquid side of the boundary, since they satisfy all the necessary continuity conditions there. On the other hand, the compound-matrix elements on the solid side of the boundary, (49), may be written as

$$Y_{ijk}^{S} = \begin{vmatrix} \hat{y}_{i1}^{S} & \hat{y}_{i2}^{S} & y_{i3}^{S} \\ \hat{y}_{j1}^{S} & \hat{y}_{j2}^{S} & y_{j3}^{S} \\ \hat{y}_{k1}^{S} & \hat{y}_{k2}^{S} & y_{k3}^{S} \end{vmatrix}.$$

By expanding the right-hand side and omitting a common factor y_{43}, we have

$$\left.\begin{array}{lll} Y_{12}^{L} = Y_{124}^{S}, & Y_{15}^{L} = -Y_{145}^{S}, & Y_{16}^{L} = -Y_{146}^{S}; \\ Y_{25}^{L} = -Y_{245}^{S}, & Y_{26}^{L} = -Y_{246}^{S}, & Y_{56}^{L} = -Y_{124}^{S}. \end{array}\right\} \tag{65}$$

6.4 Computation of eigenfunctions

Once an eigenvalue is found by integrating the compound matrix upwards, we can calculate the surface values of the corresponding eigenfunctions using (56),

$$\frac{y_3}{y_1} = \frac{Y_{346}}{Y_{146}}, \qquad \frac{y_5}{y_1} = \frac{-Y_{456}}{Y_{146}}, \tag{66}$$

when the surface layer is solid, or using (29),

$$\frac{y_5}{y_1} = \frac{Y_{56}}{Y_{16}}, \tag{67}$$

when the surface layer is liquid. Starting with these values, we integrate the original equations for the stress–motion vector, (45) or (48), *downwards,* with stepwise correction of y_2, y_4, and y_6, with the aid of (57) or its equivalent in a liquid layer (see (29)).

When we integrate the equation through a solid layer and arrive at a solid-to-liquid interface, the continuation scheme is simple because the eigenfunctions y_1, y_2, y_5 and y_6 have to be continuous across the boundary. On the other hand, when we integrate through a liquid layer and arrive at a liquid-to-solid interface, such as the outer-core–inner-core boundary, we must find y_3 at the solid side of the boundary since it is not obtained at the liquid side. From (55), we have

$$y_3 = -(Y_{125}y_1 + Y_{134}y_5)/Y_{145}, \tag{68}$$

because $y_4 = 0$ at the solid side of the interface. Since y_1, y_2, y_5 and y_6 are continuous and $y_4 = 0$ at the interface, we have all the eigenfunctions at the solid side of the interface in terms of the eigenfunctions at the liquid side of the interface.

6.5 Energy integral relations

We summarize the integral relations below:

$$\langle L\rangle = \omega^2 I_1 - I_2 = -[y_1 y_2 + y_3 y_4 + y_5 y_6]_a, \tag{69}$$

where

$$\left.\begin{aligned}
I_1 &= \int_0^a \rho(y_1^2 + y_3^2)\,\mathrm{d}r,\\
I_2 &= \int_0^a \left\{\frac{1}{C}y_2^2 + \frac{1}{L}y_4^2 + \frac{1}{r^2}\left(A - \frac{F^2}{C} - N\right)\{2y_1 - [l(l+1)]^{1/2}y_3\}^2\right.\\
&\quad + \frac{(l-1)(l+2)N}{r^2}y_3^2 - \frac{2\rho g}{r}y_1\{2y_1 - [l(l+1)]^{1/2}y_3\}\\
&\quad \left. + 4\pi G y_6^2 + \frac{2(l+1)\rho}{r}y_1 y_5 - \frac{2[l(l+1)]^{1/2}\rho}{r}y_3 y_5\right\}\mathrm{d}r,\\
U &= a^2 I_3 / c I_1,\\
I_3 &= \int_0^a \left\{\frac{1}{r^2}\left(A - \frac{F^2}{C}\right)y_3^2 + \frac{1}{r(l+\frac{1}{2})}y_5(y_6 + \rho y_1)\right.\\
&\quad + \frac{1}{r[l(l+1)]^{1/2}}\left[y_1 y_4 - \frac{F}{C}y_2 y_3 - \frac{2}{r}\left(A - \frac{F^2}{C} - N\right)y_1 y_3\right.\\
&\quad \left.\left. + \frac{\rho}{r}y_3(g y_1 - y_5)\right]\right\}\mathrm{d}r.
\end{aligned}\right\} \tag{70}$$

The rotational splitting factor β is computed using (39).

6.6 Partial derivatives of the compound matrix

Both (54) and (61) are of the same form as (41). Therefore the partial derivatives of the compounds Y_{ijk} in the solid layer and Y_{ij} in the liquid layer can be integrated in a similar manner as for the non-gravitating spheroidal mode. In a solid layer the compound and its partial derivatives together constitute a system of $14 \times 3 = 42$ first-order differential equations. At first sight, this might seem too large for stable numerical solutions to be obtained. In practice, however, we have found that the system is stable and we have been able to compute solutions even for the ${}_{20}S_{20}$ mode, with single-precision arithmethic.

When the compound and its derivatives are integrated to the free surface for an eigenvalue, we can compute the corresponding group velocity by

$$U = -\frac{\partial Y_{246}}{\partial l} \Big/ \left[\frac{\partial Y_{246}}{(\omega a)}\right]_a \tag{71}$$

when the free surface is solid, or by

$$U = -\frac{\partial Y_{26}}{\partial l} \Big/ \left[\frac{\partial Y_{26}}{\partial(\omega a)}\right]_a \tag{72}$$

when the surface is liquid.

The average kinetic energy can also be computed from the partial derivatives. By differentiating (70) with the constraint $y_4(a) = y_6(a) = 0$, and setting $y_2(a) = 0$, we obtain

$$2\omega^2 I_1 = -\left[\frac{y_1^2}{Y_{146}}\,\omega\,\frac{\partial Y_{246}}{\partial \omega}\right]_a \tag{73}$$

when the surface is solid, or

$$2\omega^2 I_1 = -\left[\frac{y_1^2}{Y_{16}}\,\omega\,\frac{\partial Y_{26}}{\partial \omega}\right]_a \tag{74}$$

when the surface is liquid.

6.7 Description of spheroidal oscillation programs

The keyword for the gravitating spheroidal mode program package is SPH. Thus SPHDSP designates a driver routine that controls SPHRKG, SPHEFR computes eigenfunctions and partial derivatives, and SPHINI, SPHDEL and SPHENG are routines called by SPHRKG and SPHEFR.

The inputs and outputs are similar to those for torsional and non-gravitating spheroidal modes, except for the gravity. It is necessary for a user of DISPER80 to specify a gravity model that is consistent with the density model. To accommodate this, a routine DSPGRV is prepared for the computation of gravity for a given density structure. The gravitational constant G is a variable in all the programs, but when a value less than zero is entered in the programs, they assume a default value $G = 6.673 \times 10^{-8}$, the Geodetic Reference System 1983 value.

7 LOVE WAVES AND RAYLEIGH WAVES

Since Love- and Rayleigh-wave algorithms are fully described in Takeuchi and Saito (1972), only the definitions of variables in the programs will be given below.

7.1 Love waves

The motion–stress vector components for Love-wave-type motion are defined by

$$\left.\begin{aligned} \boldsymbol{u} &= \frac{1}{k} y_1(z) \boldsymbol{\nabla} \times [\hat{\boldsymbol{z}} Y_k(x, y)], \\ \boldsymbol{\sigma}_z &= \frac{1}{k} y_2(z) \boldsymbol{\nabla} \times [\hat{\boldsymbol{z}} Y_k(x, y)], \end{aligned}\right\} \tag{75}$$

where $\hat{\boldsymbol{z}}$ is a unit vector in the vertical direction, k is the horizontal wavenumber and $Y_k(x, y)$ is a plane harmonic of wavenumber k or a solution of the two-dimensional Helmholtz equation

$$\left(\frac{\partial^2}{\partial x^2} + \frac{\partial^2}{\partial y^2} + k^2\right) Y_k(x, y) = 0.$$

In a Cartesian coordinate system $Y_k(x, y)$ is given by

$$Y_k(x, y) = \mathrm{e}^{\mathrm{i}(lx + my)},$$

where $k^2 = l^2 + m^2$, and in plane polar coordinates it is given by

$$Y_k(x, y) = J_m(kr)\mathrm{e}^{im\phi},$$

where $J_m(kr)$ is the Bessel function of the first kind. Substitution of the above form into the equation of motion yields the well-known equations

$$\left.\begin{aligned} \frac{\mathrm{d}}{\mathrm{d}z} y_1 &= \frac{1}{L} y_2, \\ \frac{\mathrm{d}}{\mathrm{d}z} y_2 &= k^2(N - \rho c^2) y_1, \end{aligned}\right\} \tag{176}$$

where $c = \sigma/k$ is the phase velocity. Instead of ω and k, the Love-wave programs use ω and c as independent parameters. Thus LOVRKG, for example, finds a phase velocity for a given angular frequency ω. Consistent with this choice of parameters, partial derivatives of eigenfunctions are formed with respect to ω and c, and the group-velocity formula is written in terms of the partial derivatives as

$$U = c\left(c\frac{\partial y_2}{\partial c}\right)\left(c\frac{\partial y_2}{\partial c} + \omega\frac{\partial y_2}{\partial \omega}\right)^{-1}. \tag{77}$$

The energy integral I_3 can also be written in terms of a partial derivative:

$$2k^2 I_3 = 2k^2\int_{-\infty}^{0} N y_1^2\,\mathrm{d}z = -\left[y_1^2 c\frac{\partial y_2}{\partial c}\right]_a. \tag{78}$$

Note that, while in spherical models the output of the program is the kinetic-energy integral I_1, in flat models the output is I_3.

Two sets of Love-wave programs are available in DISPER80: a Runge–Kutta–Gill version and a matrix version. In the Runge–Kutta–Gill version, in which LOVRKG finds an eigen-phase-velocity and LOVEFR computes eigenfunctions, (75) is integrated by a Runge–Kutta–Gill integrator as in the torsional programs. On the other hand, LOVMRX and LOVEFX are the matrix versions of LOVRKG and LOVEFR respectively. They integrate the equations of motion using a Thomson–Haskell-type matrix method.

7.2 Rayleigh waves

The Rayleigh-wave motion–stress vector is defined by

$$\left.\begin{aligned} \boldsymbol{u} &= y_1(z)[\hat{\boldsymbol{z}}Y_k(x,y)] + \frac{1}{k}y_3(z)\boldsymbol{\nabla}[Y_k(x,y)],\\ \boldsymbol{\sigma}_z &= y_2(z)[\hat{\boldsymbol{z}}Y_k(x,y)] + \frac{1}{k}y_4(z)\boldsymbol{\nabla}[Y_k(x,y)], \end{aligned}\right\} \tag{79}$$

and the equations of motion are given by

$$\left.\begin{aligned} \frac{\mathrm{d}}{\mathrm{d}z}y_1 &= \frac{1}{C}(y_2 + kFy_3),\\ \frac{\mathrm{d}}{\mathrm{d}z}y_2 &= -\rho\omega^2 y_1 + ky_4,\\ \frac{\mathrm{d}}{\mathrm{d}z}y_3 &= -ky_1 + \frac{1}{L}y_4,\\ \frac{\mathrm{d}}{\mathrm{d}z}y_4 &= -\frac{kF}{C}y_2 + \left[k^2\left(A - \frac{F^2}{C}\right) - \rho\omega^2\right]y_3. \end{aligned}\right\} \tag{80}$$

Again, ω and c are taken as the independent parameters, and the group velocity and one of the kinetic-energy integrals are written in terms of partial derivatives of the *y*s:

$$U = c\left(c\frac{\partial Y_{24}}{\partial c}\right)\left[c\frac{\partial Y_{24}}{\partial c} + \omega\frac{\partial Y_{24}}{\partial \omega}\right]^{-1}, \tag{81}$$

$$2k^2 I_3 = -y_1^2 \frac{1}{Y_{14}} c\frac{\partial Y_{24}}{\partial c}, \tag{82}$$

where Y_{ij} is the (i, j) element of the compound matrix.

RAYRKG and RAYEFR are the subroutines for the calculation of phase velocity and of eigenfunctions respectively, using the Runge–Kutta–Gill method of integration. The matrix versions, RAYMRX and RAYEFX, can calculate Rayleigh values in an *isotropic* layered model, but the matrix method cannot be used in a transversely isotropic model.

ACKNOWLEDGEMENTS

The earliest version of this package, DISPER76, was developed at the Earthquake Research Institute, University of Tokyo, around 1975. More recently the project DISPER80 has been partially supported by the Computer Centre of the University of Tokyo, and by the Computer Center of Kobe University.

REFERENCES

Abo-Zena, A. (1979). Dispersion function computations for unlimited frequency values. *Geophys. J. R. Astron. Soc.* **58,** 91–105.

Aki, K. and Richards, P. G. (1980). *Quantitative Seismology*. W. H. Freeman, San Francisco.

Dahlen, F. A. (1968). The normal modes of a rotating, elliptical Earth. *Geophys. J. R. Astron. Soc.* **16,** 329–367.

Dunkin, J. W. (1965). Computation of modal solution in layered, elastic media at high frequencies. *Bull. Seismol. Soc. Am.* **55,** 335–358.

Gilbert, F. and Backus, G. E. (1966). Propagator matrices in elastic wave and vibration problems. *Geophysics* **31,** 326–332.

Harvey, D. J. (1981). Seismogram synthesis using normal mode superposition: the locked mode approximation. *Geophys. J. R. Astron. Soc.* **66,** 37–69.

Ng, B. S. and Reid, W. H. (1979a). An initial value method for eigenvalue problems using compound matrices. *J. Comp. Phys.* **30,** 125–136.

Ng, B. S. and Reid, W. H. (1979b). A numerical method for linear two-point boundary-value problems using compound matrices. *J. Comp. Phys.* **33,** 70–85.

Takeuchi, H. and Saito, M. (1972). Seismic surface waves. *Methods in Computational Physics,* Vol. 11 (ed. B. A. Bolt), pp. 217–295. Academic Press, New York.

IV.2

The Calculation of Eigenfrequencies and Eigenfunctions of the Free Oscillations of the Earth and the Sun

J. H. WOODHOUSE

Department of Earth and Planetary Sciences, Harvard University, Cambridge, Massachusetts, USA

1 INTRODUCTION

The synthesis of complete long-period seismograms by the superposition of normal modes (Gilbert, 1971) forms the basis of many seismological studies. The major advantage over other methods of synthesis is that it is completely automatic and omits no part of the theoretically predictable seismic signal. To use the method, however, one must have previously calculated a complete catalogue of normal-mode eigenfrequencies and eigenfunctions, up to some preassigned limit in frequency. In this chapter we present an algorithm for the calculation of such a catalogue. This algorithm has been in use for several years at a number of institutions (a modified form called MINEOS is due to G. Masters) but, apart from a brief account for the case in which self-gravitation is neglected (Woodhouse, 1981), it has not previously been described in print. The program OBANI accompanying this book is also able to calculate eigenfrequencies and eigenfunctions of the adiabatic modes of the Sun.

The program EOS by F. Gilbert has been informally circulated since the early 1970s. This program solves the ordinary differential equations governing free oscillations by a variable order-variable step Runge–Kutta technique. Some aspects of the technique are described in a report by Gilbert and Backus (1969). The algorithm described here is a descendant of EOS in that the same finite-difference techniques are employed and

SEISMOLOGICAL ALGORITHMS
ISBN 0-12-220770-X

much of the Fortran code is a direct copy or adaptation of EOS routines. The EOS program, however, is not suitable for the calculation of complete modal catalogues, principally because of two shortcomings.

First, at moderately high frequencies the equations are "stiff", leading to grossly unstable results for some modes. In this case the direct integration of a linearly independent set of solutions acts to effectively project each of them onto the space of the most rapidly increasing solution, so that although the solutions are theoretically linearly independent, they become indistinguishable numerically from a dependent set. Later versions of the EOS program overcame this difficulty by means of a reorthogonalization procedure; the algorithm presented here circumvents the problem by solving for the minors of sets of solutions—a method discussed by Gilbert and Backus (1966, 1969). A novel result of the present paper is that it is shown how eigenfunctions may be constructed directly from sets of minors. This generalizes the result for fourth-order systems of Woodhouse (1980).

The second problem encountered by previous finite-difference algorithms lies in the difficulty of ensuring that all eigenvalues in a given interval are found. For radial modes and toroidal modes the eigenfrequencies are evenly and predictably spaced, so that locating all roots of the secular equation is not difficult. The spheroidal-mode spectrum, on the other hand, contains many closely and irregularly spaced eigenvalues, and an exhaustive search requires such a small step in frequency that the method becomes impracticable. Moreover it is not possible to specify a priori what the frequency step should be. For second-order Sturm–Liouville systems it is well known that a count of eigenvalues may be established in terms of the number of zero-crossings of some combination of the solution and its derivative. Thus it is possible to calculate how many eigenvalues lie between two trial values of the eigenvalue parameter, enabling all roots to be efficiently bracketed. Once such upper and lower bounds for all eigenvalues are known, it is comparatively simple to converge on the individual eigenvalues. The algorithm presented here establishes such a count, for a certain class of higher-order differential systems, in terms of zero-crossings of minors of sets of solutions and thus enables efficient bracketing of spheroidal-mode eigenfrequencies. We shall refer to this aspect of the procedure as the "counting algorithm".

An alternative approach using the Rayleigh–Ritz technique has been implemented by Buland and Gilbert (Buland, 1976; Buland and Gilbert, 1984; see also Buland's chapter (IV.3) in this book). In this algorithm the variational formulation of the normal-mode equations is used to reduce the problem to that of finding the eigenvalues and eigenvectors of a banded real symmetric matrix, for which efficient numerical schemes exist. One of the principal advantages of the technique is that there are known methods for bracketing eigenvalues of the matrix (Martin and Wilkinson, 1967; Peters and Wilkinson, 1969). The algorithm discussed here has a number of advantages over the Rayleigh–Ritz technique, particularly when it is desired to perform the calculation to moderately high frequencies (above 20 mHz, say), since in the variational approach the discretization interval must be chosen to achieve adequate sampling for the highest-frequency mode of interest. Another difficulty is that low-frequency but highly oscillatory modes corresponding to internal waves in the core are aliased into the seismic frequency band and need to be explicitly deleted from the resulting catalogue.

The system of ordinary differential equations governing the Earth's elastic–gravitational oscillations are of sixth order in solid regions of the Earth and of fourth order in fluid regions (outer core and ocean); the centre of the Earth is a singular point. These constitute complications that, for simplicity of presentation, we shall initially ignore. We shall consider the system

$$\frac{d\boldsymbol{y}}{dx} = \boldsymbol{A}(x, \lambda)\boldsymbol{y} \quad (x \in [x_1, x_2]), \tag{1.1}$$

where $\boldsymbol{A}$ is a given, piecewise-continuous $n \times n$ matrix function of x, also depending upon the parameter λ. We shall assume that, for each fixed $x \in [x_1, x_2]$ and for λ in a certain interval of interest, $[\lambda_1, \lambda_2]$ say, the elements of $\boldsymbol{A}$ are continuously differentiable functions of λ. Solutions $\boldsymbol{y}(x)$ will be required to be continuous and to satisfy given homogeneous boundary conditions at x_1 and x_2. The elements of $\boldsymbol{A}$ and $\boldsymbol{y}$ may, in general, be complex. The eigenvalue problem consists of finding values of λ for which non-trivial solutions exist, and determining these solutions. General homogeneous boundary conditions may be stated by specifying vector spaces $\bar{V}^{(1)} \subseteq \mathbb{C}^n$, $\bar{V}^{(2)} \subseteq \mathbb{C}^n$, of dimensions n_1 and n_2, respectively, and requiring

$$\boldsymbol{y}(x_1) \in \bar{V}^{(1)}, \qquad \boldsymbol{y}(x_2) \in \bar{V}^{(2)}. \tag{1.2}$$

The notation $\mathbb{C}^n$ represents the vector space of all ordered n-tuples of complex numbers. In the derivation of the counting algorithm we shall subsequently assume that $\bar{V}^{(1)}$ and $\bar{V}^{(2)}$ are independent of λ.

A straightforward method of solution is as follows. First choose a basis $\{\bar{\boldsymbol{\sigma}}_1^{(1)}, \bar{\boldsymbol{\sigma}}_2^{(1)} \ldots \bar{\boldsymbol{\sigma}}_{n_1}^{(1)}\}$ of $\bar{V}^{(1)}$ and a basis $\{\bar{\boldsymbol{\sigma}}_1^{(2)}, \bar{\boldsymbol{\sigma}}_2^{(2)} \ldots \bar{\boldsymbol{\sigma}}_{n_2}^{(2)}\}$ of $\bar{V}^{(2)}$ and define $\boldsymbol{\sigma}_k^{(\alpha)}(x, \lambda)$, $(\alpha = 1,\ 2;\ k = 1, 2 \ldots n_\alpha)$ to be the solutions of the initial-value problems

$$\frac{d\boldsymbol{\sigma}_k^{(\alpha)}}{dx} = \mathbf{A}(x, \lambda)\boldsymbol{\sigma}_k^{(\alpha)}, \qquad \boldsymbol{\sigma}_k^{(\alpha)}(x_\alpha) = \bar{\boldsymbol{\sigma}}_k^{(\alpha)}. \tag{1.3}$$

It is well known that if a set of solutions is linearly independent at one point then they are linearly independent everywhere; thus, at each fixed x, $\{\boldsymbol{\sigma}_1^{(\alpha)} \ldots \boldsymbol{\sigma}_{n_\alpha}^{(\alpha)}\}$ is a basis for a subspace $V^{(\alpha)}(x, \lambda)$ of dimension n_α. It follows that either $V^{(1)}(x, \lambda)$ and $V^{(2)}(x, \lambda)$ possess a non-null common subspace for all x, in which case λ is an eigenvalue and the eigenfunction for each x lies in $V^{(1)}(x, \lambda) \cap V^{(2)}(x, \lambda)$, or that they do not possess such a subspace, in which case λ is not an eigenvalue, and only the trivial solution exists. It is easily seen that if $n_1 + n_2 > n$ then a non-trivial solution always exists. The most common case met with in practice is $n_1 + n_2 = n$, since then the number of independent boundary conditions is equal to the number of integration constants for the system. Such systems normally possess, at most, discrete eigenvalues. The dimension of the space $U = V^{(1)}(x, \lambda) \cap V^{(2)}(x, \lambda)$ is independent of x and is the degree of degeneracy. A non-degenerate eigenvalue has $\dim(U) = 1$.

We assume henceforth that $n_1 + n_2 = n$. The condition for λ to be an eigenvalue is that the determinant of the $n \times n$ matrix

$$\mathbf{S} \equiv [\boldsymbol{\sigma}_1^{(1)} \quad \boldsymbol{\sigma}_2^{(1)} \quad \ldots \quad \boldsymbol{\sigma}_{n_1}^{(1)} \quad \boldsymbol{\sigma}_1^{(2)} \quad \boldsymbol{\sigma}_2^{(2)} \quad \ldots \quad \boldsymbol{\sigma}_{n_2}^{(2)}] \tag{1.4}$$

vanish for some x (and therefore for all x). In general the rank defect of this matrix is independent of x and is equal to the degree of degeneracy. The possibility of degenerate eigenfunctions significantly complicates the theory and has not been encountered in the normal-mode problem. Thus, although the case of degenerate eigenvalues is not without interest, we shall assume in this paper that all eigenvalues in the interval of interest $[\lambda_1, \lambda_2]$ are non-degenerate.

Corresponding to the eigenvalue problem formulated above, we may define the *adjoint* problem

$$\frac{d\mathbf{y}}{dx} = -\mathbf{A}^\dagger \mathbf{y}, \tag{1.5}$$

$$\mathbf{y}(x_1) \in \bar{V}^{(1)\perp}, \qquad \mathbf{y}(x_2) \in \bar{V}^{(2)\perp}, \tag{1.6}$$

where † denotes Hermitian conjugation and where $\bar{V}^{(\alpha)\perp}$ are the vector spaces of dimension $n - n_\alpha$ orthogonal to $\bar{V}^{(\alpha)}$ under the usual inner product:

$$(\mathbf{x}, \mathbf{y}) \equiv \mathbf{x}^\dagger \mathbf{y}, \qquad \mathbf{x}, \mathbf{y} \in \mathbb{C}^n.$$

It is easily shown that if $\mathbf{y}(x)$ is any solution of (1.5) that is orthogonal to the subspace $V^{(1)}(x, \lambda)$ at some point x then $\boldsymbol{y}(x)$ is orthogonal to $V^{(1)}(x, \lambda)$ for all x: we have, using (1.3) and (1.5),

$$\frac{\mathrm{d}}{\mathrm{d}x}(\boldsymbol{\sigma}_k^{(1)}(x)^\dagger \boldsymbol{y}(x)) = \boldsymbol{\sigma}_k^{(1)}(x)^\dagger \boldsymbol{A}^\dagger \boldsymbol{y}(x) - \boldsymbol{\sigma}_k^{(1)}(x)^\dagger \boldsymbol{A}^\dagger \boldsymbol{y}(x) = 0, \tag{1.7}$$

from which the stated result follows. Thus if we define $V^{(\alpha)\perp}(x, \lambda)$ to be the subspaces orthogonal to $V^{(\alpha)}(x, \lambda)$ for each x, a necessary and sufficient condition for λ to be an eigenvalue of the adjoint system is that for some x (and therefore for all x) $V^{(1)\perp}(x, \lambda)$ and $V^{(2)\perp}(x, \lambda)$ possess a non-null common subspace. The adjoint eigenfunction, for each x, lies in $\tilde{U}(x, \lambda) \equiv V^{(1)\perp}(x, \lambda) \cap V^{(2)\perp}(x, \lambda)$. The dimension of $\tilde{U}(x, \lambda)$ is equal to the dimension of the space spanned by solutions of

$$\boldsymbol{S}^\dagger \boldsymbol{x} = \boldsymbol{0}$$

(see (1.4)), namely the rank defect of $\boldsymbol{S}$. Thus the eigenvalues λ^* of the adjoint system are identical with those of the original equations, and their degree of degeneracy is also the same.

The equations governing the Earth's free oscillations may be derived from a variational principle. This imposes on the equations a property of self-adjointness in a sense to be defined here (Chapman and Woodhouse, 1981). Consider the general quadratic Lagrangian

$$L(\boldsymbol{q}, \dot{\boldsymbol{q}}, x) = \operatorname{Re}\{\boldsymbol{q}^\dagger \boldsymbol{C}\boldsymbol{q} + 2\dot{\boldsymbol{q}}^\dagger \boldsymbol{D}\boldsymbol{q} + \dot{\boldsymbol{q}}^\dagger \boldsymbol{E}\dot{\boldsymbol{q}}\}, \tag{1.8}$$

where $\boldsymbol{q}$ is a vector of m "coordinates", and where a dot signifies differentiation with respect to x. The $m \times m$ matrices $\boldsymbol{C}$, $\boldsymbol{D}$ and $\boldsymbol{E}$ may depend upon x and, without loss of generality, $\boldsymbol{C}$ and $\boldsymbol{E}$ are Hermitian. The variational problem

$$\delta \int_{x_1}^{x_2} L(\boldsymbol{q}, \dot{\boldsymbol{q}}, x)\, \mathrm{d}x = 0 \tag{1.9}$$

leads to Euler–Lagrange equations

$$(\boldsymbol{D}\boldsymbol{q} + \boldsymbol{E}\dot{\boldsymbol{q}})^{\boldsymbol{\cdot}} = \boldsymbol{C}\boldsymbol{q} + \boldsymbol{D}^\dagger \dot{\boldsymbol{q}}. \tag{1.10}$$

Thus, defining conjugate momenta $\boldsymbol{p} = \boldsymbol{D}\boldsymbol{q} + \boldsymbol{E}\dot{\boldsymbol{q}}$ and assuming $\boldsymbol{E}$ to be invertible, we may write Hamilton's equations

$$\dot{\boldsymbol{q}} = -\boldsymbol{E}^{-1}\boldsymbol{D}\boldsymbol{q} + \boldsymbol{E}^{-1}\boldsymbol{p}, \qquad \dot{\boldsymbol{p}} = (\boldsymbol{C} - \boldsymbol{D}^\dagger \boldsymbol{E}^{-1}\boldsymbol{D})\boldsymbol{q} + \boldsymbol{D}^\dagger \boldsymbol{E}^{-1}\boldsymbol{p}. \tag{1.11}$$

If we now define the vector $\boldsymbol{y}$ of $2m$ elements,

$$\boldsymbol{y} = \begin{bmatrix} \boldsymbol{q} \\ \boldsymbol{p} \end{bmatrix} \tag{1.12}$$

and the $2m \times 2m$ partitional matrix

$$\mathbf{A} = \begin{bmatrix} -\mathbf{E}^{-1}\mathbf{D} & \mathbf{E}^{-1} \\ \mathbf{C} - \mathbf{D}^\dagger\mathbf{E}^{-1}\mathbf{D} & \mathbf{D}^\dagger\mathbf{E}^{-1} \end{bmatrix}, \tag{1.13}$$

we may write (1.11) in the form (1.1). We note, however, that $\mathbf{A}$ has certain symmetries, namely that its off-diagonal submatrices are Hermitian and its diagonal submatrices are related by a change of sign and Hermitian conjugation. Introducing the skew-Hermitian $2m \times 2m$ matrix

$$\mathbf{\Sigma} = \begin{bmatrix} \mathbf{0} & \mathbf{1} \\ -\mathbf{1} & \mathbf{0} \end{bmatrix} \tag{1.14}$$

we find that $\mathbf{\Sigma A}$ is Hermitian, i.e.

$$\mathbf{\Sigma A} + \mathbf{A}^\dagger\mathbf{\Sigma} = \mathbf{0} \quad \text{or} \quad \mathbf{A}^\dagger = \mathbf{\Sigma A \Sigma}. \tag{1.15}$$

Note also that $\mathbf{\Sigma}^2 = -\mathbf{1}$. In the language of matrix algebra, $-\mathbf{A}^\dagger$ is *similar* to $\mathbf{A}$ under the transformation $\mathbf{\Sigma}$. Using this transformation in (1.5), we find that if $\mathbf{y}$ is any solution of the system (1.1) then $\mathbf{\Sigma y}$ is a solution of the adjoint system (1.5). The correspondence between the original eigenvalue problem and its adjoint becomes complete provided that the boundary conditions are correctly mapped by the transformation $\mathbf{\Sigma}$. We require that if $\{\bar{\boldsymbol{\sigma}}_k^{(\alpha)}\}$ is a basis for $\bar{V}^{(\alpha)}$ then $\{\mathbf{\Sigma}\bar{\boldsymbol{\sigma}}_k^{(\alpha)}\}$ is a basis for $\bar{V}^{(\alpha)\perp}$ ($\alpha = 1$ or 2). Since the dimension of $\bar{V}^{(\alpha)\perp}$ is $n - n_\alpha$, we require $n_\alpha = n - n_\alpha$, i.e. $n_1 = n_2 = \frac{1}{2}n$. We also require

$$\bar{\boldsymbol{\sigma}}_i^{(\alpha)\dagger}\mathbf{\Sigma}\bar{\boldsymbol{\sigma}}_j^{(\alpha)} = 0 \quad \text{for all } i, j \leqslant n_\alpha,\ \alpha = 1, 2. \tag{1.16}$$

It may be verified that if (1.15) and (1.16) hold then the vectors $\boldsymbol{\sigma}_k^{(\alpha)}(x, \lambda)$ defined by (1.3) satisfy

$$\boldsymbol{\sigma}_i^{(\alpha)\dagger}\mathbf{\Sigma}\boldsymbol{\sigma}_j^{(\alpha)} = 0 \quad \text{for all } x \in [x_1, x_2],\ i, j \leqslant n_\alpha,\ \alpha = 1, 2. \tag{1.17}$$

If the matrix $\mathbf{A}$ and the boundary conditions have these properties then the system will be said to be *self-adjoint*. It is easily shown that for such a problem if $\mathbf{y}$ is an eigenfunction of the original problem then $\mathbf{\Sigma y}$ is an eigenfunction of the adjoint problem.

2 THE ALGEBRA OF MINORS

The method of solution of the eigenvalue problem outlined above becomes numerically unstable for "stiff" differential systems because, while $\{\bar{\boldsymbol{\sigma}}_k^{(\alpha)}\}$ are chosen to be linearly independent at x_α ($\alpha = 1$ or 2), $\{\bar{\boldsymbol{\sigma}}_k^{(\alpha)}(x, \lambda)\}$ become numerically dependent for other values of x. Thus we are led to seek a characterization of vector subspaces that is independent of any basis; such a characterization is provided by the "minors" or compounds (Gantmacher, 1959; Gilbert and Backus, 1966) of the basis vectors. Here we show that if the columns of the $n \times m$ matrix $\boldsymbol{S}$ are the basis for an m-dimensional subspace $V \subseteq \mathbb{C}^n$ then the set of all $m \times m$ subdeterminants of $\boldsymbol{S}$, regarded as a vector of $\binom{n}{m}$ elements, uniquely characterizes V and is independent of the choice of basis to within a multiplying factor. Thus we may speak of the "*minor vector of* V". The questions that then naturally arise are as follows. Given the minor vector of V, what is the minor vector of $V^{\perp}$; given the minor vector of two subspaces $V^{(1)}$ and $V^{(2)}$ what is the dimension of $V^{(1)} \cap V^{(2)}$ and what is its minor vector? How are projection operators onto subspaces expressed in terms of minor vectors? Here we develop such a calculus of vector subspaces.

Let us consider a $p \times q$ matrix $\boldsymbol{X}$. For any integer s satisfying

$$1 \leqslant s \leqslant p, \qquad s \leqslant q \tag{2.1}$$

we may define a matrix, which we shall denote by $[\boldsymbol{X}:s]$, to consist of all $s \times s$ subdeterminants of $\boldsymbol{X}$. Each subdeterminant may be labelled by the s row numbers $(i_1, i_2, \ldots, i_s)$ and the s column numbers $(j_1, j_2, \ldots, j_s)$, which define a particular subdeterminant, where $(i_1, \ldots, i_s)$ is a combination of the integers $1, 2, \ldots, p$ and $(j_1, j_2, \ldots, j_s)$ is a combination of the numbers $1, \ldots, q$, with no repetitions. If these combinations are arranged in a definite order then they may be used to label the rows and columns of the matrix $[\boldsymbol{X}:s]$, which will be of dimension $\binom{p}{s} \times \binom{q}{s}$. Note that elements corresponding to different orderings of the indices are not considered as distinct; for definiteness let us specify

$$1 \leqslant i_1 < i_2 < i_3 \ldots < i_s \leqslant p, \qquad 1 \leqslant j_1 < j_2 < j_3 \ldots < j_s \leqslant q. \tag{2.2}$$

We shall refer to $[\boldsymbol{X}:s]$ as the sth *minor matrix* of $\boldsymbol{X}$; minor matrices with a single column will be termed *minor vectors.*

A fundamental result concerning minor matrices is that the operation of matrix multiplication is preserved under the operation of taking minors. That is,

$$[\boldsymbol{X}:s][\boldsymbol{Y}:s] = [\boldsymbol{XY}:s] \tag{2.3}$$

for any matrices $\boldsymbol{X}$ and $\boldsymbol{Y}$ for which matrix multiplication and the sth-order minors are defined. It is assumed, of course, that the ordering adopted for the combinations of row and column numbers is consistent throughout.

We shall use the symbol $\mathbf{1}$ for any identity matrix, relying upon the context to indicate its dimension. It is easily checked that

$$[\mathbf{1}:s]=\mathbf{1}. \tag{2.4}$$

If $\boldsymbol{X}$ is a square invertible matrix then it follows immediately from (2.3) and (2.4) that

$$[\boldsymbol{X}:s][\boldsymbol{X}^{-1}:s]=\mathbf{1}, \tag{2.5}$$

and thus

$$[\boldsymbol{X}:s]^{-1}=[\boldsymbol{X}^{-1}:s]. \tag{2.6}$$

We shall now establish the following.

Result 2.1 *Let $\boldsymbol{X}$ and $\boldsymbol{Y}$ be two matrices of dimension $n\times p$ $(n\geqslant p)$, each of full rank p. Then the columns of $\boldsymbol{X}$ and $\boldsymbol{Y}$ span the same vector space if and only if there exists a non-vanishing complex number α such that*

$$[\boldsymbol{X}:p]=\alpha[\boldsymbol{Y}:p] \tag{2.7}$$

Proof To prove this result, let us first assume that the columns of $\boldsymbol{X}$ and $\boldsymbol{Y}$ span the same space. Then the columns of $\boldsymbol{X}$ may be expressed as linear combinations of the columns of $\boldsymbol{Y}$, and since both are full rank we have

$$\boldsymbol{X}=\boldsymbol{Y}\boldsymbol{Q} \tag{2.8}$$

for some non-singular $p\times p$ matrix $\boldsymbol{Q}$. Taking minors, we obtain

$$[\boldsymbol{X}:p]=[\boldsymbol{Y}:p]\det\boldsymbol{Q},\qquad \det\boldsymbol{Q}\neq 0, \tag{2.9}$$

as in (2.7).

Now let us assume (2.7). We wish to show that the columns of $\boldsymbol{X}$ and $\boldsymbol{Y}$ necessarily span the same subspace. To do this, let us extend the basis defined by the columns of $\boldsymbol{Y}$ to be a basis for the whole n-dimensional space. If the additional basis vectors are the columns of an $n\times(n-p)$ matrix $\boldsymbol{Z}$ then we may construct the non-singular partitioned matrix

$$\boldsymbol{W}=[\boldsymbol{Y}\quad\boldsymbol{Z}], \tag{2.10}$$

and thus

$$\boldsymbol{Y}=\boldsymbol{W}\begin{bmatrix}\mathbf{1}\\ \mathbf{0}\end{bmatrix}. \tag{2.11}$$

In addition, the columns of $\boldsymbol{X}$ may be expressed in terms of the columns of $\boldsymbol{W}$:

$$\boldsymbol{X}=\boldsymbol{W}\boldsymbol{K} \tag{2.12}$$

for some matrix $\boldsymbol{K}$ of dimension $n \times p$ and rank p. Using (2.3), (2.7) and (2.11), we have

$$[\boldsymbol{X}:p] = [\boldsymbol{W}:p][\boldsymbol{K}:p] = \alpha[\boldsymbol{Y}:p] = \alpha[\boldsymbol{W}:p]\boldsymbol{m}, \tag{2.13}$$

where $\boldsymbol{m}$ is the minor vector of

$$\begin{bmatrix} \mathbf{1} \\ \mathbf{0} \end{bmatrix}. \tag{2.14}$$

Clearly $\boldsymbol{m}$ has only one non-vanishing element, namely that corresponding to the combination of rows $(1, 2, \ldots, p)$. Since, by construction, $\boldsymbol{W}$ and therefore $[\boldsymbol{W}:p]$ are non-singular, (2.13) gives

$$[\boldsymbol{K}:p] = \alpha\boldsymbol{m}. \tag{2.15}$$

From this it follows that the first p rows of $\boldsymbol{K}$ constitute a non-singular $p \times p$ matrix, $\boldsymbol{Q}$, say, and that all other rows vanish, i.e.

$$\boldsymbol{K} = \begin{bmatrix} \boldsymbol{Q} \\ \mathbf{0} \end{bmatrix}, \qquad \det \boldsymbol{Q} \neq 0. \tag{2.16}$$

Thus, from (2.12) and (2.10),

$$\boldsymbol{X} = \boldsymbol{Y}\boldsymbol{Q}, \qquad \boldsymbol{Y} = \boldsymbol{X}\boldsymbol{Q}^{-1}, \tag{2.17}$$

i.e. each column of $\boldsymbol{X}$ is expressible as a linear combination of the columns of $\boldsymbol{Y}$, and vice versa. This completes the proof of Result 2.1. ■

This result shows that the "direction" of the column vector $[\boldsymbol{X}:\boldsymbol{p}]$ uniquely defines and is uniquely defined by the p-dimensional subspace of $\mathbb{C}^n$ spanned by the columns of $\boldsymbol{X}$. Thus the minor vector provides a way of characterizing a subspace in a manner independent of the chosen basis. Thus if V is a p-dimensional subspace of $\mathbb{C}^n$ then we may speak meaningfully of "*the minor vector of* V", which is defined up to a multiplicative scalar.

For any subspace V let us now introduce the orthogonal subspace $\tilde{V}$ as the vector subspace consisting of all vectors $\boldsymbol{y}$ such that for all $\boldsymbol{x} \in V : \boldsymbol{y}^\dagger \boldsymbol{x} = 0$, where † denotes Hermitian conjugation. It is well known that a set of basis vectors of V and a set of basis vectors for $\tilde{V}$ together span $\mathbb{C}^n$, and that if V is of dimension p then $\tilde{V}$ is of dimension $n - p$. Since, by Result 2.1, V and $\tilde{V}$ may each be defined by their minor vectors, there must be a functional relationship between the minor vectors. We now show the following.

Result 2.2 *Let V be a p-dimensional subspace of $\mathbb{C}^n$, and let $\tilde{V}$ be the orthogonal space to V. If $\boldsymbol{m}$ is the minor vector of V and $\tilde{\boldsymbol{m}}$ is the minor vector of $\tilde{V}$ then*

$$\tilde{\boldsymbol{m}} = \alpha \mathbf{E}^{(n,p)} \boldsymbol{m}^* \tag{2.18}$$

for some non-vanishing α, where * denotes complex conjugation, and where $\mathbf{E}^{(n,p)}$ is the square matrix of dimension $\binom{n}{p}$ whose elements are defined by

$$E^{(n,p)}_{(i_1 \dots i_{1-p})(i_{n-p+1} \dots i_n)} = \epsilon_{i_1 \dots i_n}. \tag{2.19}$$

The alternating symbol ϵ is defined to be 1 when its indices are an even permutation of the integers $1, 2, \dots, n$, -1 for an odd permutation, and zero otherwise. Note that the first and second sets of bracketed indices on the left-hand side of (2.19) are intended to refer respectively to the rows and columns of the matrix $\mathbf{E}^{(n,p)}$. The rows and columns are assumed to be labelled by combinations of indices, in the same way as are the elements of $\tilde{\boldsymbol{m}}$ and $\boldsymbol{m}$. Result 2.2 asserts that the minor vector of $\tilde{V}$ is obtainable from the minor vector of V simply by taking the complex conjugate and reordering its elements, with appropriate changes of sign.

Proof To prove this result, let us take an $n \times p$ matrix $\mathbf{X}^{(1)}$ whose columns are a basis for V, and extend this basis to be a basis for the whole space. If the additional basis vectors are the columns of an $n \times q$ matrix $\mathbf{X}^{(2)}$ ($q = n - p$) then the columns of the partitioned matrix

$$\mathbf{X} = [\mathbf{X}^{(1)} \quad \mathbf{X}^{(2)}] \tag{2.20}$$

form a basis for the complete n-dimensional space, and $\mathbf{X}$ is invertible. Let us denote by $\mathbf{Y}^{(1)}$ and $\mathbf{Y}^{(2)}$ the subpartitions of $\mathbf{X}^{-1}$ of dimension $p \times n$ and $q \times n$, i.e.

$$\mathbf{X}^{-1} = \begin{bmatrix} \mathbf{Y}^{(1)} \\ \mathbf{Y}^{(2)} \end{bmatrix}. \tag{2.21}$$

Then

$$\mathbf{X}^{-1} = \mathbf{X} = \mathbf{1} = \begin{bmatrix} \mathbf{Y}^{(1)} \\ \mathbf{Y}^{(2)} \end{bmatrix} [\mathbf{X}^{(1)} \quad \mathbf{X}^{(2)}], \tag{2.22}$$

and we obtain the identities

$$\mathbf{Y}^{(1)} \mathbf{X}^{(1)} = \mathbf{1}, \tag{2.23a}$$

$$\mathbf{Y}^{(2)} \mathbf{X}^{(2)} = \mathbf{1}, \tag{2.23b}$$

$$\mathbf{Y}^{(1)} \mathbf{X}^{(2)} = \mathbf{0}, \tag{2.23c}$$

$$\mathbf{Y}^{(2)} \mathbf{X}^{(1)} = \mathbf{0}. \tag{2.23d}$$

Similarly,

$$\boldsymbol{X}\boldsymbol{X}^{-1} = \mathbf{1} = \boldsymbol{X}^{(1)}\boldsymbol{Y}^{(1)} + \boldsymbol{X}^{(2)}\boldsymbol{Y}^{(2)}. \tag{2.24}$$

Equation (2.23d) shows that the columns of $\boldsymbol{Y}^{(2)\dagger}$ are contained in the orthogonal space $\tilde{V}$, and (2.23b) shows that $\boldsymbol{Y}^{(2)}$ is of full rank q. Thus the columns of $\boldsymbol{Y}^{(2)\dagger}$ are a basis for $\tilde{V}$, and a minor vector of $\tilde{V}$ is given by

$$\tilde{\boldsymbol{m}} = [\boldsymbol{Y}^{(2)\dagger} : q]. \tag{2.25}$$

Thus it is sufficient to show that

$$[\boldsymbol{Y}^{(2)\dagger} : q] = \alpha \boldsymbol{E}^{(n,p)}[\boldsymbol{X}^{(1)} : q]^* \tag{2.26}$$

for some α. To prove this, we make use of the identity

$$\epsilon_{i_1 i_2 \ldots i_n} X_{i_1 k_1} X_{i_2 k_2} \ldots X_{i_n k_n} = \epsilon_{k_1 k_2 \ldots k_n} \det \boldsymbol{X}, \tag{2.27}$$

where summation over repeated indices is assumed. Let us write

$$\boldsymbol{U} \equiv \boldsymbol{X}^{-1\mathrm{T}}, \tag{2.28}$$

where T denotes matrix transposition, i.e., (2.23),

$$\boldsymbol{U} = [\boldsymbol{Y}^{(1)\mathrm{T}} \quad \boldsymbol{Y}^{(2)\mathrm{T}}]. \tag{2.29}$$

Multiplying (2.27) on both sides by

$$U_{j_1 k_{p+1}} U_{j_2 k_{p+2}} \ldots U_{j_q k_n}$$

and setting $k_i = i$ $(i = 1, 2, \ldots p)$, we obtain

$$\begin{aligned} \epsilon_{i_1 \ldots i_p j_1 \ldots j_q} X^1_{i_1} X^2_{i_2} \ldots X^p_{i_p} &= \det \boldsymbol{W} \epsilon_{12 \ldots p k_{p+1} \ldots k_n} U_{j_1 k_{p+1}} U_{j_2 k_{p+2}} \ldots U_{j_q k_n} \\ &= \det \boldsymbol{W}\, U^{p+1}_{[j_1} U^{p+2}_{j2} \ldots U^n_{j_q]} \end{aligned} \tag{2.30}$$

In the last line the square brackets have been introduced in the subscripts to indicate that the indices $j_1 j_2 \ldots j_q$ are to be completely antisymmetrized, i.e. (2.30) represents a sum over all permutations of the indices $j_1 \ldots j_q$, with a positive sign for even permutations and a negative sign for odd permutations. The superscripts are simply column indices, no significance being attached to whether they appear as subscripts or superscripts. Using a similar notation on the left-hand side, we have

$$\epsilon_{(i_1 \ldots i_p) j_1 \ldots j_q} W^1_{[i_1} W^2_{i_2} \ldots W^p_{i_p]} = \det \boldsymbol{W}\, U^{p+1}_{[j_1} U^{p+2}_{j_2} \ldots U^n_{j_q]}, \tag{2.31}$$

where the indices in parentheses ( ) are summed over all *combinations*, rather than over all permutations. The antisymmetrized quantities are simply elements of the minor vectors of $\boldsymbol{X}^{(1)}$ and $\boldsymbol{Y}^{(2)\mathrm{T}}$, and we obtain

$$\pm \boldsymbol{E}^{(n,p)}[\boldsymbol{X}^{(1)} : p] = \det \boldsymbol{W}[\boldsymbol{Y}^{(2)\mathrm{T}} : q], \tag{2.32}$$

where the lower sign is taken only if p and q are both odd. On taking the complex conjugate, this is of the required form (2.26) and thus Result 2.2 is established. ■

We shall later require the following property of the matrices $\boldsymbol{E}^{(n,p)}$, which follows straightforwardly from their definition:

$$\boldsymbol{E}^{(n,p)\mathrm{T}}\boldsymbol{E}^{(n,p)} = \mathbf{1}. \tag{2.33}$$

The next question that we address is that of expressing projection operators in terms of minor vectors. Let us suppose that $V^{(1)}$ and $V^{(2)}$ are two disjoint subspaces that together span $\mathbb{C}^n$. Thus any vector $\boldsymbol{x}$ can be expressed uniquely as

$$\boldsymbol{x} = \boldsymbol{v}^{(1)} + \boldsymbol{v}^{(2)}, \tag{2.34}$$

with $\boldsymbol{v}^{(1)} \in V^{(1)}$ and $\boldsymbol{v}^{(2)} \in V^{(2)}$. Consequently there will be projection operators $\boldsymbol{P}^{(1)}$ and $\boldsymbol{P}^{(2)}$ such that

$$\boldsymbol{v}^{(1)} = \boldsymbol{P}^{(1)}\boldsymbol{x}, \qquad \boldsymbol{v}^{(2)} = \boldsymbol{P}^{(2)}\boldsymbol{x}. \tag{2.35}$$

These operators are clearly independent of the bases chosen for the subspaces and should therefore be expressible purely in terms of the minor vectors of the subspaces.

Let us suppose that $\boldsymbol{m}$ is the minor vector of a p-dimensional subspace of the complete n-dimensional space. It will be recalled that $\boldsymbol{m}$ is of dimension $\binom{n}{p}$ and its elements are labelled by the combinations $(i_1 \ldots i_p)$ satisfying

$$1 \leq i_1 < i_2 \ldots < i_p \leq n.$$

Thus the individual elements of $\boldsymbol{m}$ may be denoted by $m_{(i_1 i_2 \ldots i_p)}$. From the elements of $\boldsymbol{m}$ we now construct a matrix $\boldsymbol{M}$ of dimension $n \times \binom{n}{p-1}$ according to

$$M_{i_1(i_2 \ldots i_p)} = \begin{cases} 0 & \text{if } i_1 = i_k \text{ for } 2 \leq k \leq p, \\ \pm m_{(j_1 j_2 \ldots j_p)} & \text{where } (j_1 \ldots j_p) \text{ is the even } (+) \\ & \text{or odd } (-) \text{ permutation of } (i_1 \ldots i_p) \text{ that} \\ & \text{places them in increasing order.} \end{cases} \tag{2.36}$$

The rows of $\boldsymbol{M}$ are labelled by i_1 $(1 \leq i_1 \leq n)$ and the rows are labelled by the combinations $(i_2, i_3 \ldots i_p)$ satisfying

$$1 \leq i_2 < i_3 < i_4 \ldots < i_p \leq n,$$

arranged in some definite order. We shall refer to $\boldsymbol{M}$ as the "*spanning matrix*" of the minor vector $\boldsymbol{m}$ and write

$$\mathbf{M} = \mathrm{sp}\,(\boldsymbol{m}). \tag{2.37}$$

We now prove the following.

Result 2.3 *Let $V^{(1)}$ and $V^{(2)}$ be two disjoint subspaces of $\mathbb{C}^n$, of dimension p and q, that together span the complete n-dimensional space ($n = p + q$), and let $\boldsymbol{m}^{(1)}$ and $\boldsymbol{m}^{(2)}$ be the minor vectors of $V^{(1)}$ and $V^{(2)}$. Furthermore, let $\tilde{\boldsymbol{m}}^{(1)}$ and $\tilde{\boldsymbol{m}}^{(2)}$ be the minor vectors of the orthogonal subspaces $\tilde{V}^{(1)}$ and $\tilde{V}^{(2)}$ given by*

$$\tilde{\boldsymbol{m}}^{(1)} = \mathbf{E}^{(n,p)}\boldsymbol{m}^{(1)*}, \qquad \tilde{\boldsymbol{m}}^{(2)} = \mathbf{E}^{(n,q)}\boldsymbol{m}^{(2)*}. \tag{2.38}$$

Then the projection operators $\mathbf{P}^{(1)}$ and $\mathbf{P}^{(2)}$ that satisfy, for all vectors $\boldsymbol{x}$,

$$\mathbf{P}^{(1)}\boldsymbol{x} \in V^{(1)}, \qquad \mathbf{P}^{(2)}\boldsymbol{x} \in V^{(2)}, \qquad \mathbf{P}^{(1)} + \mathbf{P}^{(2)} = \mathbf{1} \tag{2.39}$$

are given by

$$\mathbf{P}^{(1)} = \frac{1}{\Delta}\mathbf{M}^{(1)}\tilde{\mathbf{M}}^{(2)\dagger}, \qquad \mathbf{P}^{(2)} = \pm\frac{1}{\Delta}\mathbf{M}^{(2)}\tilde{\mathbf{M}}^{(1)\dagger} \tag{2.40}$$

where the lower sign applies only if p and q are both odd and where

$$\left.\begin{aligned} \mathbf{M}^{(1)} &= \mathrm{sp}\,(\boldsymbol{m}^{(1)}), & \mathbf{M}^{(2)} &= \mathrm{sp}\,(\boldsymbol{m}^{(2)}), \\ \tilde{\mathbf{M}}^{(1)} &= \mathrm{sp}\,(\tilde{\boldsymbol{m}}^{(1)}), & \tilde{\mathbf{M}}^{(2)} &= \mathrm{sp}\,(\tilde{\boldsymbol{m}}^{(2)}), \end{aligned}\right\} \tag{2.41}$$

and

$$\Delta = \tilde{\boldsymbol{m}}^{(2)\dagger}\boldsymbol{m}^{(1)} = \pm\tilde{\boldsymbol{m}}^{(1)\dagger}\boldsymbol{m}^{(2)}. \tag{2.42}$$

Proof To establish this result, we proceed in the same way as in proving Result 2.2. We introduce two matrixes $\mathbf{X}^{(1)}$ and $\mathbf{X}^{(2)}$ of dimensions $n \times p$ and $n \times q$ respectively, whose columns span the subspaces $V^{(1)}$ and $V^{(2)}$, and construct the matrices $\mathbf{X}$ and $\mathbf{X}^{-1}$ as in (2.20)–(2.24). Using the notation introduced there, it is easily shown that the projection operators $\mathbf{P}^{(1)}$ and $\mathbf{P}^{(2)}$ are given by

$$\mathbf{P}^{(1)} = [\mathbf{X}^{(1)} \quad \mathbf{0}]\mathbf{X}^{-1} = \mathbf{X}^{(1)}\mathbf{Y}^{(1)}, \qquad \mathbf{P}^{(2)} = [\mathbf{0} \quad \mathbf{X}^{(2)}]\mathbf{X}^{-1} = \mathbf{X}^{(2)}\mathbf{Y}^{(2)}. \tag{2.43}$$

Using the usual representation of the inverse matrix in terms of cofactors, and explicitly writing the summation involved in the matrix multiplication of (2.43), we find

$$P_{ij}^{(1)} = \frac{1}{\Delta} \epsilon_{ji_2\ldots i_p i_{p+1}\ldots i_n} \{X_i^1 X_{i_2}^2 \ldots X_{i_p}^p X_{i_{p+1}}^{p+1} \ldots X_{i_n}^n$$
$$+ X_{i_2}^1 X_i^2 X_{i_3}^3 \ldots X_{i_p}^p X_{i_{p+1}}^{p+1} \ldots X_{i_n}^n$$
$$+ \ldots$$
$$+ X_{i_2}^1 X_{i_3}^2 \ldots X_{i_p}^{p-1} X_i^p X_{i_{p+1}}^{p+1} \ldots X_{i_n}^n\}, \tag{2.44}$$

where $\Delta = \det \boldsymbol{X}$ and superscripts are column indices. This may also be written as

$$P_{ij}^{(1)} = \frac{1}{\Delta} \epsilon_{j(i_2\ldots i_p)(i_{p+1}\ldots i_n)} X_{[i}^1 X_{i_2}^2 \ldots X_{i_p]}^p X_{[i_{p+1}}^{p+1} \ldots X_{i_n]}^p \tag{2.45}$$

where brackets [ ] are used, as above, to indicate that the enclosed indices are completely antisymmetrized, where parentheses ( ) indicate summation only over all combinations of the indices, rather than over all permutations. From the definition (2.36) of the spanning matrices, (2.45) takes the required form:

$$\boldsymbol{P}^{(1)} = \frac{1}{\Delta} \boldsymbol{M}^{(1)} \tilde{\boldsymbol{M}}^{(2)\dagger}. \tag{2.46}$$

Equation (2.42) may be established by similar techniques, and the remaining results of Result 2.3 follow by symmetry between the two subspaces $V^{(1)}$ and $V^{(2)}$. ■

An expression for the orthogonal projection operators of a subspace V can be obtained as a special case of Result 2.3. If $\boldsymbol{m}$ is the minor vector of V then we may write

$$\left.\begin{aligned} &V^{(1)} = V, \qquad V^{(2)} = \tilde{V}, \\ &\boldsymbol{m}^{(1)} = \boldsymbol{m}, \\ &\boldsymbol{m}^{(2)} = \tilde{\boldsymbol{m}} = \boldsymbol{E}^{(n,p)} \boldsymbol{m}^*, \\ &\tilde{\boldsymbol{m}}^{(2)} = \boldsymbol{E}^{(n,q)} \boldsymbol{m}^{(2)*} = \boldsymbol{E}^{(n,q)} \boldsymbol{E}^{(n,p)} \boldsymbol{m} = \pm \boldsymbol{m}, \end{aligned}\right\} \tag{2.47}$$

where the last equality follows from the definition (2.19). Writing $\boldsymbol{P}_{\parallel}$ (parallel) for $\boldsymbol{P}^{(1)}$ and $\boldsymbol{P}_{\perp}$ (perpendicular) for $\boldsymbol{P}^{(2)}$, we find, simply

$$\boldsymbol{P}_{\parallel} = \frac{\boldsymbol{M}\boldsymbol{M}^\dagger}{\boldsymbol{m}^\dagger \boldsymbol{m}}, \qquad \boldsymbol{P}_{\perp} = \frac{\tilde{\boldsymbol{M}}\tilde{\boldsymbol{M}}^\dagger}{\boldsymbol{m}^\dagger \boldsymbol{m}} \tag{2.48}$$

where $\boldsymbol{M} = \mathrm{sp}\,(\boldsymbol{m})$ and $\tilde{\boldsymbol{M}} = \mathrm{sp}\,(\tilde{\boldsymbol{m}})$.

It follows from these results that the columns of $\boldsymbol{M}$ span V and the columns of $\tilde{\boldsymbol{M}}$ span $\tilde{V}$. Thus, given a minor vector, we may use the columns of $\boldsymbol{M}$ to construct a basis for the corresponding subspace; this reverses the process of constructing a minor vector from a given basis. The columns of $\tilde{\boldsymbol{M}}$ are orthogonal to those of $\boldsymbol{M}$, and thus

$$\tilde{\boldsymbol{M}}^{\dagger}\boldsymbol{M} = \boldsymbol{0}. \tag{2.49}$$

This equation constitutes a number of quadratic identities, automatically satisfied by the elements of any minor vector $\boldsymbol{m}$; thus, given a vector of appropriate dimension, it will not be the minor vector of any subspace unless (2.49) is satisfied.

The conditions of Result 2.3 guarantee that $\Delta \neq 0$; we shall later need the related result in the case when $V^{(1)}$ and $V^{(2)}$ are two subspaces, again of dimensions p and q $(p+q=n)$, but possessing a common subspace. In this case $V^{(1)}$ and $V^{(2)}$ are insufficient to span the whole n-dimensional space, and Δ vanishes. From (2.39) and (2.40) we have, for $\Delta \neq 0$,

$$\boldsymbol{M}^{(1)}\tilde{\boldsymbol{M}}^{(2)\dagger} \pm \boldsymbol{M}^{(2)}\tilde{\boldsymbol{M}}^{(1)\dagger} = \Delta\boldsymbol{1}. \tag{2.50}$$

This equation, however, may be shown to hold also in the case $\Delta = 0$; we omit the proof.

Remarkably, this result holds when the vectors $\boldsymbol{m}^{(1)}$ and $\boldsymbol{m}^{(2)}$ are not the minor vectors corresponding to two subspaces, but are simply arbitrary vectors of appropriate dimensions. We have the following.

Result 2.4 *For given n, p and q $(n=p+q)$ let $\boldsymbol{m}^{(1)}$ and $\boldsymbol{m}^{(2)}$ be arbitrary vectors of dimension $\binom{n}{p}=\binom{n}{q}$ whose elements are labelled by combinations of p indices and q indices respectively, and let $\boldsymbol{M}^{(1)}$, $\boldsymbol{M}^{(2)}$, $\tilde{\boldsymbol{M}}^{(1)}$ and $\tilde{\boldsymbol{M}}^{(2)}$ be the corresponding spanning matrices of dimensions $n\times\binom{n}{p-1}$, $n\times\binom{n}{q-1}$, $n\times\binom{n}{q-1}$ and $n\times\binom{n}{p-1}$. Then*

(i)
$$\boldsymbol{M}^{(1)}\tilde{\boldsymbol{M}}^{(2)\dagger} \pm \boldsymbol{M}^{(2)}\tilde{\boldsymbol{M}}^{(1)\dagger} = \Delta\boldsymbol{1}, \tag{2.51}$$

where

$$\Delta = \tilde{\boldsymbol{m}}^{(2)\dagger}\boldsymbol{m}^{(1)} = \pm\tilde{\boldsymbol{m}}^{(1)\dagger}\boldsymbol{m}^{(2)}, \tag{2.52}$$

and

(ii)
$$\left.\begin{aligned} \operatorname{tr}\boldsymbol{M}^{(1)}\tilde{\boldsymbol{M}}^{(2)\dagger} &= p\Delta,\\ \operatorname{tr}\boldsymbol{M}^{(2)}\tilde{\boldsymbol{M}}^{(1)\dagger} &= q\Delta. \end{aligned}\right\} \tag{2.53}$$

This is a purely algebraic result; the proof is omitted.

Note that, in general, $\boldsymbol{M}^{(1)}\tilde{\boldsymbol{M}}^{(2)\dagger}$ and $\boldsymbol{M}^{(2)}\tilde{\boldsymbol{M}}^{(1)\dagger}$ will not be the projection operators, although of course they will be in the case when

$\boldsymbol{m}^{(1)}$ and $\boldsymbol{m}^{(2)}$ are the minor vectors of disjoint subspaces having dimension p and q. It is of interest to find what conditions a vector $\boldsymbol{m}$ must satisfy in order to be a minor vector of a vector subspace. We have already shown that the quadratic relations represented by (2.49) are necessary. Using (2.51) and (2.53), they may also be shown to be sufficient.

Result 2.5 *Let* $\boldsymbol{m}$ *be an arbitrary non-vanishing vector of dimension* $\binom{n}{p}$ $(p \leq n)$ *whose elements are labelled by the combinations* $(i_1, i_2, \ldots, i_p)$ $(1 \leq i_1 \leq i_2 \ldots \leq n)$, *and let* $\mathbf{M}$ *and* $\tilde{\mathbf{M}}$ *be the spanning matrices*

$$\mathbf{M} = \mathrm{sp}\,(\boldsymbol{m}), \qquad \tilde{\mathbf{M}} = \mathrm{sp}\,(\tilde{\boldsymbol{m}}),$$

with

$$\tilde{\boldsymbol{m}} = \mathbf{E}^{(n,p)}\boldsymbol{m}^*.$$

Then $\boldsymbol{m}$ *is the minor vector of a p-dimensional subspace if and only if*

$$\tilde{\mathbf{M}}^\dagger \mathbf{M} = \mathbf{0}. \tag{2.54}$$

The proof of sufficiency depends upon using Result 2.4 to show that when $\tilde{\mathbf{M}}^\dagger \mathbf{M} = \mathbf{0}$ the operators

$$\frac{\mathbf{M}\mathbf{M}^\dagger}{\boldsymbol{m}^\dagger \boldsymbol{m}}, \qquad \frac{\tilde{\mathbf{M}}\tilde{\mathbf{M}}^\dagger}{\tilde{\boldsymbol{m}}^\dagger \tilde{\boldsymbol{m}}}$$

are orthogonal projectors onto spaces of dimension p and q respectively; the details are omitted.

We now consider the case of two subspaces $V^{(1)}$ and $V^{(2)}$ that may possess a non-null common subspace. The following result leads to a test for the existence of such a subspace and an explicit construction for the basis of this subspace in the case that it is one-dimensional. It is this construction that will later be used in the calculation of the eigenfunction corresponding to a non-degenerate eigenvalue.

Result 2.6 *Let* $V^{(1)}$ *and* $V^{(2)}$ *be two subspaces of* $\mathbb{C}^n$ *of dimensions* p *and* q, *with* $p + q = n$, *and let the corresponding orthogonal subspaces be* $\tilde{V}^{(1)}$ *and* $\tilde{V}^{(2)}$. *Further, let* $V = V^{(1)} \cap V^{(2)}$ *and* $\tilde{V} = \tilde{V}^{(1)} \cap \tilde{V}^{(2)}$. *Then:*

(i) *either* V *and* $\tilde{V}$ *are both null or they have the same dimension,* s, *say*;

(ii) *in the notation of Result 2.3,*

$$\mathbf{M}^{(1)}\tilde{\mathbf{M}}^{(2)\dagger} \pm \mathbf{M}^{(2)}\tilde{\mathbf{M}}^{(1)\dagger} = \Delta \mathbf{1}, \tag{2.55}$$

with

$$\Delta = \tilde{\boldsymbol{m}}^{(2)\dagger}\boldsymbol{m}^{(1)} = \pm\tilde{\boldsymbol{m}}^{(1)\dagger}\boldsymbol{m}^{(2)}, \tag{2.56}$$

where the lower sign is taken only if p and q are both odd, and Δ vanishes if and only if V and $\tilde{V}$ are non-null;

(iii) *If $s = 1$ then there exist non-vanishing vectors $\boldsymbol{v} \in V$ and $\tilde{\boldsymbol{v}} \in \tilde{V}$ such that*

$$\boldsymbol{M}^{(1)}\tilde{\boldsymbol{M}}^{(2)\dagger} = \mp\boldsymbol{M}^{(2)}\tilde{\boldsymbol{M}}^{(1)\dagger} = \boldsymbol{v}\tilde{\boldsymbol{v}}^{\dagger}; \tag{2.57}$$

(iv) *If $s > 1$ then*

$$\boldsymbol{M}^{(1)}\tilde{\boldsymbol{M}}^{(2)\dagger} = \boldsymbol{M}^{(2)}\tilde{\boldsymbol{M}}^{(1)\dagger} = \boldsymbol{0}. \tag{2.58}$$

Proof Part (iii) follows from the fact that the columns of $\boldsymbol{M}^{(1)}$ span $V^{(1)}$, and those of $\boldsymbol{M}^{(2)}$ span $V^{(2)}$. Thus, because of the first equality in (2.57), each column of $\boldsymbol{M}^{(1)}\tilde{\boldsymbol{M}}^{(2)\dagger}$ lies in the common subspace V, which, by hypothesis, is one-dimensional, so that $\boldsymbol{M}^{(1)}\tilde{\boldsymbol{M}}^{(2)\dagger}$ must be a dyad. Similarly, each column of $\tilde{\boldsymbol{M}}^{(2)}\boldsymbol{M}^{(1)\dagger}$ lies in $\tilde{V}$, and (2.57) is established. Part (iv) follows from the fact that each element of $\boldsymbol{M}^{(1)}\tilde{\boldsymbol{M}}^{(2)\dagger}$ is a cofactor of the matrix whose columns are the combined bases of $V^{(1)}$ and $V^{(2)}$. ∎

We shall later consider the generalization of this construction to the case in which $V^{(1)}$ and $V^{(2)}$ possess a common subspace of higher dimension (Result 3.1 below).

3 LINEAR TRANSFORMATIONS OF VECTOR SUBSPACES

Let V be a p-dimensional subspace of $\mathbb{C}^n$ and let $\tilde{V}$ be its orthogonal subspace. We shall consider here transformations of V such that for $\boldsymbol{v} \in V$ the image of $\boldsymbol{v}$ is

$$\boldsymbol{v}' = \boldsymbol{U}\boldsymbol{v}, \tag{3.1}$$

where $\boldsymbol{U}$ is a non-singular $n \times n$ complex matrix. Since the image space will be unaffected by a scaling factor, we shall assume, without essential loss of generality, that

$$\det \boldsymbol{U} = 1. \tag{3.2}$$

We shall write for the image space

$$V' = \boldsymbol{U}V, \tag{3.3}$$

and we shall use the notation $\tilde{V}'$ for the subspace orthogonal to V'; note that this is, in general, different from the image of $\tilde{V}$. It is easily shown that

$$\tilde{V}' = \tilde{\boldsymbol{U}}\tilde{V}, \tag{3.4}$$

where we define

$$\tilde{\boldsymbol{U}} \equiv \boldsymbol{U}^{\dagger-1} = \boldsymbol{U}^{-1\dagger}. \tag{3.5}$$

By analogy with vector and tensor analysis, we may say that V transforms *contravariantly* and $\tilde{V}$ transforms *covariantly* under the transformation $\boldsymbol{U}$.

We shall seek now to determine the transformation properties of the minor vectors and spanning matrices of these subspaces. As previously, we shall define the minor vector

$$\boldsymbol{m} = [\boldsymbol{X}:p], \tag{3.6}$$

where the columns of $\boldsymbol{X}$ are a basis for V and

$$\tilde{\boldsymbol{m}} = \boldsymbol{E}^{(n,p)}\boldsymbol{m}^*. \tag{3.7}$$

From the foregoing analysis, we know that there exists a basis $\tilde{\boldsymbol{X}}$ of $\tilde{V}$ such that

$$\tilde{\boldsymbol{m}} = [\tilde{\boldsymbol{X}}:q] \tag{3.8}$$

where $q = n - p$. Defining

$$\boldsymbol{m}' = [\boldsymbol{X}':p], \tag{3.9}$$

with

$$\boldsymbol{X}' = \boldsymbol{U}\boldsymbol{X}, \tag{3.10}$$

it follows immediately from (2.3) that

$$\boldsymbol{m}' = [\boldsymbol{U}:p]\boldsymbol{m}. \tag{3.11}$$

We may now define

$$\tilde{\boldsymbol{m}}' = \boldsymbol{E}^{(n,p)}\boldsymbol{m}'^* = \boldsymbol{E}^{(n,p)}[\boldsymbol{U}^*:p]\boldsymbol{m}^*. \tag{3.12}$$

To simplify this further, we need to recall the definition of $\boldsymbol{E}^{(n,p)}$. From (2.19), we see that the rows of the matrix $\boldsymbol{E}^{(n,p)}[\boldsymbol{U}^*:p]$ are labelled by combinations of indices $(i_1, i_2 \ldots i_q)$, and the columns by $(j_1 \ldots j_p)$; its elements are

$$\epsilon_{(i_1 i_2 \ldots i_q)(k_1 \ldots k_p)} U^{*j_1}_{[k_1} U^{*j_2}_{k_2} \ldots U^{*j_p}_{k_p]}, \tag{3.13}$$

where [ ] denotes complete antisymmetrization and where the indices in parentheses ( ) are summed only over all combinations. Using the identity (2.27), it may be shown that

$$\epsilon_{i_1 i_2 \ldots i_q (k_1 \ldots k_p)} U^{*j_1}_{[k_1} U^{*j_2}_{k_2} \ldots U^{*j_p}_{k_p]} = \det \boldsymbol{U}^* \, \tilde{U}^{[k_1}_{i_1} \tilde{U}^{k_2}_{i_2} \ldots \tilde{U}^{k_q]}_{i_q} \epsilon_{k_1 \ldots k_q (j_1 \ldots j_p)}, \tag{3.14}$$

i.e.

$$\boldsymbol{E}^{(n,p)}[\boldsymbol{U}^* : p] = [\tilde{\boldsymbol{U}} : q] \boldsymbol{E}^{(n,p)}. \tag{3.15}$$

Using this result, (3.12) becomes simply

$$\tilde{\boldsymbol{m}}' = [\tilde{\boldsymbol{U}} : q] \tilde{\boldsymbol{m}}. \tag{3.16}$$

Because of the transformation properties (3.11) and (3.16), we may say that $\boldsymbol{m}$ is "*a contravariant vector of type p*" and $\tilde{\boldsymbol{m}}$ is "*a covariant vector of type q*". The "*type*" signifies the order of minors to be used in the transformation. Also, from (3.15) we have

$$\boldsymbol{E}^{(n,p)} = [\tilde{\boldsymbol{U}} : q] \boldsymbol{E}^{(n,p)} [\tilde{\boldsymbol{U}}^{\mathrm{T}} : p], \tag{3.17}$$

and similarly

$$\boldsymbol{E}^{(n,p)} = [\boldsymbol{U} : q] \boldsymbol{E}^{(n,p)} [\boldsymbol{U}^{\mathrm{T}} : p]. \tag{3.18}$$

Thus we may say that $\boldsymbol{E}^{(n,p)}$ transforms both contravariantly and covariantly as an "*isotropic second-rank tensor of type* (q, p)".

Next we consider the transformation properties of the spanning matrices $\boldsymbol{M} = \mathrm{sp}\,(\boldsymbol{m})$, $\tilde{\boldsymbol{M}} = \mathrm{sp}\,(\tilde{\boldsymbol{m}})$. More generally, let us define, for any minor vector $m_{(i_1 \ldots i_p)}$ the matrix $\boldsymbol{M}^{(s,p-s)}$ with rows labelled by combinations $(i_1 \ldots i_s)$, and columns by combinations $(i_{s+1} \ldots i_p)$, according to

$$M^{(s,p-s)}_{(i_1 \ldots i_s)(i_{s+1} \ldots i_p)} = \begin{cases} 0 & \text{if there are repetitions among the indices } i_1 \ldots i_p, \\ \pm m_{(j_1 \ldots j_p)} & \text{where } (j_1 \ldots j_p) \text{ is the even } (+) \text{ or odd } (-) \text{ permutation of } (i_1 \ldots i_p) \text{ that places them in increasing order.} \end{cases} \tag{3.19}$$

Thus the spanning matrices (2.36) are the special case

$$\boldsymbol{M} = \mathrm{sp}\,(\boldsymbol{m}) = \boldsymbol{M}^{(1,p-1)}, \qquad \tilde{\boldsymbol{M}} = \mathrm{sp}\,(\tilde{\boldsymbol{m}}) = \tilde{\boldsymbol{M}}^{(1,q-1)}, \tag{3.20}$$

and we may also write

$$\boldsymbol{m} = \boldsymbol{M}^{(p,0)}. \tag{3.21}$$

We shall now show that the matrices $\boldsymbol{M}^{(s,p-s)}$ and $\tilde{\boldsymbol{M}}^{(s,q-s)}$ are respectively "*contravariant and covariant second-rank tensors of types* $(s, p-s)$

and $(s, q-s)$", i.e. that they transform according to

$$\left.\begin{aligned}\boldsymbol{M}^{(s,p-s)\prime} &= [\boldsymbol{U}:s]\boldsymbol{M}^{(s,p-s)}[\boldsymbol{U}^{\mathrm{T}}:p-s],\\ \tilde{\boldsymbol{M}}^{(s,q-s)\prime} &= [\tilde{\boldsymbol{U}}:s]\tilde{\boldsymbol{M}}^{(s,q-s)}[\tilde{\boldsymbol{U}}^{\mathrm{T}}:q-s].\end{aligned}\right\} \tag{3.22}$$

These follow directly from the transformation properties of $\boldsymbol{m}$ and $\tilde{\boldsymbol{m}}$. We have

$$\boldsymbol{m}' = [\boldsymbol{U}:p]\boldsymbol{m}, \tag{3.23}$$

i.e.

$$\begin{aligned} m'_{(i_1\ldots i_p)} &= U_{i_1}^{[j_1} U_{i_2}^{j_2} \ldots U_{i_p}^{j_p]} m_{(j_1\ldots j_p)} \\ &= U_{i_1}^{[j_1} U_{i_2}^{j_2} \ldots U_{i_s}^{j_s]} U_{i_{s+1}}^{[j_{s+1}} \ldots U_{i_p}^{j_p]} M_{(j_1\ldots j_s)(j_{s+1}\ldots j_p)}, \end{aligned} \tag{3.24}$$

as required; the second line follows from the combinatorial properties of these antisymmetrized sums. In particular, for the spanning matrices $\boldsymbol{M} \equiv \mathrm{sp}\,(\boldsymbol{m})$ and $\tilde{\boldsymbol{M}} \equiv \mathrm{sp}\,(\tilde{\boldsymbol{m}})$, we have:

$$\boldsymbol{M}' = \boldsymbol{U}\boldsymbol{M}[\boldsymbol{U}^{\mathrm{T}}:p-1], \qquad \tilde{\boldsymbol{M}}' = \tilde{\boldsymbol{U}}\boldsymbol{M}[\tilde{\boldsymbol{U}}^{\mathrm{T}}:q-1]. \tag{3.25}$$

The transformation rules derived here can be used to establish the general validity of results derived under special simplifying conditions. In particular, for any basis we may select a transformation $\boldsymbol{U}$ that maps it into the orthogonal basis

$$\begin{bmatrix} 1 & 0 & 0 & 0 & \cdots \\ 0 & 1 & 0 & 0 & \cdots \\ 0 & 0 & 1 & 0 & \cdots \\ 0 & 0 & 0 & 1 & \cdots \\ \vdots & \vdots & \vdots & \vdots & \ddots \end{bmatrix}. \tag{3.26}$$

To illustrate this, we now outline the proof of the following generalization of Result 2.6.

Result 3.1 *Let $V^{(1)}$ and $V^{(2)}$ be two subspaces of $\mathbb{C}^n$ of dimension p and q and let the corresponding orthogonal subspaces be $\tilde{V}^{(1)}$ and $\tilde{V}^{(2)}$. Further, suppose that $V^{(1)}$ and $V^{(2)}$ possess a largest common subspace $V = V^{(1)} \cap V^{(2)}$ of dimension s ($\geqslant 0$). Then $\tilde{V}^{(1)}$ and $\tilde{V}^{(2)}$ possess a largest common subspace $\bar{V} = \tilde{V}^{(1)} \cap \tilde{V}^{(2)}$ of dimension $n-p-q+s$. If* $\boldsymbol{m}$ *is the minor vector of V and* $\bar{\boldsymbol{m}}$ *is the minor vector of $\bar{V}$ then*

$$\boldsymbol{M}^{(1)(t,p-t)}\tilde{\boldsymbol{M}}^{(2)(n-p-q+t,p-t)\dagger} = \boldsymbol{M}^{(2)(t,q-t)}\tilde{\boldsymbol{M}}^{(1)(n-p-q+t,q-t)\dagger} = \mathbf{0} \tag{3.27}$$

for $0 \leqslant t < s$, *and*

$$\boldsymbol{M}^{(1)(s,p-s)}\tilde{\boldsymbol{M}}^{(2)(n-p-q+s,p-s)\dagger} = \mp\boldsymbol{M}^{(2)(s,q-s)}\tilde{\boldsymbol{M}}^{(1)(n-p-q+s,q-s)\dagger} = \alpha\boldsymbol{m}\bar{\boldsymbol{m}}^{\dagger} \tag{3.28}$$

for some non-vanishing α, *where* $\mathbf{M}^{(1)}$ *and* $\mathbf{M}^{(2)}$ *are defined by* (3.19) *for* $V^{(1)}$ *and* $V^{(2)}$ *respectively, and where the lower sign is taken only if* s *is odd and* p *and* q *are both odd, or if* s *is even and* p *and* q *are not both odd.*

Proof This may be established by first noting that, because of the transformation properties obtained above, the result is invariant under linear transformations (having unit determinant). It is clearly possible to select a basis for $\mathbb{C}^n$ such that $V^{(1)}$, $V^{(2)}$, $V^{(1)} \cap V^{(2)}$, $\tilde{V}^{(1)}$, $\tilde{V}^{(2)}$ and $\tilde{V}^{(1)} \cap \tilde{V}^{(2)}$ are spanned by certain subsets of the basis. We may then define a transformation that maps this basis into the basis (3.26). Result 3.1 is easily established for this special basis, since for $t < s$ all contributions to the matrix products vanish, and for $t = s$ there is only a single non-vanishing term. ■

Result 3.1 establishes a kind of calculus of vector subspaces in terms of their minor vectors. Given the minor vectors of two subspaces $V^{(1)}$ and $V^{(2)}$, it gives an algorithm for determining the dimension of $V^{(1)} \cap V^{(2)}$ and a formula from which the minor vector of this subspace may be determined. Let us define the union of two subspaces $V^{(1)} \cup V^{(2)}$ to be the space spanned by the vectors of $V^{(1)}$ and $V^{(2)}$ taken together. Noting that $V^{(1)} \cup V^{(2)}$ is the space orthogonal to $\tilde{V}^{(1)} \cap \tilde{V}^{(2)}$, Result 3.1 may also be used to calculate the dimension of $V^{(1)} \cup V^{(2)}$ and its minor vector.

The properties of finite transformations derived here can easily be used to find the properties of infinitesimal transformations. Here we shall consider transformations $\boldsymbol{U}$ infinitesimally close to the identity transformation:

$$\boldsymbol{U} = \mathbf{1} + \epsilon \boldsymbol{A}, \tag{3.29}$$

where $\boldsymbol{A}$ is a complex $n \times n$ matrix; ϵ is a small parameter, and terms of order ϵ^2 and higher will be neglected in subsequent analysis. The restriction

$$\det \boldsymbol{U} = 1 \tag{3.30}$$

imposes the requirement

$$\operatorname{tr}(\boldsymbol{A}) = 0. \tag{3.31}$$

It was shown above that minor vectors and other derived quantities could be transformed using matrices of the form $[\boldsymbol{U}:p]$ or $[\boldsymbol{U}^{-1\dagger}:p]$ for some p. With $\boldsymbol{U}$ given by (3.29), we may write

$$[\boldsymbol{U}:p] = \mathbf{1} + \epsilon \boldsymbol{A}^{(p)} + O(\epsilon^2), \tag{3.32}$$

which serves as a definition of $\boldsymbol{A}^{(p)}$. In addition, it is easily shown that

$$[\boldsymbol{U}^{-1\dagger}:p] = \mathbf{1} - \epsilon \boldsymbol{A}^{(p)\dagger} + O(\epsilon^2). \tag{3.33}$$

Note that $\boldsymbol{A}^{(p)}$ is a square matrix of dimension $\binom{n}{p} \times \binom{n}{p}$, with rows and columns labelled by combinations of indices, in the same way as for pth-order minor matrices. It may be shown (Gilbert and Backus, 1966) that the elements of $\boldsymbol{A}^{(p)}$ are given, in terms of the elements of $\boldsymbol{A}$, by the following rules:

$$A^{(p)}_{(i_1 \ldots i_p)(j_1 \ldots j_p)} = \begin{cases} A_{i_1 j_1} + A_{i_2 j_2} + \ldots A_{i_p j_p} & \text{if } i_1 = j_1,\ i_2 = j_2,\ i_3 = j_3, \ldots, i_p = j_p, \\ \pm A_{ij} & \text{if there exists an even } (+) \text{ or odd } (-) \text{ permutation of } (j_1 \ldots j_p) \text{ that brings it into coincidence with } (i_1 \ldots i_p) \text{ except for the two elements, } i \in \{i_1 \ldots i_p\}, j \in \{j_1 \ldots j_p\}, i \neq j, \text{ that do not match,} \\ 0 & \text{otherwise.} \end{cases} \tag{3.34}$$

Corresponding to each of the results for finite transformations, there is a result for infinitesimal transformations in terms of $\boldsymbol{A}^{(p)}$. From (3.11), we have

$$\boldsymbol{m}' = \boldsymbol{m} + \epsilon \boldsymbol{A}^{(p)} \boldsymbol{m}; \tag{3.35}$$

from (3.16)

$$\tilde{\boldsymbol{m}}' = \tilde{\boldsymbol{m}} - \epsilon \boldsymbol{A}^{(q)\dagger} \tilde{\boldsymbol{m}}; \tag{3.36}$$

from (3.17) and (3.18)

$$\boldsymbol{A}^{(q)\dagger} \boldsymbol{E}^{(n,p)} + \boldsymbol{E}^{(n,p)} \boldsymbol{A}^{(p)*} = \mathbf{0}, \tag{3.37}$$

$$\boldsymbol{A}^{(q)} \boldsymbol{E}^{(n,p)} + \boldsymbol{E}^{(n,p)} \boldsymbol{A}^{(p)\mathrm{T}} = \mathbf{0}; \tag{3.38}$$

from (3.22)

$$\boldsymbol{M}^{(s,p-s)\prime} = \boldsymbol{A}^{(s)} \boldsymbol{M}^{(s,p-s)} + \boldsymbol{M}^{(s,p-s)} \boldsymbol{A}^{(p-s)\mathrm{T}}, \tag{3.39}$$

$$\tilde{\boldsymbol{M}}^{(s,q-s)\prime} = -\boldsymbol{A}^{(s)\dagger} \tilde{\boldsymbol{M}}^{(s,q-s)} - \tilde{\boldsymbol{M}}^{(s,q-s)} \boldsymbol{A}^{(q-s)*}; \tag{3.40}$$

and from (3.25)

$$\boldsymbol{M}' = \boldsymbol{A}\boldsymbol{M} + \boldsymbol{M}\boldsymbol{A}^{(p-1)\mathrm{T}}, \tag{3.41}$$

$$\tilde{\boldsymbol{M}}' = -\boldsymbol{A}^{\dagger} \tilde{\boldsymbol{M}} - \tilde{\boldsymbol{M}} \boldsymbol{A}^{(q-1)*}. \tag{3.42}$$

4 LINEAR SYSTEMS OF ORDINARY DIFFERENTIAL EQUATIONS

Returning to the eigenvalue problem formulated in Section 1, we consider the nth-order system

$$\frac{\mathrm{d}\boldsymbol{y}}{\mathrm{d}x} = \boldsymbol{A}(x, \lambda)\boldsymbol{y}, \quad x \in [x_1, x_2], \tag{4.1}$$

and boundary conditions

$$\boldsymbol{y}(x_\alpha) \in \bar{V}^{(\alpha)} \quad (\alpha = 1, 2), \tag{4.2}$$

where $\bar{V}^{(\alpha)}$ are vector spaces of dimension n_α satisfying $n_1 + n_2 = n$. We shall also assume, without essential loss of generality, that $\mathrm{tr}\,(\boldsymbol{A}) = 0$; if this condition is not satisfied then a redefinition of the independent variable

$$\boldsymbol{y}' = \boldsymbol{y} \exp\left\{-\frac{1}{n}\int^x \mathrm{tr}\,(\boldsymbol{A})\,\mathrm{d}x\right\}, \tag{4.3}$$

leads to a system of the form (4.1) in which the system matrix $\boldsymbol{A}$ is trace-free. For a self-adjoint system it may be verified that $\mathrm{tr}\,(\boldsymbol{A})$ is imaginary and that the above transformation preserves the self adjoint property (1.15). As in Section 1, we choose bases $\{\bar{\boldsymbol{\sigma}}_k^{(\alpha)}\}$ of $\bar{V}^{(\alpha)}$ and define $\{\boldsymbol{\sigma}_k^{(\alpha)}(x, \lambda)\}$ to be the solutions of the initial-value problems (1.3). The differential equations (1.3) may be thought of as defining infinitesimal transformations on the basis vectors:

$$\boldsymbol{\sigma}_k^{(\alpha)}(x + \mathrm{d}x) = (1 + \boldsymbol{A}(x)\,\mathrm{d}x)\boldsymbol{\sigma}_k^{(\alpha)}(x). \tag{4.4}$$

Forming minor vectors $\boldsymbol{m}^{(\alpha)}(x)$ from $\{\boldsymbol{\sigma}_k^{(\alpha)}(x)\}$, the result (3.35) leads to the equations

$$\frac{\mathrm{d}\boldsymbol{m}^{(\alpha)}}{\mathrm{d}x} = \boldsymbol{A}^{(n_\alpha)}\boldsymbol{m}^{(\alpha)}, \tag{4.5}$$

where $\boldsymbol{A}^{(n_\alpha)}$ is defined in (3.34). These equations were initially derived by Gilbert and Backus (1966).

The minor vectors of the subspaces orthogonal to $V^{(\alpha)}$ are given by, (2.38),

$$\tilde{\boldsymbol{m}}^{(\alpha)} = \boldsymbol{E}^{(n,n_\alpha)}\boldsymbol{m}^{(\alpha)*}, \tag{4.6}$$

and from (3.36) we find

$$\frac{\mathrm{d}\tilde{\boldsymbol{m}}}{\mathrm{d}x} = -\boldsymbol{A}^{(n-n_\alpha)\dagger}\tilde{\boldsymbol{m}}^{(\alpha)}. \tag{4.7}$$

Thus $\tilde{\boldsymbol{m}}^{(\alpha)}$ are the minor vectors arising in the adjoint problem (1.5), (1.6).

If λ is a non-degenerate eigenvalue then the vector spaces $V^{(1)}(x)$ and $V^{(2)}(\boldsymbol{x})$, having minor vectors $\boldsymbol{m}^{(1)}(x)$ and $\boldsymbol{m}^{(2)}(x)$, possess a one-dimensional common subspace containing the eigenfunction for all x. Similarly, the subspaces $\tilde{V}^{(1)}(x)$ and $\tilde{V}^{(2)}(x)$ orthogonal to $V^{(1)}$ and $V^{(2)}$ possess a one-dimensional common subspace containing the adjoint eigenfunction. Using Result (2.6), these subspaces may be found from the matrix

$$\boldsymbol{M}^{(1)}\tilde{\boldsymbol{M}}^{(2)\dagger} = \mp \boldsymbol{M}^{(2)}\tilde{\boldsymbol{M}}^{(1)\dagger}, \tag{4.8}$$

with $\boldsymbol{M}^{(\alpha)} = \mathrm{sp}\,(\boldsymbol{m}^{(\alpha)})$ and $\tilde{\boldsymbol{M}}^{(\alpha)} = \mathrm{sp}\,(\tilde{\boldsymbol{m}}^{(\alpha)})$ (see (2.36)), where the lower sign applies only if n_1 and n_2 are both odd. The matrix (4.8) is guaranteed to be a dyad of the form

$$\boldsymbol{v}\tilde{\boldsymbol{v}}^{\dagger},$$

where

$$\left.\begin{aligned} \boldsymbol{v} &\in V^{(1)}(x) \cap V^{(2)}(x), \\ \tilde{\boldsymbol{v}} &\in \tilde{V}^{(1)}(x) \cap \tilde{V}^{(2)}(x). \end{aligned}\right\} \tag{4.9}$$

Here we show the following.

Result 4.1 *If λ is a non-degenerate eigenvalue, and $\boldsymbol{y}(x)$ and $\tilde{\boldsymbol{y}}(x)$ are the eigenfunctions of the original problem and its adjoint, then there exists a non-vanishing constant α such that*

$$\boldsymbol{M}^{(1)}\tilde{\boldsymbol{M}}^{(2)\dagger} = \mp \boldsymbol{M}^{(2)}\tilde{\boldsymbol{M}}^{(1)\dagger} = \alpha \boldsymbol{y}\tilde{\boldsymbol{y}}^{\dagger}. \tag{4.10}$$

Proof Result 2.6 guarantees that this is true for some function $\alpha(x)$, and it only remains to show that α is, in fact, independent of x. From (3.41) and (3.42), we may deduce

$$\left.\begin{aligned} \frac{\mathrm{d}\boldsymbol{M}^{(\alpha)}}{\mathrm{d}x} &= \boldsymbol{A}\boldsymbol{M}^{(\alpha)} + \boldsymbol{M}^{(\alpha)}\boldsymbol{A}^{(n_\alpha-1)\mathrm{T}}, \\ \frac{\mathrm{d}\tilde{\boldsymbol{M}}^{(\alpha)}}{\mathrm{d}x} &= -\boldsymbol{A}^{\dagger}\tilde{\boldsymbol{M}}^{(\alpha)} - \tilde{\boldsymbol{M}}^{(\alpha)}\boldsymbol{A}^{(n-n_\alpha-1)*}, \end{aligned}\right\} \tag{4.11}$$

and hence

$$\begin{aligned} \frac{\mathrm{d}}{\mathrm{d}x}(\boldsymbol{M}^{(1)}\tilde{\boldsymbol{M}}^{(2)\dagger}) &= \boldsymbol{A}\boldsymbol{M}^{(1)}\tilde{\boldsymbol{M}}^{(2)\dagger} - \boldsymbol{M}^{(1)}\tilde{\boldsymbol{M}}^{(2)\dagger}\boldsymbol{A} \\ &= \alpha(\boldsymbol{A}\boldsymbol{y}\tilde{\boldsymbol{y}}^{\dagger} - \boldsymbol{y}\tilde{\boldsymbol{y}}^{\dagger}\boldsymbol{A}) \\ &= \alpha\frac{\mathrm{d}}{\mathrm{d}x}(\boldsymbol{y}\tilde{\boldsymbol{y}}^{\dagger}); \end{aligned} \tag{4.12}$$

thus, using (4.10), we obtain

$$\frac{d\alpha}{dx} = 0, \tag{4.13}$$

as required. ■

Result 4.1 enables us to find the eigenfunctions, up to a multiplier depending upon x, by decomposing the dyad $\boldsymbol{M}^{(1)}\tilde{\boldsymbol{M}}^{(2)\dagger}$. In general, the multiplier can be found by integrating a single first-order differential equation in the interval $[x_1, x_2]$. Since $\boldsymbol{M}^{(1)}\tilde{\boldsymbol{M}}^{(2)\dagger}$ is a dyad, there exist vectors $\boldsymbol{u}$ and $\tilde{\boldsymbol{u}}$, unique up to a multiplier for each x, such that

$$\boldsymbol{M}^{(1)}\tilde{\boldsymbol{M}}^{(2)\dagger} = \boldsymbol{u}\tilde{\boldsymbol{u}}^{\dagger}. \tag{4.14}$$

If $\boldsymbol{u}(x)$ and $\tilde{\boldsymbol{u}}(x)$ are differentiable vectors satisfying (4.14) then we know from Result 4.1 that there exist scalar functions $\beta(x)$ and $\tilde{\beta}(x)$ such that $\boldsymbol{y}(x)$ and $\tilde{\boldsymbol{y}}(x)$ given by

$$\boldsymbol{y}(x) = \beta(x)\boldsymbol{u}(x), \tag{4.15}$$

$$\tilde{\boldsymbol{y}}(x) = \tilde{\beta}(x)\tilde{\boldsymbol{u}}(x) \tag{4.16}$$

are eigenfunctions. It also follows from Result 4.1 that $\tilde{\beta}^*(x)\beta(x) =$ const., and thus we may choose $\tilde{\beta}(x) = 1/\beta^*(x)$.

Substituting (4.15) into (4.1), and multiplying on the left by $\boldsymbol{u}^{\dagger}$, we find

$$\beta = \exp\left\{\int^x \frac{\boldsymbol{u}^{\dagger}\boldsymbol{A}\boldsymbol{u} - \boldsymbol{u}^{\dagger}\dot{\boldsymbol{u}}}{\boldsymbol{u}^{\dagger}\boldsymbol{u}}\,dx\right\}. \tag{4.17}$$

Thus the eigenfunctions are determined by (4.15) and (4.17) up to an overall normalization factor.

For self-adjoint systems (see Section 1) we may take $\tilde{\boldsymbol{y}} = \boldsymbol{\Sigma}\boldsymbol{y}$. Thus we find that there exists a constant α such that

$$\boldsymbol{M}^{(1)}\tilde{\boldsymbol{M}}^{(2)\dagger}\boldsymbol{\Sigma} = \alpha\boldsymbol{y}\boldsymbol{y}^{\dagger}. \tag{4.18}$$

By decomposing the left-hand side, $\boldsymbol{y}$ may be determined up to a factor $\beta(x)$ of constant modulus $|\beta(x)|$. If, in addition, the system is real—as is the case for the equations governing free oscillations—we have

$$\boldsymbol{M}^{(1)}\tilde{\boldsymbol{M}}^{(2)\mathrm{T}}\boldsymbol{\Sigma} = \alpha\boldsymbol{y}\boldsymbol{y}^{\mathrm{T}}, \tag{4.19}$$

and $\boldsymbol{y}(x)$ can be determined up to a sign for each x if the left-hand side is known. The sign must be chosen, for each x, to ensure that $\boldsymbol{y}(x)$ is continuous. Thus, in the case of a real self-adjoint system the eigenfunction $\boldsymbol{y}(x)$ can be found from the minors alone, and an integration of the form (4.17) is not required. Equation (4.19) is the basis for the construction of eigenfunctions in the normal-mode algorithm to be discussed below.

From this point we shall restrict consideration to real self-adjoint systems. As pointed out above (see Section 1), such systems are necessarily of even order n and $n_1 = n_2 = \frac{1}{2}n$. Given the basis $\{\boldsymbol{\sigma}_k^{(\alpha)}(x, \lambda)\}$ ($\alpha = 1$ or 2), the foregoing analysis leads to two different methods for the construction of the minor vector of the orthogonal space $\tilde{V}^{(\alpha)}$. First we have the result (4.6), and secondly, by virtue of (1.17), the vectors $\{\boldsymbol{\Sigma}\boldsymbol{\sigma}_k^{(\alpha)}\}$ span $\tilde{V}^{(\alpha)}$, and thus the minor vector formed from these, $[\boldsymbol{\Sigma}:n_1]\boldsymbol{m}^{(\alpha)}(x)$, is also a minor vector of $\tilde{V}^{(\alpha)}$. Since the minor vectors given by these two constructions are the minor vectors of the same space, they must be parallel (Result 2.1); i.e. there exists a non-vanising scalar β such that

$$[\boldsymbol{\Sigma}:n_1]\boldsymbol{m}^{(\alpha)} = \beta \boldsymbol{E}^{(n,n_1)}\boldsymbol{m}^{(\alpha)*}, \tag{4.20}$$

or, using (2.33),

$$\boldsymbol{E}^{(n,n_1)\mathrm{T}}[\Sigma:n_1]\boldsymbol{m}^{(\alpha)} = \beta\boldsymbol{m}^{(\alpha)}, \tag{4.21}$$

where we have also used the fact that $\boldsymbol{m}^{(\alpha)}$ is real. Equation (4.21) is an eigenvalue equation for β. Here we show that the only possible value of β is $\beta = 1$, and thus $\boldsymbol{m}^{(\alpha)}$ is restricted to lie in the subspace belonging to this eigenvalue; this restriction effectively reduces the dimension of the differential system (4.5) that must be solved to find $\boldsymbol{m}^{(\alpha)}(x)$. As we shall see below, for fourth-order systems the order of the minor differential system is reduced from six to five, and for the sixth-order systems from 20 to 14. For a system of order $n = 2p$ the order of the minor differential system is reduced from $\binom{n}{p}$ to $\frac{1}{2}\binom{n}{p} + 2^{p-1}$. To show that $\beta = 1$, we now prove the following.

Result 4.2 *Let n be an even number: $n = 2p$ and let $\boldsymbol{\Sigma}$ be the $n \times n$ matrix of the form* (1.14). *Then we have the following:*

(i) *the matrix $\boldsymbol{\Xi}$ defined by*

$$\boldsymbol{\Xi} = \boldsymbol{E}^{(n,p)\mathrm{T}}[\boldsymbol{\Sigma}:p] \tag{4.22}$$

is symmetric and satisfies

$$\boldsymbol{\Xi}^2 = \mathbf{1}; \tag{4.23}$$

consequently $\boldsymbol{\Xi}$ possesses two orthogonal eigenspaces W^+ and W^-, say, belonging to eigenvalues 1 *and* -1 *respectively;*

(ii) *if* $\boldsymbol{S}$ *is a real* $n \times p$ *matrix, of full rank* p*, satisfying*

$$\boldsymbol{S}^{\mathrm{T}}\boldsymbol{\Sigma}\boldsymbol{S} = \boldsymbol{0} \tag{4.24}$$

then its minor vector $\boldsymbol{m} = [\boldsymbol{S}:p]$ *lies in* W^+;

(iii) *the dimensions of the subspaces* W^+ *and* W^- *are* $\frac{1}{2}\binom{n}{p} \pm 2^{p-1}$.

Proof First we note that, as a consequence of their definitions, together with (2.33), $\boldsymbol{E}^{(n,p)}$ and $[\boldsymbol{\Sigma}:p]$ satisfy

$$\boldsymbol{E}^{(n,p)\mathrm{T}} = (-1)^p \boldsymbol{E}^{(n,p)}, \qquad \boldsymbol{E}^{(n,p)2} = (-1)^p \boldsymbol{1}, \tag{4.25}$$

$$[\boldsymbol{\Sigma}:p]^{\mathrm{T}} = (-1)^p[\boldsymbol{\Sigma}:p], \qquad [\boldsymbol{\Sigma}:p]^2 = (-1)^p \boldsymbol{1}. \tag{4.26}$$

Since $\det \boldsymbol{\Sigma} = 1$, (3.18) can be used to obtain

$$\boldsymbol{E}^{(n,p)} = [\boldsymbol{\Sigma}:p]\boldsymbol{E}^{(n,p)}[\boldsymbol{\Sigma}:p]^{\mathrm{T}} \tag{4.27}$$

or

$$\boldsymbol{E}^{(n,p)}[\boldsymbol{\Sigma}:p] = [\boldsymbol{\Sigma}:p]\boldsymbol{E}^{(n,p)}. \tag{4.28}$$

Hence $[\boldsymbol{\Sigma}:p]$ and $\boldsymbol{E}^{(n,p)}$ are real orthogonal matrices that commute and are either symmetric or antisymmetric according to whether p is even or odd. The results under (i) above follow immediately from these properties.

From the arguments leading to (4.21), we have

$$\boldsymbol{\Xi}\boldsymbol{m} = \beta \boldsymbol{m} \tag{4.29}$$

for some non-vanishing β, which, by virtue of (4.23), has the value 1 or -1. By direct calculation, it may be shown that the row of $\boldsymbol{\Xi}$ labelled by the combination $(1, 2, \ldots, p)$ has only the single non-vanishing element

$$\Xi_{(1,2,\ldots,p)(1,2,\ldots,p)} = 1.$$

From the $(1, 2, \ldots, p)$ element of (4.30), we find that if $\boldsymbol{m}_{(1,2,3,\ldots,p)}$ is non-zero then β is necessarily equal to 1. For many practical purposes we may be assured that $m_{(1,2,\ldots,p)} \neq 0$, or that it is so almost everywhere in the domain of certain parameters upon which $\boldsymbol{m}$ continuously depends, and in that case the inference that $\beta = 1$ is justified. In order to establish Result 4.2 in the general case, however, we consider the transformation

$$\boldsymbol{S}' = \boldsymbol{U}\boldsymbol{S}, \tag{4.30}$$

where $\boldsymbol{U}$ is the complex partitioned matrix

$$\boldsymbol{U} = \frac{1}{\sqrt{2}}\begin{bmatrix} \boldsymbol{1} & \mathrm{i}\boldsymbol{1} \\ \mathrm{i}\boldsymbol{1} & \boldsymbol{1} \end{bmatrix}. \tag{4.31}$$

It may be shown directly that det $\boldsymbol{U}=1$ and

$$\boldsymbol{\Sigma U} = \boldsymbol{U}^{-1\mathrm{T}}\boldsymbol{\Sigma}, \tag{4.32}$$

and also, from (3.18),

$$\boldsymbol{E}^{(n,p)}[\boldsymbol{U}^{-1\mathrm{T}}:p] = [\boldsymbol{U}:p]\boldsymbol{E}^{(n,p)}. \tag{4.33}$$

Thus

$$\begin{aligned}[\boldsymbol{U}:p]\boldsymbol{\Xi} &= [\boldsymbol{U}:p](-1)^p\boldsymbol{E}^{(n,p)}[\boldsymbol{\Sigma}:p]\\ &= (-1)^p\boldsymbol{E}^{(n,p)}[\boldsymbol{U}^{-1\mathrm{T}}:p][\boldsymbol{\Sigma}:p]\\ &= (-1)^p\boldsymbol{E}^{(n,p)}[\boldsymbol{\Sigma}:p][\boldsymbol{U}:p]\\ &= \boldsymbol{\Xi}[\boldsymbol{U}:p];\end{aligned} \tag{4.34}$$

i.e. $[\boldsymbol{U}:p]$ commutes with $\boldsymbol{\Xi}$. Thus, writing

$$\boldsymbol{m}' = [\boldsymbol{US}:p] = [\boldsymbol{U}:p]\boldsymbol{m}, \tag{4.35}$$

(4.29) becomes

$$\boldsymbol{\Xi m}' = \beta\boldsymbol{m}'. \tag{4.36}$$

As above, the conclusion $\beta=1$ follows, provided that $m'_{(1,2,\ldots,p)}\neq 0$; to complete the proof, we now show that the condition $m'_{(1,2,\ldots,p)}\neq 0$ is always satisfied. Writing

$$\boldsymbol{S} = \begin{bmatrix}\boldsymbol{S}_1\\ \boldsymbol{S}_2\end{bmatrix}, \tag{4.37}$$

where $\boldsymbol{S}_1$ and $\boldsymbol{S}_2$ are $p\times p$ submatrices, we have

$$m'_{(1,2,\ldots,p)} = \det\frac{1}{\sqrt{2}}(\boldsymbol{S}_1 + \mathrm{i}\boldsymbol{S}_2), \tag{4.38}$$

and, from (4.24),

$$\boldsymbol{S}_1^{\mathrm{T}}\boldsymbol{S}_2 - \boldsymbol{S}_2^{\mathrm{T}}\boldsymbol{S}_1 = \boldsymbol{0}. \tag{4.39}$$

Now if $m'_{(1,2,\ldots,p)}=0$, the matrix $\boldsymbol{S}_1+\mathrm{i}\boldsymbol{S}_2$ is singular, and hence there exists a non-vanishing p-vector $\boldsymbol{x}$ satisfying

$$(\boldsymbol{S}_1 + \mathrm{i}\boldsymbol{S}_2)\boldsymbol{x} = \boldsymbol{0}, \tag{4.40}$$

from which we obtain

$$\begin{aligned}0 &= |(\boldsymbol{S}_1 + \mathrm{i}\boldsymbol{S}_2)\boldsymbol{x}|^2\\ &= \boldsymbol{x}^\dagger(\boldsymbol{S}_1^{\mathrm{T}} - \mathrm{i}\boldsymbol{S}_2^{\mathrm{T}})(\boldsymbol{S}_1 + \mathrm{i}\boldsymbol{S}_2)\boldsymbol{x}\\ &= \boldsymbol{x}^\dagger(\boldsymbol{S}_1^{\dagger}\boldsymbol{S}_1 + \boldsymbol{S}_2^{\dagger}\boldsymbol{S}_2)\boldsymbol{x} + \mathrm{i}\boldsymbol{x}^\dagger(\boldsymbol{S}_1^{\mathrm{T}}\boldsymbol{S}_2 - \boldsymbol{S}_2^{\mathrm{T}}\boldsymbol{S}_1)\boldsymbol{x}.\end{aligned} \tag{4.41}$$

Since, by (4.39), the second term vanishes, we find

$$|\mathbf{S}_1\boldsymbol{x}|^2 + |\mathbf{S}_2\boldsymbol{x}|^2 = 0, \tag{4.42}$$

from which we deduce

$$\mathbf{S}_1\boldsymbol{x} = \mathbf{0}, \qquad \mathbf{S}_2\boldsymbol{x} = \mathbf{0}, \tag{4.43}$$

and thus

$$\mathbf{S}\boldsymbol{x} = \mathbf{0}; \tag{4.44}$$

but this violates the assumption that $\mathbf{S}$ is of full rank p, and thus $m'_{(1,2,\ldots,p)} \neq 0$, as required. This completes the proof of Result 4.2, except for (iii), of which the proof is omitted. ■

We next consider an alternative derivation of the projection operators that are relevant to real self-adjoint systems. Let us denote by $\mathbf{S}^{(\alpha)}(x, \lambda)$ $(\alpha = 1, 2)$ the $n \times n_1$ matrices having columns $\boldsymbol{\sigma}_k^{(\alpha)}(x, \lambda)$ $(k = 1, 2, \ldots, n_1)$. By virtue of (1.3), (1.15) and (1.16), we have

$$\frac{\mathrm{d}\mathbf{S}^{(\alpha)}}{\mathrm{d}x} = \mathbf{A}(x, \lambda)\mathbf{S}^{(\alpha)}, \tag{4.45}$$

$$\mathbf{S}^{(\alpha)\mathrm{T}}\boldsymbol{\Sigma}\mathbf{S}^{(\alpha)} = \mathbf{0}, \tag{4.46}$$

$$\frac{\mathrm{d}}{\mathrm{d}x}(\mathbf{S}^{(1)\mathrm{T}}\boldsymbol{\Sigma}\mathbf{S}^{(2)}) = \mathbf{0}, \tag{4.47}$$

and the minor vectors $\boldsymbol{m}^{(\alpha)}(x, \lambda)$ and $\tilde{\boldsymbol{m}}^{(\alpha)}(x, \lambda)$, introduced above, are simply

$$\boldsymbol{m}^{(\alpha)} = [\mathbf{S}^{(\alpha)} : n_1], \tag{4.48}$$

$$\tilde{\boldsymbol{m}}^{(\alpha)} = \mathbf{E}^{(n,n_1)}\boldsymbol{m}^{(\alpha)} = [\boldsymbol{\Sigma}\mathbf{S}^{(\alpha)} : n_1], \tag{4.49}$$

where the second equality is from (4.20) with $\beta = 1$. Let us assume, for the moment, that the columns of $\mathbf{S}^{(1)}$ and $\mathbf{S}^{(2)}$, taken together, span $\mathbb{C}^n$—i.e. the matrix $\mathbf{S}$ given by (1.4) is non-singular. Then we may write, for any vector $\boldsymbol{x}$,

$$\boldsymbol{x} = \boldsymbol{x}^{(1)} + \boldsymbol{x}^{(2)}, \tag{4.50}$$

where

$$\boldsymbol{x}^{(1)} = \mathbf{S}^{(1)}\boldsymbol{k}^{(1)}, \qquad \boldsymbol{x}^{(2)} = \mathbf{S}^{(2)}\boldsymbol{k}^{(2)} \tag{4.51}$$

for some n_1-vectors $\boldsymbol{k}^{(1)}$ and $\boldsymbol{k}^{(2)}$.

Multiplying (4.50) on the left by $\mathbf{S}^{(1)\mathrm{T}}\boldsymbol{\Sigma}$, and making use of (4.46), we find

$$\mathbf{S}^{(1)\mathrm{T}}\boldsymbol{\Sigma}\boldsymbol{x} = \mathbf{S}^{(1)\mathrm{T}}\boldsymbol{\Sigma}\mathbf{S}^{(2)}\boldsymbol{k}^{(2)}. \tag{4.52}$$

Thus, solving for $\boldsymbol{k}^{(2)}$ and substituting into (4.51), we find

$$\boldsymbol{x}^{(2)} = \mathbf{S}^{(2)}\boldsymbol{X}^{-1}\mathbf{S}^{(1)\mathrm{T}}\boldsymbol{\Sigma}\boldsymbol{x}, \tag{4.53}$$

and similarly

$$\boldsymbol{x}^{(1)} = -\mathbf{S}^{(1)}\boldsymbol{X}^{\mathrm{T}-1}\mathbf{S}^{(2)\mathrm{T}}\boldsymbol{\Sigma}\boldsymbol{x}, \tag{4.54}$$

where we have introduced the $n_1 \times n_1$ matrix

$$\boldsymbol{X} \equiv \mathbf{S}^{(1)\mathrm{T}}\boldsymbol{\Sigma}\mathbf{S}^{(2)}, \tag{4.55}$$

which (by (4.7)) is independent of x. The matrices that multiply $\boldsymbol{x}$ in (4.53) and (4.54) are, by definition, projection operators onto the spaces spanned by the columns of $\mathbf{S}^{(2)}$ and $\mathbf{S}^{(1)}$ respectively, and thus we may identify them with those obtained previously (Result 2.3); i.e.

$$-\mathbf{S}^{(1)}\boldsymbol{X}^{\mathrm{T}-1}\mathbf{S}^{(2)\mathrm{T}}\boldsymbol{\Sigma} = \frac{1}{\Delta}\mathbf{M}^{(1)}\tilde{\mathbf{M}}^{(2)\dagger}. \tag{4.56}$$

$$\mathbf{S}^{(2)}\boldsymbol{X}^{-1}\mathbf{S}^{(1)\mathrm{T}}\boldsymbol{\Sigma} = \pm\frac{1}{\Delta}\mathbf{M}^{(2)}\tilde{\mathbf{M}}^{(1)\mathrm{T}\dagger}, \tag{4.57}$$

$$\tag{4.58}$$

with

$$\Delta = \tilde{\boldsymbol{m}}^{(2)\mathrm{T}}\boldsymbol{m}^{(1)} = \pm\tilde{\boldsymbol{m}}^{(1)\mathrm{T}}\boldsymbol{m}^{(2)}, \tag{4.59}$$

where the lower signs apply if n_1 is odd and where

$$\mathbf{M}^{(\alpha)} = \mathrm{sp}\,(\boldsymbol{m}^{(\alpha)}), \qquad \tilde{\mathbf{M}}^{(\alpha)} = \mathrm{sp}\,(\tilde{\boldsymbol{m}}^{(\alpha)}). \tag{4.60}$$

Furthermore, taking n_1th minors of (4.55),

$$\begin{aligned} \det \boldsymbol{X} &= \boldsymbol{m}^{(1)\mathrm{T}}[\boldsymbol{\Sigma}:n_1]\boldsymbol{m}^{(2)} \\ &= \boldsymbol{m}^{(1)\mathrm{T}}\tilde{\boldsymbol{m}}^{(2)} \qquad \text{(by (4.49))} \\ &= \Delta. \end{aligned} \tag{4.61}$$

Thus we may write

$$\left.\begin{aligned} \mathbf{S}^{(1)}\bar{\boldsymbol{X}}^{\mathrm{T}}\mathbf{S}^{(2)\mathrm{T}} &= \mathbf{M}^{(1)}\tilde{\mathbf{M}}^{(2)\dagger}\boldsymbol{\Sigma}, \\ \mathbf{S}^{(2)}\bar{\boldsymbol{X}}\mathbf{S}^{(1)\mathrm{T}} &= \pm\mathbf{M}^{(2)}\tilde{\mathbf{M}}^{(1)\dagger}\boldsymbol{\Sigma}, \end{aligned}\right\} \tag{4.62}$$

where $\bar{\boldsymbol{X}}$ represents the adjoint matrix:

$$\bar{X}_{ij} = \text{cofactor of } x_{ji}. \tag{4.63}$$

Although the derivation given here relies upon the assumption $\Delta \neq 0$, it may be shown, in fact, that (4.62) is valid in general; we omit demonstration of this.

From Result 4.1, we now have the following.

Result 4.3 *If λ is a non-degenerate eigenvalue of the nth order real self-adjoint system* (1.1), (1.2) *and $\mathbf{y}(x)$ is the associated eigenfunction then there exists a constant α such that*

$$\left.\begin{aligned} &\mathbf{S}^{(1)}\bar{\mathbf{X}}^{\mathrm{T}}\mathbf{S}^{(2)\mathrm{T}} = \mathbf{S}^{(2)}\bar{\mathbf{X}}\mathbf{S}^{(1)\mathrm{T}}, \\ &\mathbf{M}^{(1)}\tilde{\mathbf{M}}^{(2)\mathrm{T}}\boldsymbol{\Sigma} = \mp\mathbf{M}^{(2)}\tilde{\mathbf{M}}^{(1)\mathrm{T}}\boldsymbol{\Sigma} = \alpha\mathbf{y}\mathbf{y}^{\mathrm{T}}, \end{aligned}\right\} \tag{4.64}$$

where $\bar{\mathbf{X}} = \operatorname{adj}(\mathbf{S}^{(1)\mathrm{T}}\boldsymbol{\Sigma}\mathbf{S}^{(2)})$ and is independent of x, and where the lower sign applies if and only if $\frac{1}{2}n$ is odd. Other notation used in (4.64) *is as defined previously.*

While the expressions for the eigenfunction dyad in terms of minors are more suitable for numerical calculation, the expressions in terms of $\mathbf{S}^{(1)}$ and $\mathbf{S}^{(2)}$ are useful in the derivation of the theoretical results that follow.

Let us again suppose that $\Delta \neq 0$, i.e. $\mathbf{S}(x)$ defined by (1.4),

$$\mathbf{S}(x) = [\mathbf{S}^{(1)}(x) \quad \mathbf{S}^{(2)}(x)], \tag{4.65}$$

is a fundamental matrix. The propagator matrix or matrizant $\mathbf{P}(x, x_0)$ (Coddington and Levinson, 1955; Gilbert and Backus, 1966) can be written, for any $x, x_0 \in [x_1, x_2]$, as

$$\mathbf{P}(x, x_0) = \mathbf{S}(x)\mathbf{S}^{-1}(x_0). \tag{4.66}$$

The properties (4.46) can be used to show that $\mathbf{P}(x, x_0)$ can be written as

$$\mathbf{P}(x, x_0) = (\mathbf{S}^{(2)}(x)\mathbf{X}^{-1}\mathbf{S}^{(1)\mathrm{T}}(x_0) - \mathbf{S}^{(1)}(x)\mathbf{X}^{-1\mathrm{T}}\mathbf{S}^{(2)\mathrm{T}}(x_0))\boldsymbol{\Sigma}, \tag{4.67}$$

where $\mathbf{X}$ is the constant $n_1 \times n_1$ matrix defined in (4.55). In order to verify this, it is necessary only to check that $\mathbf{P}(x, x_0)\mathbf{S}^{(\alpha)}(x_0) = \mathbf{S}^{(\alpha)}(x)$ for $\alpha = 1, 2$ with $\mathbf{P}(x, x_0)$ given by (4.67).

We now use this result to evaluate the derivative $\mathrm{d}\Delta/\mathrm{d}\lambda$, where λ is the eigenvalue parameter. First we evaluate $\mathbf{S}_\lambda^{(\alpha)}$, where subscript λ denotes the partial derivative at constant x. Differentiating (4.45) with respect to λ, we obtain

$$\frac{\mathrm{d}}{\mathrm{d}x}\mathbf{S}_\lambda^{(\alpha)} = \mathbf{A}\mathbf{S}_\lambda^{(\alpha)} + \mathbf{A}_\lambda\mathbf{S}^{(\alpha)}, \tag{4.68}$$

from which it follows, using standard techniques, that

$$\boldsymbol{S}_\lambda^{(\alpha)}(x) = \int_{x_\alpha}^{x} \boldsymbol{P}(x, x')\boldsymbol{A}_\lambda(x')\boldsymbol{S}^{(\alpha)}(x')\,\mathrm{d}x + \boldsymbol{P}(x, x_\alpha)\boldsymbol{S}_\lambda^{(\alpha)}(x_\alpha), \quad \alpha = 1 \text{ or } 2. \tag{4.69}$$

We shall assume that the given boundary conditions are independent of λ; thus the second term in (4.69) vanishes. Using (4.55), (4.61) and the identity

$$(\det \boldsymbol{X})_\lambda = \mathrm{tr}\,(\boldsymbol{X}_\lambda \bar{\boldsymbol{X}}), \tag{4.70}$$

we obtain, after some algebra,

$$\Delta_\lambda = -\mathrm{tr}\left\{\int_{x_1}^{x_2} \boldsymbol{S}^{(2)}\bar{\boldsymbol{X}}\boldsymbol{S}^{(1)\mathrm{T}}\boldsymbol{\Sigma}\boldsymbol{A}_\lambda\,\mathrm{d}x\right\}. \tag{4.71}$$

This result may also be shown to hold in the case $\Delta = 0$. Note, therefore, that from Result 4.2, when $\Delta = 0$,

$$\Delta_\lambda = -\alpha\int_{x_1}^{x_2} \boldsymbol{y}^{\mathrm{T}}\boldsymbol{\Sigma}\boldsymbol{A}_\lambda \boldsymbol{y}\,\mathrm{d}x, \tag{4.72}$$

where $\boldsymbol{y}$ is the eigenfunction—assumed to be non-degenerate. By virtue of (1.15) and the assumed reality of $\boldsymbol{A}$, the matrix $\boldsymbol{\Sigma}\boldsymbol{A}$, and hence $\boldsymbol{\Sigma}\boldsymbol{A}_\lambda$, is symmetric. Motivated by Sturm–Liouville theory, and also by the properties of the normal-mode equations (see below), we shall assume that for λ in the interval of interest, $[\lambda_1, \lambda_2]$, $-\boldsymbol{\Sigma}\boldsymbol{A}_\lambda$ is positive-semidefinite and, furthermore, that it has the property that if, for a solution $\boldsymbol{y}(x)$ of (1.1), $\boldsymbol{y}^{\mathrm{T}}\boldsymbol{\Sigma}\boldsymbol{A}_\lambda \boldsymbol{y} = 0$ for all $x \in [x_1, x_2]$ then $\boldsymbol{y}(x)$ necessarily vanishes for all $x \in [x_1, x_2]$, i.e. we shall consider the class of eigenvalue problems that allow us to infer from (4.72) that $\Delta_\lambda \neq 0$ and $\mathrm{sgn}\,\Delta_\lambda = \mathrm{sgn}\,\alpha$. We shall subsequently verify that the eigenvalue problem for the Earth's normal modes belongs to this class, which is a generalization of the class of Sturm–Liouville problems. Let us define a normalized eigenfunction

$$\hat{\boldsymbol{y}}(x) \equiv \left\{-\int_{x_1}^{x_2} \boldsymbol{y}^{\mathrm{T}}\boldsymbol{\Sigma}\boldsymbol{A}_\lambda \boldsymbol{y}\,\mathrm{d}x\right\}^{-1/2} \boldsymbol{y}(x). \tag{4.73}$$

Then we have, simply (Result 4.3),

$$\boldsymbol{M}^{(1)}\tilde{\boldsymbol{M}}^{(2)\mathrm{T}}\boldsymbol{\Sigma} = \Delta_\lambda \hat{\boldsymbol{y}}\hat{\boldsymbol{y}}^{\mathrm{T}}. \tag{4.74}$$

This equation has the important consequence that when λ is an eigenvalue the diagonal elements of $\boldsymbol{M}^{(1)}\tilde{\boldsymbol{M}}^{(2)}\boldsymbol{\Sigma}$ are of uniform sign in $[x_1, x_2]$ and have the same sign as Δ_λ. More generally, if $\boldsymbol{R}(x)$ is any continuous positive-definite $n \times n$ matrix function of x then the sign of $\mathrm{tr}\,(\tilde{\boldsymbol{M}}^{(2)\mathrm{T}}\boldsymbol{\Sigma R M}^{(1)})$ is uniform in $[x_1, x_2]$ and agrees with that of Δ_λ.

Let us define

$$\vartheta_R(x, \lambda) = -\frac{1}{\pi}\cot^{-1}\left\{\frac{\mathrm{tr}\,(\tilde{\boldsymbol{M}}^{(2)\mathrm{T}}(x_2)\boldsymbol{\Sigma R}(x)\boldsymbol{M}^{(1)}(x))}{\tilde{\boldsymbol{m}}^{(2)\mathrm{T}}(x_2)\boldsymbol{m}^{(1)}(x)}\right\}. \tag{4.75}$$

The ambiguity in the value of $\cot^{-1}$ is resolved by requiring that ϑ_R be continuous at points where the argument of $\cot^{-1}$ is singular. This requirement defines $\vartheta_R(x, \lambda)$ up to an additive integer that is independent of x and λ. Because, by assumption, the given boundary conditions independent of λ, $\boldsymbol{m}^{(1)}(x_1)$, $\boldsymbol{m}^{(2)}(x_2)$ etc. have this property, $\vartheta_R(x_1, \theta)$ is independent of λ. Thus the definition of $\vartheta_R(x, \lambda)$ may be made unique by specifying

$$0 \leqslant \vartheta_R(x_1, \lambda) < \pi. \tag{4.76}$$

The value of $\vartheta_R(x, \lambda)$ at other points can be uniquely determined by traversing a trajectory in the (x, λ) plane from any point on the line $x = x_1$ to the desired point and imposing continuity at points on the trajectory where the denominator in (4.75) vanishes.

The usefulness of the function $\vartheta_R(x, \lambda)$ lies in the properties of $\vartheta_R(x_2, \lambda)$ as a function of λ. When $x = x_2$ the denominator in (4.75) is simply Δ and, as has been shown above, the numerator agrees in sign with $\mathrm{d}\Delta/\mathrm{d}\lambda$ when Δ vanishes. It follows from this that if, for some λ_0, $\Delta(\lambda_0)$ vanishes then $\vartheta_R(x_2, \lambda_0)$ is an integer, k say, and that for sufficiently small $|\varepsilon|$, $\vartheta_R(x_2, \lambda_0 + \varepsilon)$ is either greater than or less than k according to the sign of ε, i.e. $\vartheta_R(x_2, \lambda)$, as a function of λ, is always increasing at points where ϑ_R is an integer. Since the condition for $\vartheta_R(x_2, \lambda)$ to be an integer is the same as that for λ to be an eigenvalue—viz $\Delta(\lambda) = 0$—we conclude that the number of eigenvalues lying between two values λ_a and λ_b ($\lambda_a < \lambda_b$; $\Delta(\lambda_a) \neq 0$; $\Delta(\lambda_b) \neq 0$) is given by

$$\nu(\lambda_a, \lambda_b) = \mathrm{ent}\,(\vartheta_R(x_2, \lambda_b)) - \mathrm{ent}\,(\vartheta_R(x_2, \lambda_a)), \tag{4.77}$$

where $\mathrm{ent}\,(\vartheta)$ is the greatest integer less than or equal to ϑ. The values $\vartheta_R(x_2, \lambda_b)$ and $\vartheta_R(x_2, \lambda_a)$ can be found by tracking the values of $\vartheta_R(x, \lambda)$ as x increases from x_1 to x_2 with λ fixed, making use of (4.76) at the point $x = x_1$. Thus (4.77) enables the number of eigenvalues between λ_a and λ_b to be calculated with only two integrations of the differential equations. This is the theoretical basis of the algorithm for bracketing eigenvalues that is implemented in the computer program accompanying this chapter. In practice it is not necessary to perform the trigonometric calculation in (4.75), but simply to count zero-crossings of the denominator, counting positively or negatively according to the product of the signs of the numerator and the derivative of the denominator. In the interests of simplicity, we have, up to now, omitted consideration of discontinuities and changes or order of the equations. In the absence of such complications, an obvious choice for $\boldsymbol{R}(x)$ is simply $\boldsymbol{R} = \mathbf{1}$; in the normal-mode problem $\boldsymbol{R}(x)$ will be chosen to ensure that $\vartheta_R(x, \lambda)$ is continuous as a function of x at the core–mantle boundary and at the ocean floor, where the solution is discontinuous and where the order of the differential equations changes from six in the solid to four in the fluid.

5 THE DIFFERENTIAL EQUATIONS GOVERNING THE EARTH'S FREE OSCILLATIONS

The normal-mode problem, including gravitation, affords examples of second-order (toroidal modes), fourth-order (radial modes, spheroidal modes in a fluid) and sixth-order (spheroidal modes in a solid) differential systems. In each case these are self-adjoint systems satisfying the positivity condition that $-\boldsymbol{\Sigma A}_\lambda$ is positive-semidefinite. Here we specialize to these particular systems the general results of previous sections. We refer to the paper by Takeuchi and Saito (1972) for a complete treatment of the differential equations governing normal modes, and to Chapman and Woodhouse (1981) for their self-adjoint forms.

Adopting a spherical coordinate system (r, θ, ϕ), the elastic displacement field $\boldsymbol{u}$ for a mode of angular order l and azimuthal order m is given, in terms of its spherical components, by (Pekeris and Jarosch, 1958; Alterman *et al.*, 1959)

$$\left.\begin{aligned} u_r &= U(r)Y_l^m(\theta, \phi), \\ u_\theta &= V(r)\partial_\theta Y_l^m(\theta, \phi) + W(r)\operatorname{cosec}\theta\,\partial_\phi Y_l^m(\theta, \phi), \\ u_\phi &= V(r)\operatorname{cosec}\theta\,\partial_\phi Y_l^m(\theta, \phi) - W(r)\,\partial_\theta Y_l^m(\theta, \phi), \end{aligned}\right\} \tag{5.1}$$

and the perturbation in gravitational potential is

$$\varphi_1 = P(r)Y_l^m(\theta, \phi), \tag{5.2}$$

where $Y_l^m(\theta, \phi)$ are spherical harmonics. For toroidal modes U, V and P are identically zero; for spheroidal modes $W = 0$, and for radial modes V and W are zero. In each case the functions among U, V, W and P that are not identically zero satisfy systems of differential equations in r that are independent of m and that depend upon the frequency ω of the oscillation; in making use of preceding results, we identify $x \equiv r$, $\lambda \equiv \omega^2$. Earth structure is specified in terms of the piecewise-continuous functions

$A(r)$, $C(r)$, $L(r)$, $N(r)$ and $F(r)$ that characterize the linear elastic constitutive law for a transversely isotropic material (see Takeuchi and Saito, 1972), together with the density $\rho_0(r)$. The equilibrium equations for the undisturbed model, under the assumption that the initial stress field $\boldsymbol{t}^0$ is isotropic,

$$t_{ij}^0 = -p_0(r)\delta_{ij}, \tag{5.3}$$

allow the initial pressure $p_0(r)$ and the initial gravitational acceleration $g_0(r)$ to be determined from $\rho_0(r)$:

$$g_0(r) = \frac{4\pi G}{r^2}\int_0^r \rho_0(r')r'^2\,\mathrm{d}r', \tag{5.4}$$

$$p_0(r) = \int_r^a \rho_0(r')g_0(r')\,\mathrm{d}r', \tag{5.5}$$

where a is the radius of the Earth or, in the presence of an ocean, of the ocean surface.

Toroidal modes

Writing

$$\boldsymbol{y} = \begin{bmatrix} r\zeta W \\ r\zeta L\dot{W} - \zeta W \end{bmatrix}, \tag{5.6}$$

where we have introduced $\zeta = [l(l+1)]^{1/2}$ and where a dot denotes the derivative with respect to r, we find that $\boldsymbol{y}$ satisfies a differential equation of the form (1.1) with the system matrix

$$\boldsymbol{A} = \begin{bmatrix} 2r^{-1} & L^{-1} \\ (\zeta^2-2)r^{-2}N - \omega^2\rho_0 & -2r^{-1} \end{bmatrix}. \tag{5.7}$$

Note therefore that

$$\boldsymbol{\Sigma A} = \begin{bmatrix} (\zeta^2-2)r^{-2}N - \omega^2\rho_0 & -2r^{-1} \\ -2r^{-1} & -L^{-1} \end{bmatrix} \tag{5.8}$$

is symmetric and

$$-\boldsymbol{\Sigma A}_\lambda = \begin{bmatrix} \rho_0 & 0 \\ 0 & 0 \end{bmatrix} \tag{5.9}$$

is positive-semidefinite. The normalization integral (see (4.73)) is

$$\int_{x_1}^{x_2} \boldsymbol{y}^{\mathrm{T}}(-\boldsymbol{\Sigma A}_\lambda)\boldsymbol{y}\,\mathrm{d}r = \int_{a_{\mathrm{cmb}}}^{a_{\mathrm{solid}}} \rho_0 l(l+1)W^2r^2\,\mathrm{d}r, \tag{5.10}$$

where we have identified $x_1 = a_{\text{cmb}}$, $x_2 = a_{\text{solid}}$, and where a_{cmb} is the radius of the core–mantle boundary and a_{solid} is the radius of the spherical surface corresponding to the ocean floor or, in the absence of an ocean, to the free surface. Note that $-\boldsymbol{\Sigma A}_\lambda$ has the required property that if $\boldsymbol{y}^{\mathrm{T}}\boldsymbol{\Sigma A}_\lambda \boldsymbol{y}$ vanishes for all $x \in [x_1, x_2]$ then $\boldsymbol{y}(x) \equiv \mathbf{0}$ (see the discussion following (4.72)).

The boundary conditions are that tractions vanish at a_{cmb} and a_{solid}:

$$y_2(x_1) = 0, \qquad y_2(x_2) = 0. \tag{5.11}$$

Thus we identify

$$\bar{\boldsymbol{\sigma}}_1^{(1)} = \bar{\boldsymbol{\sigma}}_2^{(1)} = \begin{bmatrix} 1 \\ 0 \end{bmatrix}, \tag{5.12}$$

i.e. $\bar{V}^{(1)}$ and $\bar{V}^{(2)}$ are the one-dimensional vector spaces spanned by this vector. At other boundaries in the model $\boldsymbol{y}(x)$ is required to be continuous.

For second-order systems the minor formalism is degenerate since all minors are of order 1. Thus $\boldsymbol{m}^{(1)}(x) = \boldsymbol{M}^{(1)}(x) = \boldsymbol{\sigma}^{(1)}(x, \lambda)$ and $\boldsymbol{m}^{(2)}(x) = \boldsymbol{M}^{(2)}(\boldsymbol{x}) = \boldsymbol{\sigma}^{(2)}(x, \lambda)$, where $\boldsymbol{\sigma}^{(1)}$ and $\boldsymbol{\sigma}^{(2)}$ are the solutions of the initial-value problems (1.3).

Setting $\boldsymbol{R}(x) = \mathbf{1}$ the counting variable $\vartheta_R(x, \lambda)$ is given by, (4.75),

$$\vartheta_R(x, \lambda) = -\frac{1}{\pi}\cot^{-1}\left\{\frac{\sigma_2^{(1)}(x, \lambda)}{\sigma_1^{(1)}(x, \lambda)}\right\}. \tag{5.13}$$

We note that the properties of Result 4.2 are also degenerate, since $\boldsymbol{E}^{(2,1)} = \boldsymbol{\Sigma}$ and consequently $\boldsymbol{\Xi} = \mathbf{1}$, and the eigenfunction construction (4.74) reduces to

$$\boldsymbol{\sigma}^{(1)}\boldsymbol{\sigma}^{(2)\mathrm{T}} = \Delta_\lambda \hat{\boldsymbol{y}}\hat{\boldsymbol{y}}^{\mathrm{T}}. \tag{5.14}$$

When $\Delta = 0$, $\boldsymbol{\sigma}^{(1)}(x, \lambda)$ and $\boldsymbol{\sigma}^{(2)}(x, \lambda)$ are equal up to multiplication by a constant.

Radial modes

We write

$$\boldsymbol{y} = \begin{bmatrix} rU \\ rP \\ rC\dot{U} + 2FU \\ (4\pi G)^{-1}(r\dot{P} + \dot{P}) + \rho_0 rU \end{bmatrix}, \tag{5.15}$$

where G is the universal constant of gravitation; $\boldsymbol{y}$ satisfies a differential system of the form (1.1), with system matrix of the form

$$\boldsymbol{A}=\begin{bmatrix}\boldsymbol{T} & \boldsymbol{K}\\ \boldsymbol{S} & -\boldsymbol{T}^{\mathrm{T}}\end{bmatrix}, \tag{5.16}$$

where the submatrices are

$$\boldsymbol{T}=\begin{bmatrix}r^{-1}(1-2FC^{-1}) & 0\\ -4\pi G\rho_0 & 0\end{bmatrix}, \tag{5.17}$$

$$\boldsymbol{S}=\begin{bmatrix}-\rho_0\omega^2+4r^{-2}(A-N-F^2C^{-1}-r\rho_0 g_0) & -r^{-1}\rho_0\\ -r^{-1}\rho_0 & 0\end{bmatrix}, \tag{5.18}$$

$$\boldsymbol{K}=\begin{bmatrix}C^{-1} & 0\\ 0 & 4\pi G\end{bmatrix}. \tag{5.19}$$

Note that $\boldsymbol{S}$ and $\boldsymbol{T}$ are symmetric (cf. (1.13)). It is by virtue of this structure that $\boldsymbol{\Sigma A}$ is symmetric; the matrix $-\boldsymbol{\Sigma A}_\lambda$ is given by

$$-\boldsymbol{\Sigma A}_\lambda=\begin{bmatrix}\rho_0 & 0 & 0 & 0\\ 0 & 0 & 0 & 0\\ 0 & 0 & 0 & 0\\ 0 & 0 & 0 & 0\end{bmatrix}, \tag{5.20}$$

whence the normalization integral is

$$\int_{x_1}^{x_2}\boldsymbol{y}^{\mathrm{T}}(-\boldsymbol{\Sigma A}_\lambda)\boldsymbol{y}\,\mathrm{d}r=\int_0^a \rho_0 U^2 r^2\,\mathrm{d}r, \tag{5.21}$$

where we have identified $x_1=0$, corresponding to the centre of the Earth, and $x_2=a$, the radius of the outer free surface of the Earth model. Let us adopt the following "alphabetical" ordering of the minor vectors:

$$\boldsymbol{m}^{(\alpha)}=\begin{bmatrix}m_{(12)}\\ m_{(13)}\\ m_{(14)}\\ m_{(23)}\\ m_{(24)}\\ m_{(34)}\end{bmatrix}. \tag{5.22}$$

We shall write $m_1^{(\alpha)}$, $m_2^{(\alpha)}$ etc. in place of $m_{(12)}^{(\alpha)}$, $m_{(13)}^{(\alpha)}$ etc. We find that

$\boldsymbol{E}^{(4,2)}$, (2.19), is

$$\boldsymbol{E}^{(4,2)} = \begin{bmatrix} 0 & 0 & 0 & 0 & 0 & 1 \\ 0 & 0 & 0 & 0 & -1 & 0 \\ 0 & 0 & 0 & 1 & 0 & 0 \\ 0 & 0 & 1 & 0 & 0 & 0 \\ 0 & -1 & 0 & 0 & 0 & 0 \\ 1 & 0 & 0 & 0 & 0 & 0 \end{bmatrix} \tag{5.23}$$

and

$$[\boldsymbol{\Sigma}:2] = \begin{bmatrix} 0 & 0 & 0 & 0 & 0 & 1 \\ 0 & 1 & 0 & 0 & 0 & 0 \\ 0 & 0 & 0 & 1 & 0 & 0 \\ 0 & 0 & 1 & 0 & 0 & 0 \\ 0 & 0 & 0 & 0 & 1 & 0 \\ 1 & 0 & 0 & 0 & 0 & 0 \end{bmatrix}, \tag{5.24}$$

whence, (4.22),

$$\boldsymbol{\Xi} = \begin{bmatrix} 1 & 0 & 0 & 0 & 0 & 0 \\ 0 & 0 & 0 & 0 & -1 & 0 \\ 0 & 0 & 1 & 0 & 0 & 0 \\ 0 & 0 & 0 & 1 & 0 & 0 \\ 0 & -1 & 0 & 0 & 0 & 0 \\ 0 & 0 & 0 & 0 & 0 & 1 \end{bmatrix}. \tag{5.25}$$

Result 4.2 requires that $\boldsymbol{m}^{(1)}$ and $\boldsymbol{m}^{(2)}$ (for all x and λ) lie in the eigenspace belonging to the eigenvalue 1; this is equivalent to the relation

$$m_2^{(\alpha)} = -m_5^{(\alpha)}. \tag{5.26}$$

The spanning matrices appearing in, for example, (4.74) are given by ((2.36), (2.38))

$$\boldsymbol{M}^{(1)} = \begin{bmatrix} 0 & m_1^{(1)} & m_2^{(1)} & m_3^{(1)} \\ -m_1^{(1)} & 0 & m_4^{(1)} & m_5^{(1)} \\ -m_2^{(1)} & -m_4^{(1)} & 0 & m_6^{(1)} \\ -m_3^{(1)} & -m_5^{(1)} & -m_6^{(1)} & 0 \end{bmatrix}, \tag{5.27}$$

$$\tilde{\boldsymbol{M}}^{(2)} = \begin{bmatrix} 0 & m_6^{(2)} & -m_5^{(2)} & m_4^{(2)} \\ -m_6^{(2)} & 0 & m_3^{(2)} & -m_2^{(2)} \\ m_5^{(2)} & -m_3^{(2)} & 0 & m_1^{(2)} \\ -m_4^{(2)} & m_2^{(2)} & -m_1^{(2)} & 0 \end{bmatrix}, \tag{5.28}$$

and thus (4.74) may be written as

$$\Delta_\lambda \hat{\boldsymbol{y}}\hat{\boldsymbol{y}}^{\mathrm{T}} = \boldsymbol{X} \tag{5.29}$$

where the elements of $\boldsymbol{X}$ are

$$\left.\begin{aligned}
&x_{11} = m_1^{(1)}m_3^{(2)} - m_3^{(1)}m_1^{(2)}, \qquad x_{12} = -m_1^{(1)}m_2^{(2)} + m_2^{(1)}m_1^{(2)}, \\
&x_{13} = m_1^{(1)}m_6^{(2)} - m_2^{(1)}m_5^{(2)} + m_3^{(1)}m_4^{(2)}, \qquad x_{14} = m_2^{(1)}m_3^{(2)} - m_3^{(1)}m_2^{(2)}, \\
&x_{21} = m_1^{(1)}m_5^{(2)} - m_5^{(1)}m_1^{(2)}, \qquad x_{22} = -m_1^{(1)}m_4^{(2)} + m_4^{(1)}m_1^{(2)}, \\
&x_{23} = -m_4^{(1)}m_5^{(2)} + m_5^{(1)}m_4^{(2)}, \qquad x_{24} = m_1^{(1)}m_6^{(2)} + m_4^{(1)}m_3^{(2)} - m_5^{(1)}m_2^{(2)}, \\
&x_{31} = m_2^{(1)}m_5^{(2)} - m_4^{(1)}m_3^{(2)} - m_6^{(1)}m_1^{(2)}, \qquad x_{32} = -m_2^{(1)}m_4^{(2)} + m_4^{(1)}m_2^{(2)}, \\
&x_{33} = -m_4^{(1)}m_6^{(2)} + m_6^{(1)}m_4^{(2)}, \qquad x_{34} = m_2^{(1)}m_6^{(2)} - m_6^{(1)}m_2^{(2)}, \\
&x_{41} = m_3^{(1)}m_5^{(2)} - m_5^{(1)}m_3^{(2)}, \qquad x_{42} = -m_3^{(1)}m_4^{(2)} + m_5^{(1)}m_2^{(2)} - m_6^{(1)}m_1^{(2)}, \\
&x_{43} = -m_5^{(1)}m_6^{(2)} + m_6^{(1)}m_5^{(2)}, \qquad x_{44} = m_3^{(1)}m_6^{(2)} - m_6^{(1)}m_3^{(2)},
\end{aligned}\right\} \tag{5.30}$$

and

$$\Delta = \boldsymbol{m}^{(1)\mathrm{T}}\tilde{\boldsymbol{m}}^{(2)} = m_1^{(1)}m_6^{(2)} - m_2^{(1)}m_5^{(2)} + m_3^{(1)}m_4^{(2)} + m_4^{(1)}m_3^{(2)} - m_5^{(1)}m_2^{(2)} + m_6^{(1)}m_1^{(2)}. \tag{5.31}$$

Note that, by virtue of (5.26), when Δ vanishes the right-hand sides of (5.30) are symmetric.

The free-surface boundary condition is simply

$$y_3 = y_4 = 0;$$

thus we choose

$$\bar{\boldsymbol{\sigma}}_1^{(2)} = \begin{bmatrix} 1 \\ 0 \\ 0 \\ 0 \end{bmatrix}, \qquad \bar{\boldsymbol{\sigma}}_1^{(2)} = \begin{bmatrix} 0 \\ 1 \\ 0 \\ 0 \end{bmatrix},$$

and find

$$\boldsymbol{m}^{(2)}(x_2) = \begin{bmatrix} 1 \\ 0 \\ 0 \\ 0 \\ 0 \\ 0 \end{bmatrix}.$$

The boundary condition at x_1 is replaced by the requirement that the solution be bounded at the Earth's centre, which is a regular singular point of the differential system (5.16). In practice this condition may be satisfied by setting $\boldsymbol{\sigma}_k^{(1)}(x, \lambda)$ $(k = 1, 2)$ equal to the two independent solutions of the equations that are regular at the origin in the case of a homogeneous Earth model, i.e. we assume that in some small neighbourhood of the origin, $r \in [0, r_1]$, the model is homogeneous, and make use of analytic forms for the solutions to obtain $\boldsymbol{\sigma}_k^{(1)}(r_1, \lambda)$ $(k = 1, 2)$. From these we may calculate $\boldsymbol{m}^{(1)}(r_1)$, which is then used as the starting value for the integration of the minor equations to the Earth's surface. The relevant analytic solutions are given by Pekeris and Jarosch (1958) and Takeuchi and Saito (1972).

This introduces a complication into the counting algorithm since $\boldsymbol{m}^{(1)}(r_1)$ does not have the assumed property of being independent of $\lambda\,(\equiv \omega^2)$ (see the discussion preceding (4.76)). It is therefore necessary to show that the counting variable $\vartheta_R(x, \lambda)$ has a well-defined limit, which is independent of λ, as $x\,(\equiv r)$ tends to zero. The value of r_1 must be chosen to be sufficiently small that no zeros of the denominator in (4.75) lie in the interval $(0, r_1)$, since otherwise the count will be in error. We do not pursue the details of these calculations here. In the computer program accompanying this chapter the choice of r_1 depends upon frequency, and is such that r_1 is less than one wavelength. This may be regarded as no more than a "rule of thumb" that has been found, empirically, to satisfy the aforementioned requirements. A detailed analysis of these matters remains to be carried out. Similar considerations apply to the case of spheroidal modes, to be discussed below.

Again choosing $\boldsymbol{R}(x) = \mathbf{1}$ in (4.75), the counting variable becomes

$$\vartheta_R(x, \lambda) = -\frac{1}{\pi}\cot^{-1}\left\{\frac{m_4^{(1)}(x, \lambda) - m_3^{(1)}(x, \lambda)}{m_6^{(1)}(x, \lambda)}\right\} \tag{5.32}$$

Spheroidal modes

We write

$$\mathbf{y} = \begin{bmatrix} rU \\ r\zeta V \\ rP \\ rC\dot{U} + F(2U - \zeta^2 V) \\ \zeta(r\dot{V} + U - V) \\ (4\pi G)^{-1}[r\dot{P} + (l+1)P] + \rho_0 rU \end{bmatrix}, \tag{5.33}$$

and $\mathbf{y}$ satisfies a differential system with system matrix of the form (5.16) in which

$$\mathbf{T} = \begin{bmatrix} r^{-1}(1 - 2FC^{-1}) & r^{-1}FC^{-1} & 0 \\ -r^{-1}\zeta & 2r^{-1} & 0 \\ -4\pi G\rho_0 & 0 & r^{-1}l \end{bmatrix}, \tag{5.34}$$

$$\mathbf{S} = \begin{bmatrix} -\rho_0\omega^2 + 4r^{-2}(\gamma - r\rho_0 g_0) & \zeta r^{-2}(r\rho_0 g_0 - 2\gamma) & r^{-1}\rho_0(l+1) \\ \zeta r^{-2}(r\rho_0 g_0 - 2\gamma) & -\rho_0\omega^2 + r^{-2}[\zeta^2(\gamma + N) - 2N] & r^{-1}\rho_0\zeta \\ -r^{-1}\rho_0(l+1) & r^{-1}\rho_0\zeta & 0 \end{bmatrix}, \tag{5.35}$$

$$\mathbf{K} = \begin{bmatrix} C^{-1} & 0 & 0 \\ 0 & L^{-1} & 0 \\ 0 & 0 & 4\pi G \end{bmatrix}, \tag{5.36}$$

where $\gamma = A - N - C^{-1}F^2$. We find that the normalization integral becomes

$$\int_{x_1}^{x_2} \mathbf{y}^{\mathrm{T}}(-\mathbf{\Sigma A}_\lambda)\mathbf{y}\,\mathrm{d}r = \int_0^a \rho_0[U^2 + l(l+1)V^2]r^2\,\mathrm{d}r. \tag{5.37}$$

In a fluid the sixth-order system reduces to fourth-order by virtue of the identities

$$y_5 = 0, \qquad y_2 = \zeta\omega^{-2}r^{-1}[g_0 y_1 - \rho_0^{-1}y_4 + y_3]. \tag{5.38}$$

Writing, in the fluid case,

$$\boldsymbol{y}=\begin{bmatrix} rU \\ rP \\ rC\dot{U}+F(2U-\zeta^2V) \\ (4\pi G)^{-1}[r\dot{P}+(l+1)P]+\rho_0 rU \end{bmatrix}, \tag{5.39}$$

we again obtain a system of the form (5.16), in which

$$\boldsymbol{T}=\begin{bmatrix} r^{-1}\omega^{-2}\zeta^2 g_0-r^{-1} & r^{-2}\omega^{-2}\zeta^2 \\ -4\pi G\rho_0 & -r^{-1}l \end{bmatrix}, \tag{5.40}$$

$$\boldsymbol{S}=\begin{bmatrix} -\rho_0\omega^2+\rho_0 g_0 r^{-1}(r^{-1}\omega^{-2}\zeta^2 g_0-4) & \rho_0 r^{-1}(r^{-1}\omega^{-2}\zeta^2 g_0-l-1) \\ \rho_0 r^{-1}(r^{-1}\omega^{-2}\zeta^2 g_0-l-1) & \rho_0 r^{-2}\omega^{-2}\zeta^2 \end{bmatrix}, \tag{5.41}$$

$$\boldsymbol{K}=\begin{bmatrix} C^{-1}-r^{-2}\omega^{-2}\rho_0^{-1}\zeta^2 & 0 \\ 0 & 4\pi G \end{bmatrix}. \tag{5.42}$$

Again adopting the "alphabetical" ordering of minor vectors, we have, in a solid,

$$\boldsymbol{m}^{(\alpha)}=\begin{bmatrix} m^{(\alpha)}_{(123)} \\ m^{(\alpha)}_{(124)} \\ m^{(\alpha)}_{(125)} \\ m^{(\alpha)}_{(126)} \\ m^{(\alpha)}_{(134)} \\ \vdots \\ m^{(\alpha)}_{(456)} \end{bmatrix}, \tag{5.43}$$

which is of dimension 20. We shall also write $m_1^{(\alpha)}, \ldots, m_{20}^{(\alpha)}$ for the elements of $\boldsymbol{m}^{(\alpha)}$. The matrix $\boldsymbol{\Xi}$ ((4.22), $n=6$, $p=3$) has non-vanishing elements

$$\left.\begin{aligned} \xi_{1,1}=\xi_{4,4}=\xi_{6,6}=\xi_{10,10}&=1, \\ \xi_{11,11}=\xi_{15,15}=\xi_{17,17}=\xi_{20,20}&=1, \\ \xi_{2,13}=\xi_{13,2}=\xi_{8,19}=\xi_{19,8}&=1, \\ \xi_{3,7}=\xi_{7,3}=\xi_{5,12}=\xi_{12,5}&=-1, \\ \xi_{9,16}=\xi_{16,9}=\xi_{14,18}=\xi_{18,14}&=-1, \end{aligned}\right\} \tag{5.44}$$

and thus Result 4.2 requires

$$\left.\begin{array}{lll} m_2^{(\alpha)} = m_{13}^{(\alpha)}, & m_8^{(\alpha)} = m_{19}^{(\alpha)}, & m_3^{(\alpha)} = -m_7^{(\alpha)}, \\ m_5^{(\alpha)} = -m_{12}^{(\alpha)}, & m_9^{(\alpha)} = -m_{16}^{(\alpha)}, & m_{14}^{(\alpha)} = -m_{18}^{(\alpha)}, \end{array}\right\} \tag{5.45}$$

which reduces the number of independent elements of $\boldsymbol{m}^{(\alpha)}$ from 20 to 14 (the dimension of W^+; see Result 4.2). Equation (4.74), which is used to construct the eigenfunctions, may be written in the form (5.29), in which the elements of $\boldsymbol{X}$ are:

$$\left.\begin{array}{l}
x_{11} = m_1^{(1)}m_{10}^{(2)} - m_3^{(1)}m_7^{(2)} + m_4^{(1)}m_6^{(2)} + m_6^{(1)}m_4^{(2)} - m_7^{(1)}m_3^{(2)} + m_{10}^{(1)}m_1^{(2)}, \\
x_{12} = -m_1^{(1)}m_9^{(2)} + m_2^{(1)}m_7^{(2)} - m_4^{(1)}m_5^{(2)} - m_5^{(1)}m_4^{(2)} + m_7^{(1)}m_2^{(2)} - m_9^{(1)}m_1^{(2)}, \\
x_{13} = m_1^{(1)}m_8^{(2)} - m_2^{(1)}m_6^{(2)} + m_3^{(1)}m_5^{(2)} + m_5^{(1)}m_3^{(2)} - m_6^{(1)}m_2^{(2)} + m_8^{(1)}m_1^{(2)}, \\
x_{14} = m_1^{(1)}m_{20}^{(2)} - m_2^{(1)}m_{19}^{(2)} + m_3^{(1)}m_{18}^{(2)} - m_4^{(1)}m_{17}^{(2)} + m_5^{(1)}m_{16}^{(2)} - m_6^{(1)}m_{15}^{(2)} \\
\qquad + m_7^{(1)}m_{14}^{(2)} + m_8^{(1)}m_{13}^{(2)} - m_9^{(1)}m_{12}^{(2)} + m_{10}^{(1)}m_{11}^{(2)}, \\
x_{15} = -m_5^{(1)}m_{10}^{(2)} + m_9^{(1)}m_6^{(2)} - m_8^{(1)}m_7^{(2)} - m_7^{(1)}m_8^{(2)} + m_6^{(1)}m_9^{(2)} - m_{10}^{(1)}m_5^{(2)}, \\
x_{16} = m_2^{(1)}m_{10}^{(2)} - m_3^{(1)}m_9^{(2)} + m_4^{(1)}m_8^{(2)} + m_8^{(1)}m_4^{(2)} - m_9^{(1)}m_3^{(2)} + m_{10}^{(1)}m_2^{(2)}, \\
x_{21} = m_1^{(1)}m_{16}^{(2)} - m_3^{(1)}m_{13}^{(2)} + m_4^{(1)}m_{12}^{(2)} + m_{12}^{(1)}m_4^{(2)} - m_{13}^{(1)}m_3^{(2)} + m_{16}^{(1)}m_1^{(2)}, \\
x_{22} = -m_1^{(1)}m_{15}^{(2)} + m_2^{(1)}m_{13}^{(2)} - m_4^{(1)}m_{11}^{(2)} - m_{11}^{(1)}m_4^{(2)} + m_{13}^{(1)}m_2^{(2)} - m_{15}^{(1)}m_1^{(2)}, \\
x_{23} = m_1^{(1)}m_{14}^{(2)} - m_2^{(1)}m_{12}^{(2)} + m_3^{(1)}m_{11}^{(2)} + m_{11}^{(1)}m_3^{(2)} - m_{12}^{(1)}m_2^{(2)} + m_{14}^{(1)}m_1^{(2)}, \\
x_{24} = m_{11}^{(1)}m_{16}^{(2)} - m_{12}^{(1)}m_{15}^{(2)} + m_{13}^{(1)}m_{14}^{(2)} + m_{14}^{(1)}m_{13}^{(2)} - m_{15}^{(1)}m_{12}^{(2)} + m_{16}^{(1)}m_{11}^{(2)}, \\
x_{25} = m_1^{(1)}m_{20}^{(2)} - m_2^{(1)}m_{19}^{(2)} + m_3^{(1)}m_{18}^{(2)} - m_4^{(1)}m_{17}^{(2)} - m_{11}^{(1)}m_{10}^{(2)} + m_{12}^{(1)}m_9^{(2)}, \\
\qquad -m_{13}^{(1)}m_8^{(2)} - m_{14}^{(1)}m_7^{(2)} + m_{15}^{(1)}m_6^{(2)} - m_{16}^{(1)}m_5^{(2)}, \\
x_{26} = m_2^{(1)}m_{16}^{(2)} - m_3^{(1)}m_{15}^{(2)} + m_4^{(1)}m_{14}^{(2)} + m_{14}^{(1)}m_4^{(2)} - m_{15}^{(1)}m_3^{(2)} + m_{16}^{(1)}m_2^{(2)}, \\
x_{31} = m_1^{(1)}m_{19}^{(2)} - m_6^{(1)}m_{13}^{(2)} + m_7^{(1)}m_{12}^{(2)} + m_{12}^{(1)}m_7^{(2)} - m_{13}^{(1)}m_6^{(2)} + m_{19}^{(1)}m_1^{(2)}, \\
x_{32} = -m_1^{(1)}m_{18}^{(2)} + m_5^{(1)}m_{13}^{(2)} - m_7^{(1)}m_{11}^{(2)} - m_{11}^{(1)}m_7^{(2)} + m_{13}^{(1)}m_5^{(2)} - m_{18}^{(1)}m_1^{(2)}, \\
x_{33} = m_1^{(1)}m_{17}^{(2)} - m_5^{(1)}m_{12}^{(2)} + m_6^{(1)}m_{11}^{(2)} + m_{11}^{(1)}m_6^{(2)} - m_{12}^{(1)}m_5^{(2)} + m_{17}^{(1)}m_1^{(2)}, \\
x_{34} = m_{11}^{(1)}m_{19}^{(2)} - m_{12}^{(1)}m_{18}^{(2)} + m_{13}^{(1)}m_{17}^{(2)} + m_{17}^{(1)}m_{13}^{(2)} - m_{18}^{(1)}m_{12}^{(2)} + m_{19}^{(1)}m_{11}^{(2)}, \\
x_{35} = -m_5^{(1)}m_{19}^{(2)} + m_6^{(1)}m_{18}^{(2)} - m_7^{(1)}m_{17}^{(2)} - m_{17}^{(1)}m_7^{(2)} + m_{18}^{(1)}m_6^{(2)} - m_{19}^{(1)}m_5^{(2)}, \\
x_{36} = m_1^{(1)}m_{20}^{(2)} + m_5^{(1)}m_{16}^{(2)} - m_6^{(1)}m_{15}^{(2)} - m_7^{(1)}m_{14}^{(2)} - m_{11}^{(1)}m_{10}^{(2)} + m_{12}^{(1)}m_9^{(2)} \\
\qquad -m_{13}^{(1)}m_8^{(2)} - m_{17}^{(1)}m_4^{(2)} - m_{18}^{(1)}m_3^{(2)} + m_{19}^{(1)}m_2^{(2)}, \\
x_{41} = m_2^{(1)}m_{19}^{(2)} - m_5^{(1)}m_{16}^{(2)} - m_8^{(1)}m_{13}^{(2)} + m_9^{(1)}m_{12}^{(2)} + m_{11}^{(1)}m_{10}^{(2)} + m_{14}^{(1)}m_7^{(2)} \\
\qquad -m_{15}^{(1)}m_6^{(2)} - m_{17}^{(1)}m_4^{(2)} + m_{18}^{(1)}m_3^{(2)} + m_{20}^{(1)}m_1^{(2)},
\end{array}\right\} \tag{5.46}$$

$$\left.\begin{aligned}
x_{42} &= -m_2^{(1)}m_{18}^{(2)} + m_5^{(1)}m_{15}^{(2)} - m_9^{(1)}m_{11}^{(2)} - m_{11}^{(1)}m_9^{(2)} + m_{15}^{(1)}m_5^{(2)} - m_{18}^{(1)}m_2^{(2)},\\
x_{43} &= m_2^{(1)}m_{17}^{(2)} - m_5^{(1)}m_{14}^{(2)} + m_8^{(1)}m_{11}^{(2)} + m_{11}^{(1)}m_8^{(2)} - m_{14}^{(1)}m_5^{(2)} + m_{17}^{(1)}m_2^{(2)},\\
x_{44} &= m_{11}^{(1)}m_{20}^{(2)} - m_{14}^{(1)}m_{18}^{(2)} + m_{15}^{(1)}m_{17}^{(2)} + m_{17}^{(1)}m_{15}^{(2)} - m_{18}^{(1)}m_{14}^{(2)} + m_{20}^{(1)}m_{11}^{(2)},\\
x_{45} &= -m_5^{(1)}m_{20}^{(2)} + m_8^{(1)}m_{18}^{(2)} - m_9^{(1)}m_{17}^{(2)} - m_{17}^{(1)}m_9^{(2)} + m_{18}^{(1)}m_8^{(2)} - m_{20}^{(1)}m_5^{(2)},\\
x_{46} &= m_2^{(1)}m_{20}^{(2)} - m_8^{(1)}m_{15}^{(2)} + m_9^{(1)}m_{14}^{(2)} + m_{14}^{(1)}m_9^{(2)} - m_{15}^{(1)}m_8^{(2)} + m_{20}^{(1)}m_2^{(2)},\\
x_{51} &= m_3^{(1)}m_{19}^{(2)} - m_6^{(1)}m_{16}^{(2)} + m_{10}^{(1)}m_{12}^{(2)} + m_{12}^{(1)}m_{10}^{(2)} - m_{16}^{(1)}m_6^{(2)} + m_{19}^{(1)}m_3^{(2)},\\
x_{52} &= -m_3^{(1)}m_{18}^{(2)} + m_6^{(1)}m_{15}^{(2)} - m_8^{(1)}m_{13}^{(2)} - m_{10}^{(1)}m_{11}^{(2)} - m_{12}^{(1)}m_9^{(2)} + m_{14}^{(1)}m_7^{(2)}\\
&\quad + m_{16}^{(1)}m_5^{(2)} - m_{17}^{(1)}m_4^{(2)} - m_{19}^{(1)}m_2^{(2)} + m_{20}^{(1)}m_1^{(2)},\\
x_{53} &= m_3^{(1)}m_{17}^{(2)} - m_6^{(1)}m_{14}^{(2)} + m_8^{(1)}m_{12}^{(2)} + m_{12}^{(1)}m_8^{(2)} - m_{14}^{(1)}m_6^{(2)} + m_{17}^{(1)}m_3^{(2)},\\
x_{54} &= m_{12}^{(1)}m_{20}^{(2)} - m_{14}^{(1)}m_{19}^{(2)} + m_{16}^{(1)}m_{17}^{(2)} + m_{17}^{(1)}m_{16}^{(2)} - m_{19}^{(1)}m_{14}^{(2)} + m_{20}^{(1)}m_{12}^{(2)},\\
x_{55} &= -m_6^{(1)}m_{20}^{(2)} + m_8^{(1)}m_{19}^{(2)} - m_{10}^{(1)}m_{17}^{(2)} - m_{17}^{(1)}m_{10}^{(2)} + m_{19}^{(1)}m_8^{(2)} - m_{20}^{(1)}m_6^{(2)},\\
x_{56} &= m_3^{(1)}m_{20}^{(2)} - m_8^{(1)}m_{16}^{(2)} + m_{10}^{(1)}m_{14}^{(2)} + m_{14}^{(1)}m_{10}^{(2)} - m_{16}^{(1)}m_8^{(2)} + m_{20}^{(1)}m_3^{(2)},\\
x_{61} &= m_4^{(1)}m_{19}^{(2)} - m_7^{(1)}m_{16}^{(2)} + m_{10}^{(1)}m_{13}^{(2)} + m_{13}^{(1)}m_{10}^{(2)} - m_{16}^{(1)}m_7^{(2)} + m_{19}^{(1)}m_4^{(2)},\\
x_{62} &= -m_4^{(1)}m_{18}^{(2)} + m_7^{(1)}m_{15}^{(2)} - m_9^{(1)}m_{13}^{(2)} - m_{13}^{(1)}m_9^{(2)} + m_{15}^{(1)}m_7^{(2)} - m_{18}^{(1)}m_4^{(2)},\\
x_{63} &= m_4^{(1)}m_{17}^{(2)} - m_7^{(1)}m_{14}^{(2)} + m_9^{(1)}m_{12}^{(2)} - m_{10}^{(1)}m_{11}^{(2)} + m_{13}^{(1)}m_8^{(2)} - m_{15}^{(1)}m_6^{(2)}\\
&\quad + m_{16}^{(1)}m_5^{(2)} + m_{18}^{(1)}m_3^{(2)} - m_{19}^{(1)}m_2^{(2)} + m_{20}^{(1)}m_1^{(2)},\\
x_{64} &= m_{13}^{(1)}m_{20}^{(2)} - m_{15}^{(1)}m_{19}^{(2)} + m_{16}^{(1)}m_{18}^{(2)} + m_{18}^{(1)}m_{16}^{(2)} - m_{19}^{(1)}m_{15}^{(2)} + m_{20}^{(1)}m_{13}^{(2)},\\
x_{65} &= -m_7^{(1)}m_{20}^{(2)} + m_9^{(1)}m_{19}^{(2)} - m_{10}^{(1)}m_{18}^{(2)} - m_{18}^{(1)}m_{10}^{(2)} + m_{19}^{(1)}m_9^{(2)} - m_{20}^{(1)}m_7^{(2)},\\
x_{66} &= m_4^{(1)}m_{20}^{(2)} - m_9^{(1)}m_{16}^{(2)} + m_{10}^{(1)}m_{15}^{(2)} + m_{15}^{(1)}m_{10}^{(2)} - m_{16}^{(1)}m_9^{(2)} + m_{20}^{(1)}m_4^{(2)},
\end{aligned}\right\} \tag{5.46)(cont.}$$

and

$$\begin{aligned}
\Delta &= m_1^{(1)}m_{20}^{(2)} - m_2^{(1)}m_{19}^{(2)} + m_3^{(1)}m_{18}^{(2)} - m_4^{(1)}m_{17}^{(2)} + m_5^{(1)}m_{16}^{(2)} - m_6^{(1)}m_{15}^{(2)}\\
&\quad + m_7^{(1)}m_{14}^{(2)} + m_8^{(1)}m_{13}^{(2)} - m_9^{(1)}m_{12}^{(2)} + m_{10}^{(1)}m_{11}^{(2)} - m_{11}^{(1)}m_{10}^{(2)} + m_{12}^{(1)}m_9^{(2)}\\
&\quad - m_{13}^{(1)}m_8^{(2)} - m_{14}^{(1)}m_7^{(2)} + m_{15}^{(1)}m_6^{(2)} - m_{16}^{(1)}m_5^{(2)} + m_{17}^{(1)}m_4^{(2)} - m_{18}^{(1)}m_3^{(2)}\\
&\quad + m_{19}^{(1)}m_2^{(2)} - m_{20}^{(1)}m_1^{(2)}.
\end{aligned} \tag{5.47}$$

It may be checked explicitly that, by virtue of (5.45), $\boldsymbol{X}$ is symmetric when Δ vanishes, as is required by the foregoing theoretical results (e.g. Result 4.3).

In fluid regions of the model the fourth-order system leads to the same construction for eigenfunctions as was given above for radial modes ((5.22)–(5.31)). Special consideration must be given to internal fluid–solid boundaries in order to determine how, on passing from fluid to solid, the set of minors $m_1^F, \ldots, m_6^F$ on the fluid side transforms into $m_1^S, \ldots, m_{20}^S$ on the solid side; similarly, on passing from a solid to a fluid, a rule must be found giving $m_1^F, \ldots, m_6^F$ in terms of $m_1^S, \ldots, m_{20}^S$.

Let us suppose that the minor differential equations are being integrated through a fluid region and a fluid–solid boundary is encountered. On the fluid side the minor vector $\boldsymbol{m}^F$ characterizes a two-dimensional subspace spanned, let us suppose, by vectors $\boldsymbol{y}_1^F$ and $\boldsymbol{y}_2^F$:

$$\boldsymbol{y}_1^F = \begin{bmatrix} y_{11}^F \\ y_{21}^F \\ y_{31}^F \\ y_{41}^F \end{bmatrix}, \quad \boldsymbol{y}_2^F = \begin{bmatrix} y_{12}^F \\ y_{22}^F \\ y_{32}^F \\ y_{42}^F \end{bmatrix}. \tag{5.48}$$

On the solid side continuity of radial displacement, radial traction, gravitational potential, etc. require that the corresponding solution be a linear combination of the vectors

$$\boldsymbol{y}_1^S = \begin{bmatrix} y_{11}^F \\ 0 \\ y_{21}^F \\ y_{31}^F \\ 0 \\ y_{41}^F \end{bmatrix}, \quad \boldsymbol{y}_2^S = \begin{bmatrix} y_{12}^F \\ 0 \\ y_{22}^F \\ y_{32}^F \\ 0 \\ y_{42}^F \end{bmatrix}, \quad \boldsymbol{y}_3^S = \begin{bmatrix} 0 \\ 1 \\ 0 \\ 0 \\ 0 \\ 0 \end{bmatrix} \tag{5.49}$$

where the last vector is needed to accommodate an unknown discontinuity in tangential displacement. The fifth elements of these vectors vanish by virtue of the condition of vanishing tangential traction at the boundary. The minor vector on the solid side, determined up to a multiplicative factor α, can be calculated from these three vectors:

$$m_1^S = \det \begin{bmatrix} y_{11}^F & y_{12}^F & 0 \\ 0 & 0 & 1 \\ y_{21}^F & y_{22}^F & 0 \end{bmatrix} = \alpha(y_{12}^F y_{21}^F - y_{22}^F y_{11}^F) = -\alpha m_1^F,$$

$$\vdots$$

The factor α is of no account, except that it must be chosen, along with the corresponding factor on passing from solid to fluid, in such a way that the construction for the eigenfunction in the fluid and the solid yield solutions satisfying the correct continuity conditions at the boundary.

Choosing, arbitrarily, $\alpha = -1$, the complete set of conditions is

$$\left.\begin{aligned} m_1^S &= m_1^F, & m_{11}^S &= -m_4^F, \\ m_2^S &= m_2^F, & m_{13}^S &= -m_5^F, \\ m_4^S &= m_3^F, & m_{15}^S &= -m_6^F, \end{aligned}\right\} \qquad (5.50)$$

all others being zero.

Next suppose that the minor differential equations are being integrated through a solid region and that a fluid–solid boundary is encountered. On the solid side the minor vector $\boldsymbol{m}^S$ characterizes a three-dimensional subspace, spanned, let us suppose, by vectors $\boldsymbol{y}_1^S$, $\boldsymbol{y}_2^S$ and $\boldsymbol{y}_1^S$. The allowable solution lies in the subspace of this three-dimensional space that has vanishing tangential components (i.e. vanishing fifth elements); in general, this subspace is two-dimensional. Assuming that $y_{51}^S \neq 0$, a basis for the two-dimensional subspace of allowable solutions on the fluid side is given by

$$\boldsymbol{y}_1^F = \begin{bmatrix} y_{52}^S y_{11}^S - y_{51}^S y_{12}^S \\ y_{52}^S y_{31}^S - y_{51}^S y_{32}^S \\ y_{52}^S y_{41}^S - y_{51}^S y_{42}^S \\ y_{52}^S y_{61}^S - y_{51}^S y_{62}^S \end{bmatrix}, \qquad \boldsymbol{y}_1^F = \begin{bmatrix} y_{53}^S y_{11}^S - y_{51}^S y_{13}^S \\ y_{53}^S y_{31}^S - y_{51}^S y_{33}^S \\ y_{53}^S y_{41}^S - y_{51}^S y_{43}^S \\ y_{53}^S y_{61}^S - y_{51}^S y_{63}^S \end{bmatrix}. \qquad (5.51)$$

Direct calculation of the minor vector of the spanned by these two vectors yields (omitting an overall factor $-y_{51}^S$, which is of no account):

$$\boldsymbol{m}^F = \begin{bmatrix} -m_6^S \\ -m_8^S \\ m_{10}^S \\ -m_{17}^S \\ m_{19}^S \\ m_{20}^S \end{bmatrix} \qquad (5.52)$$

which gives the required transformation on passing from a solid to a fluid. This may also be obtained by applying Result 3.1 ($n = 6$, $p = 3$, $q = 5$, $s = 2$) with $V^{(1)}$ corresponding to the space characterized by $\boldsymbol{m}^S$ and $V^{(2)}$ to the space of vectors having vanishing fifth elements. We find that the two-dimensional space $V^{(1)} \cap V^{(2)}$ is proportional simply to the fifth column of $\boldsymbol{M}^{(1)(2,1)}$, which, on omitting elements labelled by combinations (i_1, i_2) in which 2 or 5 enter, yields the result (5.52).

The boundary conditions at the free surface require vanishing tractions and a perturbation in gravitational potential that tends to zero as r tends to infinity. If the model includes an ocean then these conditions at the ocean surface lead to a six-dimensional minor vector $\boldsymbol{m}^{(2)}(x_2)$ having a single non-vanishing element that, without loss of generality, may be taken to be unity: $m_1^{(2)}(x_2) = 1$. If the model does not possess an ocean then the 20-dimensional minor vector at the free surface again possesses only a single non-vanishing element $m_1^{(2)}(x_2) = 1$. As in the case of radial modes, discussed above, the boundary condition at the Earth's centre is that the minor vector is that of the subspace spanned by the set of solutions that are non-singular at the origin (see Takeuchi and Saito, 1972). The precise form taken by this boundary condition depends, of course, upon whether or not the model possesses a solid inner core.

In solid regions we take

$$\boldsymbol{R} = \begin{bmatrix} 1 & 0 & 0 & 0 & 0 & 0 \\ 0 & 1 & 0 & 0 & 0 & 0 \\ 0 & 0 & 0 & 0 & 0 & 0 \\ 0 & 0 & 0 & 0 & 0 & 0 \\ 0 & 0 & 0 & 0 & 0 & 0 \\ 0 & 0 & 0 & 0 & 0 & 0 \end{bmatrix} \tag{5.53}$$

in (4.75), which gives for the counting variable

$$\vartheta_R(x, \lambda) = -\frac{1}{\pi}\cot^{-1}\left\{\frac{m_{15}^{(1)}(x, \lambda) - m_{10}^{(1)}(x, \lambda)}{m_{20}^{(1)}(x, \lambda)}\right\}. \tag{5.54}$$

In a fluid we use $\boldsymbol{R} = \mathbf{1}$ and find

$$\vartheta_R(x, \lambda) = -\frac{1}{\pi}\cot^{-1}\left\{\frac{m_4^{(1)}(x, \lambda) - m_3^{(1)}(x, \lambda)}{m_6^{(1)}(x, \lambda)}\right\} \tag{5.55}$$

6 DESCRIPTION OF PROGRAM

The computer program accompanying this book goes by the name OBANI (omega-bound-anisotropic). In this section we briefly describe the execution of the program for the case of spheroidal modes. It is our aim to describe, in general terms, how the theoretical results of previous sections of this chapter are employed, using spheroidal modes as an example, and making specific reference to the subroutines in which the various operations are performed. More detailed notes on the program are contained in program comments and implementation notes supplied with it. As mentioned above, many of the routines are from Freeman Gilbert's EOS program; among those that have been adopted without modification, or with only slight modification, are DG, STARTL, SPS,

SAS, FPS, BFS, STEPS and BAYLIS. Modifications of the EOS program to include transverse isotropy and to make use of a frequency-dependent model were carried out by Adam Dziewonski and were subsequently incorporated into the OBANI program. The routines so modified or created include MODEL and QCORR. A slightly earlier version of the OBANI program presented here has been modified by Guy Masters and has been renamed MINEOS (Minors-EOS). This version incorporates some improvements in the root-finding procedure and also uses an alternative construction for the eigenfunctions.

The program begins by reading an Earth model (subroutine MODEL), calculating $g_0(r)$ (subroutine DG called by MODEL) and by setting up parameters needed for the variable-order–variable-step Runge–Kutta algorithm (subroutine STEPS; see Gilbert and Backus, 1969). Subroutine WBOUND is then called and this routine retains control for the remainder of a given run. WBOUND manages the iterative process by which upper and lower bounds are established for each eigenfrequency in a specified interval. In doing this, it repeatedly calls the subroutine DETQC(WIN,KOUNT,DET,ID,IFEIF), which determines the secular determinant $\Delta = \mathrm{DET} \times 2^{\mathrm{ID}}$ for a given frequency and, optionally, also returns the value of the counting variable KOUNT, which takes a value equal to the overtone number of the mode next lower in frequency than the given argument $\omega = \mathrm{WIN}$. If $\mathrm{IFEIF} \neq 0$ the eigenfunction is calculated, this option being appropriate only when WIN is, within a specified tolerance, an eigenfrequency.

WBOUND begins by reading an input data file (unit 5) to determine the required angular order l and the range of frequencies $[\omega_{\min}, \omega_{\max}]$ in which it is required to locate eigenfrequencies. Values of frequency between these limits may also be specified in the input file to allow more rapid convergence if approximate eigenfrequencies or bounds on eigenfrequencies are known (e.g. from a prior run). After calling DETQC for each of the input frequencies and, in particular for $\mathrm{WIN} = \omega_{\min}\,(\mathrm{KOUNT} = n_{\min})$ and $\mathrm{WIN} = \omega_{\max}\,(\mathrm{KOUNT} = n_{\max})$, WBOUND proceeds, by bisection, to find a sequence of values of ω that yield all integer values of KOUNT between $n_{\min}$ and $n_{\max}$. These values serve to bracket all eigenfrequencies in the specified interval $[\omega_{\min}, \omega_{\max}]$. After this, the counting algorithm is no longer needed and WBOUND enters the mode of refining the bracketed roots; a simple but infallible scheme, based upon bisection and linear interpolation, is used to obtain upper and lower bounds differing by less than a specified tolerance on the values of ω where the secular determinant $\Delta(\omega)$ changes sign.

DETQC manages the integration of the minor differential equations, first in the direction of increasing r from the centre of the Earth to the free surface, yielding $\boldsymbol{m}^{(1)}(x)$ and allowing the secular determinant $\Delta = \tilde{\boldsymbol{m}}^{(2)\mathrm{T}}(x_2)\boldsymbol{m}^{(1)}(x_2)$ to be determined together with the counting variable $\vartheta_R(x_2, \lambda)$. If the eigenfunction is required, integration of the minor equations is then carried out in the opposite direction, yielding $\boldsymbol{m}^{(2)}(x)$; from $\boldsymbol{m}^{(1)}(x)$ and $\boldsymbol{m}^{(2)}(x)$ the eigenfunction is constructed by decomposing the dyads (subroutine DECOD) given by (5.30) and (5.47) (see subroutines SPRPMA for the construction of these dyads in a solid layer and FPROPM for a fluid layer). The subroutines SPRPMA and FPROPM are called by DETQC to perform Runge–Kutta integration of the minor differential equations; auxiliary routines DERMS and DERMF are called by SPRPMA and FPROPM to compute the solution derivatives in solid and fluid layers respectively at each evaluation that is called for by the Runge–Kutta scheme. DETQC also calls routines SFBM and FSBM to transform the minor vectors on passing from a solid to a fluid layer (SFBM, (5.52)) or from a fluid to a solid layer (FSBM, (5.50)).

DETQC determines starting solutions for the upward integration by calling SPSM in a solid layer or FPSM in a fluid layer. These, in turn, call routines SPS (which calls SAS for a solution in terms of spherical Bessel functions) and FPS, which compute sets of linearly independent solutions for a homogeneous solid or fluid sphere, either in terms of power series or in terms of spherical Bessel functions (subroutine BFS), depending upon the values of frequency, radius, etc. Although, in principle, integration should begin at the Earth's centre, considerable time is saved by starting the integration at a higher level in the model in the case where the solutions decay exponentially below a given level. A suitable starting level is determined by the subroutine STARTL, called by DETQC, which examines the minimum rate of decay of the solution to determine a level below which the value of the starting solution becomes immaterial to a given level of accuracy (as determined by the value of EPS).

When WBOUND determines that an eigenfrequency has been found to the given accuracy, DETQC is again called to evaluate the eigenfunction. On exiting DETQC, the eigenfunction $\boldsymbol{y}(x)$ is determined up to a sign for each radius. EIFOUT is called to determine the sign, which is chosen in such a way that the spline-interpolated eigenfunction between all pairs of adjacent knots is maximally smooth, and the eigenfunction is also reformatted to give $U(r)$, $\dot{U}(r)$, $V(r)$, $\dot{V}(r)$, $P(r)$ and $\dot{P}(r)$. In addition, Gauss–Legendre integration (subroutine GAUSLV) is used to calculate the normalization integral, to check the Rayleigh quotient and to determine group velocity and attenuation Q. The integrands are calculated by the auxiliary routine INTGDS.

ACKNOWLEDGEMENTS

I thank Freeman Gilbert for generously supplying me with the EOS program. This work has been supported by the United States National Science Foundation under Grants EAR80-26441, EAR82-13330 and EAR85-11400.

REFERENCES

Alterman, Z., Jarosch, H. and Pekeris, C. L. (1959). Oscillations of the Earth. *Proc. R. Soc. Lond.* **A252,** 80–95.

Buland, R. (1976). Retrieving the seismic moment tensor. Ph.D. thesis, University of California, San Diego.

Buland, R. and Gilbert, F. (1984). Computation of free oscillations of the Earth. *J. Comp. Phys.* **54,** 95–114.

Chapman, C. H. and Woodhouse, J. H. (1981). Symmetry of the wave equation and excitation of body waves. *Geophys. J. R. Astron. Soc.* **65,** 777–782.

Coddington, E. A. and Levinson, N. (1955). *Theory of Ordinary Differential Equations*. McGraw-Hill, New York.

Gantmacher, F. R. (1959). *The Theory of Matrices*. Chelsea, New York.

Gilbert, F. (1971). Excitation of normal modes of the Earth by earthquake sources. *Geophys. J. R. Astron. Soc.* **22,** 223–226.

Gilbert, F. and Backus, G. E. (1966). Propagator matrices in elastic wave and vibration problems. *Geophysics* **31,** 326–333.

Gilbert, F. and Backus, G. E. (1969). A computational problem encountered in a study of the Earth's normal modes. *Proc. Fall Joint Computer Conf., 1968, of Am. Federation of Information Processing Society,* pp. 1273–1277.

Martin, R. S. and Wilkinson, J. H. (1967). Solution of symmetric and unsymmetric band equations and the calculation of eigenvectors of band matrices. *Numer. Maths* **9,** 279–301.

Pekeris, C. L. and Jarosch, H. (1958). The free oscillations of the Earth. *Contributions in Geophysics in Honor of Beno Gutenberg* (ed. E. Ingerson), pp. 171–192. Pergamon, New York.

Peters, G. and Wilkinson, J. H. (1969). Eigenvalues of $\boldsymbol{A}\boldsymbol{x} = \lambda \boldsymbol{B}\boldsymbol{x}$ with band symmetric $\boldsymbol{A}$ and $\boldsymbol{B}$. *Comput. J.* **12,** 398–404.

Takeuchi, H. and Saito, M. (1972). Seismic surface waves. *Methods in Computational Physics,* Vol. 11 (ed. B. A. Bolt), pp. 217–295. Academic Press, New York.

Woodhouse, J. H. (1980). Efficient and stable methods for performing seismic calculations in stratified media. *Physics of the Earth's Interior*; *Proc. Int. School of Physics "Enrico Fermi"*, Course LXXVIII, pp. 127–151. North-Holland, Amsterdam.

Woodhouse, J. H. (1981). The calculation of the complete normal mode spectrum of the Earth using finite difference methods. Seismic Discrimination Semiannual Technical Summary, Lincoln Laboratory, MIT, 31 March 1981.

IV.3

Variational Methods

RAY BULAND

US Geological Survey, Golden, Colorado, USA

1 RAYLEIGH'S PRINCIPLE

In Lord Rayleigh's landmark work *The Theory of Sound* (1877) he extensively investigated the properties of mechanical systems undergoing small oscillations (simple harmonic motion). As part of this work, Rayleigh developed a special case of Hamilton's Principle, now called Rayleigh's Principle, which for the purposes of this chapter can be written as

$$\delta\int_V L(s)\,\mathrm{d}V = \delta\left[\omega^2\int_V K(s)\,\mathrm{d}V - \int_V U(s)\,\mathrm{d}V\right] = 0, \tag{1}$$

where L is the Lagrangian density of a conservative system, ω is an eigenfrequency, K is kinetic-energy density, U is potential-energy density, s is a spatial eigenfunction, $\mathrm{d}V$ is a volume element, V is a fixed volume and δ means variation with respect to s. That is, the eigenfrequencies of the system are stationary with respect to small variations in their respective eigenfunctions. Note that Rayleigh's Principle involves variations in space only. Time has been eliminated by the assumption of simple harmonic motion. A useful variant of (1) may be derived by setting ω equal to zero. In this way, an inhomogeneous static solution that is compatible with either arbitrary boundary conditions or a specified distribution of internal, equivalent body forces may be found.

Rayleigh's Principle has two important numerical applications. First, if the model of the media is perturbed then the perturbation to the eigenfrequencies, correct to first order, may be found from

$$\begin{aligned} 2\omega\,\delta\omega\int_V m(\boldsymbol{x})\tilde{K}(s,s)\,\mathrm{d}V &= -\int_V \delta m(\boldsymbol{x})\,[\omega^2\tilde{K}(s,s) - \tilde{U}(s,s)]\,\mathrm{d}V \\ &= -\int_V \delta m(\boldsymbol{x})\,\tilde{L}(s,s)\,\mathrm{d}V, \end{aligned} \tag{2}$$

where $\delta\omega$ is the perturbation to the eigenfrequency, m is the model, δm is the perturbation to the model, $\boldsymbol{x}$ is position, $m\tilde{K}=K$, $m\tilde{U}=U$ and $m\tilde{L}=L$. $\tilde{K}$, $\tilde{U}$ and $\tilde{L}$ are sometimes called "kernels". The notation $\tilde{K}(s, s)$ emphasizes that the kernel functions are quadratic in the eigenfunctions. $-\tilde{L}$ is proportional to the Frechét derivative of eigenfrequency with respect to the model. Note that, as a result of Rayleigh's Principle, the kernel functions depend only on the unperturbed eigenfunctions.

The second application involves using a given model and approximate eigenfunctions to estimate eigenfrequencies from the energy-balance equation

$$\hat{\omega}^2\int_V K(f)\,\mathrm{d}V-\int_V U(f)\,\mathrm{d}V=0, \tag{3}$$

where $\hat{\omega}$ is an approximation to ω and f is the approximate eigenfunction (see e.g. Mathews and Walker, 1965, pp. 319–320). From Rayleigh's Principle, if $|s-f|$ is $O(\epsilon)$ then $\omega^2\approx\hat{\omega}^2-O(\epsilon^2)<\hat{\omega}^2$. That is, the error in estimating an eigenfrequency will be of a higher order than the approximation of the eigenfunction, and the estimate of the eigenfrequency will be strictly an upper bound (assuming the usual case in which the stationary point corresponds to a minimum of the Lagrangian).

2 THE RAYLEIGH–RITZ ALGORITHM

The practical application of (3) is limited by the fact that choosing and refining f is, in general, both laborious and tedious and requires considerable insight into the problem (see e.g. Jeffreys and Jeffreys, 1966, pp. 302–304). A much more powerful alternative can be derived from the observation that (1) requires eigenfunctions such that

$$\frac{\partial}{\partial s}\left[\omega^2\int_V K(s)\,\mathrm{d}V-\int_V U(s)\,\mathrm{d}V\right]=0. \tag{4}$$

Choose

$$f(\boldsymbol{x})=\sum_{i=1}^{N}a_i\gamma_i(\boldsymbol{x}), \tag{5}$$

where the a_i are unknown coefficients and the γ_i are known linearly independent functions such that all possible linear combinations satisfy the boundary conditions required of solutions to (4) (see e.g. Morse and Feshbach, 1953, pp. 1117–1119). Substituting (5) into (4) and satisfying the extremum by setting derivatives with respect to the a_i equal to zero yields

$$\frac{\partial}{\partial a_i}\left[\sum_{k=1}^{N}\sum_{j=1}^{N} a_k a_j\left\{\int_V m(\boldsymbol{x})\tilde{U}(\gamma_k, \gamma_j)\,\mathrm{d}V - \hat{\omega}^2\int_V m(\boldsymbol{x})\tilde{K}(\gamma_k, \gamma_j)\,\mathrm{d}V\right\}\right]$$

$$= 2\sum_{j=1}^{N} a_j\{\mathscr{U}_{ij} - \hat{\omega}^2\mathscr{K}_{ij}\} = 0, \quad i = 1, 2, \ldots, N, \tag{6}$$

where $\mathscr{U}_{ij} = \mathscr{U}_{ji} = \frac{1}{2}(\tilde{\mathscr{U}}_{ij} + \tilde{\mathscr{U}}_{ji})$, $\mathscr{K}_{ij} = \mathscr{K}_{ji} = \frac{1}{2}(\tilde{\mathscr{K}}_{ij} + \tilde{\mathscr{K}}_{ji})$ and

$$\left.\begin{aligned}\tilde{\mathscr{U}}_{ij} &= \int_V m(\boldsymbol{x})\tilde{U}(\gamma_i, \gamma_j)\,\mathrm{d}V,\\ \tilde{\mathscr{K}}_{ij} &= \int_V m(\boldsymbol{x})\tilde{K}(\gamma_i, \gamma_j)\,\mathrm{d}V.\end{aligned}\right\} \tag{7}$$

This explicit symmetrization is necessary so that the matrices $\boldsymbol{\mathscr{U}}$ and $\boldsymbol{\mathscr{K}}$, of which $\mathscr{U}_{ij}$ and $\mathscr{K}_{ij}$ respectively are elements, will be Hermitian. Although it may not be obvious, it is quite common for some terms of the potential energy to result in $\tilde{\mathscr{U}}_{ij} \neq \tilde{\mathscr{U}}_{ji}$. When $k \neq j$, the factor of two in (6) arises from these two terms. When $k = j$, the factor of two arises from the derivative of a_j^2 with respect to a_j. Rewriting (6) in matrix notation yields

$$(\boldsymbol{\mathscr{U}} - \hat{\omega}^2\boldsymbol{\mathscr{K}})\boldsymbol{a}^{\mathrm{T}} = \boldsymbol{0}, \tag{8}$$

where T means transpose. Equations (5) and (8) define the Rayleigh–Ritz algorithm, which provides the means of simultaneously refining a suite of eigenfunctions and estimating their associated eigenfrequencies.

Equation (8) is just a generalized algebraic eigenvalue problem. In this context, "generalized" simply means that the eigenvalue problem is not in standard form (i.e. $\boldsymbol{\mathscr{K}} \neq \boldsymbol{\mathscr{I}}$, where $\boldsymbol{\mathscr{I}}$ is the identity matrix). Borrowing notation from the engineering literature, $\boldsymbol{\mathscr{U}}$ and $\boldsymbol{\mathscr{K}}$ will be referred to as the stiffness and mass matrices respectively. Owing to the quadratic nature of the potential- and kinetic-energy kernels, the stiffness and mass matrices are guaranteed to be positive-semidefinite (and usually strictly positive-definite). Therefore the system given by (8) is Hermitian and the eigenvalues and eigenvectors are real. This is to be expected, as (1) is only valid for conservative systems. If $\mathrm{D}(s) = 0$ is the differential operator derived from (1) by means of Lagrange's equations (see e.g. Goldstein, 1950, pp. 14–18), then the Hermitian nature of (8) implies that $\mathrm{D}(s)$ is also Hermitian (in a different, but, of course, related sense).

Multipling (8) by $\boldsymbol{a}$ on the left and rearranging terms yields

$$\hat{\omega}^2 = \frac{\boldsymbol{a}\boldsymbol{\mathscr{U}}\boldsymbol{a}^{\mathrm{T}}}{\boldsymbol{a}\boldsymbol{\mathscr{K}}\boldsymbol{a}^{\mathrm{T}}}, \tag{9}$$

which by analogy with (3) should provide a method of making a highly accurate estimate of an eigenvalue given an approximation to the correct eigenvector. Equation (9) is, of course, just the well-known Rayleigh quotient of linear algebra (Jeffreys and Jeffreys, 1966, pp. 143–144).

2.1 The nature of Rayleigh–Ritz solutions

In general, (1) can be rewritten as

$$\delta\left[\int_V \{\omega^2 K(s) - U(s)\}\,\mathrm{d}V\right] = \delta\int_V \mathrm{D}(s)s\,\mathrm{d}V = 0, \tag{10}$$

where $\mathrm{D}(s)$ has been Fourier-transformed with respect to time. Substituting (5) into (10) and interpreting the variation as differentiation with respect to the a_i yields (after some manipulation)

$$\int_V \mathrm{D}(f)\gamma_i\,\mathrm{d}V = 0, \quad i = 1, 2, \ldots, N. \tag{11}$$

That is, the differential operator, acting on the approximate eigenfunction, produces a spatial error that is orthogonal to all of the basis functions under the norm $(u, v) \equiv \int_V u(\boldsymbol{x})v(\boldsymbol{x})\,\mathrm{d}V$. In other words, the approximate solution is just the projection of the true solution onto the Hilbert space (with this norm) spanned by the basis functions γ_i. This implies that the approximate eigenfunction is the element of the Hilbert space spanned by the basis functions that is as close as possible to the true solution.

The N eigenvectors of (8) turn out (under all reasonable circumstances) to specify linear combinations of the N basis functions that approximate the N smoothest (lowest-wavenumber) eigenfunctions. The corresponding N eigenvalues are upper bounds on the first N squared eigenfrequencies. The precision of the eigenfrequency estimates depends on the ability to represent the corresponding eigenfunctions as linear combinations of the basis functions. As their wavenumber increases, eigenfunctions become successively more difficult to approximate with a fixed number of basis functions and the corresponding eigenfrequency estimates become successively poorer. In practice, this means that only the $R \leqslant N$ smoothest eigenfunctions will be estimated with sufficient precision. For most problems, eigenfrequency increases as wavenumber increases. This means that the $N - R$ eigenfrequency upper bounds that are not sufficiently accurate to be useful are guaranteed to be above the frequency band of interest. However, in some problems, eigenfrequency decreases as wavenumber increases. This means that the $N - R$ highest-wavenumber eigenfunctions correspond to the $N - R$ lowest eigenfrequencies. If the upper bounds on these eigenfrequencies are sufficiently poorly estimated, it is possible that the estimates may "contaminate" the eigenspectrum (frequency band) of interest. That is, a very poor estimate of an eigenfrequency that is not of interest may fall between the estimates of two eigenfrequencies that are of interest.

A significant disadvantage of the Rayleigh–Ritz technique is that it is quite difficult to estimate the accuracy to which eigenfrequencies and eigenfunctions have been estimated. In some cases it is possible to derive an alternative variational principle in which the solutions are maxima and the eigenfrequency estimates are lower bounds (see e.g. Zienkiewicz, 1967, pp. 23–24). As the calculation of lower bounds requires that an entirely new and (at least) equally complicated problem be solved, this approach is rarely used. Instead, the informal use of numerical analogues to gain a "feel" for the accuracy to expect is common. Possible analogues include: (i) comparing solutions with an existing computer program that solves some subset of the same problem using a different algorithm; (ii) comparing solutions for an analogous problem with a known analytical solution; and (iii) comparing two Rayleigh–Ritz solutions with different numbers of basis functions. The third approach is the easiest to implement. It attempts to measure the convergence of the eigenfrequencies as a function of N. This can work quite well, as eigenvalues tend to converge rapidly (as long as the new basis functions are adding significantly to the representation of the corresponding eigenfunctions) and then stabilize near the correct values. The difficulty with this method is that if the basis set is deficient with respect to some special class of eigenfunctions then convergence can be deceptively slow.

2.2 Generalization and specialization

The formulation given above, while consistent with the original derivation (Rayleigh, 1877), is not the most general statement of the Rayleigh–Ritz algorithm. Indeed, the Rayleigh–Ritz method may be applied to any problem for which a variational principle can be derived. As already pointed out, both inhomogeneous and normal-mode problems can be solved. In addition to Sturm–Liouville problems, the Rayleigh–Ritz algorithm can be applied to the Helmholtz equation, the Maxwell equations and the Schrödinger equation. It is not necessary that the extremal solutions be minima. The notion of conservative systems can be generalized somewhat to include velocity-dependent potentials (see e.g. Goldstein, 1950, p. 19). Further, the requirement of a fixed finite boundary can be relaxed to include infinite media, provided that the eigenfunctions decay sufficiently rapidly as $|\boldsymbol{x}| \to \infty$.

The Rayleigh–Ritz algorithm can be applied to discrete as well as continuous problems and to problems in one and two as well as three dimensions. Also, both the model and the eigenfunction may have multiple components. For example, in isotropic elastic problems the model consists of density and two Lamé parameters, each of which may be an independent function of position. The eigenfunction may consist of one or more scalars, vectors or tensors or some combination of these. It is convenient to think of each scalar and vector or tensor component as a separate scalar function. Each scalar is usually the solution of a separate differential equation. The relationship among the scalars is determined by coupling among the differential equations. For example, an eigenfunction of the P–SV elastic wave propagation problem has three scalars (two components of displacement and the perturbation of the gravitational potential) that are solutions of three coupled second-order differential equations.

For simplicity, the detailed development and solution of Rayleigh–Ritz problems will be discussed below in the context of one-dimensional problems. Where appropriate, multicomponent models and eigenfunctions will be discussed. This is consistent with the author's experience in the computation of free oscillations of spherically averaged non-rotating self-gravitating isotropic Earth models from which illustrations, cautions, comparisons and computer routines will be drawn. For details of this problem see Buland and Gilbert (1984). In the following section the general approach for solving static problems will be outlined. The remainder of the chapter will be concerned with eigenvalue problems. After a brief examination of natural basis functions, the discussion will be specialized to spline (finite-element) basis functions. Numerical methods for computing the basis functions, constructing the matrix problem, matching the boundary conditions and finding the eigenfrequencies and eigenfunctions will be given. Several important special cases, numerical difficulties and performance-enhancing tricks will be described. Finally, the Rayleigh–Ritz method will be compared with shooting methods and contrasted with the Galerkin and finite-element methods.

2.3 Static problems

For static problems (8) yields $\mathcal{U}\boldsymbol{a}^{\mathrm{T}} = \mathbf{0}$, which, without modification, can yield only trivial solutions. When a solution compatible with arbitrary boundary conditions is sought, a good procedure is to choose one basis function that satisfies all boundary conditions exactly and N basis functions that contribute nothing at the boundaries (see e.g. Carrier and

Pearson, 1976, pp. 174–177). The coefficient of the basis function that satisfies the boundary conditions must be fixed to unity. Repeating the steps that led to (8) yields $\mathcal{U}\boldsymbol{a}^{\mathrm{T}} = \boldsymbol{b}^{\mathrm{T}}$, where the right-hand side arises from the basis function that satisfies the boundary conditions. When a solution compatible with a specified distribution of internal, equivalent body forces is sought, no modification to the basis functions is needed. In this case, a non-trivial right-hand side arises from work done by the equivalent body forces (see e.g. Zienkiewicz, 1967, pp. 18–21). In either case the static problem only requires solving a simple system of linear equations, which can be done in a variety of ways. Owing to the symmetry of $\mathcal{U}$, the Cholesky decomposition (Martin *et al.*, 1971a) discussed below would be particularly appropriate.

3 NATURAL BASIS FUNCTIONS

Given that a particular subset of the smoothest eigenfunctions and their corresponding eigenfrequencies must be estimated to a certain precision, the choice of basis functions completely determines the methods by which (8) may be solved and the numerical performance (number of storage locations and floating-point operations required) that may be obtained. For substantial problems, the choice of basis functions will determine the minimum processor speed, memory capacity and on- and off-line storage capacity needed to solve the problem (in a realistic amount of time).

The traditional approach to choosing basis functions is to minimize the number required by utilizing functions that are "natural" for the problem in some sense. The term "natural" generally implies that each of the N basis function is either known or suspected to closely approximate one of the eigenfunctions of the N smoothest normal modes of the system. Natural basis functions are typically found either by computing the eigenfunctions of a simpler, but closely related, problem or by constructing functions that satisfy known behavior near boundaries and suspected behaviour between boundaries. A related approach is to use orthogonal polynomials that are natural for the differential equation involved (see e.g. Morse and Feshbach, 1953, p. 1118). It is, of course, much more difficult to generate natural basis functions for problems with multicomponent eigenfunctions. However, it is possible to model each scalar independently. This has two disadvantages. First, if there are J scalars then a $JN \times JN$ matrix (rather than a $N \times N$ matrix) must be solved. Secondly, choosing scalars that match boundary conditions depending on more than one scalar can be difficult.

The use of such natural basis functions will, in general, result in $\mathcal{U}$ and

$\mathcal{K}$ matrices that are dense (all elements are non-zero). However, if the basis set is well chosen, these matrices will be diagonally dominant, stabilizing subsequent numerical calculations. The method of solving (8) for dense real symmetric matrices is well known (Martin and Wilkinson, 1971a) and will only be outlined here. Note that the references, all taken from Wilkinson and Reinsch (1971), include Algol programs and test cases. We begin with a Cholesky decomposition of the kinetic-energy-density matrix, $\mathcal{K} = \mathcal{L}\mathcal{L}^{\mathrm{T}}$ (Martin *et al.*, 1971a), where $\mathcal{L}$ is lower-triangular (i.e. elements above the diagonal are zero). Multiplying (8) on the left by $\mathcal{L}^{-1}$ yields

$$(\mathcal{L}^{-1}\mathcal{U}\mathcal{L}^{-\mathrm{T}} - \hat{\omega}^2\mathcal{I})\mathcal{L}^{\mathrm{T}}\boldsymbol{a}^{\mathrm{T}} = (\mathcal{A} - \hat{\omega}^2\mathcal{I})\boldsymbol{c}^{\mathrm{T}} = \mathbf{0}, \tag{12}$$

where -1 means matrix inverse and $-\mathrm{T}$ means transpose inverse. Equation (12) is now in standard form and may be solved by a variety of standard methods. The recommended procedure is to reduce $\mathcal{A}$ to tridiagonal form (i.e. all elements except those on the main diagonal and the first sub- and superdiagonals are zero) by means of Householder rotations (Martin *et al.*, 1971b). The eigenvalues may then be found by bisection (Barth *et al.*, 1971). This procedure uses the Sturm count to control the bisection and is efficient because the Sturm count of a tridiagonal matrix can be computed very rapidly. The Sturm count is the number of eigenvalues of $\mathcal{A}$ that are less than a trial value of $\hat{\omega}^2$. Finally, eigenvectors may be found by means of inverse iteration (Peters and Wilkinson, 1971).

A Fortran routine implementing the Cholesky decomposition has been translated from Algol by R. L. Parker (subroutine CHOLIN). A routine implementing the Householder tridiagonalization, bisection for eigenvalues, and inverse iteration for eigenvectors for a real symmetric matrix has been adapted from a routine written by J. F. Gilbert (subroutine DIAG). If the Cholesky decomposition is used for the static problem, CHOLIN requires a companion subroutine, CHOLSL (also translated by R. L. Parker), to complete the solution for $\boldsymbol{a}$. Note that a brief description of the routines accompanying this chapter is given in the program index. These routines will be discussed through the text as appropriate. For the Cholesky and Householder procedures the amount of computer memory required is proportional to $(JN)^2$ and the number of floating-point operations is proportional to $(JN)^3$. In the following sections an alternative strategy for choosing basis functions, which leads to a sparse matrix problem (i.e. most matrix elements are zero), will be explored. In some cases the sparse formulation can be either faster, more convenient or both.

Both the dense and sparse problems turn out to have the following properties in common. First, the algorithm may be arranged so that the eigenvalues will be found in ascending (or descending) order and so that the bisection can be terminated at any time (for example, after the R smoothest eigenfunctions have been found). Further, eigenvectors may be found for each eigenvalue as required. That is, only the eigenvalues and eigenvectors actually desired need to be computed. Secondly, in practice, modern eigenvalue–eigenvector algorithms are amazingly stable and precise. The set of eigenvalues found is guaranteed to be complete and the eigenvectors are guaranteed to be orthogonal to machine precision. Even the eigenvectors of nearly coincident eigenvalues are properly handled.

3.1 An example

To illustrate the Rayleigh–Ritz method using natural basis functions, we consider the problem of toroidal free oscillations of a two-layer spherically symmetric non-rotating elastic isotropic Earth model with a fluid core and with density ρ and rigidity μ, which vary linearly with radius r throughout the mantle. This problem is of some geophysical interest, as it can be made to approximate the Earth rather closely. It is also simple enough to treat briefly, but complicated enough to illustrate the method in some detail. In order to derive some natural basis functions, we consider the related problem in which density and rigidity (and hence shear velocity v_s) are constant throughout the mantle. To make this model more physically meaningful, we let the core radius r_c be 3480 km, the Earth radius r_e be 6371 km, the density be 4448.8 kg m^{-3} and the rigidity be 1.6310×10^{11} nt m^{-2} (which makes the shear velocity 6.0548 km s^{-1}). These parameters have been derived from the PREM model of Dziewonski and Anderson (1981). We write the spatial portion of the eigenfunction as $s(\mathbf{x}) = {}_nW_l(r)Y_l^m(\theta, \phi)$, where $W(r)$ is the radial part of the eigenfunction, Y is a fully normalized spherical harmonic, θ is colatitude, ϕ is longitude, l is angular order, m is azimuthal order and n is radial order. In the following, l will be taken as 2, m can be neglected and n is simply a count of successively higher-frequency solutions for the same l. For convenience, these modes will be designated (in the usual way) by ${}_nT_l$.

Dropping the sub- and superscripts, W satisfies the differential equation

$$\kappa^2 r^2 \ddot{W}(\kappa r) + 2\kappa r \dot{W}(\kappa r) + [\kappa^2 r^2 - l(l+1)]W(\kappa r) = 0, \qquad (13)$$

where $\dot{W}$ and $\ddot{W}$ denote first and second derivatives with respect to radius and $\kappa = \omega\ (\rho/\mu)^{1/2} = \omega/v_s$ is wavenumber, subject to the boundary conditions that $\dot{W}(\kappa r) - W(\kappa r)/r = 0$ at r_c and r_e (there can be no toroidal motion in the fluid core). The solutions to this equation are of the form $W(\kappa r) = \beta j_l(\kappa r) + \nu y_l(\kappa r)$, where j_l and y_l are spherical Bessel functions of the first and second kind. The constants β and ν are found by satisfying the core–mantle boundary condition and normalizing the trial eigenfunction to unit kinetic energy. There are then an infinite number of discrete κs for which the free-surface boundary condition is satisfied. The spherical Bessel functions are computed by recursion (Abramowitz and Stegun, 1968, pp. 452–453). The first five eigenperiods for angular order 2 are given in Table 1.

For the linear model let $\rho(r) = \rho_0(3.4662 - 3.0481r)$ and $\mu(r) = \mu_0(1.8426 - 1.0827r)$, where ρ_0 and μ_0 are the homogeneous-model values. These functions have also been derived from the PREM model. For this problem, (1) may be written as

$$\delta\left[\tfrac{1}{2}\omega^2\int_{r_c}^{r_e} \rho W^2 r^2\,\mathrm{d}r - \frac{1}{2}\int_{r_c}^{r_e} \mu\{(l+2)(l-1)W^2 + (r\dot{W} - W)^2\}\,\mathrm{d}r\right] = 0, \quad (14)$$

where the angular coordinates have already been integrated out. Defining the perturbations to ρ and μ to be $\rho(r) - \rho_0$ and $\mu(r) - \mu_0$ respectively, (2) may be applied to compute $\delta\omega$. The integrals have been computed using a 96 point Gauss–Legendre scheme (Abramowitz and Stegun, 1968, p. 887) as discussed in Section 4.1 below. The resulting eigenperiods are also given in Table 1. A direct estimate of $\hat{\omega}^2$ may also be obtained from (3). The results of this calculation are also given in Table 1 (referred to as the quotient method). In this case the perturbation and quotient methods yield nearly the same estimates. However, these techniques are actually quite different. The perturbation is computed using the eigenfunctions of the unperturbed system. The quotient is applied to an estimate of an eigenfunction of the perturbed system. In this case the eigenfunctions of the unperturbed model can be used as

Table 1 Toroidal eigenperiods (in seconds) for the homogeneous model and for the linear model computed by the perturbation method and the quotient method.

Computation	${}_0T_2$	${}_1T_2$	${}_2T_2$	${}_3T_2$	${}_4T_2$
Homogeneous	2744.0	809.6	455.9	311.7	235.9
Perturbation	2689.3	781.3	435.3	296.9	224.5
Quotient	2684.8	783.2	436.1	297.4	224.9

Table 2 Toroidal eigenperiods (in seconds) from the Rayleigh–Ritz method using natural basis functions.

Basis functions	${}_0T_2$	${}_1T_2$	${}_2T_2$	${}_3T_2$	${}_4T_2$
2	2692.7	779.9	—	—	—
3	2694.0	795.8	436.0	—	—
4	2694.2	797.4	448.4	296.4	—
5	2694.2	797.9	450.0	306.2	223.1
6	2694.2	798.0	450.5	307.8	231.3
7	2694.3	798.0	450.6	308.4	232.8
8	2694.3	798.0	450.7	308.5	233.4
9	2694.3	798.1	450.7	308.6	233.6

estimates of the eigenfunctions of the perturbed model because they satisfy the boundary conditions of both models. The systematic shift to shorter periods between the homogeneous and linear models arises because no care was taken to ensure that the mass of the Earth remained constant.

To apply the Rayleigh–Ritz method, the lowest-frequency eigenfunctions of the homogeneous model are taken to be a set of basis functions, which, as has just been seen, already satisfy the boundary conditions of the linear model. Evaluating the integrals in (6), the eigenvalue problem given in (8) can be constructed and solved as shown in the preceding section. The resulting eigenperiods obtained using 2–9 basis functions are shown in Table 2. Note that the eigenperiod estimates for each mode increase monotonically with the number of basis functions used (i.e. the eigenfrequency estimates, which must be upper bounds, decrease monotonically). For any number of basis functions, the last (lowest) eigenperiod estimate is never as good as the perturbation result, at least for this example. However, the ith eigenperiod estimate is fairly stable by the time that $i+2$ basis functions are used (for example, the eigenperiod of ${}_1T_2$ is beginning to stabilize when four or more basis functions are used).

4 FINITE-ELEMENT BASIS FUNCTIONS

In practice, natural basis functions are often abandoned for several reasons. For complicated physical systems, it can be quite difficult to derive adequate natural basis functions either analytically or numerically. Further, when a large number of solutions are sought, it will not be economical to solve the dense system unless $R \approx JN$, which requires the natural basis functions to be very good indeed. If highly accurate eigenvalues are required, this problem will be even more severe. An alternative approach, pioneered by structural engineers, uses basis functions that are spatially localized (sometimes called finite elements). The original motivation for this approach was to identify the finite elements with physical structural elements of a building. However, finite elements may also be thought of as basis functions for piecewise functional interpolation. The complete generality of this approach can provide adequate solutions for all (even unanticipated) special cases of complicated physical systems. In addition, complicated sets of boundary conditions and internal constraints (even for multicomponent eigenfunctions) can be enforced algorithmically. Finally, even though $R \ll JN$, in general, the resulting system may still be computationally efficient because it will lead to a sparse-matrix problem.

A finite-element basis set can be formed by constructing a set of "nodes" x_i in positions such that $x_1 < x_2 < \ldots < x_N$, where (x_1, x_N) is the domain of the problem. More than one different type of basis function (I types) can be associated with each node. There will be a replica (which is identical except for translation and scaling) of each type of basis function at each node. Basis functions are chosen to be easy to evaluate and differentiate, to be sufficiently smooth to provide an acceptable interpolation of the eigenfunction, and to overlap with one another as little as possible. The degree of overlap will determine how many nearest neighbours (P on each side) each basis function will interact with. When there is no interaction between a pair of basis functions, the corresponding matrix elements will be zero. Hence the stiffness and mass matrices will be sparse. By properly ordering the simultaneous equations in (8), all of the non-zero elements will occur near the diagonal. Such a matrix is said to be banded. The number J of scalar functions sought, the number I of basis functions at each node, and the degree of overlap P will determine the bandwidth M of the matrix (the number of non-zero superdiagonals). Note that a matrix with a bandwidth of M actually has $2M+1$ non-zero diagonals (the main diagonal and, owing to the symmetry, M sub- and superdiagonals). Roughly speaking, $M = IJ(P+1)-1$. In general, both storage and computational requirements increase as M increases.

In order to estimate the scope of a particular problem, it is necessary to choose the type of basis functions to use (which determines I) and the number of basis functions to use (which determines N). For a given accuracy, N will be a function of the type and distribution of basis functions used. As smoother basis functions (more continuous derivatives) are used, the number of basis functions required decreases rapidly.

As the distribution of basis functions is changed, the ability to represent the eigenfunctions may be enhanced or degraded. In fact, N may be minimized by requiring that the worst-case structure of the eigenfunctions be represented equally well throughout the model. Buland and Gilbert (1984) give one strategy for optimizing node spacing that attempts to guarantee a minimum number of basis functions per wavelength for all eigenfunctions of interest.

The simplest finite-element basis functions are those for piecewise-linear interpolation. This choice provides a continuous function, but a discontinuous (piecewise-constant) first derivative. In this case there is one type of basis function ($I = 1$) and overlap with nearest neighbours only ($P = 1$). Given a set of nodes, each scalar is represented by

$$f(x) = \sum_{i=1}^{N} F_i \zeta_i(x), \tag{15}$$

where the F_i are coefficients and the ζ_i are the N basis functions defined as

$$\zeta_i(x) = \begin{cases} t_0/\Delta_0 & (x_{i-1} \leqslant x \leqslant x_i), \\ 1 - t_1/\Delta_1 & (x_i < x \leqslant x_{i+1}), \\ 0 & \text{otherwise}, \end{cases} \tag{16}$$

where $t_j = x - x_{i+j-1}$ and $\Delta_j = x_{i+j} - x_{i+j-1}$. Note that $\zeta_i(x_{i-1}) = \zeta_i(x_{i+1}) = 0$ to guarantee continuity. Also, the basis functions are normalized so that $\zeta_i(x_i) = 1$, which means that $f(x_i) = F_i$. The representation of $f(x)$ in terms of piecewise-linear basis functions is, in fact, identical with the piecewise-linear interpolation of the N values $f(x_i)$. This is a consequence of the fact that these basis functions are "local" (i.e. if $x_{i-1} \leqslant x \leqslant x_i$ then $f(x)$ depends only on the values of the coefficients associated with the nodes x_{i-1} and x_i). The use of piecewise-linear basis functions leads to a bandwidth of $2J - 1$.

If a continuous first derivative is desired then the basis functions for piecewise quadratic interpolation can be used. If a continuous second derivative is desired then the basis functions for piecewise-cubic interpolation (called B splines) can be used. In either case the representation is not local. In both cases there is only one type of basis function ($I = 1$). However, the quadratic basis functions overlap with their next two nearest neighbours on each side ($P = 2$), while the cubic basis functions overlap with their next three nearest neighbours on each side ($P = 3$). Also, $N + 1$ quadratic basis functions or $N + 2$ cubic basis functions are required for a complete representation. This need for additional degrees of freedom is a general property of non-local basis functions. It implies that the representation of eigenfunctions may be different from the corresponding interpolation. Another consequence of non-local basis functions is that there is no simple relationship between the coefficients and the function values at the nodes. Quadratic and cubic basis functions lead to bandwidths of $3J-1$ and $4J-1$ respectively. The quadratic basis functions for a particular set of nodes are computed by subroutine QSBF and cubic basis functions by subroutine BSBF.

It is also possible to construct piecewise-cubic basis functions that are local using the basis functions for piecewise Hermite cubic interpolation (called H splines). This type of interpolation fits not only the function at each node, but also its first derivative. Cubic Hermite interpolation provides a continuous function and first derivative, but a discontinuous second derivative. H splines are represented by two different basis functions at each node ($I=2$), but each function is local ($P=1$). Given a set of nodes, each scalar is represented by

$$f(x)=\sum_{i=1}^{N}\{F_i\chi_i(x)+\dot{F}_i\psi_i(x)\} \tag{17}$$

where the F_i and the $\dot{F}_i$ are coefficients and the χ_i and the ψ_i are the $2N$ basis functions defined as

$$\chi_i(x)=\begin{cases} 3\dfrac{t_0^2}{\Delta_0^2}-2\dfrac{t_0^3}{\Delta_0^3} & (x_{i-1}\leqslant x\leqslant x_i),\\[2ex] 1-3\dfrac{t_1^2}{\Delta_1^2}+2\dfrac{t_1^3}{\Delta_1^3} & (x_i<x\leqslant x_{i+1}),\\[2ex] 0 & \text{otherwise,}\end{cases} \tag{18}$$

and

$$\psi_i(x)=\begin{cases} -\dfrac{t_0^2}{\Delta_0}+\dfrac{t_0^3}{\Delta_0^2} & (x_{i-1}\leqslant x\leqslant x_i),\\[2ex] t_1-2\dfrac{t_1^2}{\Delta_1}+\dfrac{t_1^3}{\Delta_1^2} & (x_i<x\leqslant x_{i+1}),\\[2ex] 0 & \text{otherwise.}\end{cases} \tag{19}$$

Note that $\chi_i(x_i)=\dot{\psi}_i(x_i)=1$ and that $\dot{\chi}_i(x_i)=\psi_i(x_i)=0$, which guarantees that $f(x_i)=F_i$ and $\dot{f}(x_i)=\dot{F}_i$ (where a dot means a derivative with respect to position). H spline basis functions also lead to a bandwidth of $4J-1$. These basis functions are generated by subroutine HSBF. Given a set of $2N$ coefficients F_i and $\dot{F}_i$, the unique H spline interpolation can be evaluated at any point by function HSPLN.

If $h = \max|x_{i+1} - x_i|$ then the linear basis functions yield $O(h)$ accuracies in the eigenfunction and $O(h^2)$ accuracies in the eigenfrequency. Quadratic basis functions yield $O(h^2)$ and $O(h^4)$ accuracies respectively, while both B-spline and H-spline basis functions yield $O(h^3)$ and $O(h^6)$ accuracies respectively. Of course, H splines have twice as many basis functions for a given h as B splines and (as one would expect) can be shown to yield somewhat more accurate results (Birkhoff *et al.*, 1966). For the same number of nodes, higher-order methods have larger bandwidths and hence require more computation (and storage) than lower-order methods. However, higher-order methods may need so many fewer nodes that they may actually require less computer time than lower-order methods for the same accuracy.

4.1 Evaluation of matrix elements

As shown in (6), the variational equation is converted into an algebraic equation by substituting a linear combination of basis functions for each unknown scalar function. In general, the resulting integrals involve products of the independent variable, the model, functionals of the model, and pairs of the known basis functions (or their derivatives). When the model is multicomponent, the integrand will be a sum over model components, each multiplied by a separate kernel function. When the eigenfunction is multicomponent, each kernel will be a sum over quadratic terms, each involving products of the independent variable, functionals of the model components, a basis function for one of the eigenfunction scalars (or its derivative), and a basis function for another of the scalars (or its derivative). For example, see the formulation for the free oscillations of the Earth (Pekeris and Jarosch, 1958).

Although optimal methods of performing these integrals will depend on their exact form, for polynomial finite-element basis functions a set of algorithms has been developed that are generally fast, accurate and stable. This technique involves taking advantage of both the polynomial nature of the basis functions and their limited spatial extent. To see the motivation for this, we consider a case in which the model components are represented by piecewise polynomials that are connected at the basis-function nodes. In many cases the integrand will now be a simple, but rather high-order, piecewise polynomial in the independent variable. Therefore, between any two nodes, the integrals could be evaluated analytically. Unfortunately, this procedure is numerically unstable (owing to cancellation among the high-order terms). An alternative that is theoretically equivalent, numerically stable, and can actually be faster than evaluating the analytical solution, is to use Gauss–Legendre integration. This involves evaluating the integrand at S positions (determined by the algorithm) and summing the evaluations premultiplied by S weights (also determined by the algorithm). Gauss–Legendre integration of order S is theoretically exact when integrating polynomials of order up to $2S-1$. Gauss–Legendre integration is discussed and tables of abscissae and weights are given by Abramowitz and Stegun (1968) on pp. 887 and 916–919 respectively. Function GLQ performs Gauss–Legendre integration for a few different orders.

In some cases this procedure can be used directly (for example, some Earth models are represented by polynomials between first-order discontinuities, which must coincide with nodes in order to match boundary conditions). In some cases models are represented by values at discrete points. Such models may be satisfactorily interpolated by piecewise cubic polynomials using B splines. This interpolation is performed by subroutine BSPLN, and the piecewise polynomial can be evaluated at any point using function BSPLE. However, it is unlikely that the model sample points will coincide with the basis-function nodes. While it is possible to interpolate the model, sample it at the nodes, and then reinterpolate it, the results are not always adequate. The problem is that reinterpolation alters the polynomial representation of the model very slightly, making it difficult to reproduce results to high precision. If reinterpolation is undesirable or if the model representation is not easily converted into a polynomial or if the integrand is not exactly a polynomial (sometimes functionals of the model, occurring in the integrand, produce non-polynomial behaviour) then Gauss–Legendre integration may still be used between nodes, but it will no longer be exact. Formally, it will be equivalent to interpolating the integrand with a polynomial of order $2S-1$ and then integrating the interpolating polynomial. In practice, Gauss–Legendre integration will still yield excellent results, particularly if the order of the integration is increased slightly.

As pointed out, complicated variational integrals comprise many terms, each of which may be represented by a polynomial of different order. Despite the additional evaluations required, there are substantial efficiencies in choosing one order of integration. This is because most of the computational effort in performing the integrals is due to evaluating the basis functions and the model at the Gauss–Legendre integration points. If there is only one order of integration, these evaluations need be done only once. Further, products of the integration weights and the evaluated basis functions may be formed in advance. In fact, local basis functions scale with node spacing in the same way as the integration points. Thus only one of the basis functions need be evaluated at the integration points.

4.2 Matching boundary conditions

In most discussions of the Rayleigh–Ritz algorithm, basis functions are chosen that already match all boundary conditions. In fact, it is not necessary to match the boundary conditions exactly. It is possible to incorporate boundary conditions into the variational principle in such a way that they are matched approximately in the process of estimating the eigenfunctions (see e.g. Carrier and Pearson, 1976, pp. 176–179). By removing the constraint that boundary conditions be matched exactly, it is possible to approach the extrema more closely and hence approximate the eigenfunctions better on the average. For very simple basis functions (e.g. linear basis functions) this approach must be used when there are insufficient degrees of freedom to match all boundary conditions exactly. This technique will not be pursued further here. Instead it will be assumed that the reader desires the boundary conditions to be matched exactly, even if the eigenfunctions are not approximated quite so well.

To implement this more traditional approach for finite-element basis functions, it is necessary to construct special basis functions at the boundaries that match all boundary conditions. The simplest way to do this is to form the special basis functions from linear combinations of the original basis functions. Let us consider the boundaries at the model extremes. By construction, there will be a node exactly at each boundary. Basis functions near these boundaries will be special in the sense that they must be truncated so as not to extend beyond the domain of the model. The piecewise nature of the finite-element basis functions makes truncation at a node convenient to deal with. Only these partial basis functions will be non-zero at the boundaries. Thus the special basis functions that satisfy all constraints at a boundary will be constructed from a linear combination of the partial basis functions at the same boundary.

To see how this linear combination is formed, let us consider a typical boundary condition that requires that a linear combination of eigenfunction scalars (or their derivatives) be constant. Because each scalar is a linear combination of basis functions, this can always be transformed into the statement that a linear combination of the unknown coefficients (all associated with partial basis functions) be constant. Solving this equation, one of the coefficients can be determined in terms of the others. Substituting for this coefficient in the linear combination of basis functions that represents the affected scalar and collecting terms has the effect of eliminating one basis function and modifying others. This is illustrated by the following simple example. Suppose that x_N is a boundary node at which $\dot{f}(x_N) - \alpha f(x_N) = 0$. If the H-spline representation is used, this requires that $\dot{F}_N = \alpha F_N$. Substituting into (17) yields

$$f(x) = \sum_{i=1}^{N-1} \{F_i\chi_i(x) + \dot{F}_i\psi_i(x)\} + F_N\{\chi_N(x) + \alpha\psi_N(x)\}. \tag{20}$$

The coefficient $\dot{F}_N$ and its associated basis function have been eliminated. However, the basis function associated with the coefficient F_N is a linear combination of the two original partial basis functions that satisfies the boundary condition. By means of a sequence of similar operations, a complete set of boundary conditions (even involving different scalars) may all be satisfied simultaneously.

Algorithmically, it is more convenient to form the matrix elements from the original basis functions (which, except for truncation, are all the same) and then match the boundary conditions. This is equivalent because the integral of a linear combination of basis functions is the same as the linear combination of integrals of basis functions. Because of the quadratic nature of the kernels, each basis function affects all elements of one row and one column of the stiffness and mass matrices. Therefore the matrix equivalent of (20) is to form linear combinations of the rows and columns associated with the coefficients involved in matching a particular boundary condition. The row and column associated with the coefficient that is eliminated will also be eliminated. This procedure is described in detail by Wiggins (1976).

A problem with internal boundaries can be constructed by setting up a separate problem for each region (between two boundaries). Because the basis functions are all truncated at each boundary, each problem will be independent. The matrices for each separate problem are then combined, in the order that the boundaries are encountered, to form the blocks of two block-diagonal matrices ($\mathcal{U}$ and $\mathcal{H}$) that contain all the information. Finally, the blocks are "tied" together to form one larger system by the boundary conditions at the internal boundaries. Matching these boundary conditions is essentially identical with the procedure discussed above. The only difference is that constants in each linear combination may come from both sides of the boundary. In some cases matching boundary conditions can increase or decrease the bandwidth of the matrix locally. If bandwidth is increased, performance will suffer, even if the outermost superdiagonal contains only one element.

Linear equality constraints among the eigenfunction scalars or their derivatives may be imposed between boundaries by essentially the same procedure. In this case the constraint is imposed at each node in the region of interest. Although the procedure is the same as for matching boundary conditions, the result is to impose a linear dependence on the continuous eigenfunction scalars rather than an absolute or jump boundary condition.

4.3 Solving the banded eigenvalue problem

As has been seen, the use of finite-element basis functions will result in sparse matrices. If the equations are ordered in the matrices so that all basis functions (for all scalars) associated with each node are grouped together and so that the nodes are included in sequential order then the matrices will be block-diagonal with blocks of IJ rows by $IJ(2P+1)$ columns. Of course, this can be thought of as a banded matrix with a bandwidth of $M=IJ(P+1)-1$. Because the matrices are block-diagonal, the outermost IJ sub- and superdiagonals will become successively more sparse the further away from the main diagonal they are. This sort of problem presents two major difficulties for finding the eigenvalues and eigenvectors. First, finite-element basis functions will only be computationally useful if a method can be found for reducing the computation required to solve the problem by taking advantage of the sparse nature of the matrices. Secondly, in general, neither the stiffness nor the mass matrix will be diagonal. All known (exact) methods of diagonalizing either one will destroy the sparse nature of the problem.

A method of dealing efficiently with banded systems has been developed. In general, it parallels the method for solving dense systems except that the problem is never reduced to standard form. For any trial eigenfrequency, the banded matrix $\mathscr{C}=\mathscr{U}-\hat{\omega}^2\mathscr{K}$ is formed. Since boundary conditions have been matched in $\mathscr{U}$ and $\mathscr{K}$, they will already be matched in $\mathscr{C}$. $\mathscr{C}$ can then be upper-triangularized by a procedure that reduces the bandwidth to zero below the diagonal and increases it to $2M$ above the diagonal (Martin and Wilkinson, 1971b). This algorithm also computes the determinant and the Sturm count of $\mathscr{C}$. Each new trial eigenfrequency requires that $\mathscr{C}$ be formed and triangularized again. Thus the computation of each determinant and Sturm count is very costly (proportional to $IJNM^2$ operations and requiring $IJN(2M+1)$ storage locations). Therefore, rather than bisection only (based on the Sturm count), a combination of bisection and linear interpolation (Peters and Wilkinson, 1969) is used to estimate the eigenvalue. In detail, this process is begun by specifying an upper and a lower bound for the

Table 3 Orders of magnitude of bandwidth M, number of storage locations SL, number of floating-point operations FPO and accuracy for various basis functions.

Basis set	M	SL	FPO	$\lvert s-f \rvert$	$\lvert \omega^2 - \hat{\omega}^2 \rvert$
Linear	$2J-1$	$4J^2N$	$4J^3N$	h	h^2
Quadratic	$3J-1$	$6J^2N$	$9J^3N$	h^2	h^4
B splines	$4J-1$	$8J^2N$	$16J^3N$	h^3	h^6
H splines	$4J-1$	$8J^2N$	$32J^3N$	h^3	h^6

eigenspectrum of interest. The difference of the Sturm counts at the extremes immediately yields the number of eigenvalues that must be found. A bisection procedure (based on the Sturm count) is then used to find a set of non-overlapping intervals, each of which contains exactly one eigenvalue. Each eigenvalue is then refined by bisection and linear interpolation. The orders of magnitude of the bandwidths, numbers of storage locations, numbers of floating point computations, and accuracies of the eigenfunctions and eigenfrequencies is given in Table 3 for each of the finite-element basis functions considered.

The eigenvectors can be found by inverse iteration (Wilkinson, 1965, pp. 628–629) using the $\mathscr{C}$ matrix for the best estimate of the eigenvalue. Inverse iteration requires that the system $\mathscr{C}\boldsymbol{a}_k^{\mathrm{T}} = \boldsymbol{a}_{k-1}^{\mathrm{T}}$ be solved for $\boldsymbol{a}_k$, where $\boldsymbol{a}_k$ is the kth iterate for the eigenvector $\boldsymbol{a}$. This auxiliary problem can be solved using the same triangularization required to find the Sturm count (Martin and Wilkinson, 1971b). Once the eigenvector is found, the eigenvalue can be refined by forming a Rayleigh quotient. The elements of the eigenvector contain all information needed to completely describe the eigenfunction. If $J>1$ then coefficients corresponding to different scalars must be "demultiplexed". Finally, coefficients that were eliminated to match boundary conditions must be computed from the appropriate linear combinations of existing coefficients.

Because the performance of this algorithm is quadratic in bandwidth, the sparse outermost diagonals result in a high cost for a relatively small amount of information. If some boundary condition has increased the bandwidth at a few points then this problem will be even more severe. The effect of sparse outermost diagonals can be ameliorated by reducing the bandwidth by means of a series of Givens rotations as described by Schwarz (1971) prior to performing the upper triangularization. Properly chosen Givens rotations can eliminate one element in the outermost sub- and superdiagonals without increasing the bandwidth. Thus the performance of this algorithm is proportional to the number of non-zero elements in each diagonal eliminated. It turns out to be worthwhile to eliminate all (at least $IJ-1$) of the outermost diagonals that are systematically sparse. The series of Givens rotations must be saved and applied to "unrotate" the eigenvector before the eigenfunction is formed.

For the free oscillations of the Earth, between two and three trial eigenvalues are required per final eigenvalue during the initial bisection process. In addition, about ten trial eigenvalues are required to refine each eigenvalue by bisection and linear interpolation to a precision of about one part in 10^{11}. The set of routines for solving the generalized banded eigenvalue problem are managed by subroutine JHW. The initial bisection is supervised by subroutine EGNVL. Each eigenvalue is refined by function FZERO, which calls function BDET to do the upper triangularization and compute the determinant and Sturm count. The eigenvectors are found by subroutine SRAY, which calls subroutine BSOL to solve the system of linear equations using the triangularization. SRAY further refines the eigenvalue by inverse iteration. The Schwarz (1971) algorithm is implemented by subroutine BDRD. Note that if BDRD is used with JHW then care must be taken to apply the same transformation (subroutine TANGLE) and its inverse (subroutine UNTNGL) at appropriate places in the eigenvector computation. A convenient implementation is to do some initialization in JHW, to call BDRD in BDET and to add the appropriate vector transformations to BSOL and SRAY (see comments in the relevant routines for hints). Also, the function DOT is used in SRAY (as well as DIAG) to perform a dot product. There can be a considerable advantage in using a machine-dependent (i.e. assembler-language) version of DOT if one is available (particularly on vector machines). For static problems the use of finite-element basis functions also leads to banded systems. These problems can economically be solved by upper-triangularizing $\mathcal{U}$ using BDET and then solving the system of equations using BSOL.

The eigenvectors of any Rayleigh–Ritz problem will be orthogonal (because they are eigenvectors of a Hermitian system) and, if properly normalized, orthonormal. The natural normalization is to make the kinetic energy unity. Although it is obvious, it is easily overlooked that most of the labour necessary to perform this integral has already been done (i.e. total kinetic energy KE $\approx \hat{\omega}^2 \boldsymbol{a}\mathcal{K}\boldsymbol{a}^{\mathrm{T}}$). This same trick may be used to evaluate other related integrals if one is clever. For example, for free oscillations of the Earth it is convenient to form $\mathcal{U}$ as a linear combination of constant matrices whose coefficients are functions of angular order (see e.g. Buland and Gilbert, 1984). By taking the derivative of this linear combination with respect to angular order, a related matrix can be derived. This new matrix then yields an integral that is proportional to group velocity. Another application is in evaluating the effect of perturbing the model (i.e. (2)). This is convenient because the same computer program can be used for computing the stiffness and mass matrices for the perturbation as for the original model. For example, the quality factor Q of a free oscillation of the Earth can be computed in this way.

4.4 Another example

The toroidal free-oscillation example used above also provides a suitable example of the use of finite-element basis functions. In this case the piecewise Hermite cubic spline basis functions are used as described by Buland and Gilbert (1984). For N nodes, $2N$ basis functions are available, but two basis functions must be eliminated to match the two boundary conditions. Therefore only $2N-2$ eigenperiods may be computed in each case. Table 4 shows the results obtained using 2–7 nodes. In each case the last (lowest) eigenperiod estimate, while not bad, is less accurate than the corresponding estimate made using natural basis functions. However, convergence to the correct eigenperiod is as fast or faster than obtained using natural basis functions (even accounting for the fact that two basis functions are added for each node). This property can be attributed to the ability of finite-element basis functions to faithfully represent whatever eigenfunction shape is required by the linear model. The natural basis functions used are more similar to the eigenfunctions of the linear model than any single Hermite spline. However, information in the natural basis functions needed to represent the eigenfunctions of the linear model accurately is distributed over a large number of basis functions. In extreme cases the eigenfunctions of the system may be so different from the natural basis functions that many more natural basis functions may be required than finite-element basis functions. For example, this would occur if toroidal eigenfunctions of angular order 2 were used as basis functions for toroidal modes of

Table 4 Toroidal eigenperiods (in seconds) from the Rayleigh-Ritz method using finite element basis functions

Nodes	${}_0T_2$	${}_1T_2$	${}_2T_2$	${}_3T_2$	${}_4T_2$
2	2693.5	783.7	—	—	—
3	2694.2	797.6	447.1	289.4	—
4	2694.3	798.0	450.4	306.8	226.6
5	2694.3	798.1	450.6	308.3	232.4
6	2694.3	798.1	450.7	308.5	233.4
7	2694.3	798.1	450.7	308.6	233.6

angular order 10. The angular-order-2 eigenfunctions are oscillatory throughout the mantle. The higher frequency angular-order-10 eigenfunctions are oscillatory in the upper mantle, but nearly zero in the lower mantle. As a result, very large numbers of basis functions would be needed to achieve the requisite cancellation. No such problem arises with finite-element basis functions.

5 FREQUENCY-DEPENDENT PROBLEMS

In some problems either the model or the boundary conditions (or both) may depend on the eigenvalue being computed. If some simple method for specifying the frequency dependence can be devised and the effect on the eigenspectrum is small enough that the order of the eigenvalues is not altered then the Rayleigh–Ritz procedure may be modified to treat this case. The reason that this is possible is that, for a generalized banded eigenvalue problem, $\mathscr{C}$ is already frequency-dependent and must be recomputed for each new trial eigenvalue anyway. If the boundary conditions are not frequency-dependent then they may be matched in $\mathscr{U}$ and $\mathscr{K}$ (once) before $\mathscr{C}$ is formed. If the boundary conditions are also frequency-dependent then $\mathscr{C}$ must be computed from $\mathscr{U}$ and $\mathscr{K}$, which do not have the boundary conditions matched, and the boundary conditions must be matched directly in $\mathscr{C}$ for each trial eigenvalue. This procedure has the curious property that each eigenfunction is the solution of a slightly different problem (although all problems have exactly the same form and behaviour). Therefore the eigenfunctions of such a problem will not, in general, be orthogonal.

5.1 Inverse dispersion

While wavenumber increases with frequency for most problems, this is by no means universal. For example, in problems involving the free oscillations of a stably stratified fluid, wavenumber increases while frequency decreases. In this case the eigenfrequency associated with the smoothest eigenfunction is usually at a very low frequency. The eigenfrequencies of higher-wavenumber modes then decrease towards zero. The problem of contamination of the eigenspectrum of interest by the poorly estimated upper bounds of lower-frequency modes has been described. This problem is considerably aggravated by the use of finite-element basis functions because, generally, the number of possible eigenfrequencies is much greater than the number of desired modes (i.e. $IJN \gg R$) in order to achieve sufficient precision for the modes of interest. This can result in a large number of modes that have such high wavenumbers that even the grossest features of the eigenfunctions cannot be properly represented (i.e. there are less than two nodes per wavelength). The eigenfrequencies of such modes will be so poorly bounded that they may appear below, throughout or even above the eigenspectrum of interest. In some cases this problem may be eliminated by transforming the variational principle so that the solutions are maxima and the Rayleigh–Ritz procedure will provide lower bounds. If this cannot be done, the problem may not be a good candidate for solution by variational means.

In some problems, such as the free oscillations of the Earth, most solutions may be normally dispersed while one subset (of trapped modes) may be inversely dispersed (see e.g. Buland and Gilbert, 1984). Even if only the normally dispersed solutions are desired and the eigenspectrums of the two types of solutions are disjoint, contamination will still be a problem if the eigenspectrum of the inversely dispersed modes is at a lower frequency than the eigenspectrum of the normally dispersed modes. In some cases, it may be possible to invent a transformation that will eliminate the inversely dispersed subset of modes. If this is not possible then the only alternative is to compute all solutions in the eigenspectrum of interest, to classify the solutions according to mode type, and then to reject the poorly represented, inversely dispersed modes. The physics of the inversely dispersed subset of modes is usually quite helpful in the classification

6 INSTABILITIES

The only instability to which the Rayleigh–Ritz algorithm is commonly subject is that due to ill-conditioning in the system of differential equations. That is, at least one of the scalars sought is poorly constrained because its differential equation is only weakly coupled to the others. The instability takes the form of highly oscillatory behaviour when it is not appropriate. In complicated models the instability may be local. There are two ways in which this problem may be minimized. The first is simply to add more nodes in the unstable regions. This is observed to move the instability to a higher wavenumber (it is hoped, out of the eigenspectrum of interest). The second way to minimize instability is by the application of a constraint. Owing to the source of the instability, there will often be an approximate relation between the well-constrained scalars and the weakly coupled ones. If the approximation is sufficiently good, the use of a constraint will often completely eliminate the instability with little impact on the quality of the solution.

7. NATURAL BASIS FUNCTIONS REVISITED

There are several instances in which it may be economical to use finite-element basis function to solve one problem and then to use the eigenfunctions of this problem as natural basis functions in solving a closely related problem. The most common example is in solving exactly the same problem for a number of slightly different models. This will be economical when the eigenfunctions for one model are so similar to those for another that nearly all the eigenvalues and eigenvectors of the subsidiary dense system are useful. The only difficulty in this application is that the eigenfunctions of one problem will not necessarily satisfy the boundary conditions of the related problem. This difficulty can be solved by creating the set of natural basis functions from linear combinations of the eigenfunctions of the first problem and another set of arbitrary but smooth (linearly independent) functions. This is only worth doing when the perturbation to the eigenfunction is desired, since the perturbation to the eigenfrequency can be found with little additional computation using (2).

Another application that sometimes arises is that an approximate problem may provide natural basis functions for the problem one actually wished to solve. For example, if the perturbation to the gravitational potential is neglected then the computational labor of estimating the spheroidal free oscillations of the Earth can be greatly reduced (two coupled differential equations rather than three). Once the eigen-displacements have been found, the induced perturbation to the gravitational potential can be computed. Although these functions are not solutions of the full problem, they do provide excellent natural basis functions.

8 COMPARISON WITH OTHER METHODS

In the context of one-dimensional problems the Rayleigh–Ritz algorithm can be thought of as solving a two-point boundary-value problem. For eigenvalue problems a solution exists only for certain discrete values of the eigenvalue. Finite-difference methods of solving two-point boundary-value problems are very closely related. In this approach a set of nodes is chosen and a discrete approximation for each derivative (depending only on the scalar values at the nodes) is substituted into the differential equations. The resulting banded matrix system is identical in form with (8). Solutions of a finite-difference system may be found in the same way as for a Rayleigh–Ritz system using finite-element basis functions. However, the upper-bound and higher-order-accuracy properties of the eigenvalue estimates are not preserved.

The other means of solving two-point boundary-value problems is by direct solution of the differential equations. These so-called shooting methods involve choosing a trial eigenvalue, propagating the solution from one boundary to the other, and measuring the misfit of the final boundary condition. Zeros of this misfit correspond to eigenvalues. The behaviour of shooting methods contrasts very strongly with that of the Rayleigh–Ritz method. For example, solutions are often oscillatory in one region of the model and evanescent (exponentially decaying) in another. In evanescent regions shooting methods must solve the differential system in the direction that the solution is increasing for stability. In addition, the eigenfunction must have a finite amplitude at the final boundary. The Rayleigh–Ritz algorithm has no such restrictions. Problems with solutions that are evanescent in both directions are not well suited for shooting methods (unless they are reformulated in terms of minors). The performance of shooting methods is also dependent on the "stiffness" of the differential system. That is, they require much more computation when the exponential decay in an evanescent region is very rapid than when it is slow. All Rayleigh–Ritz solutions require the same amount of time. However, the eigenfunction of a strongly evanescent mode may not be as well represented as that of a weakly evanescent mode.

In general, shooting methods are faster for smooth eigenfunctions while variational methods are faster for rapidly changing eigenfunctions. For very complicated systems the Rayleigh–Ritz method will have fewer special cases. However, shooting methods may be more convenient for problems in which the model contains regions with radically different properties that induce radically different physical behaviours in the solutions. For example, fluid regions in Earth models result in both instabilities and an inversely dispersed subset of modes in the Rayleigh–Ritz solution. Both difficulties may be completely eliminated by the use of shooting methods. Conversely, the stable direction of integration and the point a which boundary conditions should be matched are eigenfunction-dependent for the computation of free oscillations of the Earth by shooting methods. Further, shooting methods have no way of guaranteeing either a complete set of eigenfunctions or orthogonal eigenvectors. No such problems exist for the Rayleigh–Ritz solution.

8.1 Related methods

There are many examples of problems that are similar in form to (1), but are not variational equations. For instance, the addition of anelasticity to an elastic problem destroys the Hermitian nature of the system and the extremal nature of the solutions. However, it is often possible to solve such a problem in almost exactly the manner discussed above. This technique, called the Galerkin method, can be motivated by observing that (11) does not depend on the extremal nature of the variational solutions. That is, in the absence of a variational principle, one might seek an approximate eigenfunction such that, when acted upon by the differential operator, it yields a spatial error that is orthogonal to all of the basis functions and hence to every function in the Hilbert space that they span. Thus the projection of the true solutions onto the Hilbert space is still sought (see e.g. Carrier and Pearson, 1976, pp. 177–178). For many, but not all, problems the Rayleigh–Ritz and Galerkin formulations turn out to be identical. However, when there is no variational principle, neither the upper-bound property nor the higher-order accuracy of the eigenvalue estimates is preserved.

Another closely related technique is the finite-element method. In standard engineering applications the finite-element method is often used to compute the static response of structures to external loads (see e.g. Zienkiewicz, 1967, pp. 1–10). This technique is formally identical with the Rayleigh–Ritz method using finite-element basis functions in which a static solution to loading is sought (Zienkiewicz, 1967, pp. 18–21). In practice, finite-element methods generally incorporate additional approximations not introduced above. In particular, although the elasticity of the medium is distributed, the mass is often assumed to be concentrated at the nodes (the lumped-mass approximation). A corollary of lumping the mass is that equivalent body forces act only on the nodes.

Finite-element techniques can obviously be applied equally well to the normal-mode problem. The effect of the lumped-mass approximation is to make $\mathcal{K}$ diagonal. In this case the Cholesky decomposition turns out to be diagonal and (8) can be reduced to standard form as in (12) without destroying the banding. The method of Schwarz (1971) may be used to reduce the system to tridiagonal form (again preserving the banding). The eigenvalues and eigenvectors may then be found using the same methods as for a dense system. Note that this means that all eigenvalues can be found with about the same amount of computation as one Sturm count if $\mathcal{K}$ were banded but not diagonal. This approximation can be used in any Rayleigh–Ritz problem with finite-element basis functions. Because mass is conserved, the primary effect is on the eigenfunctions rather than the eigenfrequencies.

A final application of the finite-element method that is widely used in geophysics is the modelling of transients in complicated (often three-dimensional) media. This is done by a process called time-stepping (see e.g. Zienkiewicz, 1967, pp. 166–168). The procedure is to formulate a solution for the spatial problem at a fixed point in time, treating the time derivatives as known constants (which turn out to behave in the same manner as internal body forces). Forward differences can then be substituted for the derivatives with respect to time and an iterative system developed, where each iteration leads to a new time step.

9 CLOSING REMARKS

Variational methods originally gained popularity in an era of hand calculation. By using a small number of natural basis functions, a solution of moderate precision could be found with a minimum of computation. Today, with nearly universal access to powerful super minicomputers and widespread access to supercomputers, the role of variational methods has necessarily changed dramatically. As has been implied above, variational methods are particularly well suited to complicated two-point boundary-value problems in which a large number of high-precision eigenvalues and eigenfunctions are desired. In fact, owing to the large amounts of computer memory required, such methods are more competitive today than they were a decade ago. Although there will always be highly specialized problems for which variational methods are a natural solution, only time can tell whether they will continue to be viable alternatives to the direct solution of differential equations for general eigenvalue problems. Ultimately, this will be determined by the types of problems that need to be solved, advances in numerical analysis, and especially by the evolution of the architecture of large-scale scientific computational equipment.

REFERENCES

Abramowitz, M. and Stegun, I. A. (1968). *Handbook of Mathematical Functions* US Government Printing Office, Washington, DC.

Barth, W., Martin, R. S. and Wilkinson, J. H. (1971). Calculation of the eigenvalues of a symmetric tridiagonal matrix by the method of bisection. *Linear Algebra* (ed. J. H. Wilkinson and C. Reinsch), pp. 249–256, Springer-Verlag, New York.

Birkhoff, G., De Boor, C., Swartz, B. and Wendroff, B. (1966). Rayleigh–Ritz approximation by piecewise cubic polynomials. *SIAM J. Numer. Anal.* **3,** 188–203.

Buland, R. and Gilbert, F. (1984). Computation of free oscillations of the Earth. *J. Comp. Phys.* **54,** 95–114.

Carrier, G. F., and Pearson, C. E. (1976). *Partial Differential Equations: Theory and Technique.* Academic Press, New York.

Dziewonski, A. M. and Anderson, D. L. (1981). Preliminary reference Earth model. *Phys. Earth Planet. Inter.* **25,** 297–356.

Goldstein, H. (1950). *Classical Mechanics.* Addison-Wesley, Reading, Massachusetts.

Jeffreys, H. and Jeffreys, B. S. (1966). *Methods of Mathematical Physics.* Cambridge University Press.

Martin, R. S. and Wilkinson, J. H. (1971a). Reduction of the symmetric eigenproblem $Ax = \lambda Bx$ and related problems to standard form. *Linear Algebra* (ed. J. H. Wilkinson and C. Reinsch), pp. 303–314. Springer-Verlag, New York.

Martin, R. S. and Wilkinson, J. H. (1971b). Solution of symmetric and unsymmetric band equations and the calculation of eigenvectors of band matrices. *Linear Algebra* (ed. J. H. Wilkinson and C. Reinsch), pp. 70–92. Springer-Verlag, New York.

Martin, R. S., Peters, G. and Wilkinson, J. H. (1971a). Symmetric decomposition of a positive definite matrix. *Linear Algebra* (ed. J. H. Wilkinson and C. Reinsch), pp. 9–30. Springer-Verlag, New York.

Martin, R. S., Reinsch, C. and Wilkinson, J. H. (1971b). Householder's tridiagonalization of a symmetric matrix. *Linear Algebra* (ed. J. H. Wilkinson and C. Reinsch), pp. 212–226. Springer-Verlag, New York.

Mathews, J. and Walker, R. L. (1965). *Mathematical Methods of Physics.* Benjamin, New York.

Morse, P. M. and Feshbach, H. (1953). *Methods of Theoretical Physics,* Part II. McGraw-Hill, New York.

Pekeris, C. L. and Jarosch, H. (1958). The free oscillations of the Earth. *Contributions in Geophysics in Honor of Beno Gutenberg* (ed. E. Ingerson), pp. 171–192, Pergamon, New York.

Peters, G. and Wilkinson, J. H. (1969). Eigenvalues of $Ax = \lambda Bx$ with band symmetric A and B. *Comput. J.* **12,** 398–404.

Peters, G. and Wilkinson, J. H. (1971). The calculation of specified eigenvectors by inverse iteration. *Linear Algebra* (ed. J. H. Wilkinson and C. Reinsch), pp. 418–439. Springer-Verlag, New York.

Rayleigh, J. W. S. (1877). *The Theory of Sound,* Vol. I. Dover, New York.

Schwarz, H. R. (1971). Tridiagonalization of a symmetric band matrix. *Linear Algebra* (ed. J. H. Wilkinson and C. Reinsch), pp. 273–283. Springer-Verlag, New York.

Wiggins, R. A. (1976). A fast, new computational algorithm for free oscillations and surface waves. *Geophys. J. R. Astron. Soc.* **47,** 135–150.

Wilkinson, J. H. (1965). *The Algebraic Eigenvalue Problem*. Clarendon Press, Oxford.

Wilkinson, J. H. and Reinsch, C. (eds) (1971). *Linear Algebra*. Springer-Verlag, New York.

Zienkiewicz, O. C. (1967). *The Finite Element Method in Structural and Continuum Mechanics*. McGraw-Hill, New York.

V

UTILITY ALGORITHMS AND SOURCES

V.I

Vector and Matrix Manipulation Algorithms

GARY L. PAVLIS

Department of Geology, Indiana University, Bloomington, Indiana, USA

1 INTRODUCTION

A cursory look at any book on numerical analysis shows the central importance of vector and matrix manipulations in any physical science. Given the importance of computers to seismology, it is not surprising then that this subject is of great importance to seismologists as well. Vector manipulations have risen to further importance in recent years with increased access to array processors and vector supercomputers. The architecture of these machines drives one to examine any algorithm for steps that can be expressed as a set of vector operations because of the enormous decrease in execution time that can be achieved in the process.

Matrix and vector algorithms appear in sometimes surprising contexts. For example, finite-difference solutions to differential equations can be thought of in terms of matrices (see e.g. Lanczos, 1961). This chapter, however, is strongly coloured by the author's background in geophysical inverse theory, where matrices play a central role. Therefore most concrete examples that I shall give arise from this area of seismological research.

This chapter can be thought of as a seismological footnote to the following publications.

(i) Lawson *et al.* (1979a,b) published what I consider to be a major landmark in algorithms for vector manipulation that they called the Basic Linear Algebra Subroutines (BLAS).

(ii) Dongarra *et al.* (1979) published a collection of matrix-inversion routines based on the BLAS.

SEISMOLOGICAL ALGORITHMS
ISBN 0-12-220770-X

(iii) Paige and Saunders (1982a,b) published an algorithm for solving large sparse least-squares problems by the conjugate-gradient method.

(iv) Lawson and Hanson's (1974) book describes computational aspects of solving a variety of classes of least-squares problems.

These sources are important for this chapter because they all contain portable Fortran implementations of one or more algorithms. Furthermore, all these algorithms are public-domain software. I hope to include this rather lengthy collection of software with the other algorithms distributed with this book. Alternatively, the first three sources describe how the machine readable code can be obtained for a small fee. Lawson and Hanson's code is only printed.

This chapter focuses on seismological applications of routines from these three sources. It also includes some additions that I have developed and several algorithms that use subroutines from one or more of these sources. Thus this chapter can be thought of as a review of public-domain software packages used for vector and matrix manipulations, with some special additions relevant to the seismologist.

2 BASIS VECTOR OPERATIONS

The publication of the BLAS by Lawson *et al.* (1979a,b) was a major advance for scientific algorithms that have any relationship to vectors. The functions implemented in the BLAS are summarized in Table 1. The following are, in my view, the most significant aspects of the BLAS.

(i) When properly used, the BLAS impose a valuable level of modularity. Once they have been mastered, one finds oneself automatically trying to convert an algorithm to a set of vector operations rather than a set of scalar operations.

(ii) If increased speed is of the essence in an algorithm using the BLAS then one can replace appropriate BLAS code with machine-specific assembly-language versions.

(iii) They have demonstrated portability.

(iv) They have a consistent naming convention that makes remembering the name of a given routine easy.

(v) They have a consistent argument list that is easy to remember.

The last two items and their relation to Table 1 require some additional

Table 1 BLAS functions and root names (after Lawson *et al.*, 1979a).

Function	Root name	Major prefixes
$\mathbf{x}^{\mathrm{T}}\boldsymbol{y}$	DOT	C, D, S
$\boldsymbol{x} \leftarrow a\boldsymbol{x}$	SCAL	C, D, S
$\boldsymbol{y} \leftarrow a\boldsymbol{x} + \boldsymbol{y}$	AXPY	C, D, S
$\boldsymbol{y} \leftarrow \boldsymbol{x}$	COPY	C, D, S
Swap $\boldsymbol{x}$ and $\boldsymbol{y}$	SWAP	C, D, S
$\lVert\boldsymbol{x}\rVert_2$	NRM2	SC, D, S
$\lVert\boldsymbol{x}\rVert_1$	ASUM	SC, D, S
$\lVert\boldsymbol{x}\rVert_\infty$	AMAX	IC, ID, IS
Set up and apply Givens rotation	ROTG, ROT	D, S
Set up and apply modified Givens rotation	ROTMG, ROTM	D, S

Notes:
1. $\boldsymbol{x}$ and $\boldsymbol{y}$ are vectors, a is a scalar.
2. $\leftarrow$ symbolizes an operation where one vector overwrites another.
3. $\lVert\boldsymbol{x}\rVert_p$ denotes the L_p norm (see e.g. Stewart, 1973, p. 166).
4. For complex $\boldsymbol{x}$, CASUM calculates $\sum |\mathrm{Re}\, x_i| + |\mathrm{Im}\, x_i|$ instead of $\sum [(\mathrm{Re}\, x_i)^2 + (\mathrm{Im}\, x_i)^2]^{1/2}$.

explanation. Each procedure is labelled by a "root name" that is mnemonic for its function (within the limits of six character variable names dictated by ANSI Fortran anyway). A prefix is attached to this "root name" to describe the storage structure of the vector(s) on which it is designed to operate. For example, SWAP is the root name for the operation of interchanging the locations of two vectors. There are three versions of SWAP called SSWAP, DSWAP and CSWAP. They are designed for single-precision vectors, double-precision vectors and complex vectors respectively. The S, D and C prefixes in the BLAS always carry this implication, which makes it easy to distinguish what type of object(s) the routine expects. There are two minor anomalies from this pattern.

(i) A valid vector norm by definition is a real number. To emphasize this, the authors of the BLAS chose to label the complex versions of ASUM and NRM2 as SCASUM and SCNRM2.

(ii) AMAX implements the L_∞ norm by returning the index of the largest element of $\boldsymbol{x}$. To take advantage of implicit typing, the authors of the BLAS chose to use the prefixes IS, ID, and IC for this function.

The original BLAS contained several additional routines not listed in Table 1. These were primarily some mixed-precision dot-product routines. I omit them from this discussion because I think their value for special situations does not justify the added confusion that they bring.

With the exception of the rotation routines, the arguments of the BLAS routines are consistent and easy to remember. They can be grouped into two classes:

one-vector routines: NAME(N, {A}, X, INCX)
two-vector routines: NAME(N, {A}, X, INCX, Y, INCY)

where the {A} is used to imply that this argument is not always required. Specifically, A is always a scalar. It is only used in the SCAL and AXPY routines. X and Y are vectors of length N that are to be manipulated. INCX and INCY are storage increments between the elements of X and Y. This idea is most easily understood by recognizing that in most applications INCX and INCY are one of two things. When manipulating vectors or the columns of a two-dimensional array, INC = 1. When manipulating the rows of a two-dimensional array, INC is set to the first dimension of the array. A simple example helps clarify this. Suppose that we want to calculate $\boldsymbol{b} = \boldsymbol{A}\boldsymbol{x}$, where $\boldsymbol{b} \in \mathbb{R}^m$, $\boldsymbol{x} \in \mathbb{R}^n$ and $\boldsymbol{A} \in \mathbb{R}^{m \times n}$. The following is a fragment of Fortran code for doing this that uses the BLAS:

```
      REAL A(N1A, N), B(M), X(N)
          :
      DO 100 I = 1, M
            B(I) = SDOT(N, A(I, 1), N1A, X, 1)
100   CONTINUE
```

The dot-product routine accesses $\boldsymbol{A}$ by rows and treats $\boldsymbol{x}$ as a simple vector.

For additional details on the BLAS the reader is referred to Lawson *et al.* (1979a,b). For further examples of the use of the BLAS see Dongarra *et al.* (1979).

In my opinion, there are three very basic operations that were not included in the implementation of the BLAS due to Lawson *et al.* These are the vector operations

$$y \leftarrow x + y,$$

$$y \leftarrow y - x,$$

$$y \leftarrow x - y,$$

where the arrow $\leftarrow$ means that the object on the right overwrites the object on the left. All three of these operations could be implemented with AXPY (the third would also require a call to SCAL). However, this is not very efficient since AXPY uses a floating-point multiply, which is unnecessary in the special case when $a = \pm 1$. As part of the computer code distributed with this book I have included code for these special operations. These are relatively simple programs that anyone could write in a short time. I include them purely for the reader's convenience. Because they are so simple, however, only the single-precision routines are included. The double-precision and complex versions could be easily produced by some simple changes to DAXPY and CAXPY. The routines included are named SXPY, SYMY and SXYM respectively. All have the same arguments as SDOT, SCOPY etc.

3 MANIPULATION OF NON-SPARSE MATRICES

3.1 Basic operations

There are three basic matrix operations: addition, multiplication and transposition. The first two are somewhat trivial to implement with the BLAS. Simple algorithms for both can be implemented in no more than three lines of Fortran executable code. I include two space-economizing routines for implementing these operations, as the algorithms are not quite so obvious. MTRXADD for two matrices $\boldsymbol{A} \in \mathbb{R}^{m \times n}$ and $\boldsymbol{B} \in \mathbb{R}^{m \times n}$ does the operation $\boldsymbol{B} \leftarrow \boldsymbol{A} + \boldsymbol{B}$. Similarly, a routine called MTRXMUL calculates the product $\boldsymbol{C} \in \mathbb{R}^{m \times n}$ from $\boldsymbol{A} \in \mathbb{R}^{l \times m}$ and $\boldsymbol{B} \in \mathbb{R}^{l \times n}$ as $\boldsymbol{B} \leftarrow \boldsymbol{C} = \boldsymbol{A}^{\mathrm{T}}\boldsymbol{B}$. MTRXMUL uses the product $\boldsymbol{A}^{\mathrm{T}}\boldsymbol{B}$ as a basic operation because it leads to a column-oriented algorithm that executes considerably faster on virtual memory computers. Both of these routines are good introductory examples of the utility of the BLAS. One can easily design other addition and multiplication algorithms of various flavours using the BLAS. I should hope that the versions I supply could be used as a guide in the process.

Matrix transposition is trivial to implement when storage is not at a premium by a simple application of the BLAS COPY routine. For large arrays that are nearly full, in-place transposition is preferred. I include a routine called MTRANS that does this operation for single-precision arrays. Its use is best seen by an example:

```
REAL A(M, N)
      ⋮
CALL MTRANS(A, M, N)
```

After the call to MTRANS, the array A will be stored in computer memory as if it had been declared originally as REAL A(N, M). This implies that one must be careful in using the contents of A in subsequent calculations since the actual storage after the transposition is not consistent with the declaration. The rule to remember is that if MTRANS is applied to an array then that array should not be referenced by subscripts unless it is passed with variable dimensions to another subprogram. There are two alternatives. The first is the Fortran EQUIVALENCE, but my personal prejudice is this should be avoided as I think it can produce rather obtuse code. The second is never to explicitly declare a matrix as a two-dimensional array, but only use a one-dimensional array and let the program calculate subscripts internally. In some respects this is a better programming style with the data-structure limitations of Fortran, but my experience is that this is more error-prone because one is forced to do one's own index calculations. The key thing is that any of these three approaches will work, but the main requirements of any code that uses MTRANS is a generous number of explanatory comments.

In implementing computer algorithms using matrices, there are two other basic operations that for some reason are almost always ignored in discussions of matrix algorithms. The reason, I think, is that both are mundane pure computer tasks that are unrelated to linear algebra. The first is initialization. For this purpose I am supplying a routine called MTRXO that initializes a single-precision array to zeros. Its function deserves no further comment. The second is the input and output of the data stored in matrices. There are three basic situations where I have found matrix I/O routines to be a valuable tool.

(i) Complicated matrix algorithms can often be broken down into a set of distinct tasks with well-defined brief intermediate results. In such cases, rather than writing the ultimate megaprogram that does the full job from start to finish, it may be preferable to write a series of individual programs that communicate through temporary files. The Unix concept of filters and pipelines is a simple example that encourages this approach.

(ii) Matrix input and output can be used to implement a simple type of manual, virtual memory. That is, a matrix can be written to a temporary file, after which the storage it occupied can be used for other calculations. When the matrix is needed again it can be read back into memory.

(iii) The contents of an array often need to be printed or otherwise stored in machine-retrievable form. These numbers may be either the end result of some calculation, or some intermediate result needed for debugging. Printing represents a special problem in that the result needs to be readable by a human being. In this case a large matrix is best printed by columns in multiple blocks.

The functional specification of I/O routines that I have implemented are summarized in Table 2. All are limited to single-precision arrays, but conversion to other data structures would be trivial. To assure compatibility between input and output modules, related routines are forced to be connected through the use of alternate entry points. For example, the MTRXIB code is structured as follows:

```
SUBROUTINE MTRXIB(A,...)
        ⋮
RETURN
ENTRY MTRXOB(A,...)
        ⋮
RETURN
ENTRY MTRXTOB(A,...)
        ⋮
RETURN
END
```

The transposed output routines (MTRXTOB and MTRXTOE) are useful when a later calculation is more efficiently done with the transpose. This can be used to avoid a later call to MTRANS or an equivalent.

3.2 Matrix-operator building blocks

Several operations are basic for the construction of more complex matrix operations, like matrix inversion. The operations considered here are not exhaustive, but they represent what has proved valuable to me.

Householder transforms are a valuable basic tool for some matrix-inversion algorithms. They are also sometimes called elementary reflectors because this describes geometrically what they do. From a vector $\boldsymbol{u}$ of unit length ($\boldsymbol{u}^{\mathrm{T}}\boldsymbol{u} = 1$), we can construct a Householder transformation for this vector defined as $\boldsymbol{H} = \boldsymbol{I} - 2\boldsymbol{u}\boldsymbol{u}^{\mathrm{T}}$. Geometrically, the effect of $\boldsymbol{H}$ is to reflect the coordinate system through a hyperplane with $\boldsymbol{u}$ as its normal vector. (A useful exercise to understand this is to calculate a concrete example in a two-dimensional space. Then the reflecting "hyperplane" reduces to a line through the origin.)

Table 2 Matrix input–output routines.

Routine name(s)	Function
MTRXIB, MTRXOB	Read (I) and write (O) a matrix using Fortran unformatted I/O
MTRXTOB	Equivalent of MTRXOB, but writes transpose of matrix
MTRXIE, MTRXOE	Read (I) or write (O) a matrix using formatted I/O in E format
MTRXTOE	Equivalent of MTRXOE, but writes transpose of matrix
MTRXPR	Outputs matrix in blocks of columns appropriate for hard copies on a line printer

The usefulness of Householder transforms comes from their exceptional numerical stability (see e.g. Lawson and Hanson, 1974, pp. 83–89). Because of this Householder transforms are the cornerstone of most QR and singular-value decomposition algorithms (see below). Lawson and Hanson (1974) published a Fortran routine for performing Householder transforms called H12. The reader is referred there for detailed documentation of this algorithm. For further descriptions of how Householder transforms are used the reader is referred to Lawson and Hanson's book and to another by Stewart (1973).

The BLAS contain Fortran codes for another type of building block. These are the routines for performing plane rotations known as Givens rotation matrices and modified Givens rotation matrices (Table 1). The two are functionally identical. The modified form uses a different algorithm devised by Gentleman (1973) that reduces the operation count and as a result is usually the preferred algorithm. Functionally, a plane rotation is of the form

$$\boldsymbol{Q} = \begin{bmatrix} 1 & 0 & & 0 & \dots & 0 & \dots & 0 \\ \vdots & \vdots & & \vdots & & \vdots & & \vdots \\ 0 & 0 & & \cos\theta & & \sin\theta & \dots & 0 \\ \vdots & \vdots & & \vdots & & \vdots & & \vdots \\ 0 & 0 & & -\sin\theta & & \cos\theta & \dots & 0 \\ \vdots & \vdots & & \vdots & & \vdots & & \vdots \\ 0 & 0 & \dots & 0 & & 0 & \dots & 1 \end{bmatrix}. \tag{1}$$

If the two $\cos\theta$ terms occupy columns i and j, the effect of $\boldsymbol{Q}$ is to rotate the coordinate system in the (i, j) plane through an angle θ. The main application of plane rotations is in algorithms for orthogonal decomposition of matrices. Almost any algorithm using Householder transformations has an equivalent using plane rotations. In addition, I have found plane rotations useful for simple three-space Cartesian coordinate system transformations.

A final building block I think worthy of consideration are projectors. I have found projectors to be a valuable tool in the analysis of earthquake travel-time residuals. For three different applications of projectors, the reader is referred to Jordan and Sverdrup (1981), Pavlis (1984) and Pavlis and Hokanson (1985). All of these applications centre around the joint earthquake location problem. Jordan and Sverdrup used projectors to derive what they call the hypocentroidal decomposition theorem. This states that when clusters of earthquakes are located jointly, the absolute location of the group (the hypocentroid) is decoupled, to first order, from the relative locations of events within the group. Pavlis and Hokanson (1984) took this another step. We used projectors to separate the joint location problem into two distinct operations. The absolute location of a cluster of events can only be corrected by using outside information because the hypocentral decomposition theorem shows that this information is fundamentally unavailable from the given data. We showed how this outside information can be assimilated into the solution through the use of a projector. Relative locations, on the other hand, can be obtained directly from the actual data. However, relative locations are more precise when mixed with outside information through appropriate projectors.

In Pavlis (1984) I also used projectors for a related, but somewhat different application. Residuals from two different earthquakes cannot, in general, be compared because of differences in how the residuals from the two events interact with the location process. In my 1984 paper, I derived a projector that allows one to compare only the parts of the residuals from two earthquakes that are in common with both events.

Projectors are commonly discussed in linear-algebra texts, but I am not aware of any common software package that contains any algorithm for calculating them. A projector is defined as follows: if $\boldsymbol{v}_i \in \mathbb{R}^n$ ($i = 1, 2, \ldots, r$) are a set of orthonormal vectors (i.e. $\boldsymbol{v}_i^{\mathrm{T}}\boldsymbol{v}_j = \delta_{ij}$) then

$$\boldsymbol{P} = \sum_{i=1}^{r} \boldsymbol{v}_i \boldsymbol{v}_{\mathrm{i}}^{\mathrm{T}} \tag{2}$$

is an operator called an orthogonal projector. $\boldsymbol{P}$ has an important counterpart defined as

$$\boldsymbol{P}_{\mathrm{c}} = \boldsymbol{I} - \boldsymbol{P} \tag{3}$$

and called a complementary projector. This name comes from the property that given some $\boldsymbol{b} \in \mathbb{R}^n$, $\boldsymbol{Pb}$ is the projection of $\boldsymbol{b}$ onto the r-dimensional subspace of $\mathbb{R}^n$ spanned by the basis bectors $\boldsymbol{v}_i$. In contrast, $\boldsymbol{P}_{\mathrm{c}}\boldsymbol{b}$ is the projection of $\boldsymbol{b}$ onto the $(n-r)$-dimensional subspace of $\mathbb{R}^n$ perpendicular to the subspace defined by $\boldsymbol{P}$. This subspace is called an orthogonal complement (see e.g. Ben-Israel and Greville, 1974, pp. 63–76). Orthogonal projectors are intimately related to least-squares problems. That is, if we solve

$$\boldsymbol{Ax} = \boldsymbol{b} \tag{4}$$

for $\boldsymbol{A} \in \mathbb{R}^{m \times n}$, $\boldsymbol{b} \in \mathbb{R}^m$ and $\boldsymbol{x} \in \mathbb{R}^n$ by least squares, we obtain the solution

$$\hat{\boldsymbol{x}} = \boldsymbol{A}^{+}\boldsymbol{b}, \tag{5}$$

where $\boldsymbol{A}^{+}$ is the pseudoinverse (see e.g. Lawson and Hanson, 1974, pp. 36–52). For full-rank, overdetermined problems $\boldsymbol{A}^{+}$ can be calculated from the so-called normal equations as $\boldsymbol{A}^{+} = (\boldsymbol{A}^{\mathrm{T}}\boldsymbol{A})^{-1}\boldsymbol{A}^{\mathrm{T}}$. Otherwise, it must be calculated by some other method, some of which are discussed below.

A general matrix $\boldsymbol{A}$ is associated with four fundamental vector subspaces defined as follows (Ben-Israel and Greville, 1974, p. 14):

$$\left.\begin{array}{lll} R(\boldsymbol{A}) = \{\boldsymbol{y} \in \mathbb{R}^m : \boldsymbol{y} = \boldsymbol{Ax} \quad \text{for} \quad \boldsymbol{x} \in \mathbb{R}^n\} & \text{is the } \textit{range of } \boldsymbol{A}, \\ N(\boldsymbol{A}) = \{\boldsymbol{x} \in \mathbb{R}^n : \boldsymbol{Ax} = \boldsymbol{0}\} & \text{is the } \textit{null space of } \boldsymbol{A}, \\ R(\boldsymbol{A}^{\mathrm{T}}) = \{\boldsymbol{x} \in \mathbb{R}^n : \boldsymbol{x} = \boldsymbol{A}^{\mathrm{T}}\boldsymbol{y} \quad \text{for} \quad \boldsymbol{y} \in \mathbb{R}^m\} & \text{is the } \textit{range of } \boldsymbol{A}^{\mathrm{T}}, \\ N(\boldsymbol{A}^{\mathrm{T}}) = \{\boldsymbol{y} \in \mathbb{R}^m : \boldsymbol{A}^{\mathrm{T}}\boldsymbol{y} = \boldsymbol{0}\} & \text{is the } \textit{null space of } \boldsymbol{A}^{\mathrm{T}}. \end{array}\right\} \tag{6}$$

Each of these subspaces has an associated orthogonal projector that I shall write as $\boldsymbol{P}_{R(\boldsymbol{A})}$, $\boldsymbol{P}_{N(\boldsymbol{A})}$, $\boldsymbol{P}_{R(\boldsymbol{A}^{\mathrm{T}})}$ and $\boldsymbol{P}_{N(\boldsymbol{A}^{\mathrm{T}})}$ respectively. A pseudoinverse solution (5) has the following relation to these projectors (Ben-Israel and Greville, 1974):

$$\hat{\boldsymbol{b}} = \boldsymbol{A}\hat{\boldsymbol{x}} = \boldsymbol{P}_{R(\boldsymbol{A})}\boldsymbol{b},$$
$$\boldsymbol{r} = \boldsymbol{b} - \boldsymbol{A}\hat{\boldsymbol{x}} = \boldsymbol{P}_{N(\boldsymbol{A}^{\mathrm{T}})}\mathbf{b},$$
$$\hat{\boldsymbol{x}} = \boldsymbol{P}_{R(\boldsymbol{A}^{\mathrm{T}})}\boldsymbol{x},$$
$$\boldsymbol{x} - \hat{\boldsymbol{x}} = \boldsymbol{P}_{N(\boldsymbol{A})}\boldsymbol{x}.$$

Furthermore,

$$\left.\begin{aligned} \boldsymbol{P}_{N(\boldsymbol{A}^{\mathrm{T}})} &= \boldsymbol{I} - \boldsymbol{P}_{R(\boldsymbol{A})} = [\boldsymbol{P}_{R(\boldsymbol{A})}]_{\mathrm{c}}, \\ \boldsymbol{P}_{N(\boldsymbol{A})} &= \boldsymbol{I} - \boldsymbol{P}_{R(\boldsymbol{A}^{\mathrm{T}})} = [\boldsymbol{P}_{R(\boldsymbol{A}^{\mathrm{T}})}]_{\mathrm{c}}, \end{aligned}\right\} \tag{7}$$

where the c subscript is used to emphasize the relation to (3). Equation (7) is important because it shows that the null-space projectors are easily constructed from the range projectors. The original definition of a projector (2) suggests an algorithm for calculating it based on the singular-value factorization. That is, any matrix $\boldsymbol{A} \in \mathbb{R}^{m \times n}$ can always be decomposed as

$$\boldsymbol{A} = \boldsymbol{U}\boldsymbol{\Lambda}\boldsymbol{V}^{\mathrm{T}}, \tag{8}$$

where $\boldsymbol{U} \in \mathbb{R}^{m \times r}$ and $\boldsymbol{V} \in \mathbb{R}^{n \times r}$ are orthogonal matrices whose columns span $R(\boldsymbol{A})$ and $R(\boldsymbol{A}^{\mathrm{T}})$ respectively. $\boldsymbol{\Lambda} \in \mathbb{R}^{r \times r}$ is a diagonal matrix whose r non-zero diagonal elements form a set of positive numbers called the singular values of $\boldsymbol{A}$. They can be thought of as a type of eigenvalue. The columns of $\boldsymbol{U}$ and $\boldsymbol{V}$ can similarly be thought of as a generalized set of eigenvectors called the left and right singular vectors respectively (see e.g. Lanczos, 1961). $\boldsymbol{U}$ and $\boldsymbol{V}$ can be used to construct these projectors as

$$\left.\begin{aligned} \boldsymbol{P}_{R(\boldsymbol{A})} &= \boldsymbol{U}\boldsymbol{U}^{\mathrm{T}}, \\ \boldsymbol{P}_{R(\boldsymbol{A}^{\mathrm{T}})} &= \boldsymbol{V}\boldsymbol{V}^{\mathrm{T}}. \end{aligned}\right\} \tag{9}$$

The null-space projectors can then be constructed using (7).

I have implemented two routines for calculating the projectors defined in (2) and (3). I call them PROJECT and COMPPRO respectively. PROJECT is based on (9). COMPPRO is based on (9) combined with (7). Both use the BLAS and an outer-product algorithm similar in style to Lawson and Hanson's (1974, p. 57) Householder-transformation algorithm.

3.3 Matrix inversion of even-determined systems

For solution of non-sparse equations of the form

$$\boldsymbol{A}\boldsymbol{x} = \boldsymbol{b}$$

where $\boldsymbol{A} \in \mathbb{R}^{n \times n}$, I know of no better source than the software package published by Dongarra *et al.* (1979) called LINPACK. LINPACK is based on the BLAS, and contains implementations of virtually all major modern inversion algorithms. Further discussion is unnecessary. The reader is referred to the outstanding text by Dongarra *et al.* that documents this code and gives a lot of practical advice regarding its use.

3.4 Singular-value decomposition and generalized inverses

Solution of many geophysical inverse problems centres around equations of the form of (4). Equations of this form arise in at least four ways in geophysical inverse problems.

(i) In discrete linear problems we are given m data related to n model parameters by

$$b_i = f_i(\boldsymbol{x}) = \sum_{j=1}^{n} A_{ij} x_j. \tag{10}$$

An example of a problem where this type of equation occurs is the time-term method of refraction analysis (Willmore and Bancroft, 1960).

(ii) In discrete non-linear problems we are given a similar set of equations to (10), but $f_i(\boldsymbol{x})$ is not a linear function of the parameters $\boldsymbol{x}$. One approach to solving such problems, however, is to treat them as a sequence of linear problems. This is accomplished by using a first-order expansion of f_i about some initial guess $\boldsymbol{x}_0$, which gives

$$r_i = b_i - f_i(\boldsymbol{x}_0) \approx \sum_{j=1}^{n} \frac{\partial f_i}{\partial x_j} \delta x_j. \tag{11}$$

Equations (11) and (10) are similar, with b_i equivalent to r_i, a_{ij} equivalent to $\partial f_i / \partial x_j$, and x_j equivalent to δx_j. Equation (11) is referred to as a linearization for obvious reasons. The most familiar example of this type of problem to seismologists is hypocentre location by Geiger's method.

(iii) In continuous linear inverse problems $\boldsymbol{x}$ becomes a continuous function of space, $m(\boldsymbol{r})$, that describes some physical property of Earth materials as a function of spatial position. In this case we write

$$b_i = \int_{\text{volume}} G_i(\boldsymbol{r}) m(\boldsymbol{r}) \, \mathrm{d}V, \quad i = \{1, 2, \ldots, m\}. \tag{12}$$

$G_i(\boldsymbol{r})$ is called a "data kernel" by analogy with the theory of integral equations (Parker, 1977). An example of this type of problem is the problem of estimation of the seismic Q within the Earth from attenuation measurements.

(iv) Most seismological inverse problems are properly thought of as continuous non-linear problems. In this case

$$b_i = F_i[m(\boldsymbol{r})],$$

with the functional F_i being non-linear in $m(\boldsymbol{r})$. The most common approach for dealing with these problems is the continuous equivalent of (11). That is, if we expand F_i about some initial guess of the model, $m_0(\boldsymbol{r})$, in the functional equivalent of a Taylor series and retain only the zeroth- and first-order terms, then we obtain a set of m equations of the form

$$r_i = b_i - F_i[m_0(\boldsymbol{r})] = \int_{\text{volume}} G_i(r)\, \delta m(r)\, \mathrm{d}V. \tag{13}$$

The right-hand side is the function-space equivalent of a partial-derivative operator called a Fréchet derivative and G_i is called a Fréchet kernel. $\delta m(\boldsymbol{r})$ is a model perturbation defined as $\delta m = m(\boldsymbol{r}) - m_0(\boldsymbol{r})$. As an example, travel times and seismic velocity are related by this type of equation (Backus and Gilbert, 1969).

It is not universally appreciated, I believe, that computationally there is little practical difference between discrete and continuous inversion. In continuous problems the data kernels $G_i(\boldsymbol{r})$ are seldom known explicitly and have to be calculated numerically. For instance, in Backus and Gilbert's (1969) inversion of P-wave residuals for gross Earth structure, the G_i were determined numerically using a model discretized on a set of grid points. This means that the integral in (12) is approximated as ((13) is similar except that the symbols are slightly different)

$$b_i \approx \sum_{j=1}^{n} \sum_{k=1}^{n} G_{ij} W_{jk} m_k, \tag{14}$$

where G_{ij} and m_k are the values of $G_i(\boldsymbol{r})$ and $m(\boldsymbol{r})$ at a set of n quadrature points at positions $\boldsymbol{r}_j$. The weights W_{jk} give the corresponding quadrature coefficients. For a collection of m observations, (14) can be written in matrix form as

$$\boldsymbol{b} = \boldsymbol{GWm} \tag{15}$$

For any sensible quadrature rule, the matrix $\boldsymbol{W}$ is positive definite. Then

(15) can be rewritten as

$$\boldsymbol{b} = \boldsymbol{A}\boldsymbol{x}$$

where $\boldsymbol{A} = \boldsymbol{G}/\sqrt{\boldsymbol{W}}$ and $x = \sqrt{\boldsymbol{W}}\boldsymbol{m}$. The matrix $\sqrt{\boldsymbol{W}}$ is called the square root of $\boldsymbol{W}$ (see e.g. Ben-Israel and Greville, 1974, p. 124).

With most simple one-dimensional quadrature rules, $\boldsymbol{W}$ is diagonal (cubic splines are a notable exception). For example, for the trapezoidal rule with a set of n points spaced at intervals of Δx, $W_{ij} = w_i \delta_{ij}$, where w_i is $\frac{1}{2}\Delta x$ for $i = \{1, n\}$ and w_i is Δx otherwise. When $\boldsymbol{W}$ is diagonal, calculating $\sqrt{\boldsymbol{W}}$ is trivial. For two- or three-dimensional quadrature rules, $\boldsymbol{W}$ is diagonal only in the simplest case when uniform pixels or boxes are used. However, more complicated quadrature schemes still usually lead to a $\boldsymbol{W}$ that is sparse with a well-defined structure. If $\boldsymbol{W}$ is not diagonal then the only general way I know for calculating $\sqrt{\boldsymbol{W}}$ is to use an eigenvalue–eigenvector decomposition of $\boldsymbol{W}$, but there may be better alternatives of which I am unaware. In any case, the point is that from a computational standpoint both discrete and continuous inverse problems face the need to solve the same basic set of equations (4).

A wide range of solutions can be obtained from (4) through application of appropriate weighting schemes. That is, if we are given a set of positive-definite matrices $\boldsymbol{W}_b \in \mathbb{R}^{m\times m}$ and $\boldsymbol{W}_x \in \mathbb{R}^{n\times n}$, (4) can be converted to

$$\boldsymbol{B}\boldsymbol{y} = \boldsymbol{c}, \tag{16}$$

where

$$\boldsymbol{B} = \boldsymbol{W}_b\boldsymbol{A}\boldsymbol{W}_x, \qquad \boldsymbol{y} = \boldsymbol{W}_x^{-1}\boldsymbol{x}, \qquad \boldsymbol{c} = \boldsymbol{W}_b\boldsymbol{b}. \tag{17}$$

A pseudoinverse solution of (16) yields

$$\hat{\boldsymbol{y}} = \boldsymbol{B}^{+}\boldsymbol{c} \tag{18}$$

or

$$\hat{\boldsymbol{x}} = \boldsymbol{W}_x\boldsymbol{B}^{+}\boldsymbol{W}_b\boldsymbol{b}. \tag{19}$$

The solution defined by (19) is special in that it is the unique solution that minimizes $\|\boldsymbol{r}\|_{\boldsymbol{W}_b} = [\boldsymbol{r}^{\mathrm{T}}\boldsymbol{W}_b^2\boldsymbol{r}]^{1/2}$ ($\boldsymbol{r} = \boldsymbol{b} - \boldsymbol{A}\boldsymbol{x}$) and $\|\boldsymbol{x}\|_{\boldsymbol{W}_x} = [\boldsymbol{x}^{\mathrm{T}}\boldsymbol{W}_x^{-2}\boldsymbol{x}]^{1/2}$ (see e.g. Ben-Israel and Greville, 1974, pp. 103–127; Lawson and Hanson, 1974, pp. 180–188). Note that this reduces to a simple least-squares solution when $\boldsymbol{W}_b = \boldsymbol{I}$ and $\boldsymbol{W}_x = \boldsymbol{I}$. For seismologists the role of $\boldsymbol{W}_b$ is most easily understood in reference to earthquake location. A diagonal $\boldsymbol{W}_b$ is used in every location program that I know of. These numbers are referred to as weights. They are derived from a variety of criteria, including arrival quality, distance weighting, residual weighting and azimuthal weighting. The effect of $\boldsymbol{W}_x$ is more subtle. In least-squares problems that are full-rank, $\boldsymbol{W}_x$ only indirectly affects the solution by altering the condition number of $\boldsymbol{B}$ in comparison with $\boldsymbol{A}$ (Lawson and Hanson, 1974, pp. 185–188). An example of this comes again from earthquake location. It is fairly common to scale the origin time of the hypocentre parameters to have units of length in order to be consistent with spatial coordinates. In this case $\boldsymbol{W}_x$ would be diagonal with diagonal elements equal to 1 for the spatial coordinates and $1/v$ for the origin time, where v is some representative velocity. (Note that in this case $\boldsymbol{W}_x^{-1}$ would also be diagonal, but the origin-time component would be v.) For well-constrained solutions this has no effect. However, it does tend to improve numerical stability because the scaled equations are almost always better conditioned than the corresponding unscaled equations.

For underdetermined problems or mixed determined least-squares problems $\boldsymbol{W}_x$ can have a controlling influence on the solution. Tarantola and Valette (1982) showed that such an effect on this type of solution can be interpreted in terms of an a priori model covariance.

In continuous inversion theory the operator equivalent of $\boldsymbol{W}_x$ defines what Backus (1970) calls quelling. The most common example is integration quelling (see e.g. Backus and Gilbert, 1969; Chou and Booker, 1979). In the one-dimensional case $\boldsymbol{W}_x$ is lower-triangular with all the lower-triangular elements equal to unity and

$$\boldsymbol{W}_x^{-1} = \begin{bmatrix} 1 & 0 & 0 & \dots & 0 & 0 \\ -1 & 1 & 0 & \dots & 0 & 0 \\ 0 & -1 & 1 & \dots & 0 & 0 \\ \vdots & \vdots & \vdots & \ddots & \vdots & \vdots \\ 0 & 0 & 0 & \dots & 1 & 0 \\ 0 & 0 & 0 & \dots & -1 & 1 \end{bmatrix},$$

The three-dimensional form depends on details of the quadrature scheme used to represent the model.

The pseudoinverse in (18) or (8) can be calculated in several ways. Probably the most reliable, however, is through the singular-value decomposition whose form was given in (8). From that decomposition, the pseudoinverse $\boldsymbol{A}^+$ can be calculated as

$$\boldsymbol{A}^+ = \boldsymbol{V}\boldsymbol{\Lambda}^{-1}\boldsymbol{U}^{\mathrm{T}}. \tag{20}$$

$\boldsymbol{B}^+$ can be calculated in the same way, but the factorization will, of course, yield different numbers unless $\boldsymbol{W}_x$ and $\boldsymbol{W}_b$ are identity matrices.

$\boldsymbol{\Lambda}^+$ is defined as

$$\boldsymbol{\Lambda}^+ = \begin{bmatrix} \lambda_1^{-1} & 0 & \dots & 0 \\ 0 & \lambda_2^{-1} & \dots & 0 \\ \vdots & \vdots & \ddots & \vdots \\ 0 & 0 & \dots & \lambda_r^{-1} \end{bmatrix}, \tag{21}$$

where $\{\lambda_1, \lambda_2, \dots, \lambda_r\}$ are the singular values of $\mathbf{A}$. The pseudorank r is a difficult quantity to establish. If the singular values are arranged in a decreasing sequence so that $\lambda_1 > \lambda_2 > \dots > \lambda_r$ then it follows from (21) that the smallest singular values cause the greatest magnification of experimental errors. It can be shown (see e.g. Parker, 1977) that decreasing the pseudorank r always decreases the uncertainty in the solution induced by measurement errors, but only at the expense of degrading the resolving power of the data. There is no fixed rule for dealing with small singular values. How to handle small but significant singular values in setting r is a "difficult and unsolved problem" (Dongarra *et al.*, 1977, p. 11.23). A practical approach that I have found useful in problems using a pseudoinverse solution is to find solutions for a range of r values and select one that yields a reasonable compromise between solution uncertainty and resolving power. This implies that it is prudent to separate the task of doing the singular-value decomposition from that of constructing a solution. An approach that I have found useful is to use one program whose sole purpose is to construct $\mathbf{A}$ and calculate the singular-value decomposition of $\mathbf{A}$. When this is completed, MTRXOB (or MTRXOE) is used to dump the results to a disk file. Then another program can read the results of the SVD operation and easily generate a number of solutions for almost nothing. The only computations required are some simple matrix multiplications. Furthermore, the analysis is not limited to using a pseudoinverse. Once the singular values and singular vectors are available, one can just as easily construct a set of damped solutions (see e.g. Lawson and Hanson, 1974, pp. 188–194). Resolution functions (Backus and Gilbert, 1968, 1970) are equally easy to calculate.

A number of singular-value-decomposition algorithms are available in the public domain. Two different subroutines are distributed with this book. The first is SSVDC, which appears as part of LINPACK (Dongarra *et al.*, 1977). The second is called SVDRS. It was published in the book by Lawson and Hanson (1974). Both use the same basic algorithm, which is a descendant of the revolutionary Algol procedure originally published by Golub and Reinsch (1970) that took SVD from a theoretical to a practical tool. My personal prejudice gives preference to

SVDRS largely because it uses less storage than SSVDC. It accomplishes this by never explicitly calculating $\boldsymbol{U}$. Instead, Lawson and Hanson recognized that one seldom cares about the actual form of $\boldsymbol{U}$, but all one really needs is the product with the data vector, $\boldsymbol{U}^{\mathrm{T}}\boldsymbol{b}$. Thus SVDRS only calculates $\boldsymbol{b} \leftarrow \boldsymbol{U}^{\mathrm{T}}\boldsymbol{b}$, which is all that is necessary for solving (5). This restriction is not as serious as it may first appear, however, as the algorithm allows $\boldsymbol{b}$ to be an array. Thus, if one wants to explicitly calculate $\boldsymbol{U}^{\mathrm{T}}$, one only has to set $\boldsymbol{b} = \boldsymbol{I} \in \mathbb{R}^{m \times m}$, and SVDRS will overwrite the array $\boldsymbol{b}$ with $\boldsymbol{U}^{\mathrm{T}}$. I have found this algorithm valuable for implementing what Pavlis and Booker (1980) called an annulling transform. The annulling transform has proved useful in the problem of using earthquake arrival times to determine Earth structure. The problem becomes intractable without it because every earthquake that is added to the data set not only increases the number of equations that have to be solved, but also increases the number of unknowns by four (the coordinates of the hypocentre). A brute-force solution of such a system of equations becomes completely intractable unless the number of events used to calculate the solution is kept small. The annulling transform permits one to circumvent this problem by allowing one to separate the hypocentre-location process from the velocity-inversion process. This allows larger data sets to be used in the inversion. For examples of the application of this technique the reader is referred to a recent review paper by Thurber and Aki (1987). The algorithm that I have used based on SVDRS is more stable numerically than the similar procedure developed independently by Spencer and Gubbins (1980). SVDRS uses Householder transforms and Givens rotations to calculate $\boldsymbol{U}$ and $\boldsymbol{V}$, which, as noted above, have extremely good numerical stability. Spencer and Gubbins use a normal equation algorithm, which is not as numerically stable (see e.g. Stewart, 1973, pp. 221–222).

I am also supplying two short procedures intimately connected to SVDRS called GISOL and NUMBIG. GISOL uses the output of SVDRS to calculate the pseudoinverse solution given in (5). To form this solution, it calls the function NUMBIG, which establishes the pseudo-rank r according to a user-specified criteria. That is, NUMBIG returns r given a specified minimum allowed value on λ_i. It is worth pointing out that it is usually ill advised to ever specify this cut-off in an absolute sense because r will depend upon how $\boldsymbol{A}$ is scaled. It is better to specify the cut-off as

$$\lambda_{\text{cut-off}} = \rho\lambda_{\max},$$

where $\lambda_{\max}$ is the largest singular value of $\boldsymbol{A}$ (this is λ_1 for SVDRS). ρ is chosen such that $\varepsilon < \rho < 1$, where ε is the single-precision floating-point relative precision of the computer doing the calculations ($\varepsilon \approx 10^{-6}$ for 32-bit computers). Using ρ in this way establishes a cut-off that defines the solution in terms of the condition number of the system rather than some arbitrary number.

4 SPARSE-MATRIX INVERSION ALGORITHMS

4.1 Banded matrices

A matrix is banded when a small number of rows and columns around its diagonal are non-zero. Two examples where inversion of banded matrices is required are cubic spline interpolation formulae and finite-element-analysis codes. LINPACK contains several routines useful for solving banded systems of equations. The reader is referred to Dongarra *et al.* (1977) for details.

4.2 Sparse least-squares inversion

Sparse linear-equation solvers have risen to great importance in recent years with the recognition that travel-time inversions for velocity require the inversion of large sparse systems of equations. Travel-time inversion has much in common with X-ray computer-aided tomography (CAT) scanners. Recognition of this fact led directly to the recent terminology of "seismic tomography".

Sparse systems of equations arise in tomography out of a fundamental assumption or ray theory. Under this assumption, the travel time from a source at $\boldsymbol{r}_0$ to a receiver at $\boldsymbol{r}$ is

$$t_i = \int_{\Gamma_i} u(\boldsymbol{r})\, \mathrm{d}s, \tag{22}$$

where Γ_i denotes the ray path joining source and receiver and u is the slowness as a function of $\boldsymbol{r}$. Thus, t_i can be thought of as a line integral between $\boldsymbol{r}_0$ and $\boldsymbol{r}$. Now let us suppose that we are given a cubic volume of the Earth that we divide up into N^3 cubes. If we parametrize the velocity of this volume in terms of the velocity in each of these N^3 boxes, and we assemble m equations of the form (22), we are faced with solving an $m \times N^3$ system of equations. However, these equations can be expected to always be rather sparse. Each datum is related to $u(\boldsymbol{r})$ by a line integral that will only cross about N boxes for each ray. Thus only about $N^{-2} \times 100\%$ of the matrix to be inverted will be non-zero. For N of any reasonable size the matrix to be solved is generally better than 99% zeros.

X-ray CAT scanners almost universally use a technique called the convolution method, which is basically a numerically stable algorithm for implementing an inverse Radon transform (Herman, 1980, pp. 118–179). Unfortunately, the convolution method is generally inappropriate for seismic tomography because the data-collection geometry of most seismic-tomography experiments is too heterogeneous and the convolution method assumes straight ray paths (Pavlis, 1986). As a result, seismic tomography is generally cast as a large-scale sparse-matrix-inversion problem.

I am aware of five basic classes of algorithms that have proved useful for solving large sparse systems of equations by least squares.

(i) The algebraic reconstruction technique (see Herman, 1980, pp. 180–205, and references therein) has been applied successfully for borehole seismic tomography by Peterson *et al.* (1985).

(ii) The simultaneous iterative reconstruction technique (SIRT) (Herman, 1980, pp. 206–221), which is a variant of Richardson's method for solving linear systems with positive-definite matrices, has been used successfully in X-ray tomography. Olson (1987) describes a significant variant that uses Chebyshev polynomials to speed convergence.

(iii) Dines and Lytle (1979) proposed an algorithm that is similar in its effect to the algebraic reconstruction technique. The algorithm is different, however, and has different convergence properties (Nolet, 1985). This algorithm has been used extensively by the group at the California Institute of Technology (Clayton and Comer, 1983; Fawcett and Clayton, 1984; Humphreys *et al.*, 1984).

(iv) A conjugate-gradient technique for sparse systems has been implemented by Paige and Saunders (1982a,b) for solving damped least-squares problems.

(v) It is possible to implement a sparse form of singular-value decomposition with the Lanczos algorithm (Nolet, 1985).

Little is known about the relative advantages and disadvantages of these algorithms for geophysical applications. It is my opinion that more detailed investigations are needed to address this issue. The only real comparative study of which I am aware is that by Nolet (1985). He compared the last three of the algorithms listed above and found the conjugate-gradient procedure of Paige and Saunders (1982b) to be superior in performance to both of the other algorithms that he tested.

He also described an algorithm for calculating resolution functions that use the same sparse-matrix algorithm. This is an important development since one cost of implementing a solution based on any of the above algorithms is that resolution cannot be directly calculated. Nolet's algorithm for calculating the resolution functions, however, requires the same amount of computational effort that is required to calculate a single solution. The extra effort is certainly justified, however, and is not as serious a problem as it may appear. In a recent paper (Pavlis, 1986) I used a brute-force procedure based on SVDRS to solve a 550×400 tomographic system of equations. This required slightly over four hours of CPU time on a Vax computer. Some recent calculations done by one of my students using Paige and Saunder's algorithm required one minute of CPU time to solve a 1500×1500 system on the same computer. The same system I estimate would have required several days to solve using SVDRS. Given this difference, one could calculate a large number of resolution functions for far less than the cost of a brute-force solution.

I plan to distribute Paige and Saunder's (1982a,b) algorithm with this book. I know of no other readily available working public-domain sparse-matrix least-squares equation solver. For details of this implementation the reader is referred to the above mentioned references, which serve as excellent documentation.

4.3 Sequential accumulation

Sequential accumulation (Lawson and Hanson, 1974, pp. 208–224) is a useful algorithm for solving large least-squares problems for which $m \gg n$. The algorithm given by Lawson and Hanson is not really a sparse-matrix algorithm. I discuss it in this section, however, because I have found it to be a useful algorithm for solving the multiple-event location problem. The equations one needs to solve for joint earthquake locations are very sparse with a special structure This structure can be exploited as described by Pavlis and Booker (1983). Our implementation there centres around sequential accumulation and annulling transformations (Pavlis and Booker, 1980). In this chapter I restrict the discussion to sequential accumulation as described by Lawson and Hanson, because the algorithm they describe has more general applications. The reader is referred to Pavlis and Booker (1983) for the details of how sequential accumulation can be used in multiple-event location.

For least-squares problems in which $m \gg n$ it is clear that storing the entire matrix $\boldsymbol{A}$ can become prohibitive when m is very large. The idea of sequential accumulation is that it is also totally unnecessary. To see this, it is useful to partition $\boldsymbol{A}$ and $\boldsymbol{b}$ to yield

$$\begin{bmatrix} \boldsymbol{A}_1 \\ \boldsymbol{A}_2 \\ \vdots \\ \boldsymbol{A}_q \end{bmatrix} [\boldsymbol{x}] = \begin{bmatrix} \boldsymbol{b}_1 \\ \boldsymbol{b}_2 \\ \vdots \\ \boldsymbol{b}_q \end{bmatrix}, \tag{23}$$

where $\boldsymbol{A}_i \in \mathbb{R}^{m_i \times n}$ and $\boldsymbol{b}_i \in \mathbb{R}^{m_i}$, with

$$\sum_{i=1}^{q} m_i = m.$$

The size of m_i is a matter of choice. It can be as small as 1, or large enough that q is a small number. One method for solving least-squares problems is via the normal equations

$$\hat{\boldsymbol{x}} = (\boldsymbol{A}^{\mathrm{T}}\boldsymbol{A})^{-1}\boldsymbol{A}^{\mathrm{T}}\boldsymbol{b}. \tag{24}$$

It is not hard to show that

$$\boldsymbol{A}^{\mathrm{T}}\boldsymbol{A} = \sum_{i=1}^{q} \boldsymbol{A}_i^{\mathrm{T}}\boldsymbol{A}_i, \qquad \boldsymbol{A}^{\mathrm{T}}\boldsymbol{b} = \sum_{i=1}^{q} \boldsymbol{A}_i^{\mathrm{T}}\boldsymbol{b}_i. \tag{25}$$

One approach to sequential accumulation is suggested by (25). One can simply accumulate $\boldsymbol{A}^{\mathrm{T}}\boldsymbol{A}$ and $\boldsymbol{A}^{\mathrm{T}}\boldsymbol{b}$ as given in (25). Storage is only needed for $\boldsymbol{A}^{\mathrm{T}}\boldsymbol{A}$, $\boldsymbol{A}^{\mathrm{T}}\boldsymbol{b}$, $\boldsymbol{A}_i$ and $\boldsymbol{b}_i$. The problem with this algorithm is that it is well known that the normal equation solution (24), has poor numerical stability. The algorithm described by Lawson and Hanson has superior numerical stability to the simple normal equation approach, while retaining the same storage economy. It does this by using a QR factorization equivalent of (25). The QR factorization makes use of the fact that any matrix $\boldsymbol{A}$ can be decomposed as

$$\boldsymbol{A} = \boldsymbol{Q}\boldsymbol{R}, \tag{26}$$

where $\boldsymbol{Q} \in \mathbb{R}^{m \times m}$ is orthogonal and $\boldsymbol{R} \in \mathbb{R}^{m \times n}$ is upper-triangular. Equation (4) can then be solved as

$$\hat{\boldsymbol{x}} = \boldsymbol{R}^{+}\boldsymbol{Q}^{\mathrm{T}}\boldsymbol{b}. \tag{27}$$

If $\boldsymbol{R}$ is not singular then $\boldsymbol{R}^{+}$ can be calculated by a simple back-substitution algorithm (Lawson and Hanson, 1974, pp. 63–66). If it is singular then $\boldsymbol{R}^{+}$ can be calculated by singular-value decomposition. The algorithm described by Lawson and Hanson for doing this process by sequential accumulation used Householder transforms to construct a sequence of upper-triangular matrices $\boldsymbol{R}_i$ $(i = 1, 2, \ldots, q)$, which terminate with $\boldsymbol{R}_q = \boldsymbol{R}$ in (26). The algorithm only requires space to hold the upper-triangular matrix, $\boldsymbol{R}$, the largest $\boldsymbol{A}_i$ and the largest $\boldsymbol{b}_i$. Additional storage is needed to store the vector $\boldsymbol{d} \in \mathbb{R}^{n+1}$. The first n elements of $\boldsymbol{d}$ hold the product $\boldsymbol{Q}^{\mathrm{T}}\boldsymbol{b}$. Only the first n are needed because the lower $m-n$ rows of $\boldsymbol{R}$ are forced to zero by the algorithm. These $m-n$ components of $\boldsymbol{Q}^{\mathrm{T}}\boldsymbol{b}$ are not zero, but normally individual components are not of interest. Thus the $(n+1)$th element of $\boldsymbol{d}$ is used to store the sum of the squares of these numbers, which can be shown to be the sum of squared residuals. For a detailed description of this algorithm the reader is referred to Lawson and Hanson (1974).

Lawson and Hanson (1974, p. 210) give an abbreviated sketch of an algorithm they label SEQHT for doing sequential accumulation of a QR factorization. I have implemented this algorithm in a Fortran algorithm with the same name that is included in the software distributed with this book.

ACKNOWLEDGEMENTS

This chapter was made possible through experiences in matrix-manipulation software gleaned from work over the past several years. During that time I was supported through the following grants from the National Science Foundation, to whom I am deeply indebted: EAR-7919605, EAR-8319418 and EAR-8507920.

REFERENCES

Backus, G. E. (1970). Inference from inadequate and inaccurate data, II. *Proc. Natl. Acad. Sci.* **65,** 281–287.

Backus, G. E. and Gilbert, F. (1968). The resolving power of gross Earth data. *Geophys. J. R. Astron. Soc.* **16,** 169–205.

Backus, G. E. and Gilbert, F. (1969). Constructing P-velocity models to fit restricted sets of travel time data. *Bull. Seismol. Soc. Am.* **59,** 1407–1414.

Backus, G. E. and Gilbert, F. (1970). Uniqueness in the inversion of inaccurate gross Earth data. *Phil. Trans. R. Soc. Lond.* **A266,** 123–192.

Ben-Israel, A. and Greville, T. N. E. (1974). *Generalized Inverses: Theory and Applications.* Wiley, New York.

Chou, C. W. and Booker, J. R. (1979). A Backus–Gilbert approach to inversion of travel-time data for three-dimensional velocity structure. *Geophys. J. R. Astron. Soc.* **59,** 325–344.

Clayton, R. W. and Comer, R. P. (1983). A tomographic analysis of mantle heterogeneities from body wave travel times (abstract). *EOS Trans. Am. Geophys. Union* **64,** 776.

Dines, K. A. and Lytle, R. J. (1979). Computerized geophysical tomography. *Proc. IEEE* **67,** 1065–1073.

Dongarra, J. J., Moler, C. B., Bunch, J. R. and Stewart, G. W. (1979). *LINPACK User's Guide*. SIAM, Philadelphia.

Fawcett, J. A. and Clayton, R. W. (1984). Tomographic reconstruction of velocity anomalies. *Bull. Seismol. Soc. Am.* **74,** 2201–2219.

Gentleman, W. M. (1973). Least squares computation by Givens transformations without square roots. *J. Inst. Maths Applics* **12,** 329–336.

Golub, G. H. and Reinsch, C. (1970). Singular value decomposition and least squares solutions. *Numer. Maths* **14,** 403–420.

Herman, G. T. (1980). *Image Reconstruction from Projections: The Fundamentals of Computerized Tomography,* Academic Press, New York.

Humphreys, E., Clayton, R. W. and Hager, B. H. (1984). A tomographic image of mantle structure beneath southern California. *Geophys. Res. Lett.* **11,** 625–627.

Jordan, T. H. and Sverdrup, K. A. (1981). Teleseismic location techniques and their application to earthquake clusters in the south-central Pacific. *Bull. Seismol. Soc. Am.* **71,** 1105–1130.

Lanczos, C. (1961). *Linear Differential Operators.* Van Nostrand Reinhold, New York.

Lawson, C. H. and Hanson, R. J., (1974). *Solving Least Squares Problems.* Prentice-Hall, Englewood Cliffs, NJ.

Lawson, C., Hanson, R., Kincaid, D. and Krogh, F. (1979a). Basic linear algebra subroutines. *ACM Trans. Math. Software.* **5,** 308–371.

Lawson, C., Hanson, R., Kincaid, D. and Krogh, F. (1979b). ALGORITHM 539, Basic linear algebra subroutines. *ACM Trans. Math. Software.* **5,** 308–323.

Nolet, G. (1985). solving and resolving inadequate and noisy tomographic systems. *J. Comp. Phys.* **61,** 463–482.

Olson, A. H. (1987). A Chebyshev condition for accelerating convergence of iterative tomographic methods solving large least squares problems. *Phys. Earth Planet. Inter.* (Submitted).

Paige, C. C. and Saunders, M. A. (1982a). LSQR: An algorithm for sparse linear equations and sparse least squares. *ACM Trans. Math. Software.* **8,** 43–71.

Paige, C. C. and Saunders, M. A. (1982b). ALGORITHM 582, LSQR: Sparse linear equations and least squares problems. *ACM Trans. Math. Software.* **8,** 195–209.

Parker, R. L. (1977). Understanding inverse theory. *Ann. Rev. Earth Planet. Sci.* **5,** 35–64.

Pavlis, G. L. (1984). Comparison of earthquake travel-time residuals or how to compare apples to oranges. *Bull. Seismol. Soc. Am.* **74,** 2245–2255.

Pavlis, G. L. (1986). Geotomography using refraction fan shots. *J. Geophys. Res.* **91,** 6522–6534.

Pavlis, G. L. and Booker, J. R. (1980). The mixed discrete continuous inverse problem: Application to the simultaneous determination of earthquake hypocenters and velocity structure. *J. Geophys. Res.* **85,** 4801–4810.

Pavlis, G. L. and Booker, J. R. (1983). Progressive multiple event location. *Bull. Seismol. Soc. Am.* **73,** 1753–1777.

Pavlis, G. L. and Hokanson, N. B. (1985). Separated earthquake location. *J. Geophys. Res.* **90,** 12777–12789.

Peterson, J. E., Paulsson, B. N. P. and McEvilly, T. V. (1985). Application of algebraic reconstruction techniques to crosshole seismic data. *Geophysics* **50,** 1599.

Spencer, C. and Gubbins, D. (1980). Travel-time inversion for simultaneous earthquake location and velocity structure determination in laterally varying media. *Geophys. J. R. Astron. Soc.* **63,** 95–116.

Stewart, G. W. (1973). *Introduction to Matrix Computations*. Academic Press, New York.

Tarantola, A. and Valette, B. (1982). Generalized nonlinear inverse problems solved using the least squares criterion. *Rev. Geophys. Space Phys.* **20,** 219–232.

Thurber, C. H. and Aki, K. (1987). Three-dimensional seismic imaging. *Ann. Rev. Earth Planet. Sci.* **15,** 115–139.

Willmore, D. L. and Bancroft, A. M. (1960). The time term approach to refraction seismology. *Geophys. J. R. Astron. Soc.* **3,** 419–432.

V.2

Radiation from a Moment-Tensor Source

B. L. N. KENNETT
Research School of Earth Sciences, Australian National University, Canberra, Australia

1 INTRODUCTION

A convenient and economical description of seismic sources is provided by treating them as arising from deviations from the elastically predicted stress state. The difference may be represented by a moment-tensor density $m_{ij}(\boldsymbol{x}, t)$ (Backus and Mulcahy, 1976), which will be symmetric for any source not arising from an external agency.

The seismic displacement generated at a point $\boldsymbol{x}$ by such a source lying within a volume V is given by

$$u_k(\boldsymbol{x}, t) = \int_0^t \mathrm{d}\tau \int_V \mathrm{d}^3\boldsymbol{\eta} \frac{\partial G_{kp}}{\partial \eta_q}(\boldsymbol{\eta}, \tau; \boldsymbol{x}, t) m_{pq}(\boldsymbol{\eta}, \tau), \tag{1}$$

where $G_{kp}(\boldsymbol{\eta}, \tau; \boldsymbol{x}, t)$ is the Green tensor for the medium. For a concentrated source localized near $\boldsymbol{h}$, we make a Taylor expansion of the Green tensor about the point $\boldsymbol{h}$ and write

$$u_k(\boldsymbol{x}, t) = \int_0^t \mathrm{d}\tau \frac{\partial G_{kp}}{\partial h_q}(\boldsymbol{h}, \tau; \boldsymbol{x}, t) M_{pq}(\tau) + \ldots, \tag{2}$$

where the moment tensor

$$M_{pq}(t) = \int_V \mathrm{d}^3\boldsymbol{\eta}\, m_{pq}(\boldsymbol{\eta}, t). \tag{3}$$

The higher-order terms represent the effect of successive polynomial moments of the original moment-tensor density distribution, and have

SEISMOLOGICAL ALGORITHMS
ISBN 0-12-220770-X

been exploited in schemes to improve source location via matching observed and theoretical seismograms (see e.g. Dziewonski and Woodhouse, 1983). The location of the reference point would normally be the hypocentre for small sources, but for extended faulting an improved representation is obtained by considering the centroid of the disturbance.

The specific form of the moment tensor to be used in (2) depends on the application. To consider the source radiation itself, it is convenient to envisage the point moment-tensor source lying in a uniform medium with P velocity α_s, S velocity β_s, and density ρ_s. In that case, the far-field contributions to the displacement field at a point $\boldsymbol{x}$ depend on the rate of change of the moment-tensor components with time:

$$u_k^P(\boldsymbol{x}, t) = \frac{1}{4\pi\rho_s\alpha_s^3}\frac{1}{R}\gamma_k\gamma_p\gamma_q\dot{M}_{pq}\left(t - \frac{R}{\alpha_s}\right), \tag{4}$$

$$u_k^S(\boldsymbol{x}, t) = \frac{1}{4\pi\rho_s\beta_s^3}\frac{1}{R}(\delta_{kp}\gamma_q - \gamma_k\gamma_p\gamma_q)\dot{M}_{pq}\left(t - \frac{R}{\beta_s}\right) \tag{5}$$

for P and S waves respectively. Here $R = |\boldsymbol{x} - \boldsymbol{h}|$ and the γ_i are the direction cosines of $\boldsymbol{x}$ relative to $\boldsymbol{h}$. The radiation patterns depend on the interaction of the geometrical terms and the moment-tensor components. The P radiation corresponds to the radial component of the moment tensor, and the S radiation to the sum of the tangential components in a spherical coordinate system centred on $\boldsymbol{h}$.

In this chapter we shall give a summary of the results for the radiation generated by a point source for both body waves and surface waves. We shall use a moment-tensor description and show how the conventional fault-plane description can be cast in terms of this representation. Even given a specific expression for a source, it is frequently difficult to envisage the form of the radiation pattern. We therefore show how to generate two different pictorial treatments that enable the amplitude of body-wave radiation to be assessed in three dimensions.

2 POINT SOURCES

There are two commonly encountered descriptions of point seismic sources. The first is in terms of an angular description of the nodal planes in the P radiation from a purely slip motion on a fault (the fault plant and the orthogonal auxiliary plane). The second, of increasing importance, is the description of the source by the six independent components of the moment tensor, which are assumed to have a common dependence on time. Commonly this is taken to be a step function in time, i.e.

$$M_{pq}(t) = M_{pq}H(t). \tag{6}$$

This assumption means that the far-field radiation is just a delta function in time for a uniform medium. Both representations are given for larger events in the "Preliminary Determination of Epicentres", published by the US National Earthquake Information Centre (NEIS), and these are later reprinted in the Bulletin of the International Seismological Centre (ISC).

In order to define the source parameters we have to establish a reference coordinate system. We shall follow Aki and Richards (1980) and Bullen and Bolt (1985) by choosing a right-handed system with:

- x direction to the North, unit vector $\hat{\boldsymbol{N}}$;
- y direction to the East, unit vector $\hat{\boldsymbol{E}}$;
- z direction downwards (towards nadir), unit vector $\hat{\boldsymbol{D}}$.

For a general anisotropic medium with elastic-modulus tensor c_{pqrs}, a shear dislocation with unit slip vector $\boldsymbol{e}$ lying in a fault plane with unit normal $\boldsymbol{n}$ has the (zeroth-order) moment-tensor representation

$$M_{pq} = c_{pqrs}[\bar{u}]Ae_rn_s, \qquad n_re_r = 0, \tag{7}$$

where $[\bar{u}]$ is the mean fault offset and A is the area of slip. In an isotropic medium with shear modulus μ_s this reduces to

$$M_{pq} = \mu_s[\bar{\mu}]A(e_pn_q + n_pe_q). \tag{8}$$

We note that the slip vector $\boldsymbol{e}$ and the normal $\boldsymbol{n}$ play a symmetric role in (8), so they cannot be distinguished from the character of the radiation.

We can describe the disposition of the fault system by three angles (Fig. 1):

- ϕ_s the strike angle, measured clockwise from North to the surface trace of the fault plane;
- δ the dip angle, measured down from the horizontal at right angles to the strike direction;
- λ the rake angle, the angle between the strike direction and the slip vector.

These are the angles used in the NEIS and ISC publications. A number of other conventions exist for describing the orientation of the slip relative to the strike: in particular, the angle in the vertical plane is known as the plunge ($\sin p = \sin\lambda \sin\delta$) and the angle in the horizontal plane as the trend ($\sin t = \sin\lambda \cos\delta/(1-\sin^2\lambda \sin^2\delta)^{1/2}$). Jarosch and

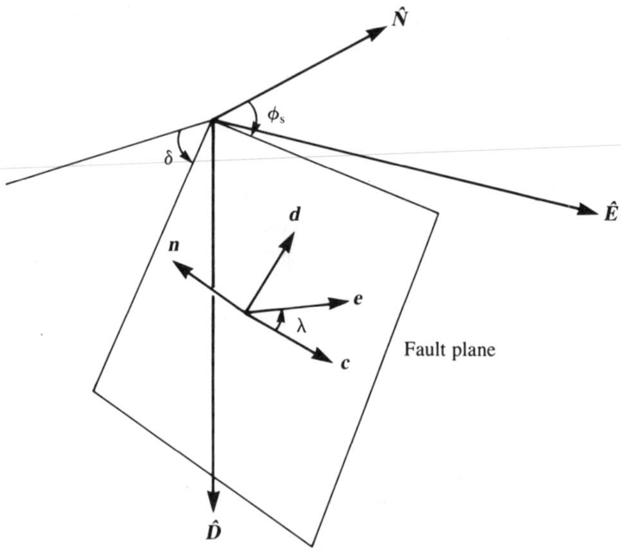

Fig. 1 Configuration of the fault plane defined by the strike angle ϕ_s and the dip angle δ, with slip defined by the rake angle λ. The orientations of the normal to the fault plane $\boldsymbol{n}$ and the slip vector $\boldsymbol{e}$ are shown, as well as the auxiliary vectors $\boldsymbol{c}$ and $\boldsymbol{d}$.

Aboodi (1970) provide a good description of conversions between the different conventions.

In the (North, East, Down) coordinate frame the normal to the fault

$$\boldsymbol{n} = -\hat{\boldsymbol{N}} \sin \delta \sin \phi_s + \hat{\boldsymbol{E}} \sin \delta \cos \phi_s - \hat{\boldsymbol{D}} \cos \delta. \tag{9}$$

The strike of the fault lies in the direction

$$\boldsymbol{n} \wedge \hat{\boldsymbol{D}} = \boldsymbol{c} \sin \delta, \tag{10}$$

where the strike vector

$$\boldsymbol{c} = \hat{\boldsymbol{N}} \cos \phi_s + \hat{\boldsymbol{E}} \sin \phi_s. \tag{11}$$

The vector in which the vertical plane intersects the fault plane lies along

$$\boldsymbol{d} = \boldsymbol{n} \wedge \boldsymbol{c} = \hat{\boldsymbol{N}} \cos \delta \sin \phi_s - \hat{\boldsymbol{E}} \cos \delta \cos \phi_s - \hat{\boldsymbol{D}} \sin \delta. \tag{12}$$

The vectors $\boldsymbol{c}$ and $\boldsymbol{d}$ provide an orthogonal basis on the fault plane, so that the unit slip vector

$$\boldsymbol{e} = \boldsymbol{c} \cos \lambda + \boldsymbol{d} \sin \lambda \tag{13}$$

in terms of the rake λ. Thus in the original basis

$$\begin{aligned} \boldsymbol{e} = \hat{\boldsymbol{N}}(\cos\lambda\cos\phi_s + \sin\lambda\cos\delta\sin\phi_s) \\ + \hat{\boldsymbol{E}}(\cos\lambda\sin\phi_s - \sin\lambda\cos\delta\cos\phi_s) - \hat{\boldsymbol{D}}\sin\lambda\sin\delta. \end{aligned} \tag{14}$$

With the identifications $x \equiv$ North, $y \equiv$ East, $z \equiv$ Down, we can now evaluate the six independent components of the moment tensor for a shear dislocation as

$$\left.\begin{aligned} M_{xx} &= -M_0(\sin\delta\cos\lambda\sin 2\phi_s + \sin 2\delta\sin\lambda\sin^2\phi_s), \\ M_{xy} &= M_{yx} = M_0(\sin\delta\cos\lambda\cos 2\phi_s + \sin 2\delta\sin\lambda\sin\phi_s\cos\phi_s), \\ M_{yy} &= M_0(\sin\delta\cos\lambda\sin 2\phi_s - \sin 2\delta\sin\lambda\cos^2\phi_s), \\ M_{xz} &= M_{zx} = -M_0(\cos\delta\cos\lambda\cos\phi_s + \cos 2\delta\sin\lambda\sin\phi_s), \\ M_{yz} &= M_{zy} = -M_0(\cos\delta\cos\lambda\sin\phi_s - \cos 2\delta\sin\lambda\cos\phi_s), \\ M_{zz} &= M_0\sin 2\delta\sin\lambda, \end{aligned}\right\} \tag{15}$$

where M_0 is the scalar moment $M_0 = \mu_s[\bar{u}]A$.

If the moment tensor is available directly then we only need to check that it is given in a form that conforms to our choice of reference axis. The convention we have adopted is common in body-wave work. However, in normal-mode studies at long period, an alternative convention based on a spherical coordinate basis ($r \equiv$ Up, $\theta \equiv$ South, $\phi \equiv$ East) is prevalent, and the solutions of the Harvard group (see e.g. Dziewonski *et al.*, 1983) are given in this form. The conversion involves some sign changes:

$$\left.\begin{aligned} M_{xx} &= M_{\theta\theta}, \quad & M_{xy} &= -M_{\theta\phi}, \quad & M_{xz} &= M_{r\theta}, \\ M_{yy} &= M_{\phi\phi}, \quad & M_{yz} &= -M_{r\phi}, \quad & M_{zz} &= M_{rr}. \end{aligned}\right\} \tag{16}$$

3 BODY-WAVE RADIATION PATTERNS

We regard the source as lying in a uniform isotropic medium and look at the distribution of radiation on a unit sphere surrounding the source point $\boldsymbol{h}$. A knowledge of the P- and S-wave radiation patterns on the focal sphere will summarize the behaviour of the source, as seen in the far field. For distant observation points we shall need to take account of the propagation paths and interactions with, for example, the free surface, which will modify the way in which the source radiation is reflected in the character of the recorded seismograms. At such locations the directly arriving energy will leave the source downwards, and attention is usually focused on the lower hemisphere of this focal sphere.

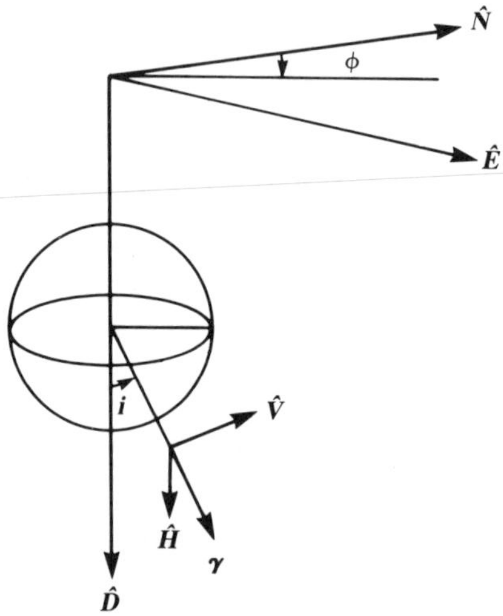

Fig. 2 The relationship of the take-off angle i and azimuth ϕ to the focal sphere surrounding a source at $\boldsymbol{h}$. The orientations of the unit vectors in the P direction γ, the SV direction $\hat{\boldsymbol{V}}$ and the SH direction $\hat{\boldsymbol{H}}$ are also indicated.

At closer ranges, as in local networks, some far-field stations will be placed so that the energy that reaches them leaves the source upwards and then the upper hemisphere of the focal sphere is of interest.

We can parametrize the points on this focal sphere by looking at the orientation of a radial unit vector γ described by the take-off angle measured up from the vertical (i) and the azimuth (ϕ) measured clockwise from north (Fig. 2)

$$\gamma = \hat{\boldsymbol{N}} \sin i \cos \phi + \hat{\boldsymbol{E}} \sin i \sin \phi + \hat{\boldsymbol{D}} \cos i. \tag{17}$$

P waves will be radiated along the direction of γ; to describe the S waves we introduce two vectors orthogonal to γ: $\hat{\boldsymbol{V}}$ lying in a vertical plane and $\hat{\boldsymbol{H}}$ lying horizontally, which allows us to polarize the S waves into SV and SH components at the source.

$$\left.\begin{aligned} \hat{\boldsymbol{V}} &= \hat{\boldsymbol{N}} \cos i \cos \phi + \hat{\boldsymbol{E}} \cos i \sin \phi - \hat{\boldsymbol{D}} \sin i, \\ \hat{\boldsymbol{H}} &= -\hat{\boldsymbol{N}} \sin \phi + \hat{\boldsymbol{E}} \cos \phi. \end{aligned}\right\} \tag{18}$$

From (4), the P-wave contribution to the far-field displacement at a point on the unit focal sphere is given by

$$4\pi\rho_s\alpha_s^3\boldsymbol{u}^{\mathrm{P}}(\gamma, t) = \gamma(\gamma_p\gamma_q\dot{M}_{pq}), \tag{19}$$

where the moment-rate components are evaluated at time $t - \alpha_s^{-1}$. When all the moment-tensor components have the same time dependence, we can extract the P-wave radiation pattern

$$\begin{aligned}\mathcal{F}^{\mathrm{P}} &= \gamma_p\gamma_q M_{pq}\\ &= \sin^2 i\,(\cos^2\phi\, M_{xx} + \sin 2\phi\, M_{xy} + \sin^2\phi\, M_{yy} - M_{zz})\\ &\quad + 2\sin i\cos i\,(\cos\phi\, M_{xz} + \sin\phi\, M_{yz}).\end{aligned} \tag{20}$$

For use in generalized ray theory, the horizontal slowness $p = \sin i/\alpha_s$ may be a more useful variable (see e.g. Aki and Richards, 1980, Box 9.10).

The shear-wave displacement (5) has to be polarized into the SV and SH components. In a vertical plane the far-field SV contribution is

$$4\pi\rho_s\beta_s^3\boldsymbol{u}^{\mathrm{SV}}(\gamma, t) = \hat{\boldsymbol{V}}(V_p\gamma_q\dot{M}_{pq}), \tag{21}$$

where the time argument of the moment rate components is now $t - \beta_s^{-1}$. The SV radiation pattern

$$\begin{aligned}\mathcal{F}^{\mathrm{SV}} &= V_p\gamma_q M_{pq}\\ &= \sin i\cos i\,(\cos^2\phi\, M_{xx} + \sin 2\phi\, M_{xy} + \sin^2\phi\, M_{yy} - M_{zz})\\ &\quad + \cos 2i\,(\cos\phi\, M_{xz} + \sin\phi\, M_{yz}).\end{aligned} \tag{22}$$

We note that both the P and SV body-wave radiation terms depend on the same combinations of moment-tensor components. In the horizontal plane the SH contribution has a similar structure:

$$4\pi\rho_s\beta_s^3\boldsymbol{u}^{\mathrm{SH}}(\gamma, t) = \hat{\boldsymbol{H}}(H_p\gamma_q\dot{M}_{pq}), \tag{23}$$

again with a shear-wave delay term. The SH radiation pattern has a rather different dependence on both the take-off angle i and the moment-tensor components:

$$\begin{aligned}\mathcal{F}^{\mathrm{SH}} &= H_p\gamma_q M_{pq}\\ &= \sin i\,[\sin\phi\cos\phi(M_{yy} - M_{xx}) + \cos 2\phi\, M_{xy}]\\ &\quad + \cos i\,[\cos\phi\, M_{yz} - \sin\phi\, M_{xz}].\end{aligned} \tag{24}$$

Having now obtained these geometrical radiation patterns, the problem remains as to how they may be displayed to best effect. For dislocation sources, the conventional answer is to use the "beachball" with a projection of the lower focal hemisphere on which the dilatational segments in the P wave radiation are shaded. This representation is useful because of the simplicity of the form of the nodal surfaces, which are the projections of the fault plane and the auxiliary plane. However, even in this simple case, the variation of the radiation across the focal sphere is not adequately mapped. For general moment-tensor sources, the shapes of the nodal surfaces are useful for judging pictorially the extent of departures from double-couple behaviour, but are not very helpful in allowing an assessment of the radiation in a particular direction.

An effective scheme for displaying the radiation patterns, which can be used for P, SV and SH radiation, is to display a projection of one focal hemisphere (usually the lower) with a regular grid of symbols, scaled to represent the size of the local amplitude of the radiation and oriented to show the sense of motion. This may be done for either the simple dislocation model or for the general moment tensor.

As an example we show in Fig. 3 three radiation patterns for the centroid moment-tensor solution determined by the Harvard group for an event in Mexico in July 1984, where the parameters in Table 1 are taken from the ISC Bulletin. This diagram has been produced using a 10° increment in both azimuth and take-off angle and is plotted using an

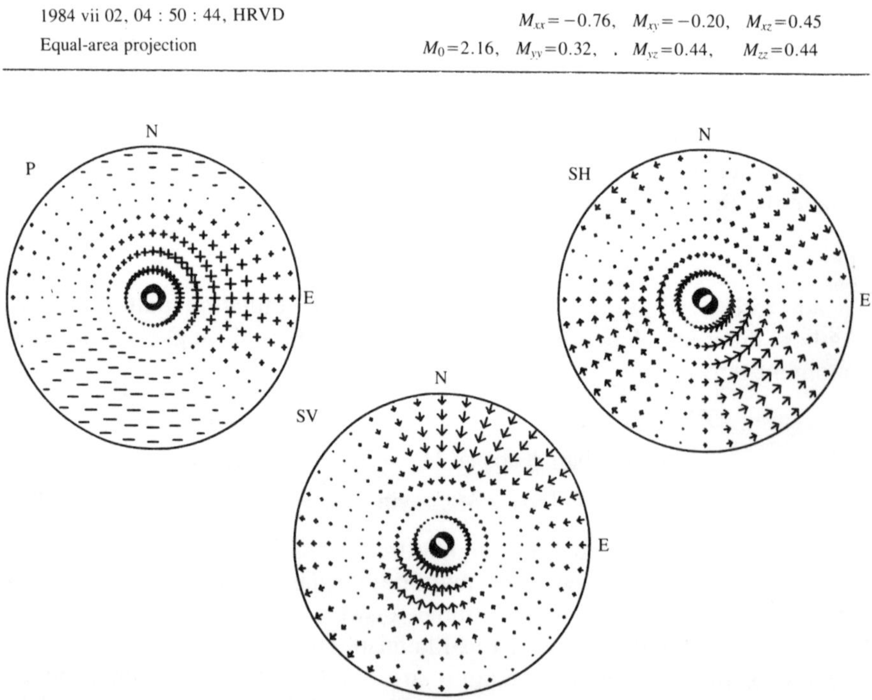

Fig. 3 Radiation patterns for P, SV and SH waves for the event of Table 1. The size and configuration of the symbols indicate the sense of motion (pluses indicate compression and minuses dilatation for P waves). The symbols are plotted on a 10° grid in both take-off angle i and azimuth ϕ.

Table 1 Parameters of the Mexican earthquake used in Figs. 3–5.

Date:	1984 vii 02
Origin time:	04:50:44
Latitude:	16.28°N
Longitude:	98.46°W
Depth:	46 km (ISC), 39 km (HRVD)

Centroid moment-tensor solution

Scale: 10^{18} N m

$M_{xx} = M_{\theta\theta} = -1.65,$	$M_{xy} = -M_{\theta\phi} = -0.44,$	$M_{xz} = M_{r\theta} = 0.97$
$M_{yy} = M_{\phi\phi} = 0.69,$	$M_{yz} = -M_{r\phi} = 0.95,$	$M_{zz} = M_{rr} = 0.95$

Principal axes

T	1.85,	plunge 52°,	azimuth 76°
B	0.31,	plunge 30°,	azimuth 299°
P	−2.16,	plunge 21°,	azimuth 196°

Best double couple: $M_0 = 2.0 \times 10^{18}$ N m

NP1:	strike 246°,	dip 36°,	rake 32°
NP2:	strike 129°,	dip 72°,	rake 72°

The moment-tensor components are scaled by the largest principal value for plotting in Figs 3–5.

equal-area projection (as is commonly used in source studies). For a display of radius r, at take-off angle i a symbol is placed at a distance $r \sin \frac{1}{2}i$ at azimuth ϕ. If a stereographic representation is preferred, the mapping point is placed at $r \tan \frac{1}{2}i$ at the same azimuth. In order to get a reasonable pattern of symbol sizes, it is convenient to normalize the moment-tensor components by the size of the largest principal value so that the sizes of the terms are comparable to the angularly dependent portions of those in the dislocation model (15). The principal values are frequently published along with the moment-tensor solution, but can be readily found from the tensor components (see Appendix).

One of the features of this display of the radiation pattern is that the significance of nodal and near-nodal regions are clearly displayed. Although the event shown in Fig. 3 has sufficient non-double-couple component to have no crossing nodal surfaces, the difference from the best fitting double couple is mostly in the regions of low amplitude in the north-west quadrant. We note that these regions are up to 20–30° wide and are in many ways the dominant features of the patterns; in the presence of noise, practical assignment of polarity in such regions may be very difficult. Thus the most stable estimates of source mechanism come from waveform studies at long period (Dziewonski and Woodhouse, 1983). For a general moment-tensor model the radiation patterns do not have the four-lobed symmetries of the dislocation model, and therefore for surface-reflected phases, which leave the source upwards, it can be advantageous to also draw an upper focal-hemisphere projection, plotting at $r \sin(\frac{1}{2}\pi - \frac{1}{2}i)$ to avoid excessive distortion.

Although the symbol displays of Fig. 3 give some insight into the nature of the source radiation, it can still be very difficult to visualize what is happening in three dimensions. An alternative approach to displaying a flat projection of a hemisphere is therefore to generate projected views of the total P- and S-wave radiation in three dimensions

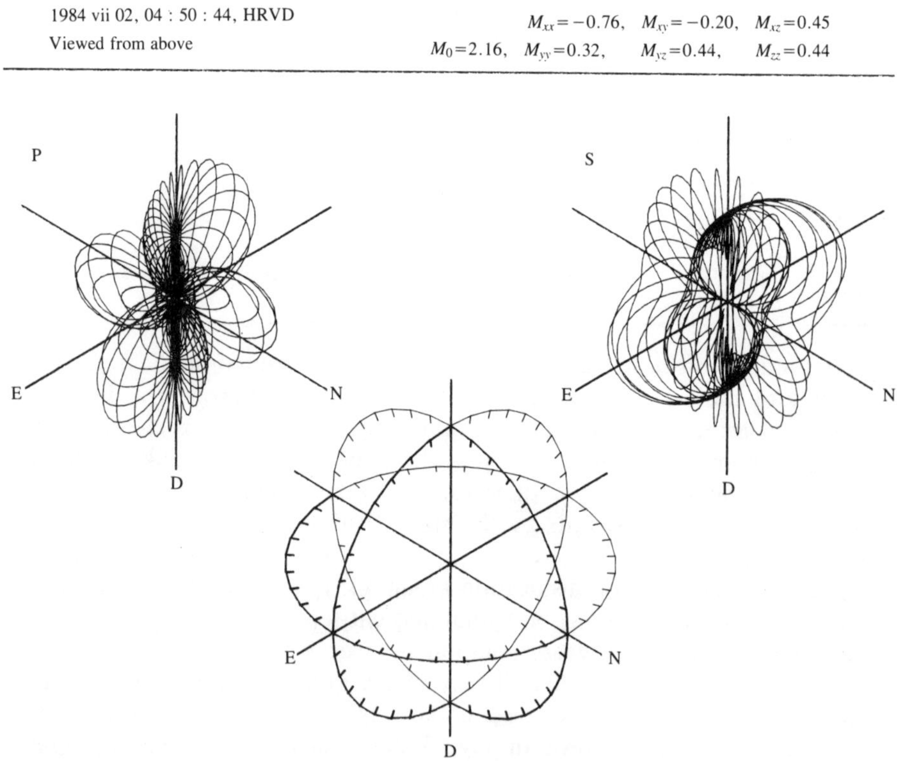

Fig. 4 Projective display of the total P and S radiation for the same event as in Fig. 3. The key diagram shows how the three principal great circles behave under projection.

as a function of angular position. This can be done by drawing the outline of the pattern as a function of take-off angle $0° < i < 360°$ at 10° intervals in azimuth. Such a relatively simple scheme does not have the benefits of hidden line removal but can still give a useful picture. In Fig. 4 we show the P and S radiation for the centroid moment-tensor solution of Table 1, together with a key diagram designed to help with viewing the three-dimensional relations. The moment-tensor solution is dominated by the major double couple and so the patterns are not far removed from the familiar four-lobed pattern of the dislocation source.

This projective plot is not difficult to obtain. First a viewpoint needs to be selected and defined in terms of its direction cosines

$$(l, m, n) = (\sin\theta\cos\psi, \sin\theta\sin\psi, \cos\theta) \tag{25}$$

for some specified view angle θ and azimuth ψ. Then, looking from infinity, a point at (x, y, z) is to be plotted onto axes perpendicular to the line of view with coordinates

$$\left.\begin{aligned} x_p &= (-mx + ly)/k, \\ y_p &= \left[-lx - my + \left(n - \frac{1}{n}\right)z\right]\frac{n}{k}, \end{aligned}\right\} \tag{26}$$

where $k = (1 - n^2)^{1/2} = \sin\theta$. For Fig. 4 we have taken $(l, m, n) = (-1, -1, -1)/\sqrt{3}$, so that the North and East axes lie in the same configuration as in Fig. 3. The radiation patterns $\mathscr{F}^{\mathrm{P}}$, $[(\mathscr{F}^{\mathrm{SV}})^2 + (\mathscr{F}^{\mathrm{SH}})^2]^{1/2}$ have been represented by scaling the radius vector for each combination of azimuth and take-off angle:

$$(x, y, z) = \mathscr{F}(\sin i\cos\phi, \sin i\sin\phi, \cos i). \tag{27}$$

4 SURFACE-WAVE RADIATION PATTERNS

For a point source described by a moment tensor, relatively simple expressions exist for the far-field surface-wave displacement in the frequency domain. The displacement can be represented as a sum of contributions corresponding to individual surface-wave modes (see e.g. Aki and Richards, 1980, Chapter 7; Kennett, 1983, Chapter 11). In cylindrical coordinates (r, ϕ, z) the far-field displacements for Rayleigh waves couple the vertical and radial components, and the Love waves appear only on the tangential component. Each modal contribution is built up from the modal eigenfunctions as a function of wavenumber, depth and frequency, a wavenumber-dependent radial phase term and a source-depth-dependent radiation term.

For Rayleigh waves

$$\begin{bmatrix} u_z^{\mathrm{RAYL}}(r, \phi, z, \omega) \\ u_r^{\mathrm{RAYL}}(r, \phi, z, \omega) \end{bmatrix} = \sum_j \begin{bmatrix} U_E(k_j, z, \omega) \\ iV_E(k_j, z, \omega) \end{bmatrix} \times \frac{\mathcal{F}^{\mathrm{R}}(k_j, \phi, h, \omega)}{4c_j U_j I_j^{\mathrm{R}}} \left(\frac{2}{\pi k_j r}\right)^{1/2} \exp\left(\mathrm{i}k_j r + \frac{\mathrm{i}\pi}{4}\right), \quad (28)$$

where U_{E} and V_{E} are the vertical and horizontal displacement components of the modal eigenfunction ($-r_2$ and r_1 in Aki and Richards' notation), c_j and U_j are the phase and group velocities of the mode, and

$$I_j^{\mathrm{R}} = \int_0^\infty \mathrm{d}z\, \rho(z)[U_{\mathrm{E}}^2(k_j, z, \omega) + V_{\mathrm{E}}^2(k_j, z, \omega)], \quad (29)$$

which comes from a kinetic-energy term. The Rayleigh-wave radiation term depends on the eigendisplacements and their derivatives evaluated at the source depth, as well as the specific source excitation terms

$$\begin{aligned} \mathcal{F}^{\mathrm{R}}(k_j, \phi, h, \omega) &= k_j V_{\mathrm{E}}(k_j, h, \omega)(\cos^2\phi\, M_{xx} + \sin 2\phi\, M_{xy} + \sin^2\phi\, M_{yy}) \\ &\quad + \mathrm{i}k_j\left[U_{\mathrm{E}}(k_j, h, \omega) + \frac{\partial V_{\mathrm{E}}}{\partial z}(k_j, h, \omega)\right](\cos\phi\, M_{xz} + \sin\phi\, M_{yz}) \\ &\quad - \frac{\partial U_{\mathrm{E}}}{\partial z}(k_j, h, \omega) M_{zz}, \end{aligned} \quad (30)$$

which depends on frequency through the frequency dependence of the eigendisplacements and the moment-tensor components. $\mathcal{F}^{\mathrm{R}}$ depends on the same combinations of moment-tensor terms as the body-wave radiation terms $\mathcal{F}^{\mathrm{P}}$ and $\mathcal{F}^{\mathrm{SV}}$, (20) and (22). The angular dependence of the moment-tensor combinations appearing in the Rayleigh-wave radiation pattern (30) is shown in polar plots in Fig. 5, for the same event as before. The total radiation is dictated by the interaction of the excitation terms with the structure through the eigendisplacements.

For Love waves

$$u_\phi^{\mathrm{LOVE}}(r, \phi, z, \omega) = -\sum_l W_{\mathrm{E}}(k_l, z, \omega) \frac{\mathcal{F}^{\mathrm{L}}(k_l, \phi, h, \omega)}{4c_l U_l I_l^{\mathrm{L}}} \times \left(\frac{2}{\pi k_l r}\right)^{1/2} \exp\left(\mathrm{i}k_l r + \frac{\mathrm{i}\pi}{4}\right), \quad (31)$$

where W_{E} is the eigendisplacement in the Love mode and

$$I_l^{\mathrm{L}} = \int_0^\infty \mathrm{d}z\, \rho(z) W_{\mathrm{E}}^2(k_l, z, \omega). \quad (32)$$

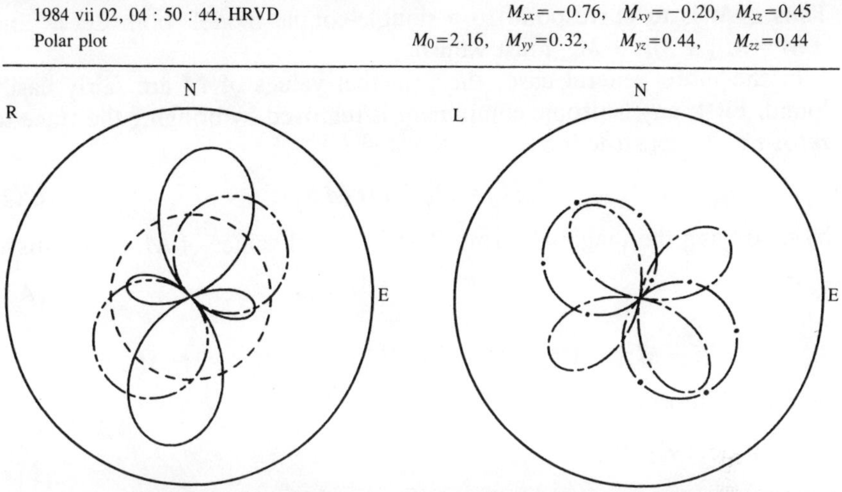

Fig. 5 The azimuthal dependence of the moment-tensor combinations appearing in the surface-wave radiation patterns (30) and (33) shown as polar plots: ——, $\cos^2\phi\, M_{xx} + \sin 2\phi\, M_{xy} + \sin^2\phi\, M_{yy}$; — – —, $\cos\phi\, M_{xz} + \sin\phi\, M_{yz}$; - - - -, M_{zz}; ——–, $\sin\phi\cos\phi\,(M_{yy} - M_{xx}) + \cos 2\phi\, M_{xy}$; —·—, $\cos\phi\, M_{yz} - \sin\phi\, M_{xz}$.

The Love-wave radiation

$$\mathscr{F}^{\mathrm{L}}(k_l, h, h, \omega) = \mathrm{i}k_l W_{\mathrm{E}}(k_l, h, \omega)[\sin\phi\cos\phi\,(M_{yy} - M_{xx}) + \cos 2\phi\, M_{xy}] - \frac{\partial W_{\mathrm{E}}}{\partial z}(k_l, h, \omega)(\cos\phi\, M_{yz} - \sin\phi\, M_{xz}), \tag{33}$$

which incorporates the same moment-tensor combinations as the SH body-wave term $\mathscr{F}^{\mathrm{SH}}$, (24). The angular dependence of the moment-tensor terms in the Love-wave radiation pattern is also shown in Fig. 5.

APPENDIX: PRINCIPAL VALUES OF THE MOMENT TENSOR

We have seen above that we can construct the moment tensor corresponding to a point shear-dislocation source, but what are the conditions for a given moment tensor to correspond to a shear dislocation? We require (i) one principal value of $\boldsymbol{M}$ to vanish and (ii) the trace of $\boldsymbol{M}$ to be zero. These are necessary and sufficient conditions for M_{pq} to be of the form $e_p n_q + n_p e_q$, since this has principal directions

$$\left.\begin{array}{ll} \hat{\boldsymbol{p}} = 2^{-1/2}(\boldsymbol{n} - \boldsymbol{e}) & \text{principal value } 1, \\ \hat{\boldsymbol{t}} = 2^{-1/2}(\boldsymbol{n} + \boldsymbol{e}) & \text{principal value } -1, \\ \hat{\boldsymbol{b}} = \boldsymbol{n} \wedge \boldsymbol{e} & \text{principal value } 0, \end{array}\right\} \tag{A1}$$

Thus if $\boldsymbol{M}$ is to correspond to a double-couple model, both $\det \boldsymbol{M}$ and $\operatorname{tr} \boldsymbol{M} = M_{xx} + M_{yy} + M_{zz}$ must vanish.

In the more general case, the principal values of $\boldsymbol{M}$ are fairly easily found. First, any isotropic component is removed by bringing the trace to zero, i.e. by constructing

$$M'_{ij} = M_{ij} - \tfrac{1}{3} \operatorname{tr} \boldsymbol{M}\, \delta_{ij} \tag{A2}$$

by modifying the diagonal terms. The principal values of $\boldsymbol{M}'$ then satisfy

$$\lambda^3 - K\lambda - \det \boldsymbol{M}' = 0, \tag{A3}$$

where

$$K = M_{xy}^2 + M_{xz}^2 + M_{yz}^2 - M_{xx}M_{yy} - M_{yy}M_{zz} - M_{xx}M_{zz},$$

with the following solutions:

(a) if $\det \boldsymbol{M}' = 0$,

$$\lambda = 0, \quad K^{1/2}, \quad -K^{-1/2}; \tag{A4}$$

(b) if $\det \boldsymbol{M}' \neq 0$,

$$\lambda = q \cos \theta, \quad q \cos(\theta + \tfrac{2}{3}\pi), \quad q \cos(\theta + \tfrac{4}{3}\pi), \tag{A5}$$

where

$$q = 2(\tfrac{1}{3}K)^{1/2}, \qquad \theta = \tfrac{1}{3} \cos^{-1}\left(\frac{3 \det \boldsymbol{M}'}{Kq}\right).$$

The principal directions may then be readily found by solving

$$(M'_{ij} - \lambda\delta_{ij})u_j = 0, \quad i = 1, 2, 3, \tag{A6}$$

for the components of the vector $\boldsymbol{u}$.

If the trace of the original tensor $\boldsymbol{M}$ is non-zero, then $\tfrac{1}{3} \operatorname{tr} \boldsymbol{M}$ has to be added to each principal value.

REFERENCES

Aki, K. and Richards, P. G. (1980). *Quantitative Seismology*. W. H. Freeman, San Francisco.

Backus, G. E. and Mulcahy, M. (1976). Moment tensors and other phenomenological descriptions of seismic sources I—Continuous displacements. *Geophys. J. R. Astron. Soc.* **46,** 341–371.

Bullen, K. E. and Bolt, B. A. (1985). *An Introduction to the Theory of Seismology*. Cambridge University Press.

Dziewonski, A. M. and Woodhouse, J. H. (1983). Studies of the seismic source using normal-mode theory, *Earthquakes: Observation, Theory and*

Interpretation; *Proc. Intl School of Physics "Enrico Fermi"*, Course LXXXV, (ed. H. Kanamori and E. Boschi), pp. 45–137. North-Holland, Amsterdam.

Dziewonski, A. M., Friedman, A., Giardini, D. and Woodhouse, J. H. (1983). Global seismicity of 1982: centroid moment tensor solutions for 308 earthquakes. *Phys. Earth Planet. Inter.* **33,** 76–90.

Jarosch, H. and Aboodi, E. (1970). Towards a unified notation for source parameters. *Geophys. J. R. Astron. Soc.* **21,** 513–529.

Kennett, B. L. N. (1983). *Seismic Wave Propagation in Stratified Media.* Cambridge University Press.

V.3

Single and Joint Fault-Plane Solutions from First-Motion Data

A. UDÍAS and E. BUFORN
Cátedra de Geofísica, Universidad Complutense de Madrid, Spain

1 INTRODUCTION

Although other methods, such as those based on waveform analysis or moment-tensor inversion of body and surface waves, have been developed recently for the retrieval of source parameters, fault-plane solutions from first-motion polarities of P waves are still widely used. In many cases they constitute the only method available for obtaining information about the orientation of the focal mechanism. Moreover, in many cases such as microearthquake surveys or aftershock detection using a small number of portable stations, joint or composite fault-plane solutions are the only possibility. Most often, fault plane solutions are determined graphically with a major consumption of time and effort and no provision of formal error margins for the solutions. In the case of joint solutions, graphical analysis becomes an extremely difficult task. For both single earthquakes and joint solutions of groups of events algorithms have been developed that allow not only rapid determination of the orientation of fault-planes and their graphical representation, but also the use of polarities of SV and SH first-motion and above all the estimation of the standard deviations of the angles involved. For joint solutions the algorithms allow for the separation of events into groups of different mechanism, and as a consequence provide the orientation of regional stresses.

2 GEOMETRY OF THE PROBLEM

In the algorithms the point-source problem is reduced to the focal sphere and solved in spherical coordinates. Observations are projected on the

SEISMOLOGICAL ALGORITHMS
ISBN 0-12-220770-X

surface of the focal sphere by conveniently following the ray back to the source. Position of the observations on the focal sphere are given by i, the take-off angle at the source, measured from the vertical (positive downward) and ϕ, the azimuth from North. For teleseismic distances, values of i are obtained from travel-time tables, assuming a certain velocity at the focal region. For near shocks, precise knowledge of the depth and crustal structure is needed to determine i, from stratified or continuously variable velocity models.

The source is assumed to be a pure shear dislocation or double couple and it is represented by the normal to the shear plane and direction of slip (X and Y axes) or the direction of the principal axes of stress, pressure and tension (P and T axes). The X and Y axes correspond to the directions of the two couples of the double-couple model or normals to the two nodal planes. In both cases, owing to the orthogonality condition, only three angles are necessary to represent the source (e.g. Φ_T, Θ_T, Φ_P, or Φ_X, Θ_X, Φ_Y). The axes are defined by the trend Φ measured from North and the complement of the plunge Θ measured from the vertical (positive downward or nadir) (Fig. 1). The relations between this set of axes and those used to define the orientation of the fault plane, i.e. azimuth ϕ, dip δ and slip λ, if X is normal to plane A,

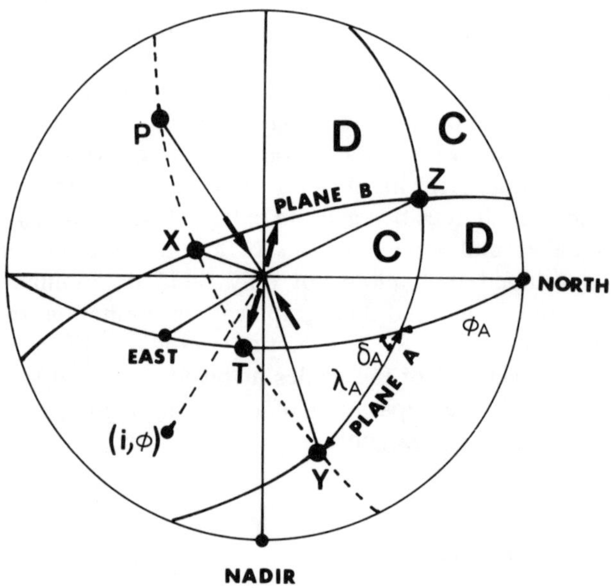

Fig. 1 Representation of P, T, X, Y and Z axes and nodal planes of P (planes A and B) on the focal sphere.

are given by

$$\phi_A = \Phi_X + 90°, \qquad 0° \leqslant \phi \leqslant 360°, \tag{1}$$

$$\delta_A = \Theta_X, \qquad 0° \leqslant \delta \leqslant 90°, \tag{2}$$

$$\lambda_A = \sin^{-1}\left(\frac{\cos \Theta_Y}{\sin \Theta_x}\right), \qquad -180° \leqslant \lambda \leqslant 180°. \tag{3}$$

The P and T axes can be obtained from X and Y or vice versa by a simple rotation of 45° around the Z axis, normal to both.

3 PROBABILITY MODEL AND LIKELIHOOD FUNCTION

The algorithm is based on the probability model of Brillinger, Udías and Bolt (1980). For a single earthquake, this model assumes that the probability of observing a compression or dilatation at a particular station is a function of the theoretical amplitude expected at that point. The probability of reading a compression in station j may then be expressed in the form

$$\pi_j = \gamma + (1 - 2\gamma)\Phi[\rho A_j(\hat{\Theta})]. \tag{4}$$

In this expression γ $(0 \leqslant \gamma \leqslant \frac{1}{2})$ is a parameter that allows for readers' errors. Its extreme values are $\gamma = \frac{1}{2}$, which reduces the probability to total uncertainty $(\pi_j = \frac{1}{2})$ and $\gamma = 0$, which implies that there are no readers' errors present. Since the performance of the readers is not known a priori, in the applications of the algorithm γ is assigned a small constant value, $\gamma = 10^{-6}$ which implies small reading errors. This parameter does not enter the maximization process. Φ is the normal cumulative function, $\rho = \alpha/\sigma$, where α is a proportionality constant and σ the standard deviation of the noise. A_j are the normalized theoretical amplitudes corresponding to a particular orientation of the source $\hat{\Theta}(\Theta_T, \Phi_T, \Phi_P)$. For one earthquake, since there is no a priori knowledge of the noise at each station, ρ is given a constant value (e.g. $\rho = 1$). Introducing the theoretical amplitudes $A_j(\hat{\Theta})$ in the definition of the probability, the method gives lower weight to data near nodal planes, where expected amplitudes are smaller.

The probability that a particular set of first-motion observations Y_j corresponds to a determined orientation of the source $\hat{\Theta}$ can be written as

$$F = \prod_{j=1}^{N} [\pi_j^{(1+Y_j)/2}(1 - \pi_j)^{(1-Y_j)/2}]. \tag{5}$$

In this expression it is assumed that probabilities for each observation are independent. Y_j is equal to plus or minus one, corresponding to observations of compressions or dilatations, and the π_j are given by (4).

The solution of the problem consists in finding the orientation of the source $\hat{\Theta}$, defined by the three angles (either Θ_T, Φ_T, Φ_P or Θ_X, Φ_X, Φ_Y), that maximize the probability that this orientation corresponds to the set of observations Y_j. This is done by maximizing the likelihood function (represented by (5)) that in an equivalent logarithmic form making use of the symmetry of π_j as a function of A_j is given by

$$L = \sum_{j=1}^{N} \log \{\tfrac{1}{2}[1 + (2\pi_j - 1)Y_j]\}. \tag{6}$$

It can be seen that L is a differentiable function of the parameters involved, so that there are a variety of efficient optimization routines available to solve the problem. In their present version, the programs utilize the subroutine VA09A of the Harwell (UKARE) library, which is based on the gradient method. Relevant maximization routines and related statistical details are discussed in Chambers (1977). Further, because the procedure is maximum-likelihood, there exist expressions for the standard errors of the estimates and methods for testing hypotheses concerning parameter values. From the function L the information matrix $\boldsymbol{I}(\hat{\Theta})$ can be determined as

$$[\boldsymbol{I}]_{\ln} = -E\left\{\frac{\partial^2 L}{\partial\Theta_l\, \partial\Theta_n}\right\}. \tag{7}$$

The inverse of $\boldsymbol{I}$ is the covariance matrix $\boldsymbol{C}(\hat{\Theta})$. The diagonal elements of $\boldsymbol{C}$ provide estimates of the variances of the parameters $\hat{\Theta}$. Although this procedure only directly calculates the standard errors of the three angles used for the solution, those of the other angles defining the orientation of the axes or nodal planes can be easily determined. This point is important, because no other method using polarities provides formal estimates of the standard errors of the parameters that define the orientation of the source.

Although the absolute value of L gives an overall measure of the error of the solution, this is difficult to compare from one case to another. For this reason it was decided to use as a measure of the goodness of a solution the score or proportion of correct polarities:

$$p = \frac{1}{N} \sum_{j=1}^{N} \tfrac{1}{2}(1 + Y_j \operatorname{sgn} A_j). \tag{8}$$

This value, together with the value of N, the number of data used, and the standard deviations of the parameters, gives the necessary information about the statistical precision of the solutions. It should be noted, however, that in the estimation of the standard deviation of the parameters, problems may arise for cases with perfect score ($p = 1$). In those cases some diagonal entry of $\boldsymbol{I}(\hat{\Theta})$ may be zero, indicating that, in a certain sense, no information is available concerning that particular parameter value.

3.1 Applications

For single solutions, the parameters of the problem are the three angles Θ_T, Φ_T and Φ_P that define the orientation of the P and T axes. From them, we obtain X and Y, the normals to the nodal planes A and B, as well as the three angles ϕ, δ and λ which define the motion at each plane. It is also possible to enter as independent parameters the angles Θ_X, Φ_X and Φ_Y that define the orientation of the X and Y axes, and from them calculate that of P and T.

Since the method is an iterative process, an initial solution must be provided. Depending on the number of observations, this solution need not be a good approximation, since practice has shown that the program, in most cases, is sufficiently robust to converge to the best solution from a relatively bad first choice (Brillinger *et al.*, 1980).

Signs of first motion of SV and SH components of S waves can also be used (Buforn and Udías, 1984). In this case the criterion followed to

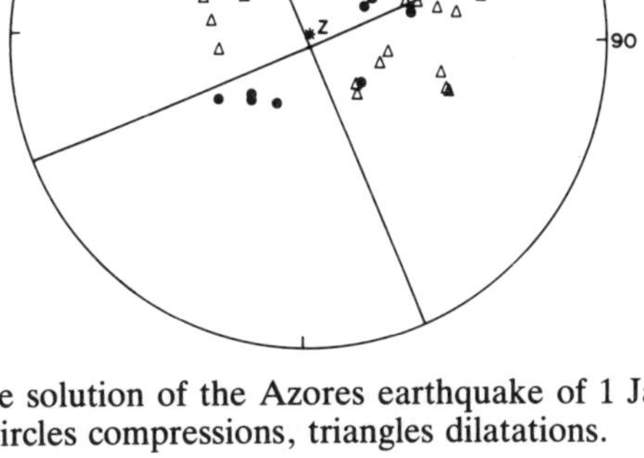

Fig. 2 Fault-plane solution of the Azores earthquake of 1 January 1980. Based on P-wave data. Circles compressions, triangles dilatations.

Table 1 Fault-plane solution for the 1 January 1980 earthquake based on P and P and S first-motion data.

	T and *P* axes			Nodal planes			
N		Θ	Φ	ϕ	δ	λ	*p*
36 (P)	*T*:	88 ± 4	21 ± 5	*A*: 66 ± 5	88 ± 4	7 ± 5	0.89
	P:	89 ± 2	291 ± 19	*B*: 156 ± 5	89 ± 4	1 ± 5	
46 (P + S)	*T*:	88 ± 3	194 ± 5	*A*: 239 ± 5	88 ± 3	5 ± 3	0.85
	P:	85 ± 6	284 ± 10	*B*: 149 ± 5	85 ± 4	2 ± 2	

establish the positive or negative character of motion is that given by Stauder (1960). Observations of polarities of S first motion can be used to complement the P observations, so that the number of independent equations is increased, thereby reducing the uncertainties of the solutions and, in some cases, the ambiguities of P data alone.

An example for an earthquake with strike–slip mechanism is given in Fig. 2. This solution corresponds to the Azores earthquake, 1 January 1980, 16 h 42 m 42 s, 38.8°N, 27.8°W, $M = 7$. The number of P data is 36. The standard errors of both planes are small (less than 6°). Adding SV and SH polarities from 10 stations does not appreciably change the solution or the standard errors. Both solutions are given in Table 1.

An example of an earthquake with dip–slip mechanism is given in Fig. 3. This solution corresponds to the Gulf of Cádiz earthquake, 15 March

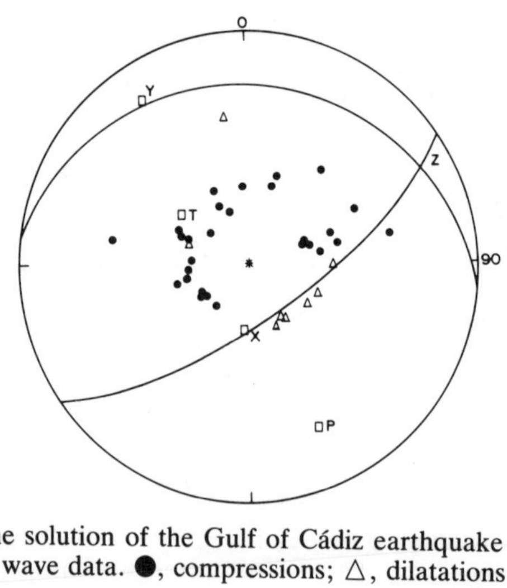

Fig. 3 Fault plane solution of the Gulf of Cádiz earthquake of 15 March 1964. Based on P and S wave data. ●, compressions; △, dilatations.

Table 2 Fault-plane solution for the 15 March 1964 earthquake based on P and P and S first-motion data.

	T and P axes		Nodal planes				
N	Θ	Φ	ϕ	δ	λ	p
38 (P)	T: 30 ± 9	355 ± 43	A: 220 ± 120	19 ± 3	60 ± 30	0.92
	P: 62 ± 9	154 ± 46	B: 71 ± 18	74 ± 6	81 ± 8	
56 (P + S)	T: 29 ± 4	304 ± 19	A: 276 ± 37	24 ± 9	53 ± 10	0.89
	P: 65 ± 4	158 ± 9	B: 56 ± 11	71 ± 4	75 ± 6	

1964, 22 h 30 m 26 s, 36.2°N, 7.6°W, $M = 6.2$. The number of P data is 38. Only the dips of planes and the plunges of the axes are well defined. In particular, the azimuth of the nearly horizontal plane is very poorly estimated. By addition of 18 SV and SH polarities, the errors of the azimuths of the planes and the trends of the axes are considerably reduced. The largest difference between the two solutions is for the azimuth of the nearly horizontal plane. The results can be seen in Table 2.

4 JOINT SOLUTIONS

The algorithm presented here proves to be more useful as a means of analysing composite or joint data from many earthquakes (Udías *et al.*, 1982). In a joint or composite solution a set of M earthquakes in the same region is investigated under the starting assumption that most of them will have the same mechanism. For each event the observations may be small in number, so they do not allow well-determined individual solutions. The number of observations for each earthquake is N_i. The probability of reading a compression in station j from event i is now given by an expression similar to (4), in the form

$$\pi_{ij} = \gamma + (1 - 2\gamma)\Phi[\rho_i A_{ij}(\hat{\Theta})]. \tag{9}$$

In this expression ρ_i is a set of parameters assigned to all observations of each event as a weight. During the process of maximization of the likelihood function these parameters enter as variables. Their final value shows the degree of agreement of the data of each event with the joint solution. The logarithmic likelihood function corresponding to (6) is

$$L = \sum_{i=1}^{M} \sum_{j=1}^{N_i} \log \{\tfrac{1}{2}[1 + (2\pi_{ij} - 1)Y_{ij}]\}. \tag{10}$$

The number of variables in the maximization process of L is $M+3$; that is, the orientation of the joint solution Θ_T, Φ_T, Φ_P and the values of ρ_i, $i=1,\ldots,M$. To initiate the iterative process, besides an initial solution, a set of initial values of ρ_i must be given. In order not to bias the solution, these values must all be equal. Practice has shown that $\rho_i=6$ is a good choice. During the process ρ_i may be assigned negative values for events whose observations are predicted to have opposite sign by the joint solution. In order to know to what extent events agree with the joint solution, one examines the scores for each event p_i and the plots of p_i versus ρ_i. In these plots, for perfect agreement $p_i=1$ and ρ_i tends to infinity; for $p_i=0.5$, $\rho_i=0$, as this represents pure chance; and for $p_i<0.5$, ρ_i takes negative values, increasing and tending to minus infinity for $p_i=0$ (perfect opposite agreement).

To separate events, a threshold value of p is used (e.g. $p=0.7$), below which events are not considered as having the same mechanism as that obtained for the joint solution. Events are then separated into two groups according to values of p_i less or greater than the threshold value. The second group corresponds to the events satisfied by the joint solution. The procedure is continued by looking for successive joint solutions.

4.1 Application

An example of the procedure for joint solutions is presented for a set of 49 aftershocks of the Arudy earthquake (29 February 1980) in the Pyrenees recorded by temporary stations. Magnitudes of the aftershocks range between 1.5 and 4. The location of the epicentres is given in Fig. 4. The procedure is better understood by considering three successive plots of p_i versus ρ_i, as shown in Fig. 5. In the first step all 49 events and 728 observations are used with equal values of ρ_i ($\rho_i=6$). The program converges to a solution for which the plot of ρ_i versus p_i is given in Fig. 5(a). Using $p=0.7$ as threshold value, 27 events ($p_i\geqslant 0.7$) are separated, forming group I. These 27 events are run again in the program to refine the obtained joint solution for group I. The other 22 earthquakes are run again, resetting $\rho_i=6$ for all of them. A new solution is obtained that, for $p\geqslant 0.7$, results in group II with seven events. The ρ_i versus p_i plot and mechanism are shown in Fig. 5(b). The process is repeated again with the remaining 15 events forming group III with six events (Fig. 5c). For the nine events not included in the three groups, individual solutions have been obtained. In this case this was possible, since there are about 18 observations for each event.

The final solutions for the three groups are given in Table 3. The

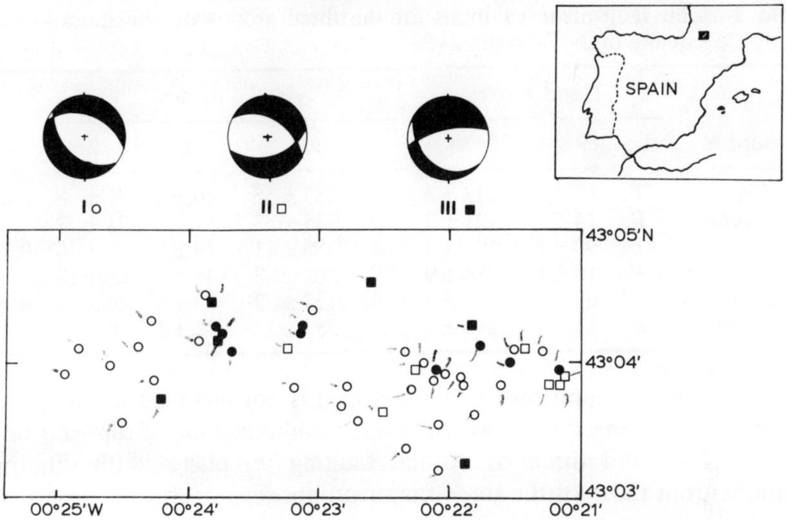

Fig. 4 Epicentres of 49 aftershocks of the Arudy (France) earthquake of 29 February 1980 and joint fault-plane solutions for groups I, II and III. Solid black circles correspond to events not belonging to any group.

mechanisms for the three groups are quite similar, corresponding to predominantly dip–slip normal faulting with horizontal tension axes trending in a general North–South direction. The standard deviations of the parameters are small (between 1° and 10°). The geographic distribution of the mechanisms is given in Fig. 4 and does not show any special

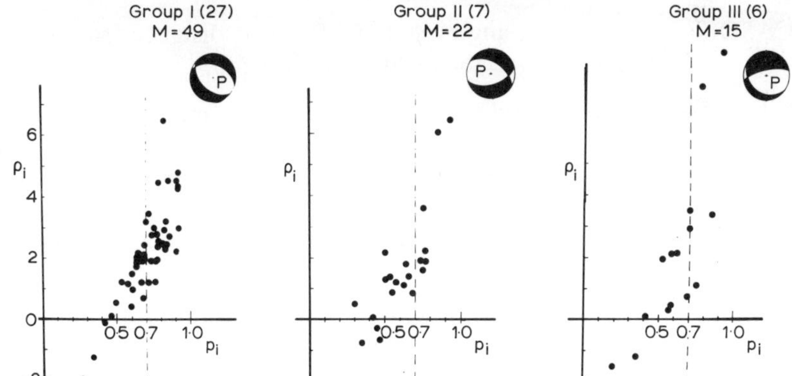

Fig. 5 Plots of p_i versus ρ_i for the 49 aftershocks of the Arudy (France) earthquake of 29 February 1980, showing procedure of separation of the three groups of events and their corresponding joint solutions.

Table 3 Joint fault-plane solutions for the three groups of aftershocks of the Arudy earthquake of 29 February 1980.

	T and P axes			Nodal planes				
Group, N		Θ	Φ		ϕ	δ	λ	p
I, 398	T:	90 ± 2	212 ± 3	A:	289 ± 5	46 ± 1	71 ± 8	0.80
27 events	P:	14 ± 3	121 ± 7	B:	135 ± 3	47 ± 2	71 ± 7	
II, 106	T:	90 ± 4	192 ± 5	A:	85 ± 11	48 ± 2	65 ± 10	0.78
7 events	P:	18 ± 4	284 ± 9	B:	301 ± 3	48 ± 3	65 ± 10	
III, 82	T:	67 ± 1	4 ± 2	A:	127 ± 2	28 ± 1	46 ± 4	0.78
6 events	P:	31 ± 1	140 ± 4	B:	259 ± 3	71 ± 1	70 ± 1	

grouping. The mechanisms of the nine events not included in any group are of the following types, two of reverse faulting, four of dip–slip on a vertical plane and three of normal faulting on planes with different azimuths from those of the three major groups.

5 COMPUTER PROGRAMS

The algorithms described in this chapter have been programmed in Fortran IV for a Univac-1100/80. A package of programs has been prepared entitled as follows.

- PAPUSI Determines angles ϕ and i for teleseismic distance from travel-time tables and a given focal-depth velocity.
- CASOL Determines angles ϕ and i for near distance from crustal models of variable velocity with depth.
- MECSTA Joint or single solutions for P first motions (for single solutions $M = 1$). Final solution is given in terms of orientation (Θ, Φ) of P, T, X, Y and Z axes and (ϕ, δ, λ) of A and B nodal planes and their standard errors. Intermediate results and situation of the observations and axes are printed.
- ONDAPS Joint or single solutions for P, SV and SH first motions. Same characteristics as MECSTA.
- EJESPT Plots final solution (axes and planes) and observations (maximum 500) on an equal-area (Schmidt) projection of the lower hemisphere of the focal sphere (Figs 2 and 3) written for a CALCOMP plotter.

ACKNOWLEDGEMENTS

The authors thank Dr A. Cisternas, Institut de Physique du Globe, Strasbourg for providing the Arudy earthquakes data, and Professor B. A. Bolt, University of California, Berkeley, and Dr J. Mezcua, Instituto Geográfico Nacional, Madrid, for helpful comments. Computations have been carried out in the Centro de Cálculo de la Junta de Energía Nuclear. Part of this work has been supported by the Comisión Asesora de Investigación Científica y Técnica 1666/82, C3 and the Comité Conjunto Hispano-Norteamericano para la Cooperación Científica y Tecnológica 83–009. This chapter is Publication 274 of the Cátedra de Geofísica, Universidad Complutense de Madrid.

REFERENCES

Brillinger, D. R., Udías, A. and Bolt, B. A. (1980). A probability model for regional focal mechanism solutions. *Bull. Seismol. Soc. Am.* **70,** 149–170.

Buforn, E. and Udías, A. (1984). An algorithm for focal mechanism determination using signs of first motion of P, SV and SH waves. *Rev. Geofísica* **40,** 11–26.

Chambers, J. M. (1977). *Computational Methods for Data Analysis.* Wiley, New York.

Stauder, W. (1960). S waves and focal mechanisms: The state of the question. *Bull. Seismol. Soc. Am.* **50,** 333–346.

Udías, A., Buforn, E., Brillinger, D. and Bolt, B. A. (1982). Joint statistical determination of fault plane parameters. *Phys. Earth Planet. Inter.* **30,** 178–184.

Program Index

All programs are arranged in files. Each file may contain more than one program. Each file is identified by a file name. Files are located on the program tape in the sequence given by the file number. Programs are in Fortran.

No.	Name	Contents	Chapter
1	CONTENTS	Contents and general information	
2	SCHAIRY	Spectral synthesis in Earth models composed of vertically varying layers. Output is the body-wave complex spectrum.	I.1
3	WIGGLES	Inverts complex spectra into the time domain by fast Fourier transform. Takes output from SCHAIRY.	I.1
4	VALUES	Assembles the integrand of the complex-ray-parameter integral in the SCHAIRY program package.	I.1
5	RAYLEG	A set of routines to calculate delay time τ by Romberg and three-point integration	I.1
6	TRAPEZ	Performs trapezoidal integration for a specified segment of complex ray parameter. Also assembles source and receiver component factors for the integrand.	I.1

No.	Name	Contents	Chapter
7	COEFF	Calculates 25 reflection/transmission coefficients at an interface. Also routines to calculate reflectively from a stack of piecewise smooth layers.	I.1
8	CVELOC	Complex velocity functions for the SCHAIRY program. Velocities may be frequency-dependent. Includes also standard earth models.	I.1
9	TURN	Finds the complex turning point (radius) in a given layer, for a given complex ray parameter. Also routines for SH and P–SV in a transversely isotropic layer.	I.1
10	VSL	Computes generalized vertical slowness.	I.1
11	LANGER	Given radian frequency ω and delay time τ, returns generalized slowness and vertical wave functions using Langer's uniformly asymptotic approximation.	I.1
12	BLOCK DATA	Data statements specifying the velocity–density–Q model for use in the SCHAIRY program package.	I.1
13	RAWIO	IO routine in the C language for use with the SCHAIRY program package. Subroutines in RAWIO are used to write temporary binary storage files.	I.1
14	SCHAIRY D	Instructions how to compile and link routines in Chapter I.1 (MAKEFILE), example main program (VCMAIN2), and example input for PKP waves and for S, SKS and ScS waves.	I.1
15	SZEGO	Szego asymptotic formula for Legendre functions near the antipode or source point.	I.1
16	WKBJ PRG	WKBJ program (and subroutine RAYIN).	I.2
17	WKBJ DAT	Example input data for WKBJ program.	I.2
18	WKBJ OUT	Resultant output of WKBJ program.	I.2
19	WKBJ FOR	Subroutines THETAC and CCSQRT for the WKBJ algorithm.	I.2
20	RAY1D FOR	Subroutines PXINT, COEFF and COMPON for ray tracing in one-dimensional model.	I.2
21	RAY1DQ FOR	Subroutines PXINT, COEFF and COMPON for ray tracing in one-dimensional model with attenuation.	I.2

No.	Name	Contents	Chapter
22	COEFF TST	Programs to test subroutines COEFF with and without attenuation.	I.2
23	EXMAR	Routines for reading and parameterizing radial and aspherical models including ellipticity.	I.3
24	EXRAY	Ray tracing in radial models, plus asphericity corrections to first order.	I.3
25	EXPER	Travel-time terms in spherical-harmonics expansion of $\delta v/v$ and $\delta r/r$.	I.3
26	EXASP	Example application program for asphericity and ellipticity corrections.	I.3
27	EXTRA	Sample input, output and model files for asphericity and ellipticity corrections.	I.3
28	METRIC	Computation of metric tensor and Christoffer symbols (Section 2).	II.1
29	MODEL	Reads and stores model input data (Section 3.2.1).	II.1
30	SRFC	Routines for specification of smooth surfaces (Sections 3.2.2 and 5.3).	II.1
31	PARM	Routines for specification of medium parameters (Section 3.1.3).	II.1
32	BLOCK	Determines position in simple and complex blocks (Section 3.3.1).	II.1
33	VELOC	Transforms medium parameters into wave velocity and loss factor (Section 3.3.2).	II.1
34	CODE	Transforms elementary wave code into instructions at the ray's initial point and intersections with interfaces (Section 4).	II.1
35	RAYCB	Complete ray tracing through a complex block (Section 5.8).	II.1
36	TRANS	Complete ray tracing across a curved interface (Section 5.9).	II.1
37	INIT	Routines for specification of quantities at the initial point of the ray.	II.1
38	INIT1	Reads and preprocesses input data for initial points of rays (Section 6).	II.1
39	INIT2	Evaluates quantities at the initial point of the ray (Section 6.1).	II.1

No.	Name	Contents	Chapter
40	REVIEW	Review of the main programs for three-dimensional ray tracing in Chapter II.2.	II.2
41	USER	List of user-supplied subprograms defining the model for three-dimensional ray tracing in Chapter II.2.	II.2
42	ITH	Two-point ray tracing in homogeneous isotropic blocks. Head-wave rays included.	II.2
43	ITHEX	Constant-offset version of ITH for near-trace configuration in seismic surveying	II.2
44	ITHGEN	Two-point ray tracing in homogeneous isotropic blocks. Head-wave rays included. Generates first-approximation ray.	II.2
45	GENLIB	Subprogram library for ITHGEN	II.2
46	ITHDIF	Two-point ray tracing in homogeneous isoropic blocks. Allows for edge-diffracted rays.	II.2
47	ITNIPH	Iterative solution of the boundary-value problem for NIP rays in isotropic and homogeneous blocks.	II.2
48	ITHIPDH	Iterative solution of the boundary-value problem for NIP rays. Allows for edge diffraction at the NIP reflector.	II.2
49	HOMLIB	Subprogram library for ITH, ITHEX, ITHGEN, ITHDIF, ITNIPH, ITNIPDH.	II.2
50	ITIL	Two-point ray tracing in inhomogeneous and/or anisotropic media. Parallel ray tracing. Allowance for at most one edge-diffracted ray.	II.2
51	INHLIB	Subprogram library for ITIL.	II.2
52	TRGRDI	Initial-value ray tracing in isotropic homogeneous and constant-gradient blocks.	II.2
53	TRGRDC	Initial-value ray tracing in inhomogeneous and/or anisotropic blocks.	II.2
54	TRGRDB	Initial-value ray tracing in inhomogeneous and/or anisotropic blocks. Alternative ray code.	II.2
55	TRNIPH	Initial-value ray tracing orthogonal to reflecting interface in homogeneous isotropic blocks.	II.2
56	TRILIB	Subprogram library for TRGRDI, TRGRDC, TRGRDB, TRNIPH.	II.2

No.	Name	Contents	Chapter
57	DYNTRAC	Dynamic ray tracing. Mixed methods: linear, regular and singular DRT.	II.2
58	DYNTRAC L	Exclusively linear dynamic ray tracing.	II.2
59	DYNLIB	Subprogram library for DYNTRAC, DYNTRAC L.	II.2
60	DYNEXAMP	Examples and comparison of DYNTRAC and DYNTRAC L.	II.2
61	DYNPROP	Tracing of a propagator matrix along the ray	II.2
62	CONTOUR	Tracing diffracting edges and wire diagrams of a model specified in terms of interfaces.	II.2
63	EDGLIB	Subprogram library for CONTOUR.	II.2
64	MODLIB	Examples of model libraries for testing purposes.	II.2
65	REFER	Reference to formulas used in algorithms of Chapter II.2.	II.2
66	SNERTMAN	Description of the program for calculating theoretical seismograms for layered attenuative media, for a moment-tensor source, with control of approximations to the response.	III.1
67	SNERTRESP	Main program and routine to calculate the response of a layered structure in the frequency-slowness domain by choice of reflection and transmission terms.	III.1
68	SNERTMMRF	Calculation of slowness-dependent but frequency-independent parts of the response, i.e. source-excitation terms and interface reflection/transmission coefficients.	III.1
69	SNERTCOEFF	Calculation of reflection and transmission matrices for portions of the stratification including free-surface effects.	III.1
70	SNERTSINT	Simple integration scheme based on trapezium rule in slowness and FFT over frequency, to get theoretical seismograms at specified ranges.	III.1
71	SHPSV	A package for solving two-point problems in vertically varying media to construct the complete response in the frequency-slowness domain.	III.2
72	SPDRIVE	An example driver for the package SHPSV.	III.2

No.	Name	Contents	Chapter
73	TRAPEZON	Trapezoidal Filon quadrature formula (equation (6)).	III.3
74	DISPER	Summary information of program package DISPER 80.	IV.1
75	TOROID	Calculation of torsional oscillations. Contains subroutines TOREX1, TORDSP, TORRKG, TOREFR, TORENG, TORINI, DSPZN, DSPREV, DSPDEP, DSPRNT.	IV.1
76	NOGRAV	Calculation of non-gravitating spheroidal oscillations. Contains subroutines NOGEX1, NOGDSP, NOGRKG, NOGEFR, NOGENG, NOGINI, DSPZN, DSPREV, DSPDEP, DSPRNT.	IV.1
77	SPHERO	Calculation of gravitating spheroidal oscillations. Contains subroutines SPHEX1, SPHDSP, SPHRKG, SPHEFR, SPHENG, SPHINI, SPHDEL, DSPZN, DSPGRV, DSPDEP, DSPRNT.	IV.1
78	LOVE	Calculation of Love waves using Runge–Kutta–Gill scheme (LOVE 1) and using matrix method (LOVE 2). Contains subroutines LOVEX1, LOVEX2, LOVDSP, LOVRKG, LOVMRX, LOVEFR, LOVEFX, LOVENG, DSPREV, DSPDEP, DSPRNT, DSPRNX.	IV.1
79	RAYLEI	Calculation of Rayleigh waves using Runge–Kutta–Gill scheme (RAYLE 1) and using matrix method (RAYLE 2) Contains subroutines RAYEX1, RAYEX2, RAYDSP, RAYRKG, RAYMRX, RAYEFR, RAYEFX, RAYENG, DSPDEP, DSPRNT, DSPRNX.	IV.1
80	OBANI	Main program and all subroutines for the calculation of free oscillations in Chapter IV.2.	IV.2
81	DOC	Notes on implementation of program OBANI.	IV.2
82	PREM	Earth model used in producing sample output (files BOUNDS and EIFS)	IV.2
83	DATA	Sample input file for OBANI	IV.2
84	BOUNDS	Sample output file of eigenfrequency bounds. Output of OBANI.	IV.2
85	EIFS	Sample output file containing eigenfunctions. Output of OBANI	IV.2

No.	Name	Contents	Chapter
86	CHOL	Choleski decomposition routines: CHOLIN does the decomposition $\boldsymbol{A} = \boldsymbol{L}\boldsymbol{L}^{\mathrm{T}}$. CHOLSL solves the system $\boldsymbol{A}\boldsymbol{x} = \boldsymbol{L}\boldsymbol{L}^{\mathrm{T}}\boldsymbol{x} = \boldsymbol{b}$.	IV.3
87	DIAG	Eigenvalues and eigenvectors of $(\boldsymbol{A} - \lambda \boldsymbol{I})\boldsymbol{x} = \boldsymbol{0}$, with $\boldsymbol{A}$ real, symmetric and dense. Also contains DOT (dot product of two vectors).	IV.3
88	SPLNBF	Spline basis-function routines, given N node positions: QSBF ($N+1$ quadratic splines), BSBF ($N+2$ cubic B splines), HSBF ($2N$ cubic H splines).	IV.3
83	INTERP	Interpolation routines: BSPLN (determines piecewise cubic B splines), BSPLE (B-spline interpolation using BSPLN), HSPLE (H-spline interpolation).	IV.3
89	GLQ	Gauss–Legendre quadrature of selectable order.	IV.3
90	JHW	Eigenvalues and eigenvectors of $(\boldsymbol{A} - \lambda \boldsymbol{B})\boldsymbol{x} = \boldsymbol{0}$, with $\boldsymbol{A}$ and $\boldsymbol{B}$ real, symmetric and banded. Contains JHW (interface routine), EGNVL, FZERO, SRAY, BDET, BSOL and DOT.	IV.3
91	BDRD	Givens rotations for a banded system. Contains BDRD, TANGLE and UNTNGL.	IV.3
92	BLAS	Basic linear-algebra subroutines from Lawson *et al.* (1979). These routines do basic vector operations: addition, dot products, norms, move, copy and elementary rotation operations.	V.1
93	BLAS SUP	Basic vector algorithms supplementary to BLAS. Single-precision, basic vector add and subtract routines.	V.1
94	MTRXADD	Space-economizer matric addition routine.	V.1
95	MTRXMUL	Space-economizer matrix multiplication routine.	V.1
96	MTRANS	In-place matrix transposition routine.	V.1
97	MTRXO	Matrix initialization routine.	V.1
98	MTRXIO	Matrix input, output and print routines.	V.1
99	LLSQ	Linear least-squares analysis routines from Lawson and Hanson (1974). Includes, among others, routines for QR factorization, singular-value decomposition (SVDRS), Householder transforms (H12).	V.1

No.	Name	Contents	Chapter
100	PROJECTOR	Routines to implement orthogonal projectors (PROJECT) and their complements (COMPPRO).	V.1
101	GISOL	Routines GISOL and NUMBIG for solving systems of linear equations by pseudoinverse solution. Uses routine SVDRS (Lawson and Hanson, 1974) supplied in file LLSQ.	V.1
102	LSQR	Sparse linear least-squares solution algorithm using conjugate-gradient method of Page and Saunders (1982a,b).	V.1
103	SEQHT	QR factorization using sequential accumulation based on Householder transforms.	V.1
104	LINPACK	Information about how to obtain a copy of the matrix-manipulation software package of Dongarra *et al.* (1979).	V.1
105	RADMAN	Description of the programs for calculating radiation patterns for a moment-tensor source, including specification of input and output files.	V.2
106	RADPAT	Main routine for presentation of body-wave radiation patterns for P, SV and SH in equal-area projection.	V.2
107	RADPROJ	Main routine for presentation of body-wave radiation patterns for P and S in perspective view.	V.2
108	RADSW	Main routine for presentation of surface-wave radiation terms for Rayleigh and Love waves.	V.2
109	RADPLT	Library of subroutines used by the other radiation programs: routines for manipulating moment-tensor components, and plot routines.	V.2
110	CASPAP	Ray direction angles from a source (programs CASSOL and PAPUSI)	V.3
111	MECOND	Joint or single solutions from first-motion data (programs MECSTA and ONDAPS).	V.3
112	EJESPT	Calcomp plot of fault-plane solutions and data, in equal-area projection.	V.3

Subject Index

Absorption, 158–159
  band, 32
Addition rule
  for reflection matrices, 256–258
  for vectors, 407
Addition theorem
  for Legendre functions, 82
Adjoint, 324–326
  self, 326, 343, 345, 349, 354
Airy function, 214
Algebraic reconstruction, 421
Amplitude
  of ray, 90, 110, 148–149, 154–158, 173, 205, 213, 216
  paraxial approximation, 156–157
Anisotropy, 239, 242, 429
  transverse isotropy, 34–35, 175, 218–221, 293–294, 355
Annulling transform, 419
Antipode, 36–38
Asphericity, 75–77, 79–81
Asymptotic ray theory, 169
  body wave spectrum, 5
Attenuation, 31–33, 69–72, 158–159, 262

Banded matrix, 382, 420
Bandwidth of matrix, 382
BLAS, 404–406
Basis functions
  finite element, 381–385
  local, 383–384
  natural, 377–379, 392, 395
  orthogonal polynomial, 377
  piecewise cubic B-spline, 383–385
  piecewise cubic H-spline, 384–385
  piecewise linear, 383–385
  piecewise quadratic, 383–384
Basis vectors, 107, 118, 140–141, 145, 197, 430, 444
Bending, 181
Bessel function, 283, 317
  spherical, 369, 380
Bisection, 368, 378, 389–391
Block, 174, 179
  complex, 94–105, 113, 118, 121–122, 133–134, 143
  isotropic, 175, 218
  simple, 94–105, 113, 123, 127–128, 134, 142, 143
Body force, 53
Body wave, 3, 47, 159, 169, 252, 268–269, 431
  amplitude, 170, 195
  far field, 428–437
  pulse, 213–217
Boundary conditions, 263, 297, 301, 323, 372, 387–389
  for ray spreading Jacobian, 207
B-spline
  basis functions, 383
  interpolation, 386

Cagniard-de Hoop, 237, 279
Cartesian coordinate system, 93
Causality, 31, 70, 158, 280
Caustic, 47, 91, 110, 114, 126, 129, 153, 154, 163, 172, 183, 195, 196, 201, 212
  cusp, 216, 226, 229
  diffraction, 169
  phase shift, 110, 128–130, 154–155, 163, 195, 201, 212
Characteristic function, 297, 304, 312, 313

Christoffel symbols, 92–94, 106, 125, 134
Cholesky decomposition, 377, 378
Clenshaw–Curtis quadrature, 286, 287
Code
  of elementary wave, 105, 113, 115, 117, 121, 133, 134, 159
  of ray, 170, 180, 181
  of NIP ray, 194
  reduced raycode, 179
Coefficients
  conversion, 110, 118, 120
  reflection/transmission, 54–55, 110, 139, 140, 239, 251, 256, 257
  generalized reflection/transmission, 8, 11, 20–24
Collocation, 268–269
Complete
  response, 243–247
  seismogram, 262, 273–276
Composite pulse, 195, 214, 215
Compound-matrix method, 302–304, 309–313, 322, 326–337
Conjugate gradient, 421
Continuation method, 188
Contravariant
  transformation, 338
  vector, 339
Contour integration, 5, 15–24, 285, 286
Conversion
  coefficients, 110, 118, 120
  P-SV, 238, 241–243, 251, 256
Coordinate system
  Cartesian, 93
  cylindrical, 49, 261, 317
  geographic spherical, 93–94
  ray centred, 107–108, 196–198
  source reference, 429, 444
  spherical polar, 76, 93, 296, 354
Core, 17–24, 36–37
Core–mantle boundary, 17, 82–85
Covariant
  transformation, 338
  vector, 339
Curvature
  of interface, 136
  of wavefront, 149, 150, 156, 198
Cylindrical coordinate system, 49, 261, 317
  vector harmonics, 50
Delay function, 7, 15, 29–30, 34, 56
Differential equations, 263, 267, 270, 294, 299, 306, 310–313, 317, 318, 343–354, 368, 369
Diffraction, 16–18, 167
  edge, 169, 181, 190, 210–212
  poles, 16
Dines and Lytle algorithm, 421
Dip, 6, 429, 430, 444
Dip-slip, 448
Dispersion relation, 297, 301, 304, 308, 312, 313
  inverse, 393–394
Displacement
  vector, 107, 110, 156, 158–159, 161, 163
  matrix, 239–241, 243–247, 251
Dissipative medium, 32–33, 69–70, 144, 158–160
Double couple, 429–431, 444
Dynamic group, 53
Dynamic ray tracing, 90, 106–112, 137, 146, 172
  initial conditions, 108, 146–147
  linear, 190, 203–205
  regular, 190–199
  singular, 196, 201–203

Earth
  model, 3–4, 50–51, 76–77, 252–253, 293–294, 368
  flattening transform, 48
Earthquake
  location, 145, 167, 416, 419, 422
  fault plane solution, 443, 447, 449–452
Edge
  diffraction, 169, 181, 190, 210–212
  tracing, 191–193
Eigenfrequency, 321, 368
Eigenfunction, 305, 314–319, 321, 324, 344–346, 351–353, 368, 369
  multicomponent, 376, 385
Eigenproblem
  banded matrices, 389–392
  dense matrices, 377–379
Eigenvalue, 305, 314, 322–324, 344, 346, 351–353, 377–379, 389–392

Elementary
  reflector, 410
  synthetic seismogram, 159
  wave, 91, 105, 117–121, 140, 159–162, 164
Ellipticity, 81–82
  correction, 81–82
Energy integrals, 298, 306, 315, 318, 319, 355, 357, 361, 369
  kinetic, 372, 378
  potential, 372
Equal-area projection, 435–452

Far-field
  body waves, 50, 428–437
  surface waves, 437–439
Fault plane, 6
  solution, 443, 447, 452
  joint solution, 443, 449–452
Filon integration, 24–25, 280–286
Finite difference method, 395–396
  collocation, 265–268
  Runge–Kutta, 294–295, 318–319, 321, 368–369
Finite element method, 397–398
Finite source, 167
First motion, 443, 447, 452
First order perturbation, 78
Focal sphere, 431, 432, 443, 444
Fourier–Bessel transform, 261, 262
Frechét derivative, 372
Free oscillations, 293, 296–300, 307–316, 323, 325, 354–367, 379–381, 392–393
Fresnel volume, 152–153
Full-wave theory, 4–5, 238
Fundamental matrix, 9, 34, 38–41

Galerkin algorithm, 397
Gaussian
  beams, 91, 163–165
  packets, 163–165
Gauss–Legendre quadrature, 369, 380, 386
Generalized
  eigenvalue problem, 373
  inverse, 414–420
  ray method, 237, 279
  ray spreading Jacobian, 202
  reflection/transmission coefficients, 8, 11, 20, 24
Geographic spherical coordinate system, 93–94
Geometrical spreading, 109, 110, 148, 156, 163, 171, 195
  matrix, 109, 128–129, 141, 147–148, 154
Givens rotations, 390, 391
  rotation matrices, 410
Gravitational potential, 308, 354
Green function, 91, 161, 263
  ray theoretical, 161
  tensor, 427
Group velocity, 298, 299, 306, 316–318, 369
Guided waves, 248

Hamiltonian
  wave, 176, 182, 188
Head wave
  interference, 18–24
  ray, 186, 187
Hermitian
  matrix, 373
  operator, 373
Hilbert transform, 172
Homotopy of models, 188
Householder
  rotations, 378
  transform, 410, 423
H-spline
  basis functions, 384–385
  interpolation, 384–385
Hybrid method, 261, 271, 273
Hypocentre, 428
  location, 145, 167, 416, 419, 422

Impulse response, 160
Initialization of a matrix, 408
Initial value ray tracing, 176–177
Integration
  Filon, 24–25, 280–286
  Gauss-Legendre, 369, 380, 386
  Runge-Kutta, 294, 295, 318, 319, 321, 368, 369
  Thomson–Haskell, 318, 319
  trapezoidal, 42–43, 282

Interference head wave, 18–24
Inverse
  dispersion, 374, 393–394
  iteration, 378, 390, 391
Inversion
  banded matrices, 420
  continuous linear, 414
  continuous nonlinear, 415
  discrete linear, 414
  discrete nonlinear, 414
  even determined system, 413–414
  generalized, 414–420
  least squares, 416, 417
  maximum likelihood, 445–447
  sparse least squares, 420–422

Jacobian
  ray spreading, 171, 195
  boundary conditions, for, 207

Kernel
  kinetic energy, 372
  Lagrangian, 372
  potential energy, 372
Kinematic
  group, 53
  analogous rays, 187
Kinetic energy, 298, 306, 307, 315, 316, 372, 378
KMAH index, 114, 129, 153–154, 195
Kramers–Krönig, 31
Kirchhoff integral, 167, 280
Kirchhoff–Helmholtz, 280, 281

Lagrangian, 298, 306, 315, 325, 371–372
Lanczos algorithm, 421
Langer approximation, 4–5, 8, 13–15, 25–28, 30–31, 40, 55
Leaky mode, 270–271
Least squares, 416, 417
  sparse inversion, 420–422
Legendre function, 5, 8, 36, 38
  addition theorem, 82
Levin's method, 287, 288
Likelihood function, 445–447, 449
Linear
  constraints, 388, 394
  interpolation, 383
  interpolation for eigenvalues, 389–391
  inversion, 414
LINPACK, 414, 418, 420
Local
  model, 190
  variations method, 184
Location of hypocentre, 145, 167, 416, 419, 422
Love waves, 317–318
Lumped mass approximation, 397

Mantle, 26, 252–253
Maslov
  index, 114, 129, 153, 154, 195
  method, 48, 69, 91, 164
Matrix
  addition, 407
  banded, 382, 420
  fundamental, 9, 34, 38–41
  initialization, 408
  inversion, 413–422
  mass, 373
  minor, 302–304, 309–313, 322, 326–337
  multiplication, 407
  paraxial approximation, 156–157
  propagator, 38–41
  ray propagator, 89, 106, 108, 111, 146, 147, 151, 153, 163
  reflection/transmission, 38–41, 239–242, 244, 247, 249–258
  sparse, 420–422
  stiffness, 373
  Thomson–Haskell method, 318–319
  transposition, 408
Maximum likelihood, 445–447
Metric tensor, 92–94, 106, 124, 125, 134
Midpoint scheme, 266, 268
Minor
  matrix method, 302–304, 309–313, 322, 326–337
  algebra of, 326–337
  differential equation, 368, 369

vector, 327, 329–337, 340–343, 346, 349, 362, 367
Modal poles, 243
Mode
leaky, 270–271
normal, 293, 321, 346, 354, 379
radial, 322, 356–360
spheroidal, 300–316, 322, 361–367
toroidal, 296–300, 322, 355–356, 379–381, 392–393
Model
block construction, 94–105, 174–176
box, 174–176
multicomponent, 376, 385
piecewise smooth, 3–4, 362
plane layered, 50–51
PREM, 3–4, 72
transversely isotropic, 293–294
upper mantle, 26, 252–253
velocity, 76–77
Moment tensor, 6, 161–162, 252, 261, 264, 427–440
Multifold path integral method, 280
Multiple reflection, 240, 245–250, 254–258

NIP
profile, 194
ray, 193
ray, edge diffracted, 193
Nodal
plane, 444, 447
region, 435
Node
finite element, 382
optimal spacing, 383
Normalization of eigenvectors, 391
Normal mode, 293, 321, 346, 354, 379

Parallel ray tracing, 182–185
paraxial approximation
matrix, 156–157
rays, 108–110, 114, 151–152
two-point ray tracing, 152
travel times, 145, 150
Particle motion diagram, 162
Pearcy function, 214
Perturbation
ray theoretical, 89, 106, 108, 111, 146, 147, 151, 153, 163
theory, 165
travel time to first order, 78
Phase integral, 280
Phase velocity, 298, 317, 318
Plane harmonic, 317
Plunge, 429, 444
Polarity, 443, 446, 448
Polarization
angle, 205
vector, 106, 107, 111, 120, 126, 135, 145, 146, 148, 170
Potential energy, 372
PREM, 3–4, 72
Principal axes, 444, 445
Probability model, 445–447
Projection operator, 349, 350
Projector, 411
Propagator matrix, 38–41
ray theoretical, 89, 106, 108, 111, 146, 147, 151, 153, 163
Pseudoinverse, 412, 416, 418
Pseudorank, 418

Q, 31–33, 69–72
Quadrature, 415
Gauss–Legendre, 369, 380, 386
Filon, 24–25, 280–286
Quasiphotons, 163
Quasilongitudinal, 220
Quasitransverse, 220
Quelling, 417
QR factorization, 423

Radial mode, 322, 356–360
Radiation pattern, 428, 431–439
Rake, 6, 429–431
Rational approximation, 67–68
Ray
amplitude, 90, 110, 148–149, 154–158, 173, 205, 213, 216
centred coordinates, 107–110, 113, 118, 125, 137, 145–149, 151, 154, 155, 196–200
code, 170, 179–181, 194
coordinates, 109, 147, 148

definition, 52
dynamic tracing, 90, 106–112, 137, 146–147, 172, 190–199, 201–205
expansion, 160
generalized spreading Jacobian, 202
Green function, 161
parallel tracing, 182–185
parameter, 5, 50, 262, 280
paraxial, 108–110, 114, 151–152
propagator matrix, 89, 106, 108, 111, 146, 147, 151, 153, 163
spreading Jacobian, 171, 195, 207
synthetic seismogram, 159–160
tracing, 77–79, 89, 90, 106, 108, 122, 152
two-point tracing, 180–195
Rayleigh
principle, 371–372
quotient, 374
wave, 318–319
Rayleigh–Ritz algorithm, 372–374
Reflection, multiple, 240, 245–250, 254–258
Reflection/transmission
coefficients, 54–55, 110, 139, 140, 239, 251, 256, 257
generalized coefficients, 8, 11, 20–24
matrices, 38–41, 239–242, 247, 249, 251, 256–258
matrix addition rule, 240, 242, 244, 249–258
matrix recursive construction, 256–258
Reflectivity, 38–41
method, 238, 252, 283
Reverberation operator, 240–242, 245–246, 248–249
Rotation
Givens matrices, 410
Householder, 378
of coordinates with spherical harmonics, 79–81
Rotational splitting, 306
Runge–Kutta, 294, 295, 318, 319, 321, 368, 369

Saddle point, 15–17
Sequential accumulation, 422–424
Shadow boundary, 214, 216–217
Shear dislocation, 6, 428, 444
Shooting method, 181, 396
Simultaneous iterative reconstruction, 421
Singular
dynamic ray tracing, 196, 201–203
region, 200, 201
Singular value decomposition, 413
algorithms, 418–419
singular values, 413
singular vectors, 413
Slip vector, 429
Slowness, 170, 251
method, 4, 24–25, 279
surface, 206–207, 213
vertical, 6, 30–31
Source
moment tensor, 6, 161–162, 252, 261, 264, 427–440
radiation pattern, 5–6, 431–439
reference coordinate system, 429, 444
vectors, 243–245, 248, 250
Sparse least squares inversion, 420–422
algebraic reconstruction, 421
conjugate gradient, 421
Dines and Lytle algorithm, 421
Lanczos algorithm, 421
simultaneous iterative reconstruction, 421
Spectral method, 3–43, 279
Spherical
Bessel function, 369, 380
harmonics, 76, 79–81, 296, 354
polar coordinate system, 76, 93, 296, 354
vector harmonics, 254
Spherical modes, 300–316, 322, 361–367
Spline
B-spline, 383
H-spline, 384–385
interpolation, 369, 384–386
Split stratification, 244–247
Spreading
geometrical, 109, 110, 148, 156, 163, 171, 195
matrix, 109, 128–129, 141, 147–148, 154

Jacobian, 171, 195
Static problem, 371–372, 376–377, 391, 397
Stress-displacement vector, 263, 296, 300–301, 307, 317, 318, 354–356, 361–362
Strike, 6, 429–431
Strike–slip, 448
Sturm count, 322, 352–354, 378, 389, 391
Sturm–Liouville, 322, 352
Surface wave
  Love, 317–318
  Rayleigh, 318–319
Synthetic seismogram, 15–31, 49–56
  ray theoretical, 159–160, 213–215

tau function, 7, 15, 29–30, 34, 56
Thomson–Haskell matrix method, 318–319
Time stepping, 398
Tomography, 420–422
Toroidal mode, 296–300, 322, 355–356, 379–381, 392–393
Transform
  annulling, 419
  Fourier–Bessel, 261, 262
  Hilbert, 172
  Householder, 410, 423
  of spherical harmonics, 79–81
Transmission, *see* Reflection/transmission
Transposition, 408
Transverse isotropy, 34–35, 175, 218–221, 293–294, 355
Trapezoidal integration, 42–43, 282
  Filon, 280–286
Travel time, 19, 27, 106, 111, 123, 134, 140, 144, 170, 196
  complex-valued, 106, 111, 123, 134, 140, 144, 159, 160
  first order perturbation, 78
  paraxial, 145, 150
  residual, 83, 84
  second derivatives, 109, 115, 149
Triplication, 26–28
Turning point, 7, 28–29, 34, 52, 270–271
Two-point problem, 261, 262, 264–271
  ray tracing, 180–195
t*, 70–72, 144, 158

Upper mantle
  model, 26, 252–253
  propagation, 252–256

Variational principle, 325, 371–377
Vector algorithms
  addition, 407
  BLAS, 404–406
  subtraction, 407
Vectorize, 282, 286
Vector spherical harmonics, 254
Velocity models, 76–77

Wave
  diffracted, 17–18
  elementary, 91, 105, 117–121, 140, 159–162, 164
  Hamiltonian, 176, 182, 188
  head, 18–24, 186, 187
Wavefront curvature, 149, 150, 156, 198
Wavefuntion
  horizontal, 8
  vertical, 5–8, 30–31
Wavenumber, 262, 279–280
WKBJ approximation, 7–8, 12–13, 50
  seismogram, 47–56